T0226123

Beam-Solid Interactions
and Phase Transformations

MATERIALS RESEARCH SOCIETY SYMPOSIA PROCEEDINGS

ISSN 0272 - 9172

Volume 1—Laser and Electron-Beam Solid Interactions and Materials Processing, J. F. Gibbons, L. D. Hess, T. W. Sigmon, 1981

Volume 2—Defects in Semiconductors, J. Narayan, T. Y. Tan, 1981

Volume 3—Nuclear and Electron Resonance Spectroscopies Applied to Materials Science, E. N. Kaufmann, G. K. Shenoy, 1981

Volume 4—Laser and Electron-Beam Interactions with Solids, B. R. Appleton, G. K. Celler, 1982

Volume 5—Grain Boundaries in Semiconductors, H. J. Leamy, G. E. Pike, C. H. Seager, 1982

Volume 6—Scientific Basis for Nuclear Waste Management, S. V. Topp, 1982

Volume 7—Metastable Materials Formation by Ion Implantation, S. T. Picraux, W. J. Choyke, 1982

Volume 8—Rapidly Solidified Amorphous and Crystalline Alloys, B. H. Kear, B. C. Giessen, M. Cohen, 1982

Volume 9—Materials Processing in the Reduced Gravity Environment of Space, G. E. Rindone, 1982

Volume 10—Thin Films and Interfaces, P. S. Ho, K.-N. Tu, 1982

Volume 11—Scientific Basis for Nuclear Waste Management V, W. Lutze, 1982

Volume 12—In Situ Composites IV, F. D. Lemkey, H. E. Cline, M. McLean, 1982

Volume 13—Laser Solid Interactions and Transient Thermal Processing of Materials, J. Narayan, W. L. Brown, R. A. Lemons, 1983

Volume 14—Defects in Semiconductors II, S. Mahajan, J. W. Corbett, 1983

Volume 15—Scientific Basis for Nuclear Waste Management VI, D. G. Brookins, 1983

Volume 16—Nuclear Radiation Detector Materials, E. E. Haller, H. W. Kraner, W. A. Higinbotham, 1983

Volume 17—Laser Diagnostics and Photochemical Processing for Semiconductor Devices, R. M. Osgood, S. R. J. Brueck, H. R. Schlossberg, 1983

Volume 18—Interfaces and Contacts, R. Ludeke, K. Rose, 1983

Volume 19—Alloy Phase Diagrams, L. H. Bennett, T. B. Massalski, B. C. Giessen, 1983

Volume 20—Intercalated Graphite, M. S. Dresselhaus, G. Dresselhaus, J. E. Fischer, M. J. Moran, 1983

Volume 21—Phase Transformations in Solids, T. Tsakalakos, 1984

Volume 22—High Pressure in Science and Technology, C. Homan, R. K. MacCrone, E. Whalley, 1984

Volume 23—Energy Beam-Solid Interactions and Transient Thermal Processing, J. C. C. Fan, N. M. Johnson, 1984

Volume 24—Defect Properties and Processing of High-Technology Nonmetallic Materials, J. H. Crawford, Jr., Y. Chen, W. A. Sibley, 1984

MATERIALS RESEARCH SOCIETY SYMPOSIA PROCEEDINGS

Volume 25—Thin Films and Interfaces II, J. E. E. Baglin, D. R. Campbell, W. K. Chu, 1984

Volume 26—Scientific Basis for Nuclear Waste Management VII, G. L. McVay, 1984

Volume 27—Ion Implantation and Ion Beam Processing of Materials, G. K. Hubler, O. W. Holland, C. R. Clayton, C. W. White, 1984

Volume 28—Rapidly Solidified Metastable Materials, B. H. Kear, B. C. Giessen, 1984

Volume 29—Laser-Controlled Chemical Processing of Surfaces, A. W. Johnson, D. J. Ehrlich, H. R. Schlossberg, 1984.

Volume 30—Plasma Processsing and Synthesis of Materials, J. Szekely, D. Apelian, 1984

Volume 31—Electron Microscopy of Materials, W. Krakow, D. Smith, L. W. Hobbs, 1984

Volume 32—Better Ceramics Through Chemistry, C. J. Brinker, D. E. Clark, D. R. Ulrich, 1984

Volume 33—Comparison of Thin Film Transistor and SOI Technologies, H. W. Lam, M. J. Thompson, 1984

Volume 34—Physical Metallurgy of Cast Iron, H. Fredriksson, M. Hillerts, 1985

Volume 35—Energy Beam-Solid Interactions and Transient Thermal Processing/1984, D. K. Biegelsen, G. Rozgonyi, C. Shank, 1985

Volume 36—Impurity Diffusion and Gettering in Silicon, R. B. Fair, C. W. Pearce, J. Washburn, 1985

Volume 37—Layered Structures, Epitaxy and Interfaces, J. M. Gibson, L. R. Dawson, 1985

Volume 38—Plasma Synthesis and Etching of Electronic Materials, R. P. H. Chang, B. Abeles, 1985

Volume 39—High-Temperature Ordered Intermetallic Alloys, C. C. Koch, C. T. Liu, N. S. Stoloff, 1985

Volume 40—Electronic Packaging Materials Science, E. A. Giess, K.-N. Tu, D. R. Uhlmann, 1985

Volume 41—Advanced Photon and Particle Techniques for the Characterization of Defects in Solids, J. B. Roberto, R. W. Carpenter, M. C. Wittels, 1985

Volume 42—Very High Strength Cement-Based Materials, J. F. Young, 1985.

Volume 43—Coal Combustion and Conversion Wastes: Characterization, Utilization, and Disposal, G. J. McCarthy, R. J. Lauf, 1985

Volume 44—Scientific Basis for Nuclear Waste Management VIII, C. M. Jantzen, J. A. Stone, R. C. Ewing, 1985

Volume 45—Ion Beam Processes in Advanced Electronic Materials and Device Technology, F. H. Eisen, T. W. Sigmon, B. R. Appleton, 1985

Volume 46—Microscopic Identification of Electronic Defects in Semiconductors, N. M Johnson, S. G. Bishop, G. D. Watkins, 1985

MATERIALS RESEARCH SOCIETY SYMPOSIA PROCEEDINGS

Volume 47—Thin Films: The Relationship of Structure to Properties, C. R. Aita, K. S. SreeHarsha, 1985

Volume 48—Applied Material Characterization, W. Katz, P. Williams, 1985

Volume 49—Materials Issues in Applications of Amorphous Silicon Technology, D. Adler, A. Madan, M. J. Thompson, 1985

Volume 50—Scientific Basis for Nuclear Waste Management IX, L. O. Werme, 1985

Volume 51—Beam-Solid Interactions and Phase Transformations, H. Kurz, G. L. Olson, J. M. Poate, 1986

Volume 52—Rapid Thermal Processing, T. O. Sedgwick, T. E. Siedel, B. Y. Tsaur, 1986

Volume 53—Semiconductor-on-Insulator and Thin Film Transistor Technology, A. Chiang. M. W. Geis, L. Pfeiffer, 1986

Volume 54—Interfaces and Phenomena, R. H. Nemanich, P. S. Ho, S. S. Lau, 1986

Volume 55—Biomedical Materials, M. F. Nichols, J. M. Williams, W. Zingg, 1986

Volume 56—Layered Structures and Epitaxy, M. Gibson, G. C. Osbourn, R. M. Tromp, 1986

Volume 57—Phase Transitions in Condensed Systems—Experiments and Theory, G. S. Cargill III, F. Spaepen, K. N. Tu, 1986

Volume 58—Rapidly Solidified Alloys and Their Mechanical and Magnetic Properties, B. C. Giessen, D. E. Polk, A. I. Taub, 1986

Volume 59—Oxygen, Carbon, Hydrogen, and Nitrogen in Crystalline Silicon, J. W. Corbett, J. C. Mikkelsen, Jr., S. J. Pearton, S. J. Pennycook, 1986

Volume 60—Defect Properties and Processing of High-Technology Nonmetallic Materials, Y. Chen, W. D. Kingery, R. J. Stokes, 1986

Volume 61—Defects in Glasses, Frank L. Galeener, David L. Griscom, Marvin J. Weber, 1986

Volume 62—Materials Problem Solving with the Transmission Electron Microscope, L. W. Hobbs, K. H. Westmacott, D. B. Williams, 1986

Volume 63—Computer-Based Microscopic Description of the Structure and Properties of Materials, J. Broughton, W. Krakow, S. T. Pantelides, 1986

Volume 64—Cement-Based Composites: Strain Rate Effects on Fracture, S. Mindess, S. P. Shah, 1986

Volume 65—Fly Ash and Coal Conversion By-Products: Characterization, Utilization and Disposal III, G. J. McCarthy , D. M. Roy, 1986

Volume 66—Frontiers in Materials Education, G. L. Liedl, L. W. Hobbs, 1986

MATERIALS RESEARCH SOCIETY SYMPOSIA PROCEEDINGS   VOLUME 51

Beam-Solid Interactions
and Phase Transformations

Symposium held December 2-4, 1985, Boston, Massachusetts, USA

EDITORS:

H. Kurz
Technical University Aachen, Aachen, Federal Republic of Germany

G. L. Olson
Hughes Research Laboratories, Malibu, California, U.S.A.

J. M. Poate
AT&T Bell Laboratories, Murray Hill, New Jersey, U.S.A.

M⁅R⁆S MATERIALS RESEARCH SOCIETY
Pittsburgh, Pennsylvania

CAMBRIDGE UNIVERSITY PRESS
Cambridge, New York, Melbourne, Madrid, Cape Town,
Singapore, São Paulo, Delhi, Mexico City

Cambridge University Press
32 Avenue of the Americas, New York NY 10013-2473, USA

Published in the United States of America by Cambridge University Press, New York

www.cambridge.org
Information on this title: www.cambridge.org/9781107405721

Materials Research Society
506 Keystone Drive, Warrendale, PA 15086
http://www.mrs.org

First published 1986
First paperback edition 2012

Single article reprints from this publication are available through
University Microfilms Inc., 300 North Zeeb Road, Ann Arbor, MI 48106

CODEN: MRSPDH

ISBN 978-0-931-83716-6 Hardback
ISBN 978-1-107-40572-1 Paperback

Contents

PREFACE   xv

ACKNOWLEDGMENTS   xvii

PART IA: PLENARY REVIEWS I
Laser-Solid and Surface Interactions

*PULSED LASER INTERACTIONS WITH CONDENSED MATTER
N. Bloembergen   3

*FEMTOSECOND DYNAMICS OF HIGHLY EXCITED SEMICONDUCTORS
Charles V. Shank and Michael C. Downer   15

*DYNAMICS OF ELEMENTARY PROCESSES AT SURFACES:
NITRIC OXIDE SCATTERED FROM A GRAPHITE SURFACE
H. Vach, J. Häger, B. Simon, C. Flytzanis,
and H. Walther   25

*NONLINEAR OPTICS AND SURFACE SCIENCE
Y.R. Shen   39

PART IB: PLENARY REVIEWS II
Ion-Solid Interactions and Phase Transformations

*COLLISION CASCADES, IONIZATION SPIKES, AND ENERGY
TRANSFER
W.L. Brown   53

*TRANSITION BETWEEN CONDENSED PHASES IN Si and Ge
David Turnbull   71

*AMORPHIZATION, CRYSTALLIZATION AND RELATED PHENOMENA
IN SILICON
J.S. Williams   83

PART II: LASER-INDUCED PHASE
TRANSITIONS IN SEMICONDUCTORS

*ANOMALOUS BEHAVIOR DURING THE SOLIDIFICATION OF
SILICON IN THE PRESENCE OF IMPURITIES
Michael O. Thompson and P.S. Peercy   99

OVERHEATING IN SILICON DURING PULSED-LASER
IRRADIATION?
B.C. Larson, J.Z. Tischler, and D.M. Mills   113

TIME-RESOLVED X-RAY ABSORPTION OF AN AMORPHOUS Si
FOIL DURING PULSED LASER IRRADIATION
Kouichi Murakami, Hans C. Gerritsen, Hedser
Van Brug, Fred Bijerk, Frans W. Saris, and
Marnix J. Van der Wiel   119

*Invited Paper

NUCLEATION OF INTERNAL MELT DURING PULSED LASER
IRRADIATION
    P.S. Peercy, Michael O. Thompson, J.Y. Tsao
    and J.M. Poate   125

DIRECT IMAGING OF "EXPLOSIVELY" PROPAGATING BURIED
MOLTEN LAYERS IN AMORPHOUS SILICON USING OPTICAL, TEM,
AND ION BACKSCATTERING MEASUREMENTS
    D.H. Lowndes, G.E. Jellison, Jr., S.J. Pennycook,
    S.P. Withrow, D.N. Mashburn, and R.F. Wood   131

PHASE DIAGRAM OF LASER INDUCED MELT MORPHOLOGIES IN
SILICON
    John S. Preston, John E. Sipe, and Henry
    Van Driel   137

TIME-RESOLVED REFLECTIVITY MEASUREMENTS OF SILICON
AND GERMANIUM USING A PULSED EXCIMER LASER
    G.E. Jellison, Jr., D.H. Lowndes, D.N. Mashburn,
    and R.F. Wood   143

RAMAN MICROPROBE ANALYSIS OF LASER-INDUCED MICRO-
STRUCTURES
    P.M. Fauchet   149

CHARACTERISTICS OF LASER/ENERGY BEAM-MELTED SILICON
MECHANICAL DAMAGES
    El-Hang Lee   155

PULSED CO_2-LASER INDUCED MELTING AND NONLINEAR
OPTICAL STUDIES OF GaAs
    R.R. James, W.H. Christie, B.E. Mills,
    and H.L. Burcham, Jr.   161

IN SITU TIME-RESOLVED REFLECTIVITY MEASUREMENTS OF
GROWTH KINETICS DURING SOLID PHASE EPITAXY: A TOOL
TO ESTIMATE INTERFACE NON PLANARITY DURING GROWTH
    C. Licoppe and Y.I. Nissim   167

LASER INDUCED OXIDATION OF HEAVILY DOPED SILICON
    F. Fogarassy, C.W. White, D.H. Lowndes, and
    J. Narayan   173

RAMAN STUDIES OF PHASE TRANSFORMATIONS IN PULSE LASER
IRRADIATED DIELECTRIC FILMS
    Gregory J. Exarhos   179

FORMATION OF $Al_xGa_{1-x}As$ ALLOY ON THE SEMI-INSULATING
GaAs SUBSTRATE BY LASER BEAM INTERACTION
    N.V. Joshi and J. Lehman   185

OSCILLATORY MORPHOLOGICAL INSTABILITIES DURING RAPID
SOLIDIFICATION - THE ROLE OF DIFFUSION IN THE SOLID
    Atul Bansal and Arijit Bose   191

PART III: ULTRAFAST LASER
EXCITATION OF SEMICONDUCTORS

*TIME-RESOLVED SPECTROSCOPY OF PLASMA RESONANCES IN
HIGHLY EXCITED SILICON AND GERMANIUM
A.M. Malvezzi, C.Y. Huang, H. Kurz, and
N. Bloembergen   201

CROSS-SECTIONAL TEM CHARACTERIZATION OF STRUCTURAL
CHANGES PRODUCED IN SILICON BY ONE MICRON PICOSECOND
PULSES
Arthur L. Smirl, Ian W. Boyd, Thomas F. Boggess,
Steven C. Moss, and R.F. Pinizzotto   213

OBSERVATION OF SUPERHEATING DURING PICOSECOND LASER
MELTING?
N. Fabricius, P. Hermes, D. von der Linde,
A. Pospieszczyk, and B. Stritzker   219

TIME-RESOLVED OPTICAL STUDIES OF PICOSECOND LASER
INTERACTIONS WITH GaAs SURFACES
J.M. Liu, A.M. Malvezzi and N. Bloembergen   225

PART IV: LASER-INDUCED PHASE
TRANSFORMATIONS IN GRAPHITE

*PULSED LASER MELTING OF GRAPHITE
G. Braunstein, J. Steinbeck, M.S. Dresselhaus,
G. Dresselhaus, B.S. Elman, T. Venkatesan,
B. Wilkens, and D.C. Jacobson   233

TIME-RESOLVED PICOSECOND OPTICAL STUDY OF LASER-
EXCITED GRAPHITE
C.Y. Huang, A.M. Malvezzi, J.M. Liu, and
N. Bloembergen   245

NANOSECOND TIME RESOLVED REFLECTIVITY MEASUREMENTS AT
THE SURFACE OF PULSED LASER IRRADIATED GRAPHITE
T. Venkatesan, J. Steinbeck, G. Braunstein,
M.S. Dresselhaus, G. Dresselhaus, D.C. Jacobson,
and B.S. Elman   251

DYNAMICAL LASER DESORPTION OF IONIC CLUSTERS IN
GROUP IV ELEMENTS
A. Kasuya and Y. Nishina   257

MICROSTRUCTURAL STUDIES OF PULSED-LASER IRRADIATED
GRAPHITE SURFACES
J.S. Speck, J. Steinbeck, G. Braunstein,
M.S. Dresselhaus, and T. Venkatesan   263

*Invited Paper

PART V: LASER-INDUCED PHASE
TRANSFORMATIONS IN METALS

*AMORPHOUS GALLIUM PRODUCED BY PULSED EXCIMER LASER
IRRADIATION
J. Fröhlingsdorf and B. Stritzker   271

PICOSECOND TRANSIENT REFLECTANCE MEASUREMENTS OF
CRYSTALLIZATION IN PURE METALS
C.A. MacDonald, A.M. Malvezzi, and F. Spaepen   277

TRANSIENT CONDUCTANCE MEASUREMENTS AND HEAT-FLOW
ANALYSIS OF PULSED-LASER-INDUCED MELTING OF
ALUMINUM THIN FILMS
J.Y. Tsao, S.T. Picraux, P.S. Peercy, and
Michael O. Thompson   283

TIME-RESOLVED LASER INDUCED TRANSFORMATIONS IN
CRYSTALLINE Te THIN FILMS
E.E. Marinero, W. Pamler, and M. Chen   289

NUCLEATION OF ALLOTROPIC PHASES DURING PULSED
LASER ANNEALING OF MANGANESE
J.H. Perepezko, D.M. Follstaedt, and P.S.
Peercy   297

LASER-MATERIAL INTERACTIONS IN PHASE CHANGE OPTICAL
RECORDING
Roger Barton and Kurt A. Rubin   303

Au CLUSTER REDISTRIBUTION DURING NANOSECOND
LASER-ANNEALING OF METAL/INSULATOR MATRICES
W. Pamler, E.E. Marinero, M. Chen, and
V.B. Jipson   309

PART VI: ION-SEMICONDUCTOR
INTERACTIONS AND THIN FILM GROWTH

*KINETICS, MICROSTRUCTURE AND MECHANISMS OF ION
BEAM-INDUCED EPITAXIAL CRYSTALLIZATION OF SEMI-
CONDUCTORS
R.G. Elliman, J.S. Williams, D.M. Maher, and
W.L. Brown   319

SELF AND ION BEAM ANNEALING OF P, Ar, AND Kr IN
SILICON
S. Cannavo, A. La Ferla, S.U. Campisano,
E. Rimini, G. Ferla, L. Gandolfi, J. Liu,
and M. Servidori   329

ENHANCEMENT OF GRAIN GROWTH IN ULTRA-THIN GERMANIUM
FILMS BY ION BOMBARDMENT
Harry Atwater, Henry I. Smith, and Carl V.
Thompson   337

*Invited Paper

MICROSTRUCTURE OF NIOBIUM FILMS ORIENTED BY NON-NORMAL
INCIDENCE ION BOMBARDMENT DURING GROWTH
    J.M.E. Harper, D.A. Smith, L.S. Yu, and
    J.J. Cuomo   343

THE CRYSTALLINE-TO-AMORPHOUS PHASE TRANSITION IN
IRRADIATED SILICON
    D.N. Seidman, R.S. Averback, P.R. Okamoto,
    and A.C. Baily   349

TRAPPING OF INTERSTITIALS DURING ION IMPLANTATION IN
SILICON
    R.J. Culbertson and S.J. Pennycook   357

MEV ION INDUCED MODIFICATION OF THE NATIVE OXIDE OF
SILICON
    R.L. Headrick and L.E. Seiberling   363

ION BEAM DEPOSITION OF MATERIALS AT 40-200 eV:
EFFECT OF ION ENERGY AND SUBSTRATE TEMPERATURE ON
INTERFACE, THIN FILM AND DAMAGE FORMATION
    N. Herbots, B.R. Appleton, S.J. Pennycook,
    T.S. Noggle, and R.A. Zuhr   369

TEMPERATURE AND DOSE DEPENDENCE OF AN AMORPHOUS
LAYER FORMED BY ION IMPLANTATION
    Eliezer Dovid Richmond and Alvin R. Knudson   375

TEMPERATURE DEPENDENT AMORPHIZATION OF SILICON DURING
SELF IMPLANTATION
    W.P. Maszara, G.A. Rozgonyi, L. Simpson,
    and J.J. Wortman   381

IMPURITY DIFFUSION, CRYSTALLIZATION AND PHASE
SEPARATION IN AMORPHOUS SILICON
    R.G. Elliman, J.M. Poate, J.S. Williams,
    J.M. Gibson, D.C. Jacobson, and D.K. Sood   389

AMORPHIZATION AND RECRYSTALLIZATION PROCESSES IN
MONOCRYSTALLINE BETA SILICON CARBIDE THIN FILMS
    J.A. Edmond, S.P. Withrow, H.S. Kong, and
    R.F. Davis   395

            PART VII: ION BEAM MIXING
          AND METASTABLE PHASE FORMATION

*METASTABLE ALLOY FORMATION BY ION BEAM MIXING
    F.W. Saris, J.F.M. Westendorp, and A. Vredenberg   405

*FORMATION OF SURFACE LAYERS OF ICOSAHEDRAL Al(Mn)
    J.A. Knapp and D.M. Follstaedt   415

*Invited Paper

xi

ION IRRADIATION INDUCED AMORPHOUS TO QUASICRYSTALLINE
TRANSFORMATION: COMPOSITION DEPENDENCE IN THE AlMn
SYSTEM
    D.A. Lilienfeld, M. Nastasi, H.H. Johnson,
    D.G. Ast, and J.W. Mayer   427

PHASE TRANSFORMATIONS IN NICKEL–ALUMINUM ALLOYS DURING
ION BEAM MIXING
    James Eridon, Lynn Rehn, and Gary Was   433

SILICIDE FORMATION BY HIGH DOSE TRANSITION METAL
IMPLANTS INTO Si
    F.H. Sanchez, F. Namavar, J.I. Budnick,
    A. Fasihudin, and H.C. Hayden   439

PULSED LASER AND ION BEAM SURFACE MODIFICATION OF
SINTERED ALPHA–SiC
    K.L. More, R.F. Davis, B.R. Appleton, D. Lowndes,
    and P. Smith   445

METASTABLE PHASES PRODUCED IN NICKEL BY HIGH DOSE La
AND O IMPLANTATION AND PULSED LASER MELT QUENCHING
    D.K. Sood, A.P. Pogany, G. Battaglin, A. Carnera,
    G. Della Mea, V.L. Kulkarni, P. Mazzoldi, and
    J. Chaumont   455

SOME CHARACTERISTICS OF $Al_{12}Mo$ IN ALUMINUM ANNEALED
AFTER IMPLANTATION WITH MOLYBDENUM
    L.D. Stephenson, J. Bentley, R.B. Benson, Jr.,
    G.K. Hubler, and P.A. Parrish   461

ION IMPLANTATION AT ELEVATED TEMPERATURES
    Nghi Q. Lam and Gary K. Leaf   467

BINARY AND TERNARY AMORPHOUS ALLOYS OF ION-IMPLANTED
Fe–Ti–C
    D.M. Follstaedt and J.A. Knapp   473

ROLE OF CHEMICAL DISORDERING IN ELECTRON IRRADIATION
INDUCED AMORPHIZATION
    D.E. Luzzi, H. Mori, H. Fujita, and M. Meshii   479

ELECTRON BEAM-INDUCED PHASE TRANSFORMATION IN
$MgAl_2O_4$ SPINEL
    S.J. Shaibani and S.N. Buckley   485

ION BEAM MIXING IN BINARY AMORPHOUS ALLOYS
    Horst Hahn, R.S. Averback, T. Diaz de la
    Rubia, and P.R. Okamoto   491

A COMPARATIVE STUDY OF THE CRYSTALLIZATION OF Ni-P
AMORPHOUS ALLOYS PRODUCED BY ION-IMPLANTATION AND
MELT SPINNING
    James Hamlyn-Harris, D.H. St. John, and
    D.K. Sood   497

AUTHOR INDEX   503

SUBJECT INDEX   505

Preface

This volume contains papers presented at the Materials Research Society symposium on "Beam-Solid Interactions and Phase Transformations" which was held in Boston, Massachussetts, December 2-4, 1985. This symposium was devoted to the fundamental aspects of beam-solid interactions and beam-induced phase transformations in materials. It was the eighth in a series of MRS symposia which have evolved from the original symposium on "Laser Annealing." The symposium began with a full day plenary session organized jointly with the symposium on "Beam-Induced Chemical Processes." This session consisted of eight invited papers by internationally recognized scientists who reviewed recent work in the fields of laser-solid and surface interactions, ion-solid interactions and phase transformations. The final two days of the symposium consisted of both oral and poster sessions on the interaction of photon, ion and electron beams with solids and the thermodynamics and kinetics of rapid phase transformations between metastable and stable states.

Except for the section devoted to plenary reviews, the papers published in this volume are ordered according to content and do not reflect the order of presentation at the meeting. All papers in this volume were refereed using the peer review process. An important element in this symposium was the emphasis on fundamental issues relating to interactions between a wide variety of energy beams and materials. For example, in addition to research results in the field of laser interactions with metals, semiconductors, and insulators, new developments in the areas of ion-induced epitaxy, metastable phase formation by ion irradiation, and ion beam mixing for production of new phases were also presented at this meeting. The symposium revealed not only how the understanding of beam-solid interactions has matured over the last eight years, but also showed how such understanding is leading to important new developments in this expanding field.

Symposium Co-Chairmen

H. Kurz   G.L. Olson   J.M. Poate

January 1986

Acknowledgments

We wish to thank all of the participants and contributors who made this symposium so successful. We particularly wish to acknowledge the invited speakers who provided excellent summaries of specific areas and set the overall tone of the meeting. They are:

N. Bloembergen	C.V. Shank
G. Braunstein	Y.R. Shen
W.L. Brown	B. Stritzker
R.G. Elliman	M.O. Thompson
J.A. Knapp	D. Turnbull
A.M. Malvezzi	H. Walther
J.W. Mayer	J.S. Williams
F.W. Saris	

We are also indebted to the session chairs who directed the sessions, guided the discussions, and gave invaluable help in having the papers refereed. They are:

W.L. Brown	P.S. Peercy
A.G. Cullis	S.T. Picraux
G.K. Hubler	C.W. White
J. Narayan	R.J. von Gutfeld
M.-A. Nicolet	

It is our pleasure to acknowledge with gratitude the financial support provided by the Air Force Office of Scientific Research (Dr. K.J. Malloy), Army Research Office (Dr. J. Hurt), Office of Naval Research (Dr. L.R. Cooper) and the Hughes Aircraft Company.

Finally, special thanks are due to Ms. H.P. Weston (AT&T Bell Laboratories) and Ms. P.E. Carnell (Hughes Research Laboratories) for the secretarial and administrative support that they provided during the organization of the symposium and the preparation of this volume.

PLENARY REVIEWS I

Laser-Solid and Surface Interactions

PULSED LASER INTERACTIONS WITH CONDENSED MATTER

N. BLOEMBERGEN
Division of Applied Sciences, Harvard University, Cambridge, MA   02138

ABSTRACT

The primary interaction is the absorption of photons by electrons. In metals free-free transitions increase the energy of the electron gas. In semi-conductors and insulators electron-hole pairs are created, if the photon energy exceeds the band gap. If it is less, only multiphoton processes can initiate energy transfer from the light beam. In nearly all solid materials Auger processes and electron-phonon interactions occur on a picosecond time scale for the high density and energy of the carrier gas created by intense short laser pulses. Thus melting and evaporation of the material can occur on this time scale. These processes may be considered as the initial phases in the creation of laser produced plasmas. They have been studied by time-resolved measurements of the complex index of refraction, by electron and ion emission, by second harmonic generation, by electrical conductivity and other techniques. Fast time resolution is essential. The dynamic behavior of atoms and phase transitions in the picosecond and femtosecond regime has been opened up for experimental investigation.

1. INTRODUCTION

The basic dissipative interaction process of light with matter is the absorption of light by electrons. In the infrared the light may be absorbed by optical phonons. The electrons in excited states transfer their energy to other electrons and lattice vibrational modes. The infrared excited optical phonons will also share their energy with other phonons. Heating of the material will result. In metals the dominant absorption mechanism is by the excitation of conduction electrons in inelastic free-free transitions. Such transitions in a collisional plasma are sometimes described as inverse brems-strahlung. At long wavelengths and electron energies at a few eV, the process is also well described by a high frequency conductivity.

In semiconductors and insulators, the dominant absorption process is the creation of an electron-hole pair, if the photon energy exceeds the band gap. For photons below the band gap, carriers may be created by two-photon (or multi-photon) processes. Stepwise excitation through discrete levels in the band gap is also possible. Once the carrier density has become sufficiently high, additional absorption by the same mechanism as in metals will take place.

In transparent insulators, the absorption is so low that usually no signi-ficant absorption and heating occurs, until the light intensity is raised to such high values that a plasma is formed by dielectric breakdown. Then further absorption by the carriers in the plasma takes place. The initial electrons are either created by ionization of impurities, by multiphoton absorption or, at extremely high intensities, by tunneling during one half light cycle. At lower intensities nonlinear optical interactions in a transparent dielectric will occur. These are usually of a purely parametric nature and do not give rise to energy dissipation. A hybrid form is presented by the stimulated Raman effect, in which energy is deposited in Raman active phonon modes.

This introductory overview is restricted to strongly absorbing materials, in which the energy is deposited in a relatively thin layer of thickness α^{-1}, where α is the absorption coefficient. The light intensity decays as $I = I_0 \exp(-\alpha z)$ as it penetrates into the medium normal to the boundary at $z = 0$. The results of the interaction depend strongly on the power level and on the pulse duration of the irradiation. The salient features of the interaction in each regime are reviewed, when surfaces of metals or semiconductors are irradiated at increasing power flux densities with millisecond, microsecond, nanosecond, picosecond and femtosecond duration pulses, respectively.

2. INTERACTION WITH MILLISECOND TO NANOSECOND PULSES

For these relatively long irradiation times a quasi-steady state develops. The absorption depth $d_{abs} = \alpha^{-1}$ is generally small compared to the heat diffusion length, $d_{diff} \approx (t_p \kappa)^{\frac{1}{2}}$, during the pulse duration t_p. Here $\kappa = K/\rho C$ is the thermal diffusivity, K is the thermal conductivity, ρ the density, and C the specific heat per unit mass. For metals κ lies in the range of 0.1-1 in cgs units. So, d_{diff} is about 10^{-2} cm for $t_p = 10^{-3}$ s and 10^{-5} cm at $t_p = 10^{-9}$ s. In metals $d_{abs} \approx (3-5) \times 10^{-6}$ cm, the London penetration depth in a strongly collisional plasma, largely independent of the wavelength for $\omega\tau << 1$. For silicon $d_{abs} \approx 10^{-5}$ cm for green light at room temperature. It is shorter for shorter wavelengths and at higher temperatures. Liquid silicon is metallic.

At low intensities the temperature rise at the surface will be small, and the rise will be determined mostly by convective cooling in the atmosphere. For power flux densities exceeding 1 kW/cm², neither convective cooling nor radiative cooling plays a significant role. For a black body at 3347°K, the radiative flux equals 1 kW/cm², while the convective cooling at a Mach number of one by a flowing gas over a surface at this temperature is less than 100 W/cm². The heat balance is largely determined by heat diffusion into the interior of the material or by evaporative losses. Since the thickness of the sample d is usually more than a few mm, one has $d > d_{diff}$ and the back face remains essentially at room tempreature. A one-dimensional heat balance problem results if the beam diameter is also large compared to d_{diff}. The average temperature rise near the front surface may be estimated by assuming that a volume of depth d_{diff} is heated uniformly

$$(kt_p)^{1/2} \rho C_v (T-T_0) \approx (1-R) I_0 t_p \tag{1}$$

where RI_0 is the reflected intensity. Thus the temperature rise is proportional to the power flux density and the square root of the pulse duration, or proportional to the energy fluence and inversely proportional to the square root of the pulse duration. It is easy to reach the melting point, or indeed the boiling point, of any material with laser flux densities of 10 kW/cm² or more. Thus laser welding, drilling, etc., have found widespread applications. Figure 1 is taken from an early standard text [1] to remind the reader what happens in the regime of relatively long irradiation times. Detailed numerical solutions of the heat transport equation can take account of the variation of the material constants with temperature, of the latent heat on melting, etc.

When the temperature is elevated so that significant vaporization occurs, a steady state regime may develop for microsecond pulses where the energy absorbed per unit time per unit area is carried away by the evaporated mass. The surface recedes by ablation at a constant speed v_s given by

$$\rho L_v v_s = (1-R) I_0 \tag{2}$$

Here L_v is the heat of evaporation per unit mass. This behavior is illustrated in Fig. 2, where the surface temperature of an Al alloy target, irradiated by a laser flux of 10^7 W/cm² is plotted vs time [2]. The mass flow rate ρv_s for evaporation into vacuum may be crudely estimated by taking the saturated vapor pressure at the surface temperature $p(T_s)$ and assuming the flow normally outward from the surface to be described by a one-sided Maxwell-Boltzmann distribution at temperature T_s. This procedure yields

$$\rho v_s = p(T_s)/(2\pi kT_s/m)^{1/2} \tag{3}$$

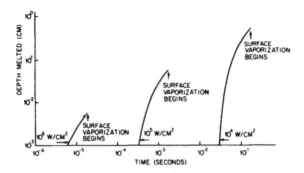

Fig.1. Melted depth in yellow brass as a function of
time for various laser flux densities (after ref. 1).

Combination of Eqs. (2) and (3) yields an equation for the steady state surface
temperature.

$$\frac{p(T_s)L_v}{(2\pi kT_s/m)^{1/2}} = (1-R) I_0 \qquad (4)$$

Since the saturated vapor pressure depends exponentially on the temperature
according to the Clausius-Clapeyron equation, the surface temperature will
rise only logarithmically with increasing intensity above the boiling point,
while the surface ablation rate increases linearly with I_0.

It may be possible to heat a substance above the critical point, when
the intensity is so high that the pressure required to yield the steady state
evaporation rate exceeds the pressure at the critical point. In this situation
the following model description may be used. The solid material is superheated
under pressure to a high temperature. Individual particles are held in a
potential well determined by an average bond strength. If the normally
directed kinetic energy exceeds the potential well depth, the binding energy
per atom in the condensed phase, E_{coh}, the particle will escape. The number
of particles escaping per unit area per unit time is given by

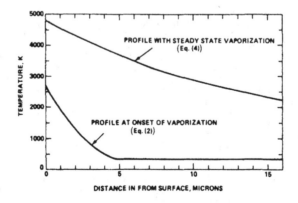

Fig.2. Temperature profiles of an aluminum target absorbing
a laser flux $(1-R)I_0 = 10^7$ W/cm^2 (after ref. 2).

$$N_0 \frac{1}{(2\pi k T_s m)^{1/2}} \int_0^\infty e^{-\frac{1}{2}(m v_x^2)/(kT_s)} v_x \, dv_x .$$

$$v_x = \sqrt{\frac{2E_{coh}}{m}}$$

Multiplication by the mass of particle leads to

$$\rho v_s = \rho_s \left(\frac{kT_s}{m}\right)^{1/2} e^{-E_{coh}/kT_s} \tag{5}$$

where ρ_s is the solid density and T_s the surface temperature. This model is, of course, entirely analogous to that of thermionic emission, which describes the evaporation of electrons in the Maxwell-Boltzmann tail of the Fermi-Dirac distribution. In this case E_{coh} is replaced by the work function. It is reassuring that Eq. (3) with the saturation vapor pressure has a similar exponential character as Eq. (5).

It is possible to make these considerations more quantitative. The one-dimensional model of heat diffusion is valid if the laser spot size is large compared to the thickness of the sample and large compared to the thermal diffusion length. The diffusion equation is

$$\rho C \frac{\partial T}{\partial t} = \frac{\partial}{\partial z}\left(K \frac{\partial T}{\partial z}\right) + \alpha(1-R)I_0 \exp - \int_0^z \alpha \, dz . \tag{6}$$

In this equation the material constants for the density ρ, the specific heat C, the thermal diffusion constant K, the absorption coefficient α and the reflection coefficient R are all functions of the temperature. The boundary condition at the front surface, $z = 0$, is

$$-K \frac{\partial T}{\partial t} + \rho v_s \Delta H_v + \bar{\varepsilon} \sigma T^4 + C'(T-T_0) = 0 . \tag{7}$$

Here $\bar{\varepsilon}$ is the average emissivity of "grey-body" radiation, $C'(T-T_0)$ represents the heat loss by convection at the front surface, ΔH_v is the enthalpy of vaporization. The occurrence of melting or another phase transition can be incorporated in the model by adding a delta function to the specific heat, $L_m \delta(T-T_m)$, on the left-hand side of Eq. (7), where L_m is the latent heat of melting.

The evaporation process into an atmospheric environment may also be analyzed more precisely by considering a Knudsen-type boundary layer adjacent to the surface. Through this boundary layer the vapor expands into a "rarefaction-wave." This process has been analyzed by Anisimov [3], Krokhin [4] and Knight [5], among others. The result is that the recession rate at a given surface temperature T_s is about 18 percent smaller for a monatomic gas than that given by Eq. (3) or (5). This can be ascribed to gas kinetic collisions in the boundary layer, so that some atoms which have escaped according to those equations are turned back toward the interface.

A more important effect from a practical point of view is that the temperature of the evaporating gas may be so high that a considerable fraction of the evaporating atoms is ionized. If molecules evaporate at high temperature, they may be dissociated. Neutral particles in excited electronic states play an important role in the dynamics of plasma formation by photo-ionization processes. A dense plasma in front of the surface has the consequence that the laser light may now be absorbed by the plasma, rather than by the condensed phase target. These plasma effects have been investigated in great detail both theoretically and experimentally [1,4,6]. They become generally important when the incident power flux density exceeds 10^8 W/cm^2. Clearly the plasma formation processes will also depend on the presence of an atmosphere in front of the irradiated target.

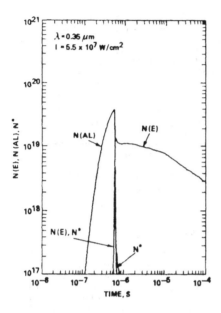

Fig.3. Calculated species densities
(Al atoms, excited Al atoms and
electrons) vs time for an incident
laser intensity of 5.5×10^7 W/cm^2.
The reflection coefficient of the
alloy at $\lambda = 0.35$ μm is assumed to
be R = 0.75 (after ref. 2).

Figure 3 illustrates the initial stages of plasma formation [2] in front
of an aluminum alloy in vacuo, which is irradiated by a microsecond pulse from
an excimer laser at $\lambda = 0.35$ μm. It may be concluded that the interaction of
laser pulses with condensed matter is well understood in the regime where the
pulse length is longer than 10^{-9} s and the maximum power flux density is less
than 10^8 W/cm^2.

3. IRRADIATION BY PICOSECOND PULSES

When t_p lies in the range 10^{-12}-10^{-9} s, we enter the regime where the
heat diffusion length becomes comparable to or smaller than the absorption
depth $\sqrt{Kt_p} \leq \alpha^{-1}$. When $d_{diff} \ll d_{abs}$, the heat is deposited in a layer of
constant depth, d_{abs}, and the temperature rise increases linearly with t_p.
The average temperature in the surface layer is proportional to the fluence,
and Eq. (1) is replaced by

$$C_v(T-T_0)d_{abs} = (1-R) \int_0^{t_p} I_0 dt . \qquad (8)$$

Since all the light energy is absorbed in a thin layer, the melting point or
the evaporation temperature of any metallic material can be reached at modest
energy fluence. For example, the melting point of tungsten is reached for
0.1 J/cm^2 at $\lambda = 1.06$ μm in a 30 ps pulse, and the melting point of silicon
is reached for 0.2 J/cm^2 at $\lambda = 0.53$ μm and $t_p \approx 2 \times 10^{-11}$ s. On melting,
silicon becomes metallic, the absorption depth at $\lambda = 0.53$ μm decreases
abruptly from about 2×10^{-5} cm to 3×10^{-6} cm. It is estimated that at 1 J/cm^2
the silicon surface is heated above the critical point temperature of 5000°K.
Significant surface damage due to evaporation may be observed after the irradiation.

It is important to emphasize that practically no evaporation occurs *during*
a picosecond pulse of 10^{-11} s duration, or less. There simply is not enough
time for the atoms to move and to establish a steady state evaporation regime.

Consider an elevated surface temperature T_s, where the average thermal velocity in the normal direction is $(kT_s/m)^{\frac{1}{2}}$. The model of evaporation out of a potential well of depth E_{coh}, given by Eq. (5) should describe the situation rather well, as the average time between collisions is longer than t_p, even if the vapor pressure were one thousand atmospheres. The maximum rate of evaporation is $\rho_g(T_s)(kT_s/m)^{\frac{1}{2}}$. During the light pulse the surface of the condensed matter recedes by an amount $d_{evap} = (\rho_g/\rho_s)(kT_s/m)^{\frac{1}{2}}t_p$. Let us take $v_{th} \approx 2 \times 10^{+5}$ cm/s, $t_p \sim 10^{-11}$ s. Even if ρ_g/ρ_s were as large as $\sim 10^{-2}$, which would correspond to a pressure of about six hundred atmospheres for atomic silicon vapor at about 5000°K, one would find $d_{evap} \approx 2 \times 10^{-8}$ cm, or less than one atomic layer of Si would have evaporated during the pulse. Thus the optical thickness of the vapor during the irradiation is negligible. Furthermore, no collisions have taken place and no plasma has developed during the pulse.

Still another way to look at the same situation is as follows. Only atoms at the surface can escape from the condensed phase. These atoms vibrate with a typical Debye frequency, characteristic for phonons at the boundary of the Brillouin zone. This frequency is on the order of 10^{13} s^{-1}. During each vibrational period the atom has a probability for escape equal to $\exp[-E_{coh}/kT_s]$ to overcome the potential barrier. This language is identical to that of a simple Eyring model for molecular dissociation. For $E_{coh} \sim 2.5$ eV and $T_s \sim 0.5$ eV corresponding to 6000°K, the probability for escape on each try is $e^{-5} \approx 0.01$. Again not more than one surface layer will escape during 10^{-11} s.

It is quite likely that a significant fraction of the relatively small number of particles escaping during the pulse is ionized. After the picosecond pulse the surface layer cools very rapidly by heat diffusion into the cold interior substrate. Since the vapor pressure depends exponentially on the temperature, the evaporation rate also diminishes very rapidly on cooling. It may be estimated that the number of particles escaping from the metal after the light pulse is of the same order of magnitude as the number escaping during the light pulse. Thus the picosecond pulse regime of heating of strongly absorbing materials is significantly different from that by pulses of a nanosecond or longer, because of the following two inequalities,

$$(\kappa t_p)^{1/2} < d_{abs}, \quad \text{and} \quad (\rho_g/\rho_s)(kT_s/m)^{1/2}t_p \ll d_{abs}.$$

In the limit $\rho_g/\rho_s = 1$ the last condition becomes $v_{ac}t_p \ll d_{abs}$, where v_{ac} is the sound velocity. It is possible to raise the condensed matter in less than 10^{-11} s to temperatures well above the boiling point or critical point. Since $d_{abs} \gg v_{th}t_p$, the atoms have no time to move over significant distance during the pulse, and any absorbing condensed material may be raised to the temperatures in the range of $0.5-1 \times 10^4$ K and pressures of about 10^3 atmospheres. The incident energy fluence for a 10 ps pulse required to reach this state in materials with an absorption depth $d_{abs} \sim 50$ nm lies typically in the range between 0.3 and 2 J/cm². This corresponds to average power flux densities of about 10^{11} W/cm². Because of the short duration, plasma effects or absorption in the vapor are usually absent.

In laser generation of fusion plasmas with inertial confinement, much higher power flux densities are used, of about 10^{14} W/cm². Then the electric field amplitude is comparable to the Coulomb field, which is responsible for the binding of the valence electrons. During one light cycle considerable numbers of electrons can tunnel out of their orbits. Any material, including transparent dielectrics, is converted into plasma at the solid density during the early stages of the pulse. Thus a dense plasma with initial electron temperature of 10^2-10^3 eV is assumed as the initial state. This plasma gets further heated by the bulk of the laser pulse [6].

The present paper focuses on the "pre-plasma" conditions, which are realized for pulses with between 10^{-12} and 10^{-10} s duration and peak power levels below $10^{10}-10^{11}$ W/cm². Then the material is characterized by electron temperatures on the order of a few eV or less, and the phonon temperature lies in the range 10^3-10^4 K.

9

In metallic materials the electron-phonon as well as the phonon-phonon interaction times under these conditions are generally shorter than 1 ps, so that local thermodynamic equilibrium prevails in the absence of phase transitions. The very rapid heating rates during the pulse and the very rapid cooling rates after the pulse, which may amount to 10^{14} K/s, may produce considerable superheating of the solid phase and undercooling of the liquid phase. Phase front boundaries may be driven to velocities of 10^4 cm/sec and molten material may resolidify in amorphous phases. Alloys may retain a composition characteristic of the liquid phase, as there is no time for chemical segregation.

4. PICOSECOND IRRADIATION OF SEMICONDUCTORS

The proceedings of the MRS symposia on laser-beam interactions from 1978 onward provide a detailed history of this topic [7-14]. In a similar general introductory lecture seven years ago, I described the phenomenon of heating on a picosecond time scale [15]. After considerable controversy and discussion, a general consensus has developed that the hot carriers and the lattice in silicon reach a common temperature on a picosecond time scale, and that a silicon surface layer may melt on this same time scale. A threshold value for the fluence inducing melting at the surface of 0.2 J/cm^2 was established for 10^{-11}s pulses at λ = 0.53 μm. Since molten silicon is metallic, an abrupt change in the reflectivity occurs, as shown in Fig. 4. The most detailed information about the nature of the hot layer during and following the pump pulse is obtained by an independent optical picosecond probe pulse [16]. The complex index of refraction is measured by reflection and transmission of a small area in the center of the heated area. Data have been obtained over a wide parameter space, including variation of the pump pulse fluence, time delay between pump and probe, and variation of the probe wavelength from the far-infrared to the near ultraviolet. At long wavelengths the index is predominantly determined by the density of the carrier plasma; at shorter wavelengths the index is determined by indirect band transitions involving the lattice temperature. An illustrative example of the pertinent experimental data on the surface reflectivity as a function of pump fluence 15 ps after the pump pulse at three wavelengths is shown in Fig. 4. With similar data at other delay times, and with data on transmission, the temporal dependence of the

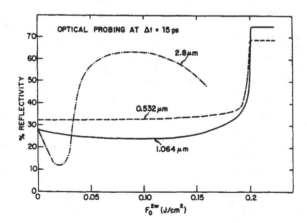

Fig.4. The reflectivity of a silicon crystalline surface at three different wavelengths, 15 ps after a green pump pulse, as a function of pump fluence.

10

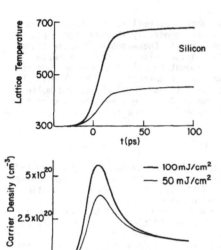

Fig.5. The carrier density and the
lattice temperature of a crystalline
silicon surface irradiated at t = 0
by a green pump pulse of 20 ps
duration for fluences of 0.05 and
0.1 J/cm^2, respectively.

carrier density and lattice temperature in crystalline silicon is obtained,
as shown in Fig. 5, for two fluences below the melting threshold. The carrier
density is limited during and after the pump pulse by Auger recombination.
Just before melting, a maximum carrier density of about 1.2×10^{21} /cm^3 is
obtainable with green pulses of 10^{-11} s duration just before melting sets in.
This density is insufficient to induce a transition to another hypothetical
solid phase. The lattice temperature falls off slowly, because a rather thick
layer of 10^{-5} cm is heated by green light in crystalline silicon.

    In the metallic liquid phase the absorption depth is reduced by an order
of magnitude. The temperature rises rapidly to the boiling point and beyond
the critical point as the fluence is increased to over 1 J/cm^2. Emission of
positive ions and electrons is observed. The velocity distribution of the
emitted neutral particles has been measured and is in agreement with a thermal
Maxwell-Boltzmann distribution [17]. The velocity distribution of positive
ions has also been measured [18]. The average kinetic energy increases with
increasing fluence corresponding to a surface temperature of 5000°K, beyond
which space charge and dense plasma effects become important. Here one enters
a regime characteristic of the picosecond time domain where emission of neutral
particles, electrons and positive ions proceeds at comparable rates. The
process of the plasma formation is distinct from that discussed earlier, in
the microsecond regime.

    There are numerous other experimental methods which support the picture
of melting and subsequent resolidification. In certain ranges of pump fluence
and wavelength,the resolidification occurs so rapidly with such large undercooling of the liquid phase that amorphous silicon may be formed [19,20], as
was reported in early 1979. The velocity of the liquid-solid interface has
been measured by dc conductivity experiments with nanosecond time resolution.
X-ray diffraction has also been accomplished on this time scale, confirming
thermal expansion and melting. The post-pulse examination of the surface
structure by LEED, Rutherford back scattering and the change in impurity distribution are also powerful tools [7-14]. It may be difficult to improve the
temporal resolution of most of these techniques to the picosecond domain. Picosecond electron pulses can, however, be obtained by picosecond photoelectric
emission, and picosecond electron diffraction from surfaces has been accomplished
[21].

The semiconductors Ge and GaAs show qualitatively similar behavior to Si. The melting of GaAs on a picosecond time scale is very convincingly demonstrated by the technique of second harmonic reflection [22]. The integrated reflected second harmonic signal from a single crystal increases as the square of the energy fluence pulses of fixed duration, but it saturates when the energy fluence exceeds the threshold for melting. In the isotropic liquid phase the second harmonic generation is negligible. Note that on a picosecond time scale the transition is studied at constant composition, since the evaporation of As is negligible.

Graphite shows an abrupt change in the reflectivity at all wavelengths, when the pump fluence is raised above a threshold value [23] at which the lattice temperature at the surface exceeds 4000°K. Obviously a strong disruption of the graphite structure takes place. Cross linking between planes and other bond deformations occur to such an extent that perhaps a liquid-like structure evolves. Attempts have also been made to describe the distorted structure as carbyne, a random conglomeration of C_n fragments. It is well known from the existence of both graphite, a semimetal, and diamond, a large band-gap insulator, that the electronic structure is very sensitive to the atomic arrangement. Predictions have been made that liquid carbon should be metallic, in analogy with silicon. It is also possible to argue that a strongly perturbed spatial arrangement of carbon atoms, starting from either the graphite or the diamond structure, would exhibit a band gap. The advantage of picosecond irradiation is that no vapor density builds up in front of the surface during the pulse. Experimental data presented elsewhere at this conference [23] indicate that the reflectivity in the high temperature phase is lower, suggesting a decrease in both the real and imaginary part of the dielectric constant compared to that of graphite at wavelengths extending from 1.9 μm to 0.53 μm. Thus the high temperature phase of carbon becomes more transparent and the incident pulse intensity can penetrate through it and melt the next layer. A high fluence picosecond pulse could thus transform a layer of considerable depth to the transparent high temperature phase. The cooling rate is slower than for metallic high temperature phases, and more evaporation may occur after the pulse. The picosecond pump and probe techniques clearly provide a key to the solution of the problem of the structure of matter at high density, characteristic of condensed phases, and high temperatures in the range of 3,000-10,000°K. This range, corresponding to 0.5-1 eV, has hardly been explored. It lies well below the hot plasma range with temperatures between 10^2-10^4 eV, where very extensive work on laser-plasma interactions has been carried out.

5. FEMTOSECOND PULSE INTERACTIONS

The characteristic time for electron-phonon and phonon-phonon interactions at the densities and temperature ranges under consideration appears to lie in the range of 10^{-13} to 10^{-12} s for most materials. With the recent development of femtosecond pulse techniques, it has become possible to supply energy to the carriers in a time shorter than the characteristic time for energy exchange with the lattice. The carriers among themselves will still attain a common temperature, T_e, as the characteristic time for collisions between carriers in a dense plasma is less than 10^{-14} s. Femtosecond irradiation would thus initially create an extremely hot carrier gas in a cold lattice. Subsequently energy would be transferred to phonons, perhaps in several stages before a common lattice temperature would be reached. Thermal evaporation of particles would thus occur after the femtosecond pulse.

Evidence for the existence of a higher electron temperature has been found in experiments on photoelectric emission from tungsten induced by femtosecond pulses [24]. Optical pump and probe experiments with femtosecond resolution are clearly very important and promising. The experimental techniques require further refinement to obtain reliable data for theoretical interpretation.

Progress in this exciting field is reviewed at this conference in a separate paper by Shank [25].

ACKNOWLEDGMENTS

I wish to thank Professor H. Kurz, Dr. A. M. Malvezzi and Dr. C. Y. Huang for many helpful discussions. This work was supported by the U.S. Office of Naval Research under contract no. N00014-83K-0030.

REFERENCES

1. J.F. Ready, Effects of High-Power Laser Radiation (Academic Press, New York, 1971).
2. D.I. Rosen, J. Mitteldorf, G. Kothandaraman, A.N. Pirri and E.R. Pugh, J. Appl. Phys. 53, 3190 (1982).
3. S. Anisimov, Soviet Physics JETP 27, 182 (1968).
4. O.N. Krokhin, in Physics of High Energy Density, edited by P. Caldirola and H. Knoepfel (Academic Press, New York, 1971), pp. 278-305.
5. C.J. Knight, J. Fluid Mech. 75, 469 (1976); AIAA Journal 17, 519 (1979).
6. T.P. Hughes, Plasmas and Laser Light (Wiley, New York, 1975).
7. Laser Solid Interactions and Laser Processing, edited by S.D. Ferris, H.J. Lemay and J.M. Poate (American Institute of Physics, New York, 1979).
8. Laser and Electron Beam Processing of Materials, edited by C.W. White and P.S. Peercy (Academic Press, New York, 1980).
9. Laser and Electron Beam Solid Interactions and Materials Processing, edited by J.F. Gibbons, L.D. Hess and T.W. Sigman (Elsevier North-Holland, New York, 1981), Mat. Res. Soc. Symp. Proc. 1, (1981).
10. Laser and Electron Beam Interactions with Solids, edited by B.R. Appleton and G.K. Celler (Elsevier North-Holland, New York, 1982), Mat. Res. Soc. Symp. Proc. 4, 49 (1982).
11. Laser-Solid Interactions and Transient Thermal Processing of Materials, edited by J. Narayan, W.L. Brown and R.A. Lemons (Elsevier North-Holland, New York, 1983), Mat. Res. Soc. Symp. Proc. 13 (1983).
12. Laser Annealing of Semiconductors, edited by J.M. Poate and J.W. Mayer (Academic Press, New York, 1982).
13. Energy Beam-Solid Interactions and Transient Thermal Processing, edited by J.C.C. Fan and N.M. Johnson (Elsevier North-Holland, New York, 1984), Mat. Res. Soc. Symp. Proc. 23 (1984).
14. Energy Beam-Solid Interactions and Transient Thermal Processing, edited by D.K. Biegelsen, G.A. Rozgonyi and C.V. Shank (Elsevier North-Holland, 1985), Mat. Res. Soc. Symp. Proc. 35 (1985).
15. N. Bloembergen, reference 7, p. 1.
16. H. Kurz and N. Bloembergen, reference 14, p. 3, and other references quoted therein.
17. N. Fabricius, P. Hermes, D. von der Linde, A. Pospieszczyk and B. Stritzker, Phys. Rev. , (1985) to be published.
    B. Danielzik, N. Fabricius, M. Rowekamp and D. von der Linde, Appl. Phys. Lett. (1985) to be published.
18. A.M. Malvezzi, H. Kurz and N. Bloembergen, reference 14, p. 75.
19. P.L. Liu, R. Yen, N. Bloembergen and R.T. Hodgson, Appl. Phys. Lett. 34, 864 (1979).
20. R.Tsu, R.T. Hodgson, T.Y. Tan and J.E. Baglin, Phys. Rev. Lett. 42, 1356 (1979).
21. S. Williamson, G. Mourou and J.C.M. Li, Phys. Rev. Lett. 52, 2364 (1984).
22. A.M. Malvezzi, J.M. Liu and N. Bloembergen, Appl. Phys. Lett. 45, 1019 (1984).
23. C.Y. Huang, A.M. Malvezzi and N. Bloembergen, in Beam-Solid Interactions and Phase Transformations, edited by H. Kurz, G.L. Olson and J.M. Poate, Mat. Res. Soc. Symp. Proc., to be published (1986).

24. J.G. Fujimoto, J.M. Liu, E.P. Ippen and N. Bloembergen, Phys. Rev. Lett. 53, 1837 (1984).
25. C.V. Shank, in Beam-Solid Interactions and Phase Transformations, edited by H. Kurz, G.L. Olson and J.M. Poate, Mat. Res. Soc. Symp. Proc. (1986) to be published.

FEMTOSECOND DYNAMICS OF HIGHLY EXCITED SEMICONDUCTORS

CHARLES V. SHANK AND MICHAEL C. DOWNER*
AT&T Bell Laboratories
Holmdel, New Jersey 07733

ABSTRACT

The advent of femtosecond optical pulse techniques has provided new opportunities for the investigation of the dynamical properties of highly excited semiconductors. In this paper we describe recent investigations of delayed Auger heating in Si.

INTRODUCTION

The power and the range of investigations into the dynamical properties of matter have been greatly enhanced by the development of new measurement techniques. Progress in ultrashort pulse generation has continued to advance at a very brisk pace where currently the generation of optical pulses as short as 8 femtoseconds [1] presses the limits of current technology.

The impact of short pulse measurements on the study of highly excited semiconductors has added greatly to our understanding of these materials under intense illumination. The interaction of ultrashort pulses with semiconductors has been studied by a variety of approaches including time resolved transmission, reflectivity, photoluminescence, ellipsometry, surface second harmonic generation, and femtosecond imaging. [2-13]

DYNAMICS OF AUGER HEATING

The processes by which the energy of an intense short laser pulse is transferred to heat in a solid upon absorption have been investigated extensively on time scales ranging from nanoseconds to femtoseconds. Femtosecond time resolved measurements have revealed the dynamics of melting and evaporation at a highly photo-excited silicon surface. These experiments have shown that a silicon surface becomes highly reflective [11] and disordered [12] on a time scale of less than one picosecond following femtosecond pulsed excitation above a threshold fluence of approximately $0.1 \ J/cm^2$.

---

* Present address: Dept. of Physics, Univ. of Texas at Austin

In the work reported here we investigate heating of the crystalline silicon lattice by laser pulses having an energy below the threshold for melting. Lattice heating arises primarily from two processes. The first process, direct heating, results from the emission of phonons by photo-excited carriers as they relax toward the band edge. This process takes place in less than a picosecond. At higher densities when the bottom of the conduction bands begins to be occupied, carriers relax by phonon emission until the Pauli exclusion principle prevents further direct heating. In this case, the second process, phonon emission by carriers which have been re-excited by Auger recombination, becomes the dominant lattice heating mechanism. The delayed reheating of the carriers by Auger recombination thus results in a delay of the transfer of energy to the crystal lattice. Kurz and Bloembergen [10] have calculated that the temperature rise of a Si surface excited by a 30 psec laser pulse is not completed for tens of picoseconds following the peak of the pulse.

In order to observe these effects, a thin film of Si deposited on a sapphire substrate is used as a sample. The thin film geometry permits the measurement of transmittance and absorbance spectra under conditions of uniform illumination. In addition, the thin film geometry eliminates the complications that arise from diffusion. Using thin film optics equations, n and k can be determined as a function of time following excitation.

In our experiments optical pulses of 100 femtoseconds duration and approximately 0.3 mJ energy at 620 nm were provided by a colliding pulse mode-locked ring dye laser [14] followed by a four stage optical amplifier operating at a repetition rate of 10 Hz. [15] A 50/50 beam splitter was used to divide output beam equally into excitation and probe beams. The excitation beam was focused to a spot size of approximately 100 μm on the SOS sample and attenuated to a fluence below the damage threshold of 0.1 J/cm^2. The probe beam passed through a variable optical delay line, and was then focused into a cell containing water to generate a white light continuum. [16] The green-blue portion of this continuum beam was spectrally filtered and attenuated by a factor of 100 before being focused at near normal incidence onto a spot illuminating the photo-excited area of the sample.

In order to ensure that we studied a uniformly excited sample area, we selected probe light from only a 10 micron diameter circular area near the center of the photo-excited region using an imaging technique, in which transmitted and reflected probe beams were collected with microscope objectives, then imaged onto irises placed in the image planes. The selected portions of these transmitted and reflected beams, along with a reference probe beam, were then collected at the entrance slit of a spectrometer equipped with a vidicon detector. The normalized spectra between 480 nm and 580 nm. were stored and analyzed with an optical multi-channel analyzer.

Our sample was a 0.5 micron silicon film on a 0.5 mm thick sapphire substrate. The sample was mounted on an X-Y electronic translation stage and rastered over a 2 mm square area during data collection in order to guard against cumulative damage resulting from occasional intense pulses or hot spots in the excitation pulse. Variations in sample thickness over this area were negligible. Spectra from approximately 300 consecutive shots were collected and averaged for each pump-probe time delay. Figure 1 shows (a) reflectivity and (b) transmission spectra at time delays $\Delta t = -1$ ps, 0.5 ps, and 200 ps. following photo-excitation of the silicon film with a fluence of 0.02 J/cm^2. Shifts in the strong Fabry-Perot interference fringes in these spectra allow sensitive measurement of the small index changes which occur at this fluence at visible wavelengths. In Figure 1a, the initial blue shift at $\Delta t = 0.5$ ps results from a decrease in the real index n caused by the presence of a dense electron-hole plasma. Note that the shift is more pronounced in the red, as expected from a Drude model description of the plasma-induced index change. The slight decrease in fringe contrast results from a corresponding increase in the imaginary index k. At later time delays, the fringes in the reflectivity spectrum drift back towards the red, eventually passing the original fringe positions in a few tens of picoseconds, before reaching a final steady-state red shifted position evident in the curve at $\Delta t = 200$ ps. There is also a slight recovery in fringe contrast. This red shift, which now is more pronounced in the blue, results from the lattice heating which accompanies the relaxation of carriers which have been re-excited through Auger recombination.

The corresponding transmission spectra in Figure 1b shows qualitatively similar shifts in fringe positions. The effect of changes in k, or absorption, however, is now more clearly evident as a drop in overall transmission and fringe contrast. The strong initial photo-induced absorption evident in the curve at $\Delta t = 0.5$ ps., followed by a slow partial recovery, suggests that plasma-induced absorption, along with heating, contributed to the changes in k.

We analyzed the data in Figure 1 by extracting values of the index of refraction from the reflectivity (R) and transmission (T). A program was written which read the experimental R and T values for each wavelength and time delay, then solved the thin film optics equations [17] iteratively to find n and k values which yielded the experimental R and T within 0.2 percent. The results of this analysis of the temporal evolution of n at probe wavelength 485 nm. and pump fluence 0.02 J/cm^2 are shown in Figure 2. A wavelength as far in the blue as is possible, consistent with good signal-to-noise ratio, was selected because the heating effect is largest and the plasma effect smallest in this part of the spectrum.

The real index n (filled circles, middle curve in Figure 2) drops immediately by 1.2 percent upon photoexcitation because of the presence of a dense electron-hole plasma (approximately $5 \times 10^{20} cm^{-3}$). The value of n then rises as the lattice temperature rises and the plasma density decreases. The index n returns to its original value at $\Delta t = 25$ ps., primarily through Auger recombination, and then continues rising toward a final constant value nearly 1 percent higher than its initial value within 200 ps. At longer probe wavelengths, the magnitude of the initial drop in n increased in good agreement with the prediction of a Drude model for the electron-hole plasma, while the rise in n above its initial value at later time delays became progressively smaller. At infrared wavelengths, Δn never rose above zero in the time scale investigated, as shown by the data for probe wavelength 1.0 micron also plotted in Figure 2 (x's, lower curve). The value of $\Delta n/n_0$ in this

18

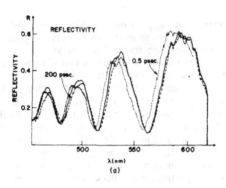

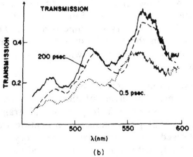

Figure 1

Normalized experimental time-resolved (a) reflectivity and (b) transmission spectra of $0.5\mu m$ silicon-on-sapphire film before excitation (solid curves), and 0.5 psec (dotted) and 200 psec (dashed) after excitation. Note strong Fabry-Perot interference fringes. In (a), note initial blue shift of fringes, stronger at longer wavelengths, caused by presence of dense electron-hole plasma. At later times, fringes red shift, more strongly at shorter wavelengths, because of lattice heating. In (b), note strong initial photo-induced absorption (transmission decrease), followed by partial recovery.

data has been scaled down by a factor of approximately four for direct comparison with the data at $\lambda = .485\ microns$, as discussed in more detail below.

The open circles and upper (dashed) curve in Figure 2 result from subtracting the lower data points and curve from the middle data points and curve, for reasons also discussed below. We now analyze the photo-induced

$\lambda_{PROBE} = 0.485 \, \mu m$

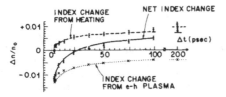

Figure 2

Photo-induced fractional changes in real part of refractive index (n). Curves result from theoretical calculations based on Auger heating (see text).

index changes shown in Figure 1 in terms of a simple model which includes plasma and lattice heating effects. We assume that the effect of the plasma on the refractive index at this wavelength is reasonably well described by a Drude model because of the observed wavelength dependence of Δn at $\Delta t = 0$ noted above, and because the probing wavelength 485 nm lies well below the direct edge of silicon. The plasma-induced index change therefore depends on the electron-hole pair density N exactly as at infrared wavelengths, apart from a scaling factor of λ^2. At $\lambda = 1.0$ micron, the heat-induced index change is three times smaller than at while the plasma-induced index change is more than four times greater. Thus, within our experimental error, the infrared data represents a pure plasma-induced index change and, when scaled by λ^2, models the plasma-induced index change at 485 nm. To an excellent approximation, this fractional index change is given by:

$$\Delta n/n_o = - (2\pi e^2/\epsilon_o m^* \omega^2) \, N, \qquad (1)$$

where silicon refractive index $n_o = 4.3$ at 485 nm, dielectric constant $\epsilon_o = 19$, and $m^* = 0.2 m_e$ is the reduced effective mass for electrons and holes. Infrared optical data from n-doped silicon, [19] experiments [10,20] on plasmas produced by picosecond laser pulses in undoped silicon, and calculations, [21,22] support an effective mass value only twenty percent higher than the low density value for $N = 5 \times 10^{20} cm^{-3}$, although for $N > 10^{21} cm^{-3}$, m^* is expected to increase rapidly with N. The space-time evolution of the electron-hole pair density N(x,t) is described by a one-dimension continuity equation. In a film with thickness less than the optical absorption depth, however, diffusion is suppressed, so that Auger recombination dominates the temporal evolution of N according to the formula:

$$dN/dt = - (C_e + C_h)N^3, \qquad (2)$$

where C_e, C_h are the Auger coefficients for e-e-h and e-h-h processes, respectively. For carrier densities up to $10^{20} cm^{-3}$ these coefficients have been measured [23] to be $C_e, C_h = 2.8, 0.99 \times 10^{-31} cm^6 sec.^{-1}$. At higher densities, screening may reduce the Auger coefficients somewhat. [24] As a first approximation, we neglect any density dependence of C_e, C_h, allowing analytic solution of (2). The simultaneous solution of equations (1) and (2) yields an expression for the fractional plasma-induced index change at 485 nm, which can be used to fit the scaled infrared data in Figure 2. Neglecting carrier and heat diffusion on the time scale of interest, the Auger heating can be described by the simple energy balance equation:

$$\rho C_p \, dT/dt = (C_e + C_h) \, N^3 \, (E_g + kTH), \qquad (3)$$

H is a degeneracy factor equal to 3 for a non-degenerate plasma and which increases with degeneracy and carrier density. Although H strictly depends on density N, we can approximate H as a constant, thus allowing analytic solution of Eq. (3) using N(t) as derived from the solution of Eq. (2). The temporal evolution of the positive fractional index change caused by Auger heating can be related to the temperature evolution by:

$$\Delta n/n_o = \beta \Delta T(t) n_o, \qquad (4)$$

where $\beta - 8.0 \times 10^{-4\circ} K^{-1}$ at 485 nm. [18]

The lower (dotted) theoretical curve in Figure 2, which describes the effect of the e-h plasma alone on the index, independent of heating, was obtained from simultaneous solution of Eqns. (1) and (2) using initial density $N_o = 5 \times 10^{20} cm^{-3}$ and combined Auger coefficient $C_e + C_h = 3.0 \times 10^{-31} cm^6 sec.^{-1}$. The upper (dashed) theoretical curve, which describes the effect of heating alone on the index, independent of the plasma, was obtained by solving Eqns. (3) and (4) using these same initial values. The middle curve, which fits the original data at 485 nm, is the sum of the upper and lower curves, and describes the combined effect of heating and e-h plasma on the index. The value of N_o at the central (probed) area of the Gaussian pump spot intensity profile was determined independently by measuring the self-reflectivity and self-transmission of the excitation pulse at .02 J/cm^2. The value of $C_e + C_h$ is somewhat lower than values measured at densities under $10^{20} cm^{-3}$, indicating that some screening may be present.

The only adjustable parameter used was an instantaneous fractional index change at $\Delta t = 0$ from direct heating, for which the value $+.001$ yielded the best fit. Clearly, this simple model satisfactorily explains the magnitude as well as the temporal evolution of the index changes shown in Figure 2.

At higher excitation fluences, we have observed that the index change rises considerably more rapidly. At a fluence of $.06J/cm^2$, $\Delta n/n_o$ crosses zero within 10 psec. This result is consistent with the faster Auger recombination and Auger heating rates expected at higher carrier densities,as described by the N^3 dependence shown in Eqns. (2) and (3). More sophisticated calculations of the Auger heating process would take into account the density dependence of the degeneracy factor H and the Auger coefficients C_e, C_h and make use of numerical methods to solve Eqns. (1) through (5). We believe, however, that the simpler calculations outlined above capture the essential physics of the optical heating process.

More detailed studies at higher excitation levels will be necessary to study the heating rate at the melting threshold, and thus to elucidate the relationship between heating rate and lattice disordering observed above E_{TH}. Equations (3) and (5), obtained from measurements on samples in thermal equilibrium, [18] appear to remain valid even on the short time scale involved in the present experiment. Detailed studies of the plasma evolution on a femtosecond time scale may provide information on the approach to thermal equilibrium, as well as on Coulombic screening of the Auger coefficient.

REFERENCES

1. W. H. Knox, R. L. Fork, M. C. Downer, R. H. Stolen, and C. V. Shank, Appl. Phys. Lett., *46*, 1120 (1985).

2. *Laser Solid Interactions and Laser Processing*, edited by S. D. Ferris, H. J. Lemay and J. M. Poate, (American Institute of Physics, New York, 1979).

3. *Laser and Electron Beam Processing of Materials*, edited by C. B. White and P. S. Peercy, (Academic Press, New York, 1980).

4. *Laser and Electron Beam Solid Interactions and Materials Processing*, edited by J. F. Gibbons, L. D. Hess and T. W. Sigmon, (Elsevier North-Holland, New York, 1981), Matl. Res. Soc. Symp. Proc., *1*, (1981).

5. *Laser and Electron Beam Interactions with Solids*, edited by B. R. Appleton and G. K. Celler, (Elsevier North-Holland, New York, 1982), Matl. Res. Soc. Symp. Proc., *4*, 49 (1982).

6. *Laser-Solid Interactions and Transient Thermal Processing of Materials*, edited by J. Narayan, W. L. Brown and R. A. Lemons, (Elsevier North-Holland, New York, 1983), Matl. Res. Soc. Symp. Proc., *13*, (1983).

7. *Laser Annealing of Semiconductors*, edited by J. M. Poate and J. W. Mayer (Academic Press, New York, 1982).

8. *Energy Beam-Solid Interactions and Transient Thermal Processing*, edited by J. C. C. Fan and N. M. Johnson, (Elsevier North-Holland, New York, 1984) Matl. Res. Soc. Symp. Proc., *23*, (1984).

9. J. M. Liu, Ph.D. Dissertation, Harvard University, Cambridge, Massachusetts.

10. H. Kurz and N. Bloembergen, Matl. Res. Soc. Symp. Proc., *35*, 3 (1985) and references cited therein.

11. C. V. Shank, R. Yen, and C. Hirlimann, Phys. Rev. Lett., *50*, 454 (1983).

12. C. V. Shank, R. Yen, and C. Hirlimann, Phys. Rev. Lett., *51*, 900 (1983).

13. M. C. Downer, R. L. Fork, and C. V. Shank, J. Opt. Soc. Am B, *2*, 595 (1985).

14. R. L. Fork, B. I. Greene, and C. V. Shank, Appl. Phys. Lett., *38*, 671 (1981).

15. R. L. Fork, C. V. Shank, and R. Yen, Appl. Phys. Lett., *41*, 223 (1982).

16. R. L. Fork, C. V. Shank, C. Hirlimann, R. Yen, and W. J. Tomlinson, Opt. Lett., *8*, 1 (1983).

17. O. S. Heavens, "Optical Properties of Thin Solid Films," (New York, Dover Publications, 1985).

18. G. E. Jellison, Jr. and F. A. Modine, Appl. PHys. Lett., *41*,120*(1982)*; Phys. Rev. B 27, 7466 (1983).

19. M. Miyao, T. Motooka, N. Natsuaki, and T. Tokuyama, Solid State Comm., *37*, 605 (1981).

20. L. A. Lompre, J. M. Liu, H. Kurz, and N. Bloembergen, Appl. Phys. Lett., *44*, 3 (1984).

21. H. M. van Driel, Appl. Phys. Lett., *44*, 617 (1984).

22. G. Yang and N. Bloembergen, to be published.

23. J. Dziewior and W. Schmid, Appl. Phys. Lett., *31*, 346 (1977).

24. E. J. Yoffa, Phys. Rev. B., *21*, 2415 (1980).

DYNAMICS OF ELEMENTARY PROCESSES AT SURFACES:
NITRIC OXIDE SCATTERED FROM A GRAPHITE SURFACE

H. VACH, J. HÄGER, B. SIMON, C. FLYTZANIS*, AND H. WALTHER
Max-Planck-Institut für Quantenoptik, D8046 Garching and
Sektion Physik der Universität München, Fed. Rep. of Germany

ABSTRACT

Molecular beam scattering from solid surfaces has long been recognized
as a powerful means for investigation of gas-surface reaction dynamics. With
the help of the recently developed laser-induced fluorescence and ionization
techniques for state-selective detection, one can now measure the angular and
velocity distributions of the scattered molecules together with their internal energy distributions. Such measurements fully describe the average energy
and momentum exchanges between molecules and surfaces and give thus full
information on the dynamics of the interaction. Recently, also the scattering
of vibrationally excited NO molecules was investigated. The paper gives a
review of new experiments with emphasis on the investigation of the scattering of NO molecules from a pyrographite surface. A simple model using
transport properties of the solid is presented which accounts surprisingly
well for the observed features.

INTRODUCTION

In recent years a large amount of detailed information has been obtained
on the interaction of gases with clean single-crystal surfaces. The knowledge
on binding sites of adsorbed molecules and their vibrational motion has been
provided by infrared absorption [1] and electron energy loss spectroscopy
[2]. Binding energies and residence times can be investigated by temperature
dependent desorption [3] and modulated molecular beam techniques [4]. Furthermore, angular and velocity distributions of scattered particles provide
information on the corrugation and the repulsive part of the interaction po-

*
Permanent address: Laboratoire d'Optique Quantique, Ecole Polytechnique,
91128 - Palaiseau, Cedex, France

tential responsible for the translational energy exchange [5].

Further progress was made when molecular beam scattering from surfaces was combined with subsequent analysis by laser techniques leading to the distribution of the internal degrees of freedom of the scattered molecules as well as their angle and state resolved velocity distributions [6 - 14]. These laser measurements provide new and surprising results on the scattering process, not available by conventional methods. It could be observed that both trapping/desorption and inelastic scattering processes show similar behaviour for the rotational state distribution: the molecules were fully accommodated to the surface at low surface temperatures, but not at higher ones. The observed results do not depend on the kinetic energy of the incoming particles for the case of adsorption/desorption and show a small change in the case of inelastic scattering. This is the case as long as the velocities of the incoming particles are in the range of thermal energies. Higher translational energies (> 0.4 eV) lead to rotational rainbows and to a polarization of the scattered particles [7].

The laser investigation of the vibrational motion of the particles shows excitation of the molecules during the scattering process that is dependent on both the incoming translational energy and the surface temperature [7e]. In general, it has been observed that there is only a very inefficient vibrational energy transfer to the scattering surface, i.e. vibrational accommodation coefficients are small. This was predicted theoretically [15] and observed experimentally [10b].

The study of NO molecules desorbing from a Pt(111) surface showed accommodation of the velocity distributions to the surface temperature for intermediate temperature ranges [9c]. This result could be expected according to the mainly isotropic angular distributions which were found experimentally [6c, 9c]. In the case of the inelastic NO/graphite system, however, a more complicated velocity distribution was measured [6d]. At low surface temperatures two different scattering channels exist: one corresponding to quasi-specularly reflected molecules with a velocity slightly smaller than that of the incoming beam, while the other one represents a diffusively scattered part with particles with a much smaller velocity. The peak velocity of the quasi-specularly scattered molecules increases with increasing surface temperature, increasing incidence and decreasing scattering angle. This effect can be so substantial that some of the scattered molecules show a velocity larger than that of the incoming ones, especially for high surface temperatures, large incidence and small scattering angles.

In the following the results of our investigations into the scattering of NO molecules from a graphite surface will be discussed in more detail.

EXPERIMENTAL

An ultrahigh vacuum (UHV) chamber with a pulsed or optionally continuous molecular beam was developed for measuring the angular, rotational,
vibrational and velocity distributions of surface scattered NO molecules.
The details of this set-up have already been described [6]. The NO molecules of an (optionally seeded) supersonic molecular beam are scattered from a
cleaved pyrographite surface in the centre of an UHV chamber (background gas
pressure 10^{-10} mbar). The molecular beam is crossed by the frequency doubled radiation of a tunable excimer pumped dye laser (226 nm). The
laser beam can be positioned in such a way that it probes either the incoming
or the scattered molecules. The NO molecules are excited electronically in a
one-photon process or ionized by resonantly enhanced two-photon absorption.
The fluorescence intensity or the number of ionized molecules is recorded as
a function of the laser wavelength. The population density of the rotational
and vibrational levels can be deduced from the line intensities of the resulting spectra. In addition, information on the angle and state resolved
velocity distributions of the scattered molecules can be obtained [6d]. For
this purpose a cylindrical cage of metal wire mesh is set in front of the
sample, with the focused laser beam propagating along the cage axis. Scattered
molecules passing through the centre of the cage are ionized, starting from a
specific rotational-vibrational state, by a resonantly enhanced two-photon
transition. The ions produced have the same velocity as the parent molecules.
Changing the position of the surface relative to the cage leads to a detection of molecules with different scattering angles.

For the investigation of the scattering behaviour of vibrationally
excited molecules part of the incoming particles are excited into the first
vibrational state by a CO-laser pumped Spin-Flip Raman laser (SFL), tuned to
the R(1/2) NO line. The SFL beam is focused into a multireflection cell
(White cell) properly adjusted for efficient excitation of the incoming
molecular beam (Fig. 1).

The degree of vibrational excitation of the incoming NO molecules by the
SFL is limited owing to both the small absorption linewidth of the incoming
molecular beam and the relatively high number of populated rotational states.
We estimate, however, that about 2 per cent of the incoming particles are
excited into the $v = 1$ ($J = 3/2$) state. In order to probe the vibrational
excitation of the scattered molecules, the frequency-doubled dye laser was
tuned over the $^2\Pi$ ($v = 1$) $\rightarrow$ $^2\Sigma$ ($v = o$) transition (around 236 nm), leading to
fluorescence or ionization spectra and, consequently, to both rotational and
angular distributions of vibrationally excited molecules.

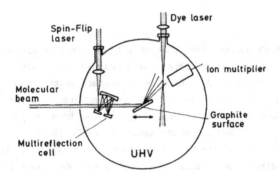

Fig. 1: Schematic set-up for the investigation of the scattering of vibrationally excited NO molecules.

The NO molecules were scattered from the surface of a cleaved pyrolytic graphite crystal or from a carbon-covered Pt(111) surface. The graphite
crystal consisted of microcrystals with highly oriented basal planes and a
mosaic spread of $0.4^\circ \pm 0.1^\circ$ (Union Carbide, grade ZYA).

RESULTS

The adiabatic expansion of the NO molecular beam in the nozzle causes an
increase of the average velocity in the beam direction and, simultaneously, a
strong reduction of the rotational temperature and the velocity spread. The
unseeded molecules in our supersonic beam had an average velocity of ~ 750
m/s, a velocity spread (FWHM) of ~ 140 m/s and a rotational population distribution corresponding to about 35 K [6]. The beam impinged on the surface at a
selectable incidence angle between 30° and 70°. The NO/graphite system has a
very low interaction potential depth (0.1 eV), leading to a very short residence time of the molecules on the surface. This time is short compared with
the duration of the molecular beam pulses (~ 5 ms FWHM).

The angular distributions of NO molecules scattered from the graphite
surface clearly indicate the weakly inelastic character of the interaction
[6b]. At low T_s, the distributions are generally composed of a quasi-specular
scattering or lobular part peaked at a scattering angle θ_s significantly less
than the specular reflection angle and a diffusive scattering or cosine
distribution part. With increasing surface temperature, the diffusive part
decreases relative to the specular part together with a shift of the specular
lobe to lower scattering angles. This behaviour results from a weakly inelastic scattering process with an increasing trapping/desorption contribution

for decreasing temperatures. The change of the angle of the specular lobe is correlated with the behaviour of the velocity distributions of the scattered NO molecules, as will be discussed later.

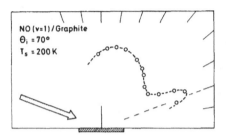

NO(v=1)/Graphite
$\Theta_i = 70°$
$T_s = 200 K$

Fig. 2: Angular distribution of vibrationally excited NO molecules scattered from a 200 K graphite surface.

The vibrationally excited molecules show a similar behaviour. For the measurements shown in Fig. 2, the incoming molecules had a velocity of 1400 m/s (NO seeded in He), an incidence angle of 70°, and were excited to the v = 1 (J = 3/2) state. The laser probed only NO molecules which were scattered from the 200 K graphite surface into the v = 1 (J = 11/2) state, that means molecules with preserved vibrational and changed rotational state. The vibrationally excited molecules exhibit a specular scattering lobe and a diffusive part very similar to the results obtained for the molecules in the vibrational ground state [6b]. Within the given experimental accuracy the angular distributions of the scattered molecules seem to be independent of the rotational and of, in addition, the vibrational state.

The rotational distributions of both the vibrational ground state and the first vibrational state were derived from the spectra of the $^2\Pi$ (v = 0) → $^2\Sigma$ (v = 0) transition (~ 226 nm) and the $^2\Pi$ (v = 1) → $^2\Sigma$ (v = o) transition (~ 236 nm), respectively. In both cases, the rotational state population could be fitted approximately to Boltzmann distributions with a characteristic rotational temperature T_{rot}. For the ground state measurements, it was found that the population distribution in the electronic states $^2\Pi_{1/2}$ and $^2\Pi_{3/2}$ could be described by the same T_{rot}. Even the overall population ratio $N(^2\Pi_{1/2})$: $N(^2\Pi_{3/2})$ is given roughly by $\exp(-\Delta E/kT_{rot})$, where ΔE is the fine-structure splitting.

The dependence of T_{rot} on the surface temperature for the NO/graphite system is shown in Fig. 3 for both the vibrational ground state (left) and the vibrational excited state (right). The solid lines correspond to complete accommodation of the rotational degree of freedom to the surface temperature. For the vibrational ground state of the unseeded molecules the experimental points follow this line up to a surface temperature of about 190 K, while at higher temperatures they deviate. At T_s higher than 350 K, T_{rot} approaches a

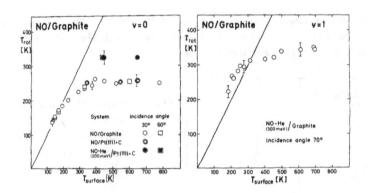

Fig. 3: Scattering of NO molecules from a pyrographite crystal and a carbon-covered Pt(111) surface. The rotational temperature is plotted versus the
surface temperature for the vibrational ground state (left) and vibrationally
excited molecules (right).

constant value of about 250 K. The results characterized by open symbols were
obtained at an incoming energy of about 80 meV; increasing the average kinetic energy of the incoming NO molecules to about 200 meV (NO seeded in He)
led to a somewhat higher rotational temperature (solid points).

Most of the rotational distributions were measured for NO molecules with
an incidence angle of 30°. In this case, molecules predominantly in the
specular direction were analyzed. Nearly the same rotational temperatures
were obtained at a second incidence angle of 60°, where mainly molecules in
the surface normal direction were investigated. Because of the bad angular
resolution of this measurement a clear distinction between the fluorescence
signal of diffusively and specularly scattered particles and, therefore,
between their respective rotational temperatures was not possible. Later,
however, we will see that the effective rotational temperature is slightly
different for diffusively and quasi-specularly scattered molecules.

A similar dependence of T_{rot} on the surface temperature is measured for
vibrationally excited NO molecules (Fig. 3, right). In this case, seeded NO
molecules in $v = 1$ with a velocity of about 1400 m/s ($\sim$ 300 meV) impinged on
the surface at an incidence angle of 70°. Only specularly scattered particles
(scattering angle 64°) were analyzed. As with seeded molecules in the vibrational ground state, the rotational temperature approaches a constant value
of about 335 K for high surface temperatures. For low surface temperatures
our (preliminary) results show rotational temperatures higher than the corresponding surface temperatures, which could be a result of either the vibrational excitation or the high incoming translational energy. An evaluation

of vibrational ground state measurements with seeded NO molecules will clarify this effect. From these results we can conclude that the survivability of
vibrational states is high, and that the rotational degree of freedom behaves
independently of the vibrational excitation, which means a decoupling of
rotation and vibration for the NO/graphite system.

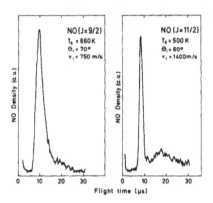

Fig. 4: Time-of-flight spectra
of unseeded (left) and
seeded (right) NO mole-
cules scattered from a
high temperature gra-
phite surface in the
specular direction.

For the measurement of state and angle resolved velocity distributions
(in the vibrational ground state) the scattered molecules are ionized from a
specific rotational state [6d]. Since the translational energy is conserved
during photoionization, the time-of-flight spectra of the ions correspond
directly to the velocity distributions of the parent neutral molecules. These
time-of-flight measurements show that the spectra are nearly independent of
the rotational states for all scattering angles. In the specular direction,
they are composed of a very high quasi-specular peak of very fast molecules
and a longer tail resulting from particles with lower velocities; see Fig. 4.'
In the left part, unseeded NO molecules with an incoming velocity v_i of 750
m/s were scattered from a high temperature graphite surface in the specular
direction. The outgoing time-of-flight spectrum exhibits an overlapping of
specularly (fast) and diffusively (slow) scattered molecules. For seeded
incoming molecules, however, the flight time is shorter and the distribution
has a smaller width. Therefore, specularly and diffusively scattered
particles appear clearly separated as a result of their different arrival
times at the ion multiplier (right side of Fig. 4).

Comparing the time of-flight spectra of molecules in different rotational states, we find the diffusive part to be stronger - relative to the
specular part - for higher rotational states than for lower J-states. Consequently the diffusively scattered molecules gain more rotational energy
during the scattering process than the specular ones. The diffusive and

specular parts can be independently characterized by two rotational temperatures, with T_{rot} of the diffusively scattered molecules being about 20 %
higher than that of the specularly scattered ones [6d].

Changing the position of the surface relative to the cage leads to
velocity distributions of molecules with different scattering angles. Molecules scattered in the specular direction show a behaviour similar to Fig. 4,
molecules scattered at a smaller angle and at low surface temperatures exhibit a velocity distribution with a smaller average velocity corresponding
to the diffusive part of the angular distribution. This slow diffusive part
disappears when the particles are scattered from a sufficiently hot surface.
In this case, the scattered molecules show a clean lobular angular distribution where the peak velocity is dependent on both the incidence and the scat-

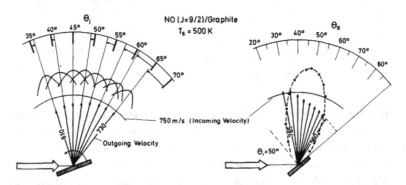

Fig. 5: Average velocity of scattered NO molecules (arrows) as a function
of the incidence angle θ_i (left) and the scattering angle θ_s (right).
The maximum of the scattering lobe is shifted to smaller scattering
angles for $\theta_i > 45°$ and to larger angles for $\theta_i < 45°$ (left).

tering angle. This is demonstrated in Fig. 5. Increasing the incidence angle
leads to an increasing shift of the angular distribution maximum to lower
scattering angles. Simultaneously, the average velocity in the direction of
the angular maximum increases. However, the velocities inside a scattering
lobe are not constant: they increase with decreasing scattering angle (Fig.
5, right).

With inceasing surface temperature, the average velocity of the scattered particles increases. Figure 6 summarizes this effect as a function of
the scattering angle for NO (J = 9/2) molecules with an incidence angle of
70°. It is also obvious that the variation of the velocities inside the
scattering lobe increases with the surface temperature. These effects can be
so substantial that some of the scattered molecules show a larger velocity

than the incoming ones, especially for high surface temperature, large incidence and small scattering angles.

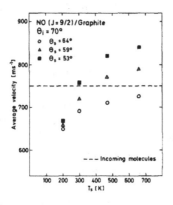

Fig. 6: Average velocity of
scattered NO molecules versus
surface temperature for different scattering angles.

DISCUSSION

Certainly the above results reflect the microscopic energy exchange processes that take place during different stages of the molecule surface interaction. Extensive numerical integration of the classical motion of the gas
molecule and the nearby surface and bulk atoms using an empirical interaction
potential for the scattering and desorption of NO molecules from Ag(111) and
Pt(111) yielded satisfactory agreement with the experimental observations
[16]. Such calculations, however, can only be performed for simple systems
and even there the number of parameters to be adjusted in the assumed interaction potential is not to be minimized; furthermore only a few features of
this potential seem to be relevant as the gas molecule interacts with a large
number of surface and bulk atoms.

On the other hand it is well established that a relatively small number
of the constituant atoms quite closely reproduces the properties of the infinite surface and bulk material and, in particular, the density of states for
all degrees of freedom, electronic or ionic. Hence an approach that from the
outset introduces the collective dynamical properties of the surface and the
bulk should be appropriate and in practice very useful for a systematic study
of different gas molecule/surface systems. From the fundamental point of view
it is of course of considerable interest to extent the applicability of macroscopic solid state aspects down to a microscopic scale and study the
accommodation between individual molecular motion and collective one. As it
will be shown below transport properties can be used and account surprisingly
well for the observed features. Combined with the extensive calculations [16]

that use detailed molecular dynamics such an approach may provide a more unifying reformulation of the microscopic energy transfer processes.

From the previous observations one may assume that to a first approximation the translation, rotation and vibration of the molecule are decoupled and accordingly each motion obtains its own thermal equilibrium separately from the others, in particular because of the different energy spacings. We concentrate our attention on the rotational motion and we admit that the observed rotational temperature T_{rot} is reached during the separation stage of the gas molecule from the surface. We proceed to determine this T_{rot}.

During the short duration of the separation stage, of the order of a few femtoseconds, a nonequilibrium situation occurs and the thermodynamic equilibrium that is expected to exist between the surface and the bulk in the stationary regime is destroyed; this is due to the vastly different dynamical features of the collective surface and bulk motions respectively [17]. The gas molecule together with the nearby surface maintains during this stage a local equilibrium at a temperature T_{rot} which is different from the bulk temperature T_B; the latter will be taken as the reservoir. The resulting nonequilibrium is restored through heat conduction from the bulk to the molecule complex which during the interaction undergoes energy redistribution and relaxation. Let us admit that this can be described by a rate equation for the rotational energy

$$dE_{rot}/dt = - E_{rot}/\tau + j_T \qquad (1)$$

where τ is a relaxation or memory time of the rotational energy redistribution; j_T is the thermal energy per unit time transmitted accross the effective cross section of the molecule S which depends on its orientation on the surface; let us put $S = d^2$ where d is a characteristic molecular dimension projected on the surface. We set

$$j_T = - \kappa S(T_{rot} - T_B)/a \qquad (2)$$

where κ is the thermal conductivity [18,19] at the bulk temperature T_B and a is a characteristic surface depth. At local equilibrium $dE_{rot}/dt = 0$ and $E_{rot} = kT_{rot}$ for a linear molecule so that

$$T_{rot} = \tau d^2 \kappa (T_B - T_{rot})/ka = \alpha(T_B - T_{rot}) \qquad (3)$$

or

$$T_{rot} = T_B \alpha/(1 + \alpha). \qquad (4)$$

The main temperature dependence in α comes through κ, the latter behaves quite differently for metals and dielectrics, respectively.

For crystalline dielectrics phonons are responsible [18] for heat con-
duction and for $T_B \gtrsim \Theta_D$, the Debye temperature [18], $\kappa \sim 1/T_B^{1+\varepsilon}$ where roughly
$0 < \varepsilon < 1$; for an ideal crystalline dielectric $\varepsilon = 0$. For metals the electrons are mainly responsible for heat conduction and for $T_B \ll \Theta_F$, the Fermi
temperature [18], $\kappa \sim 1/T_B^{1+\varepsilon}$ where now $-1 < \varepsilon < 0$ and $\varepsilon = -1$ for an ideal
Fermi metal; for $T_B \gg \Theta_D$ or impure metals or disordered alloys the heat
conduction becomes again dominated by phonons.

From these considerations on κ one may write in general $\alpha = \Lambda/T_B^{1+\varepsilon}$
with $-1 \leqq \varepsilon \leqq 1$ or

$$T_{rot} = T_B/(1 + T_B^{1+\varepsilon}/\Lambda) \qquad (5)$$

which accounts for the observed behaviour of T_{rot} as a function of T_B in all
studied cases with $-1 < \varepsilon < 0$ for metals and $0 < \varepsilon < 1$ for crystalline dielectrics or dielectric layers on metals. It is instructive to have an order
of magnitude estimation. Thus for the scattering of a NO molecule from an
ideal dielectric ($\varepsilon = 0$) like NaCl where $\kappa = \chi/T_B$ with $\chi \cong 40$ J cm^{-1}s^{-1}, a $\cong$
10 Å, d $\cong 2 \cdot 10^{-8}$cm and $\tau \cong 3 \cdot 10^{-14}$s we obtain $\Lambda = T_{sat} \cong 300$ K and (5)
can be written

$$T_{rot} = T_B / (1 + T_B/T_{sat}) \qquad (6)$$

where T_{sat} is a saturation temperature for the rotational energy distribution. In contrast for an ideal metal $\varepsilon = -1$ and one obtains a straight line
with a slope $\Lambda/(1 + \Lambda)$ from eq. (4).

The above considerations are valid for $T_B > \Theta_D$ where almost all phonons
are available for thermal conduction. For $T_B \ll \Theta_D$ (Debye regime) the number
of phonons available for conduction is more restricted and $\kappa \sim (T_B/\Theta_D)^3$ and
according to (3) T_{rot} may even be larger than T_B for dielectrics in the Debye
regime; as $\Theta_D \cong 100 - 200$ K for most cases this situation should occur below
30 - 50 K; in this regime, however, quantum effects may become partially
important.

The above model should apply both for desorption as well as for scattering since T_{rot} is essentially determined during the separation stage which
has a very short memory. In the short times involved in the scattering interaction, however, partial coupling with translation or vibration mediated
through the surface may occur. Such a situation can be accounted for by
additional energy flows in (1). Thus coupling to translational motion to a
first approximation leads again to T_{rot} described by (6) with a slightly
larger T_{sat} in the case of an ideal dielectric and a slope > 1 for $T_B <$
T_{sat}.

The use of this model for the rotational motion can be justified on
account of the small rotational energy spacing of all molecules heavier than

deuterium. Expressed in a temperature scale this spacing is θ_{rot} and is smaller than 3 K so that classical heat transport is valid. The situation is expected to be quite different in the case of vibrational motion where the energy spacing corresponds to a temperature $\theta_{vib} \cong 2700$ K which is a huge energy amount unlikely to be transferred rapidly; molecules vibrationally excited in the bulk are expected to lose at most ~ 300 K per ps [20] and thus a very long time with respect to the duration of the separation stage is required to channel out the energy $k\theta_{vib}$ unless resonant quantum transfer effects with the phonons or other collective modes are present. Thus a vibrationally excited molecule in many respects can be expected to show the same behaviour as in its ground vibrational state in particular as far as the rotational temperature is concerned.

In conclusion within the stated assumptions and approximations the above model provides a remarkably straightforward explanation of the observed behaviour and relates it to macroscopic parameters. Extensive application to a large number of cases should allow one to better define its range of applicability and establish its link with the models based on detailed molecular trajectory dynamics.

ACKNOWLEDGEMENT

One of the authors (C.F.) sincerely thanks the Alexander von Humboldt Foundation for a senior Humboldt award.

References

1. C.R. Brundle and H. Morawitz, Vibrations at Surfaces, (Elsevier, Amsterdam, 1983).
2. H. Ibach and D.L. Mills, Electron Energy Loss Spectroscopy and Surface Vibrations (Academic, New York, 1982).
3. R.J. Madix and J. Benziger, Annu. Rev. Phys. Chem. 29, 285 (1978).
4. M.J. Cardillo, Annu. Rev. Phys. Chem. 32, 331 (1981).
5. C.W. Muhlhausen, J.A. Serri, J.C. Tully, G.E. Becker, M.J. Cardillo, Isr. J. Chem. 22, 315 (1982).
6. (a) F. Frenkel, J. Häger, W. Krieger, H. Walther, C.T. Campbell, G. Ertl, H. Kuipers, J. Segner, Phys. Rev. Lett. 46, 152 (1981);

(b) F. Frenkel, J. Häger, W. Krieger, H. Walther, G. Ertl, J. Segner, W. Vielhaber, Chem. Phys. Lett. 90, 225 (1982); (c) J. Segner, H. Robota, W. Vielhaber, G. Ertl, F. Frenkel, J. Häger, W. Krieger, H. Walther, Surf. Sci. 131, 273 (1983); (d) J. Häger, Y.R. Shen, H. Walther, Phys. Rev. A 31, 1962 (1985); (e) J. Häger and H. Walther, J. Vac. Sci. Technol. B 3, 1490 (1985).

7. (a) A.W. Kleyn, A.C. Luntz, D.J. Auerbach, Phys. Rev. Lett. 47, 1169 (1981); (b) A.W. Kleyn, A.C. Luntz, D.J. Auerbach, Surf. Sci. 117, 33 (1982); (c) A.C. Luntz, A.W. Kleyn, D.J. Auerbach, Phys. Rev. B 25, 4273 (1982); (d) J.E. Hurst, L. Wharton, K.C. Janda, D.J. Auerbach, J. Chem. Phys. 78, 1559 (1983); (e) C.T. Rettner, F. Fabre, J. Kimman, D.J. Auerbach, Phys. Rev. Lett. 55, 1904 (1985).

8. (a) G.M. McClelland, G.D. Kubiak, H.G. Rennagel, R.N. Zare, Phys. Rev. Lett. 46, 831 (1981); (b) G.D. Kubiak, J.E. Hurst, Jr., H.G. Rennagel, G.M. McClelland, R.N. Zare, J. Chem. Phys. 79, 5163 (1983).

9. (a) M. Asscher, W.L. Guthrie, T.H. Lin, G.A. Somorjai, Phys. Rev. Lett. 49, 76 (1982); (b) H. Asscher, W.L. Guthrie, T.H. Lin, G.A. Somorjai, J. Chem. Phys. 78, 6992 (1983); (c) W.L. Guthrie, T.H. Lin, S.T. Ceyer, G.A. Somorjai, J. Chem. Phys. 76, 6398 (1982).

10. (a) H. Zacharias, M.M.T. Loy, P.A. Roland, Phys. Rev. Lett. 49, 1790 (1982); (b) J. Misewich, H. Zacharias, M.M.T. Loy, Phys. Rev. Lett. 55, 1919 (1985).

11. J.S. Hayden and G.J. Diebold, J. Chem. Phys. 77, 4767 (1982).

12. (a) J.W. Hepburn, F.J. Northrup, G.L. Ogram, J.C. Polanyi, J.H. Williamson, Chem. Phys. Lett. 85, 127 (1982); (b) D. Ettinger, K. Honma, M. Keil, and J.C. Polanyi, Chem. Phys. Lett. 87, 413 (1981).

13. J.B. Cross and J.B. Lurie, Chem. Phys. Lett. 100, 174 (1983).

14. (a) D.A. Mantell, S.B. Ryali, G.L. Haller, J.B. Fenn, J. Chem. Phys. 78, 4250 (1983); (b) D.A. Mantell, Y.F. Maa, S.B. Ryaly, G.L. Haller, J.B. Fenn, J. Chem. Phys. 78, 6338 (1983).

15. R.R. Lucchese and J.C. Tully, J. Chem. Phys. 80, 3451 (1984).

16. C.W. Muhlhausen, L.R. Williams, J.C. Tully, J. Chem. Phys. 83, 2594 (1985).

17. R.F. Wallis, Progr. Surf. Sci. 4, Ed. S.G. Davinson, 233, (1973)

18. N.W. Ashcroft and N.D. Mermin, Solid State Physics, (Holt-Saunders, Tokio, 1981), chp. 25.

19. C. Kittel, Introduction to Solid State Physics, (John Wiley, N.Y., 1976), chps. 5 and 6.

20. A. Seilmeier and W. Kaiser, (private communication)

NONLINEAR OPTICS AND SURFACE SCIENCE

Y. R. SHEN
Department of Physics, University of California, Berkeley, California and
Materials and Molecular Research Division, Lawrence Berkeley Laboratory,
Berkeley, California 94720

ABSTRACT

The recent status of applications of nonlinear optics to surface
science is reviewed. The basic theory of wave mixing on a surface layer,
and the possibility of using various nonlinear optical processes for surface probing are briefly discussed. Emphasis is on surface second harmonic
generation, which is shown with many illustrations to be a rather unique
and versatile tool for surface studies.

INTRODUCTION

In recent years, laser applications to surfaces have created much ex-
citement and many unique opportunities in surface science and technology
[1]. On the other hand, numerous laser techniques have been invented for
surface material processing. Laser annealing, alloying, ablation, and photochemical etching or deposition are among the well-known examples. On the
other hand, novel laser methods have been developed for surface probing.
These include photoacoustic [2], photothermal [3], and photodesorption [4]
spectroscopy, as well as state-selective detection of molecules collided
with or desorbed from a surface [5]. The recent advances in the applications of nonlinear optical techniques to material studies have naturally
brought our attention to the possible applications of nonlinear optics to
surfaces. One might question whether nonlinear optical techniques are sensitive enough for surface probing. Actually, the surface sensitivity of
nonlinear optical effects was already demonstrated more than 15 years ago
[6], but applications of nonlinear optics to surface probing were only recently realized [7].

As optical probes, nonlinear optical techniques have many intrinsic
merits. They are nondestructive, capable of in-situ remote sensing with
high spatial and temporal resolution, and applicable to any interface accessible by light. In comparison with linear optical methods, nonlinear
optical techniques have the further advantages of being more versatile and
more surface sensitive and specific. The nonlinear signal usually increases with increasing pump laser intensity, but is limited eventually by optical damage. As we shall see later, submonolayers of adsorbates on a surface are often readily detectable by the nonlinear techniques [7].

The nonlinear optical effects that are capable of probing surfaces via
measurements of their optical properties generally fall into the class of
wave mixing. Both four-wave mixing and three-wave mixing (including second
harmonic generation) have been considered as viable surface probes. The
former has the advantage that a wide range of resonances can be resonantly
excited and probed [8]. However, being a third-order effect, it suffers
from a relatively weak signal strength and a poor discrimination of surface
against bulk. Three-wave mixing could discriminate surface against bulk
and yields a fair signal, but its spectroscopic capability is severely limited by the existing laser tuning range.

In the following, we shall first briefly review the basic understanding
of surface nonlinearity and associated surface nonlinear optics, and then
describe the applications of various nonlinear optical effects to surface
studies.

SURFACE NONLINEARITY

The abrupt termination of a bulk at a surface makes the surface layer structurally very different from the bulk. Consequently, the surface and bulk are expected to have significantly different optical properties. In addition, the normal component of an optical field varies rapidly across the surface layer such that the response of the surface layer to the field is extremely nonlocal. All these indicate that the surface layer should be characterized by a different set of optical constants than the bulk.

We can define the surface layer as the layer where both the structure of the medium and the optical field vary significantly [9]. It is well-known that such a variation generally occurs in a distance of few atomic or molecular layers. Thus, the surface layer thickness is always much smaller than an optical wavelength. As a result, in discussing optical effects resulting from the optical response of the surface layer, we can treat the layer as infinitely thin. We can then use an nth-order surface nonlinear susceptibility $\overleftrightarrow{\chi}_S^{(n)}$ to describe the nth-order surface polarization $\vec{P}_S^{(n)}$ induced in the surface layer [9].

$$P_{S,i}^{(n)}(\omega = \omega_1 + \ldots + \omega_n) = \chi_{S,ij_1\ldots j_n} F_{j_1}(\omega_1)\ldots F_{j_n}(\omega_n)$$

$$F_j = E_j \quad \text{if } j = x,y$$

$$= D_j \quad \text{if } j = z \tag{1}$$

assuming $\hat{x} - \hat{y}$ is the surface plane. Here, E and D denote the electric field and displacement current, respectively. Since both the parallel field and the normal displacement current components are continuous across the surface layer, $\overleftrightarrow{\chi}_S^{(n)}$ so defined is unique, unaffected by the different values of the normal field components on the two sides of the layer.

As it stands, $\overleftrightarrow{\chi}_S^{(n)}$ appears as a local quantity, although it contains all the nonlocal response of the surface layer. Symmetry of the surface layer dictates the symmetric form of $\overleftrightarrow{\chi}_S^{(n)}$. Because the inversion symmetry is necessarily broken at a surface, the second-order susceptibility $\overleftrightarrow{\chi}_S^{(2)}$ is nonvanishing even if $\overleftrightarrow{\chi}^{(2)}$ vanishes in the bulk under the electric-dipole approximation. This result is a clear indication of the surface specificity of the second-order processes.

The various components of $\overleftrightarrow{\chi}_S^{(n)}$ for a surface layer can in principle be obtained from measurements of nonlinear reflection at the surface, as will be discussed in the next section. Typically, $\overleftrightarrow{\chi}_S^{(2)}$ is of the order of $10^{-14} - 10^{-16}$ esu and $\overleftrightarrow{\chi}_S^{(3)}$ is of the order of $10^{-20} - 10^{-23}$ esu. How $\overleftrightarrow{\chi}_S^{(n)}$ is related to the microscopic structure of the surface layer is of course a matter of great concern. Unfortunately, the theory is not yet well developed, and the limited understanding depends very much on the particular system under consideration.

NONLINEAR OPTICAL GENERATION FROM A SURFACE

We consider the system in Fig. 1. The plane interface at $z = 0$ is formed by an isotropic medium at $z < 0$ and vacuum at $z > 0$. Laser excitation of the medium induces a bulk nonlinear polarization $\vec{P}_v^{(n)} = \vec{\mathscr{P}}_v^{(n)} \times \exp(iK_x x + iK_{sz} z - i\Omega t)$ in the bulk and a surface nonlinear polarization $\vec{P}_S^{(n)} = \vec{\mathscr{P}}_S^{(n)} \exp(iK_x x - i\Omega t)$ on the surface. We are interested in the output generated by $\vec{P}_v^{(n)}$ and $\vec{P}_S^{(n)}$ on the vacuum side.

The wave equation for the output field takes the form

$$\left[\nabla^2 + \left(\frac{\Omega}{c}\right)^2 \varepsilon\right]\vec{E}(\Omega) = -4\pi\left(\frac{\Omega}{c}\right)^2\left[\vec{P}_S^{(n)}\delta(z) + \vec{P}_v^{(n)}\right]. \tag{2}$$

The solution of this equation has been obtained in various forms by differ-

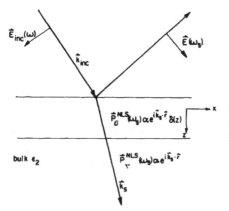

Fig. 1 An interface system with a surface nonlinear polarization $\vec{P}_S^{(n)}$ and a bulk nonlinear polarization $\vec{P}_V^{(n)}$.

ent authors [10,11]. We find

$$E_y(\Omega) = \frac{i4\pi(\Omega/c)^2}{K_{1z} + K_{2z}} [\mathscr{P}_{Sy}^{(n)} - \frac{\mathscr{P}_{vy}^{(n)}}{K_{sz} + K_{2z}}] e^{i\vec{K}_1 \cdot \vec{r} - i\Omega t}$$

$$E_x(\Omega) = (K_{1z}/K_x)E_z(\Omega)$$

$$= \frac{i4\pi K_{1z}K_{2z}}{K_{2z} + \varepsilon_2 K_{1z}} [\mathscr{P}_{Sx}^{(n)} + \frac{\varepsilon_2 K_x}{K_{2z}} \mathscr{P}_{Sz}^{(n)} + i \frac{K_{2z}\mathscr{P}_{vx}^{(n)} - K_x\mathscr{P}_{vz}^{(n)}}{K_{2z}(K_{sz} + K_{2z})}], \tag{3}$$

where $\vec{K}_1$ and $\vec{K}_2$ denote the wavevectors of the free waves at frequency Ω in media 1 and 2, respectively, with $K_{1x} = K_{2x} = K_{sx} \equiv K_x$.

It is readily seen, from Eq. (3), that both surface and bulk nonlinear polarizations contribute to the output. For $(K_{sz} + K_{2z}) \sim K_2$, the ratio of the two contributions to $E(\Omega)$ is of the order of $(2\pi/\lambda)|\mathscr{P}_S^{(n)}/\mathscr{P}_V^{(n)}|$ [11]. If the surface layer has a structure not very different from the bulk, we expect $|\mathscr{P}_S^{(n)}| \sim |\mathscr{P}_V^{(n)}|d$, where d is the surface layer thickness, then the ratio becomes $2\pi(d/\lambda)$ which is much less than 1. The output signal would completely be dominated by the bulk contribution unless one could resort to special techniques to enhance the surface contribution and/or suppress the bulk contribution. This is the case with four-wave mixing and with three-wave mixing in noncentrosymmetric media. In the case of a centrosymmetric medium, the second-order bulk polarization $P_V^{(2)}$ is zero in the electric dipole approximation. It is nonzero only if electric-quadrupole and magnetic-dipole contributions are taken into account, but is reduced by a factor of $\sim 2\pi a/\lambda$ in comparison with that for a noncentrosymmetric medium, where a denotes the atomic size. The ratio of surface contribution to bulk contribution in this case becomes [11] $\sim (2\pi/\lambda)[d/(2\pi a/\lambda)] = d/a > 1$. The surface contribution can be significantly modified by the presence of adsorbates, as we shall see. With special arrangement of beam polarizations and directions, it is also possible to yield a vanishing $P_V^{(2)}$ so that only the surface is responsible for the nonlinear output [9].

Assuming that $\vec{P}_S^{(n)}$ can be separately deduced from the measurements, we can then obtain $\chi_S^{(n)}$ from the results using Eq. (1). The approximate output signal from $P_S^{(n)}$ can be calculated from Eq. (3). We find

$$S(\Omega) \sim (4\pi\Omega/c\hbar)|P_S^{(n)}|^2 AT \text{ photons/pulse}, \tag{4}$$

where A is the beam cross-section on the surface and T is the pulsewidth.

In many cases, we may be interested only in the change due to surface modification such as adsorption of molecules to the surface. Relative measurements to deduce $\Delta\vec{\chi}_S^{(n)}$ are always much simpler and more straightforward.

SURFACE COHERENT RAMAN SPECTROSCOPY

Among the many possible four-wave mixing processes, we shall consider only coherent Raman effects here because they are potentially useful as spectroscopic tools for surface analysis [11]. In the so-called antiStokes Raman scattering (CARS) [8], two incident laser beams at frequencies ω_1 and ω_2 induce a nonlinear polarization $\vec{P}^{(3)}(\Omega = 2\omega_1 - \omega_2)$ in a medium. If $\omega_1 - \omega_2$ is near a Raman resonance at ω_v, then $\vec{P}^{(3)}$ is resonantly enhanced. The corresponding $\chi^{(3)}$ can be decomposed into a resonant and nonresonant part

$$\chi^{(3)}(\Omega = 2\omega_1 - \omega_2) = \chi_R^{(3)} + \chi_{NR}^{(3)}$$

$$\chi_R^{(3)} = A/[(\omega_1 - \omega_2) - \omega_{ex} + i\Gamma_{ex}]. \tag{5}$$

Thus, by scanning $(\omega_1 - \omega_2)$ and observing the resonant enhancement of the output at Ω, we can learn about the Raman resonance. The very attractive feature of this technique is that an extremely wide range of $(\omega_1 - \omega_2)$, from 0 to $\sim 10^5$ cm^{-1}, can be easily covered by beating a tunable dye laser with a fixed-frequency laser.

Obviously, CARS should also be applicable to surfaces. The difficulty lies in the strong bulk contribution to the signal. As a third-order nonlinear effect, there is no intrinsic symmetry rule that discriminates a surface from a bulk. Although $\vec{\chi}_S^{(3)}$ may have a different resonant spectrum than $\vec{\chi}_V^{(3)}$, spectral discrimination by the detection system to suppress the bulk contribution is often not sufficient. Polarization discrimination must be used in addition to further reduce the bulk contribution. This is based on the fact that $(\vec{\chi}_S^{(3)})_R$ generally yields an output polarized in a different direction than $\vec{\chi}_V^{(3)}$ and $(\vec{\chi}_S^{(3)})_{NR}$. There is also the question whether the output signal from $(\vec{\chi}_S^{(3)})_R$ is strong enough to be detected. Using $(\vec{\chi}_S^{(3)})_R \sim 10^{-20}$ esu and picosecond excitation pulses with $I \sim 10^{10}$ W/cm^2, $T \sim 1$ psec, and $A \sim 0.1$ cm^2 in Eq. (4), we find $S \sim 10^3$ photons/pulse, which should be easily detectable.

The surface CARS signal can be greatly enhanced if surface [12] or guided [13] optical waves are used to enhance the pump laser intensity and the effective interaction length. This has been demonstrated experimentally. The disadvantage of such a scheme is that the polarization discrimination method is not easily applicable [13].

Another coherent Raman technique which automatically eliminates the nonresonant surface and bulk contribution is stimulated Raman gain spectroscopy [14]. The input beams are again at ω_1 and ω_2 with $(\omega_1 - \omega_2) \sim \omega_{ex}$, but the output resulting from the induced $\vec{P}^{(3)}(\omega_2 = \omega_1 - \omega_1 + \omega_2)$ is measured as a change superimposed on the input $E(\omega_2)$. Assuming that only $(\chi_S^{(2)})_R$ is resonant at $(\omega_1 - \omega_2) \sim \omega_{ex}$, one would find

$$\Delta|E(\omega_2)|^2 = G|E(\omega_2)|^2$$

$$G = (4\pi\omega_2/c)\text{Im}(\chi_S^{(2)})_R|E(\omega_1)|^2. \tag{6}$$

If $\text{Im}(\chi_S^{(2)})_R \sim 10^{-20}$ esu and $I \sim 10^9$ W/cm^2, we find $G \sim 5 \times 10^{-8}$. It has been demonstrated that with CW mode-locked laser pulses and a synchronous detection system, G as small as 10^{-8} can be detected. Therefore, surface Raman spectra are indeed measurable [14]. Unfortunately, the limited laser intensity, the residual adsorption of the substrate, and the limited sta-

bility of the laser and detection systems make the measurements quite difficult.

SURFACE SECOND HARMONIC GENERATION

As we mentioned earlier, second-order nonlinear effects favor the surface against the bulk in a medium with inversion symmetry. It is therefore
ideal as a tool for surface studies. Second harmonic generation (SHG) is
particularly attractive because of the simplicity in the experimental arrangement, as shown in Fig. 2. Unlike the third-order effects, the signal

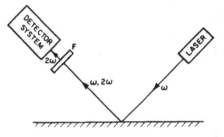

Fig. 2 Experimental arrangement
for second harmonic generation from
a surface.

from SHG is generally much stronger [11]. Even without resonance, $\chi_S^{(2)}$ can
be as large as $\sim 10^{-15}$ esu. Using Eq. (4) with $I \sim 10^{10}$ W/cm^2, $T \sim 1$ psec,
and $A \sim 0.1$ cm^2, we find $S \sim 4 \times 10^5$ photons/pulse. With $T \sim 10$ nsec, $I \sim$
10^6 W/cm^2, and $A \sim 1$ cm, we still have $S \sim 400$ photons/pulse. The signal
is actually only limited by optical damage. It then happens that because
of the higher energy damage threshold in the CW case, even a CW laser is
intense enough for generating detectable SHG from a surface monolayer. In
a recent experiment using a 20-mW CW diode laser, we could indeed use SHG
to monitor adsorption and desorption of monolayers of molecules on an electrode in an electrolytical cell [15].

In the past several years, it has been firmly established that SHG is
a viable tool for surface studies. The technique has been successfully applied to many different surfaces and interfaces to study either bare surface properties or molecular adsorbates. We describe here a few representative cases to illustrate the broad range of applications of the technique.
Surface SHG can be used to study bare surfaces [16]. A freshly cleaved
Si(111) surface is known to reconstruct to form a (2 × 1) structure. Annealing changes it into (7 × 7). The (2 × 1) structure should yield a 2-
fold rotational symmetry in the SHG about the surface normal, while the (7
× 7) structure should lead to an isotropic variation. This was indeed observed by Heinz et al. [16], as shown in Fig. 3. The result confirmed the
π-bonded chain model description for the (2 × 1)-Si(111) surface. A real
time monitoring of the surface transformation from (2 × 1) to (7 × 7)
structure during annealing was also possible using SHG. In this case, with
the selected polarization combination, the bulk contribution to the SH signal was negligible. Surface melting can also be monitored by SHG [17].
Using fsec-laser pulses, it was found that the laser-induced melting of
Si(111) takes place in ~ 1 psec. Aside from probing surface symmetry, SHG
should also be useful for spectroscopic studies of surface states.

Surface SHG is generally more useful for studies of adsorbates at an
interface. A monolayer of adsorbates changes the surface susceptibility
from its bare value $\overset{\leftrightarrow}{\chi}_S^{(2)}$ to the modified value $\overset{\leftrightarrow}{\chi}_S'^{(2)}$, with

$$\overset{\leftrightarrow}{\chi}_S'^{(2)} = \overset{\leftrightarrow}{\chi}_S^{(2)} + \overset{\leftrightarrow}{\chi}_{AS}^{(2)} + \overset{\leftrightarrow}{\chi}_A^{(2)}$$

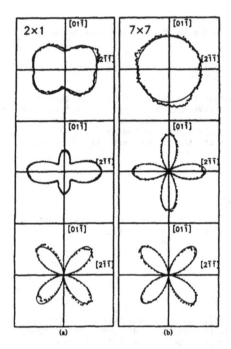

Fig. 3 SH intensity from Si(111)-(2 × 1) and (7 × 7) surfaces as a function of pump polarization. Different curves correspond to different SH polarizations. See Ref. 16.

where $\overleftrightarrow{\chi}_{AS}^{(2)}$ and $\overleftrightarrow{\chi}_{A}^{(2)}$ denote, respectively, the susceptibility changes arising from the adsorbate/surface interaction and from the intrinsic nonlinearity of the layer of adsorbates. If the interaction between adsorbate molecules is negligible, then both $\chi_{AS}^{(2)}$ and $\chi_{A}^{(2)}$ are proportional to the surface density of adsorbate molecules, N/cm^2. Typically, for metal and semiconductor surfaces, $\chi_S^{(2)} \sim 10^{-14}$ esu, which is quite appreciable. A monolayer of atoms or molecules chemisorbed on metal or semiconductor is expected to yield a $\chi_{AS}^{(2)}$, which is an appreciable fraction of $\chi_S^{(2)}$, and therefore should be easily detectable even if $\chi_A^{(2)}$ is small. On the other hand, for insulators, $\chi_S^{(2)} \sim 10^{-15}$ esu is relatively small, and $\chi_{AS}^{(2)}$ is often even smaller. A monolayer of molecular adsorbates is detectable only if the intrinsic nonlinearity of the molecules, $\chi_A^{(2)}$, is sufficiently large.

Indeed, experiments showed that even submonolayers of adsorbates on metals [18,19] and semiconductors [19] are easily detectable. A number of examples are given in Figs. 4-6. These are results obtained from well defined sample surfaces in ultrahigh vacuum. Figure 4 describes how the SH signal from Ni(111) varies with surface exposure to CO [19]. Adsorption of CO on Ni(111) is known to be on sites of the same type, and hence the adsorption kinetics could obey the simple Langmuir model. The data in Fig. 4 are indeed in excellent agreement with prediction from the Langmuir model. In the case of CO adsorption on Rh(111) [18], the result in Fig. 5 shows that the variation of the SHG with CO coverage θ, as calibrated by LEED, exhibits a sudden change in slope at θ = 1/3 monolayer. This agrees with the fact that for θ < 1/3, only adsorption sites of a single type can be occupied by CO on Rh(111), but for $1/3 \le \theta \le 3/4$, new sites of a different type also becomes open for CO adsorption. The data in Fig. 5 can actually be described very well by such a two-site model. Oxidation of a semiconductor surface can also be monitored by SHG [20]. Figure 6 shows the results obtained from Si(111). The decrease in the SH signal is dominated by the adosrption of the first monolayer of oxygen on Si, and can be fit by

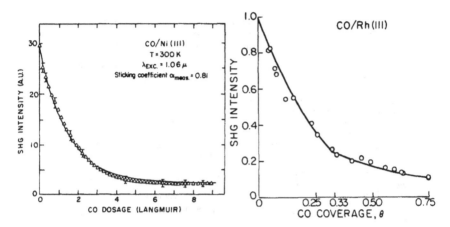

Fig. 4 SH signal as a function of surface coverage of CO on Ni(111) at 300K. The solid theoretical curve derived from the Langmuir kinetic model is used to fit the data.

Fig. 5 SH signal from CO/Rh(111) as a function of CO fractional coverage. The solid curve is a theoretical fit.

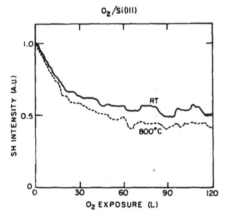

Fig. 6 SHG from Si(111) during oxidation of the surface at room temperature (RT) and at 800°K (---).

the simple Langmuir model assuming a single type of adsorption site, as indicated in Fig. 6. In all the cases presented here, the SH signal appears to decrease with increase of surface coverage of adsorbates. This is however not generally true. It depends on the adsorbate/surface interaction which dictates the relative sign of $\chi_S^{(2)}$ and $\chi_{AS}^{(2)}$. For example, adsorption of alkali atoms on metals would actually lead to a strong increase in the surface SHG [18].

As a tool to monitor adsorbates, one may need to calibrate the SH signal for absolute measurement of surface coverage of adsorbates. This can be done by correlating the SH signal with the results from thermal desorption spectroscopy (TDS) [19]. In comparison with TDS and other surface probes, SHG has the advantage of being almost instantaneous in response. It therefore could be an ideal tool for studies of surface dynamics.

Adsorption of molecules on insulator surfaces can also readily be de-

tected by SHG, although the sensitivity is relatively low if the molecules
are not highly nonlinear. This has been demonstrated in a number of cases
[15,21-24]. An example is given in Fig. 7, where it is shown that a small
hole in a monolayer of rhodamine 6G molecules adsorbed on fused quartz can
be clearly observed by SHG [15]. The technique is therefore potentially
useful for surface microscopy. With a better focusing lens, a spatial resolution close to 1 μm should be achievable.

Spectroscopy of molecular adsorbates is also possible with surface SHG
[21]. Figure 8 shows the spectra of the $S_0 \rightarrow S_2$ transition of half monolayers of rhodamine 6G and rhodamine 110 on fused quartz. The spectral
lines of the two dye molecules are well resolved even though structurally
the two molecules are quite similar. This indicates that SHG can be used
to identify or distinguish different adsorbates on a surface.

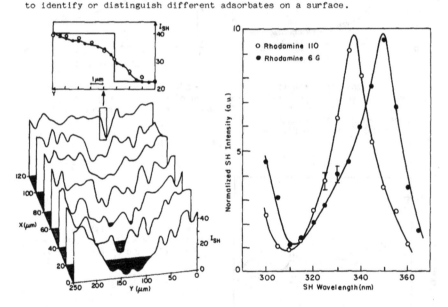

Fig. 7 SH image of an ablated hole
in a Rhodamine 6G dye monolayer.
Insert: a high resolution scan of
the region bracketed by the rect-
angle.

Fig. 8 SH spectra of the $S_0 \rightarrow S_2$
transition for submonolayers of rhodamine 6G and 110 dye molecules adsorbed
on fused silica.

Surface SHG is rather unique as a tool that is applicable to essentially all interfaces. For example, it has been used to study molecules at li-
quid/solid [22] and air/liquid [23] interfaces. Figure 9 shows the adsorption isotherm of p-nitrobenzoic acid (PNBA) on fused quartz from an ethanol
solution obtained by SHG [22]. From the adsorption isotherm, the adsorption free energy of ~ 8 KCal/mole can be deduced. Time-dependent measurement of adsorption and desorption of molecules at a liquid/solid interface
is also possible.

When the intrinsic molecular nonlinearity $\overset{\leftrightarrow}{\chi}_A^{(2)}$ dominates in the surface
nonlinear susceptibility $\overset{\leftrightarrow}{\chi}_S^{(2)}$ or can be extracted from $\overset{\leftrightarrow}{\chi}_S^{(2)}$, surface SHG
with different polarization combinations can yield information about the
orientation of molecules on a surface [22]. This is because $\chi_S^{(2)}$ in the
lab coordinates is related to the nonlinear polarizability $\overset{\leftrightarrow}{\alpha}^{(2)}$ of the molecules by a simple coordinate transformation assuming negligible interac-

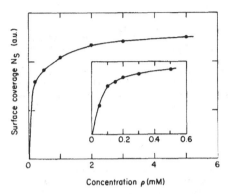

Concentration ρ (mM)

Fig. 9 Adsorption isotherm of PNBA
on fused quartz out of an ethanol
solution as deduced from SHG.

tion between molecules. We illustrate it with the simple case where $\vec{\alpha}^{(2)}$
is dominated by a single element $\alpha^{(2)}_{\xi\xi\xi}$ along a molecular coordinate $\hat{\xi}$, and
the molecules are arranged with an azimuthal symmetry. Then one can easily
show that

$$(\chi_A^{(2)})_{zzz} = N\langle\cos^3\theta\rangle\alpha^{(2)}_{\xi\xi\xi}$$

$$(\chi_A^{(2)})_{zxx} = (\chi_A^{(2)})_{xzx} = \frac{1}{2}N\langle\cos\theta\sin^2\theta\rangle\alpha^{(2)}_{\xi\xi\xi} , \tag{7}$$

where θ is the angle $\hat{\xi}$ makes with the surface normal $\hat{z}$. The weighted average orientation of the molecules, $\langle\cos^3\theta\rangle/\langle\cos\theta\sin^2\theta\rangle$, can be deduced from
the ratio of the measured $(\chi_A^{(2)})_{zzz}$ and $(\chi_A^{(2)})_{zxx} + (\chi_A^{(2)})_{xzx}$. The method
has been applied to different molecules at various interfaces [22-24].

Application of the above technique to pentadecanoic acid molecules
floating on a water surface led to the result in Fig. 10 [24]. It is seen

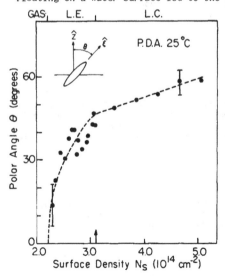

Fig. 10 Tilt angle θ between the
molecular axis and the surface normal as a function of surface density of PDA on water at 25°C.

that at the surface density where the so-called liquid-expanded to liquid-
compressed phase transition first takes place, the slope of molecular ori-

entation versus surface density changes suddenly. Analysis of the result
in Fig. 10 shows that the phase transition must be of first-order and the
two phases differ not only by a density change but also by a change in the
molecular orientation.

There are many other possible applications of surface SHG one can think
of. Studies of surface dynamics, polymerization, and catalytical reactions
are just a few examples. The strength of the technique clearly lies in its
great time-resolving capability and applicability to the large variety of
interfaces.

SURFACE SUM-FREQUENCY GENERATION

The major shortcoming of SHG as a surface analytical tool is its poor
spectral selectivity. The electronic transitions that SHG can probe are
often too broad to be useful for distinguishing species with similar structures. The vibrational transitions can better characterize molecular species, but surface SHG in the infrared is difficult to detect because of insensitivity of infrared photodetectors. A possible solution is to employ,
instead of SHG, infrared-visible sum-frequency generation. A tunable infrared laser is used to probe the vibrational transitions and the resonances are displayed in the visible by up-conversion. Such a scheme is presently being attempted.

OTHER NONLINEAR OPTICAL TECHNIQUES

A number of other nonlinear optical effects have been employed for surface studies. They are however not all-optical methods. Multiphoton ionization has been used to map out the energy distribution in molecules after
collision with or desorption from a surface [5]. Multiphoton photoemission
has been used to probe the surface states of semiconductors [25]. Finally,
infrared multiphoton absorption of molecular adsorbates on a surface could
lead to desorption of the molecules, and can therefore be used as a surface
spectroscopy technique [4].

CONCLUDING REMARKS

Applications of nonlinear optical techniques to surface studies are
still in the developing stage. However, it is already obvious that they
possess some advantages unique among conventional surface techniques. The
possibility of studying surface dynamics with ultrashort laser pulses is
one example. As complementary tools to the existing surface probes, the
nonlinear optical techniques are likely to bring surface science into a new
dimension.

ACKNOWLEDGEMENT

This work was supported by the Director, Office of Energy Research,
Office of Basic Energy Sciences, Materials Sciences Division of the U.S.
Department of Energy under Contract No. DE-AC03-76SF00098.

REFERENCES

1. See, for example, Surface Studies with Lasers, eds. F. R. Aussenegg,
    A. Leitner, and M. E. Lippitsch (Springer-Verlag, Berlin, 1983); pa-
    pers presented in Topical Meeting on Microphysics on Surfaces, Beams,

and Adsorbates (Santa Fe, February, 1985) (to be published in J. Vac. Sci. Techn. B).

2. F. Trager, H. Coufal, and T. J. Chuang, Phys. Rev. Lett. 49, 1720 (1982).

3. A. C. Boccara, D. Fournier, and J. Badoz, Appl. Phys. Lett. 36, 130 (1980); M. A. Olmstead and N. M. Amer, Phys. Rev. Lett. 52, 1148 (1984).

4. T. J. Chuang and H. Seki, Phys. Rev. Lett. 49, 382 (1982).

5. See, for example, H. Zacharias, M. M. T. Loy, and P. A. Roland, Phys. Rev. Lett. 49, 1790 (1982); J. Haeger, Y. R. Shen, and H. Walther, Phys. Rev. A 31, 1962 (1984); and references therein.

6. F. Brown, R. E. Parks, and A. M. Sleeper, Phys. Rev. Lett. 14, 1029 (1965); N. Bloembergen, R. K. Chang, and C. H. Lee, Phys. Rev. Lett. 16, 986 (1966); N. Bloembergen, R. K. Chang, S. S. Jha, and C. H. Lee, Phys. Rev. 174, 813 (1968); C. C. Wang and A. N. Duminski, Phys. Rev. Lett. 20, 668 (1968); F. Brown and R. E. Parks, Phys. Rev. Lett. 16, 507 (1966); J. M. Chen, J. R. Bower, C. S. Wang, and C. H. Lee, Optics Commun. 9, 132 (1973).

7. See, for example, T. F. Heinz, H. W. K. Tom, and Y. R. Shen, Laser Focus 19, 101 (1983).

8. See, for example, Y. R. Shen, The Principles of Nonlinear Optics (J. Wiley, New York, 1984), Chapter 15.

9. P. Guyot-Sionnest, W. Chen, and Y. R. Shen (to be published).

10. N. Bloembergen, R. K. Chang, S. S. Jha, and C. H. Lee, Phys. Rev. 174, 813 (1968); C. C. Wang, Phys. Rev. 178, 1457 (1969).

11. Y. R. Shen, The Principles of Nonlinear Optics (J. Wiley, New York, 1984), Chapter 25.

12. C. K. Chen, A. R. B. de Castro, Y. R. Shen, and F. DeMartini, Phys. Rev. Lett. 43, 946 (1979).

13. W. M. Hetheringon, N. E. Van Wyck, E. W. Koenig, G. I. Stegeman, and R. M. Fortenberry, Optics Lett. 9, 89 (1984); J. Chem. Phys. (to be published).

14. J. P. Heritage and D. L. Allara, Chem. Phys. Lett. 74, 507 (1980).

15. G. T. Boyd, Y. R. Shen, and T. W. Hansch (submitted to Optics Lett.)

16. T. F. Heinz, M. M. T. Loy, and W. A. Thompson, Phys. Rev. Lett. 54, 63 (1985).

17. C. V. Shang, R. Yen, and C. Hirlimann, Phys. Rev. Lett. 51, 900 (1983); S. A. Akhmanov, N. I. Koroteev, G. A. Paitian, I. L. Shumay, M. F. Galjautdinov, I. B. Khaibullin, and E. I. Shtyrkov, Optics Commun. 47, 202 (1983); J. Opt. Soc. Am. B 2, 283 (1985); A. M. Malvezzi, J. M. Lire, and N. Bloembergen, Appl. Phys. Lett. 45, 1019 (1984).

18. H. W. K. Tom, C. M. Mate, X. D. Zhu, J. E. Crowell, T. F. Heinz, G. Somorjai, and Y. R. Shen, Phys. Rev. Lett. 52, 348 (1984).

19. X. D. Zhu, R. Carr, and Y. R. Shen, Surf. Sci. (to be published).

20. H. W. K. Tom, X. D. Zhu, Y. R. Shen, and G. A. Somorjai, Proc. XVII Internatl. Conf. on Physics of Semiconductors (Springer-Verlag, Berlin, 1984), p.99; T. F. Heinz, M. M. T. Loy, and W. A. Thompson, J. Vac. Soc. Tech. (to be published).

21. T. F. Heinz, C. K. Chen, D. Ricard, and Y. R. Shen, Phys. Rev. Lett. 48, 478 (1982).

22. T. F. Heinz, H. W. K. Tom, and Y. R. Shen, Phys. Rev. A 28, 1883 (1983).

23. Th. Rasing, Y. R. Shen, M. W. Kim, P. Valint, and J. Bock, Phys. Rev. A 31, 537 (1985).

24. Th. Rasing, Y. R. Shen, M. W. Kim, and S. Grubb (to be published).

25. R. Haight, J. Bokor, J. Stark, R. H. Storz, R. R. Freeman, and P. H. Bucksbaum, Phys. Rev. Lett. 54, 1302 (1985).

PLENARY REVIEWS II

Ion-Solid Interactions
and Phase Transformations

COLLISION CASCADES, IONIZATION, SPIKES AND ENERGY TRANSFER

W. L. BROWN
AT&T Bell Laboratories, Murray Hill, NJ 07974

ABSTRACT

The collisions of an individual energetic ion in a solid set atoms of the solid in motion and excite electronic states. The moving atoms collide with others to form a collision cascade and the electronic excitation also forms a branched trail around the track of the ion. If the densities of the energy deposition in either of these modes is high the result is called a "spike". The low density regimes of collision and ionization cascades and evidence for the transition to non-linear high density spikes are summarized. The understanding of the latter is still quite incomplete as is also understanding of the transfer of energy between the kinetic and the electronic modes, particularly at high energy densities.

INTRODUCTION

An energetic ion entering a solid interacts with the atoms of the solid in two principal ways. Atoms are set in motion through momentum transferring collisions with the nuclei of the solid via the screened repulsive coulomb potential between the ion and those nuclei. This is often referred to as the "nuclear" energy loss of the ion, not because there are nuclear reactions but because the collisions are between the nuclei of the ion and atom. In addition, electrons are excited by coulomb interactions with the moving ion. These excitations may either be to bound states or to continuum, ionization, states. This type of interaction is referred to as the "electronic" energy loss of the ion. Figure 1 shows the partitioning of the energy loss of the ion into the nuclear, S_n and electronic S_e stopping power. The plot is as a function of $\epsilon^{1/2}$ where ϵ is Lindhard's dimensionless energy [1] which is proportional to the ion energy with a scaling factor depending on the mass and nuclear charge of the ion and the atoms of the solid.

In general, along the path of an energetic ion in a solid both nuclear and electronic energy loss occur: a succession of momentum transferring collisions is intermingled with a succession of electronic excitations. Individual collisions may involve both. The relative importance of the two energy loss modes clearly depends on ϵ, as Figure 1 indicates, but both always occur. Furthermore, atoms set in motion by nuclear

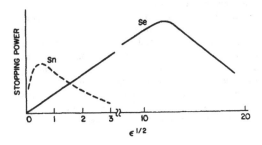

Fig. 1 The partitioning of energy loss of an ion into nuclear S_n and electronic S_e stopping power. The abscissa is $\epsilon^{1/2}$ where ϵ is Lindhard's dimensionless energy parameter [1] which is proportional to the ion energy with a scaling factor depending on ion and target atom masses and atomic numbers.

collisions of the primary ion may collide with other atoms and set them in motion and these in turn, still others. This branched sequence of energy sharing collisions is a collision cascade. Special types of electronic excitation form somewhat analogous branched sequences of electronic energy sharing, for example an inner shell electronic vacancy leading to a shower of Auger and secondary electrons or an energetic electron (delta ray) producing a cascade of further ionizations. There may also be transfer of energy between the collisional and electronic regimes.

This paper will qualitatively discuss the current understanding of:

1. Processes going on when the linear density of momentum transferring collisions or of electronic excitations along a single ion path is low;

2. Changes in the phenomena as the linear density increases; and,

3. Interchanges in energy between the kinetic and the electronic realms.

It is important to recognize what measurement diagnostics are available to indicate what is happening. Figure 2 sketches a collision cascade due to a 20 keV P$^+$ ion and an ionization track due to a 1MeV He$^+$ ion in silicon. The region affected by a single

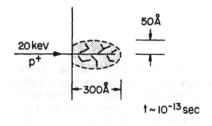

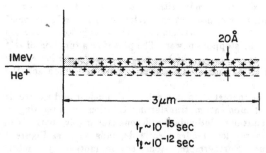

Fig. 2 A collision cascade due to a 20 keV P$^+$ ion in silicon and an ionization track due to a 1 MeV He$^+$ ion in silicon.

ion in either case is small and the time, short. As a result, there are no time-resolved probes available. The time scales are in principal accessible by current femto second laser techniques, [2] but the volumes involved are much too small. The regions affected are not just thin as they are, for example, in picosecond and subpicosecond time-resolved studies of reflectivity of laser excited materials, but laterally small as well. Furthermore, individual particle events are not time-synchronized on a fast time scale, with any external trigger. As a result, experimental information about collision cascades and ionization track processes all come from "postmortem" diagnostics. Some of these are prompt, but not measurable on a picosecond time scale, and some are available through permanent material changes measurable days later. In the two energy loss regimes these diagnostics are listed in Table I with asterisks to indicate those that are prompt.

TABLE I

COLLISIONAL	ELECTRONIC
Sputtering*	Secondary and Auger Electrons*
Damage	Hole-Electron Pairs*
Ion Beam Mixing	Luminescence*
	Sputtering*
	Damage

Section 2 and 3 below consider respectively the collisional and electronic cases.

THE COLLISIONAL REGIME

When the separation between primary momentum transferring collisions along an ion path is large, the branched tree of secondary, tertiary and higher order collisions that proceed from these primary events has a low density and the cascade is linear. Moving atoms only collide with atoms at rest. The average energy of atoms in the volume encompassing all struck atoms is low, much less than the binding energy of the atoms. This situation is represented in part c of Figure 3 [3]. The elliptical boundary indicates the volume which approximately encompasses statistically varying trajectories and their progeny. The number of collisionally displaced atoms is small and the volume relatively large. It is clear that the situation in Figure 3a is quite different. The volume is an order of magnitude smaller and the number of collisionally displaced atoms much larger. In cases of this kind moving atoms may collide with

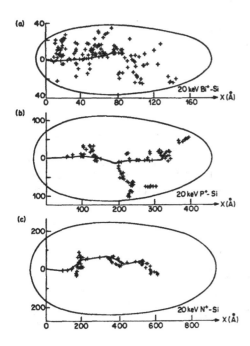

Fig. 3 Calculated collision cascades due to a) 20 keV Bi$^+$, b) 20 keV P$^+$ and c) 20 keV N$^+$ ions in silicon. The boundary ellipses represent contours at which the deposited collisional energy is 10% of the maximum. The primary ion trajectory is indicated by the continuous line. Cascade atoms receiving more than the displacement energy are indicated by $^+$. (after Walker and Thompson [3])

moving atoms, the average collisional energy per atom in the encompassed volume may be comparable to or even larger than the binding energy between atoms and the cascade is no longer linear. This case is referred to as a spike; an energy spike or a collision spike. It is also sometimes referred to as a thermal spike. This last term suggests that there are so many collisions among the atoms that they can be considered to have come to statistical equilibrium and hence their energy distribution can be described by a "temperature". The energy distribution is changing extremely rapidly with time as atoms at the boundary of the cascade region share their energy with those outside and the gradient in the energy distributions are extremely steep because of the small size of the cascade volume. At best, the description of a dense cascade as a thermal spike can only be a convenient simplification for a very complex fast transient state.

Sputtered Particle Energy Distributions

The time evolution of the energy distribution among atoms in the cascade is important but unmeasurable. As indicated in Table I above, sputtering is the only prompt diagnostic in the collisional regime and sputtered particles move much too slowly through distances in which they can be probed to provide time-resolved information. However, the energy distribution of sputtered particles can be measured and compared with theoretical predictions. Figure 4 shows results for sputtering of gold by various 20 keV ions [4]. In the linear collision cascade regime Thompson [5] has predicted the energy distribution should be

$$N(E) \propto E/(E+E_b)^3 \qquad (1)$$

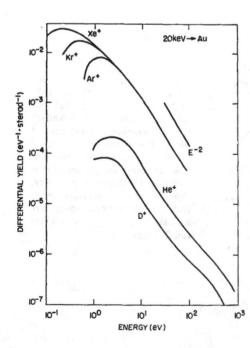

Fig. 4 Energy distributions of sputtered particles from gold bombarded by 20 keV D^+, He^+, Ar^+, Kr^+ and Xe^+ ions. (after Ahmad and Thompson [4])

where E_b is a surface binding energy of atoms to the solid. This expression has a E^{-2} dependence for $E \gg E_b$. Such a slope is shown in the figure and it describes the experimental results in the 10-100 eV regime quite satisfactorily. The deviations from this slope at higher energies for sputtering by deuterium and helium ions are believed to be due to backscattered incident ions that produce forward sputtering as they pass back out through the surface.

The other characteristic feature of expression 1 is the prediction that the energy distribution will have a maximum at $E_b/2$. That expectation is adequately borne out by the D and He results and also approximately by these for Ar. However, sputtering by the heavier ions is clearly very different below ~ 10 eV. The maximum of the distribution shifts to lower and lower energy as the mass of the projectile increases. The maximum no longer has any simple connection with the surface binding energy of gold. This is a regime of collision spikes and linear cascade theory should not be expected to apply. There is still a high energy tail with an E^{-2} shape but most of the sputtered particles have very much lower energy. This is qualitatively consistent with the picture of a much greater sharing of the collisional energy in a spike and a move toward a more "thermal" energy distribution inside the solid and toward a more evaporative energy distribution for sputtered particles.

Total Sputtering Yields

Another measure of the transition from linear cascade to collision spike regimes is contained in the variation of total sputtering yields with the energy density in the cascade. Such results are shown in Figure 5 [6]. At low values of F_D, the collisional energy density deposited at the surface, the yields scale linearly with F_D as expected

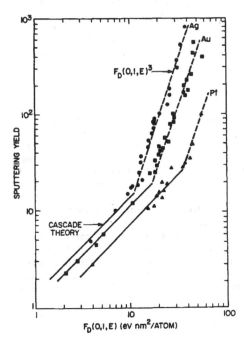

Fig. 5 Total sputtering yields for Ag, Au and Pt as a function of F_D, the collisional energy deposited at the surface by different incident ions and ion energies. (after Thompson [6])

58

from linear collision cascade theory. However, beyond critical energy densities which vary with the material being sputtered, the yield deviates sharply to much steeper, approximately cubic, dependences. The onset of collective effects due to collision spikes at energy densities in the 1-4 eV/atom range is clear. However, the cubic dependence is not easily accounted for. Sputtering that results from thermal evaporation from the intersection of a small heated volume with the surface is predicted to be quadratic in the deposited energy density [7,8]. Other effects must be contributing. For yields larger than the ~ 20 sputtered atoms per ion, at which the breaks in slopes occur, the surface is heavily perturbed and the effective binding energy may be reduced since many neighbors are simultaneously undergoing ejection.

Damage Production

The production of damage by heavy ion bombardment provides a different view of collision cascades. Results for silicon are shown in Figure 6 [9] as a function of the total energy deposited in collisional processes, ν (E). The effective number of atoms

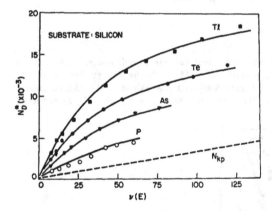

Fig. 6 Total number of displaced silicon atoms N_D per incident ion vs ν (E), the total energy deposited in nuclear collisions, for different incident ions and incident ion energies. The dashed line is a displaced atom prediction due to Kinchin and Pease. (after Thompson and Walker [9])

displaced per incident ion was deduced in each case from channeling and Rutherford backscattering measurements using MeV helium ions. None of the curves start with the slope predicted according to the Kinchin and Pease theory of defect production by collisional displacements [10]. All curves have curvature that tends to bring them toward that slope at high ν (E). The Kinchin and Pease theory assumes isolated collisions of atoms, essentially the assumption of linear cascade theory. A high energy density cascade, a collisional spike, severely violates this assumption. All of the ions for which data is shown in Figure 6 produce relatively high energy density cascades at low energies. The heavier ions produce denser cascades than the lighter ones. On the other hand, higher energy ions produce relatively low energy density cascades along the early part of their tracks. As they slow down, of course, the later parts of their tracks are equivalent to the tracks of ions incident at low energy. Thus the added displacements produced at the higher energies fall more nearly in the Kinchin and Pease regime and the defect production slopes approach the Kinchin and Pease prediction.

The interpretation of the defect production slope at low energies is not yet certain but one way of considering it is from the point of view of the quenching of a thermal spike. The small regions highly energized in a collision cascade lose their energy to surrounding atoms very quickly. Figure 7 shows calculations of the quenching time and the average energy per atom, θ_0, in the cascade for cascades formed by tellurium ions in silver [11]. Notice that θ_0 is high for low energies and low for higher energies as discussed above. Calculations of the quenching times require simplifying assumptions concerning "thermal" conductivity at enormous gradients in energy density, but for small cascades (produced with lower energy ions and higher θ_0) Sigmund [11] calculates them to be in the 10^{-13} to 10^{-12} second range. This is much faster than the quenching times that would be expected for thin "hot" layers of materials such as are formed in sub-picosecond laser heating of silicon [12]. The difference is geometrical as indicated in Figure 8. A small hot sphere shares energy with a much larger volume of surrounding material when the heat has diffused outward a distance Δr than does a hot sheet when heat has diffused a comparable distance Δx.

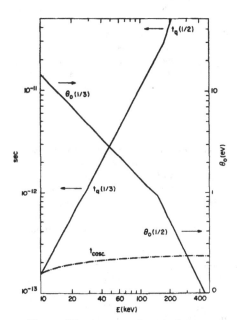

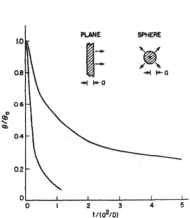

Fig. 7 The average deposited energy per atom (θ_0) in cascades produced by Te$^+$ ion bombardment of Ag at different energies. The 1/2 and 1/3 notations are for different power law interatomic potential regimes. The quenching time, t_q, is the time for the average energy to be reduced to 1/2 of θ_0. (after Sigmund [11])

Fig. 8 The time dependence of the maximum energy per atom in a spherical "thermal spike" of initial radius a in comparison with the time dependence of the maximum energy per atom in an energized plate of thickness a.

It is known from time-resolved conductivity studies with nanosecond laser pulse heated silicon that molten silicon frozen at an interface velocity >15 meters per second is amorphous [13]. The interfacial cooling rates appropriate to Figure 6 are hundreds or even thousands of meters per second. The thermal quenching interpretation of the initial slopes in Figure 6 is then the following. Small, highly energized cascade regions "melt" and quench too rapidly to regrow epitaxially from the surrounding crystalline material. As a result the melted region associated with each cascade becomes a small amorphous zone. Many more atoms are included in such a zone than could have been energetically displaced in individual Kinchin and Pease damaging events. In fact, the highest initial slope in Figure 6 corresponds to ~ 1 eV per atom, approximately the latent heat of melting of silicon. However, it is not at all clear that there is ever a phase transition to molten silicon in a collision spike that survives only 10^{-13} to 10^{-12} seconds. Furthermore, the subsequent low temperature thermal annealing of isolated "amorphous" zones formed in that way does not produce the same defect free material as results from thermal annealing of material that has been driven amorphous by many overlapping defect zones [14] or by quenching of a molten layer [15] or that has been formed by ultrahigh vacuum deposition of silicon [16]. As a result, it is not certain that the highly defective region of individual cascades is "amorphous", at least in the same sense as the planar regions formed in any of the three ways above.

Whether the dense energized cascade regions are thought of as having a "temperature" and hence being in some sort of transient local thermal equilibrium and referred to as "thermal spike" is a real issue only in the sense that unless they have a temperature it is hard to even discuss concepts of phases and phase changes. The time scales are indeed very short.

Electronic Excitation

There have been a wide range of experiments designed to measure the electronic state of atoms sputtered by collision cascades. It might be hoped that such measurements would serve as independent diagnostics of cascade parameters through energy interchange between the kinetic energy of motion of atoms and the electronic state of excitation. They might be used to describe a "temperature" if the system is indeed in equilibrium. Studies of charge states of sputtered atoms, critical to Secondary Ion Mass Spectroscopy, SIMS, are among these. In the last few years laser fluorescence has enabled studies to be made of different excited states of sputtered particles and of the velocity of atoms in these states as well. Results for the ground state and a metastable excited state of iron are shown in Figure 9 [17]. The relevant part of the energy level diagram is shown in Figure 10. The solid curves of Figure 9 are Thompson distributions (expression 1). The results for the ground state are in reasonable agreement with expression (1) with $E_b = 4.3$ eV. The most striking feature of the results however, is the major difference in the velocity distributions of atoms in the two states. The metastable atom distribution implies a completely unreasonable binding energy of ~ 20 eV. If these two states reflected equilibrium in the collision cascade (in this case produced by 10 keV Ar^+ bombardment) they should have the same velocity distributions but the density of atoms in the metastable state should be lower by a Boltzmann factor reflecting the $6928 cm^{-1}$ (0.9 eV) energy difference between the two and the "temperature" of the cascade which should also be measured by the velocity distribution. This is clearly not the case. Although there are cases in which the relative population of states with higher quantum numbers falls off in an approximately Boltzmann fashion from which a temperature can be deduced (and it tends to be thousands of degrees) the inference that that temperature reflects the cascade temperature is highly speculative. The electronic interactions of atoms as they leave the surface of a solid have been recognized for many years to be extremely important in determining their final electronic state. This has been observed directly

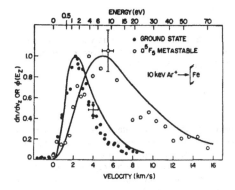

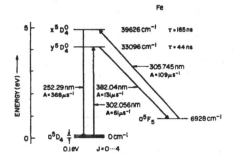

Fig. 9 Velocity distributions of sputtered ground state and metastable excited state neutrals from polycrystalline Fe. The curves are of the form of expression (1). The distribution of ground state atoms corresponds to a binding energy of 4.3 eV. The distribution of metastable atoms would imply a binding energy of ~ 20 eV. (after Schweer and Bay [17])

Fig. 10 The energy levels of Fe relevant to the laser fluorescence experiments of Figure 9. (after Schweer and Bay [17])

through Auger neutralization of incoming ions [18] and also in the velocity dependence of luminescence from excited atoms leaving the surface [19]. These final electronic interactions as sputtered particles pass out through the selvage of the rapidly decreasing electron density at the surface so strongly dominate the final states that information on electronic excitation inside the solid is lost. The fact that well-defined atomic states for an atom in the vacuum outside a surface may not even exist inside the solid is a further contributor to ineffectiveness of this potential diagnostic indication of the condition of the cascade inside. Inner shell excitations and the x-rays produced from them carry information about collisions in the highest energy collisions in the cascade but they have nothing to say about the cascade when it is at all fully developed.

Summary of the Collisional Regime

The evidence from direct measurements of sputtered particle energy distributions and total sputtering yields is completely consistent with linear cascade theory when the average energy per atom in the cascade region is low as required by that theory. Furthermore, there is clearly a transition to a collective mode when the average energy per atom in the cascade is high. However, details of this extremely short-lived collective mode are not yet clear and it seems generally more appropriate to refer to it as a collision spike than as a thermal spike since the latter implies equilibrium conditions for which there is very little definitive evidence.

THE ELECTRONIC REGIME

Electronic excitation of a solid is evident in a number of different ways as shown in Table I. Some of these are directly electronic while others result from conversion of electronic excitation to energy of motion of the atoms of a solid.

Electronic responses

Direct electronic responses to electronic excitation are important detectors of incident ions and of their energy and sometimes also their mass. Secondary electron emission is essential to detection of ions by electron multipliers. Luminescence is the heart of scintillation detection. Hole-electron pair generation is central to semiconductor particle detectors. For detectors these electronic excitation processes are used as much as possible in ranges where the measurable responses are linear, that is, proportional to the number of electronic excitations or ionization events produced by the incident ions. This is analogous to the linear cascade regime of collisions discussed above. For ion tracks with high densities of ionization, the measured responses may fail to be linear even if one believes the basic excitations themselves continue to be proportional to the electronic energy loss of the ions. Even for 5 MeV alpha particles the luminescence response in scintillation detectors drops below the energy proportionately obtained for low stopping power ions [20] because of nonradiative processes that tend to compete at high excitation density with the normal luminescent decay. High energy heavy ions in semiconductor detectors also give signals, from charge collection of holes and electrons, that fall below the energy proportional results obtained for low ionization density ions [21]. In this case the high density of hole-electron pairs results in accentuating higher order hole-electron pair recombination processes that reduce the number of pairs that are collected. The high density of hole-electron pairs also contributes to this loss by delaying the collection processes. The pairs represent a plasma which temporarily excludes the collecting electric field from all but the outer boundaries of their charge cloud.

It is interesting that in both of these cases high electronic excitation density results in less than an expected linear response. This is in contrast to the cases of collisional sputtering and damage discussed in Section 2 above for which high collisional energy density resulted in larger effects than expected based on extrapolation of the response to low density cascades.

Damage and Sputtering

Electronic excitation of clean metals does not result in creation of damage or in sputtering of atoms from the surface (although desorption of atoms from oxidized or otherwise chemically modified metallic surfaces under electronic excitation by photons (PSD, photostimulated desorption,) or electrons (ESD, electron stimulated desorption) are active scientific fields [22]. Electronic excitation is so rapidly and efficiently shared among the continuum electronic states in the conduction band of metals that it is neither spatially localized nor retained in sufficiently large units to produce atomic displacements. However, electronic excitation of insulators results in both damage and sputtering. Defect production in alkali halides is classic. In these materials a wide variety of color centers is produced by electronic excitation with UV, x-rays, electrons or ions [23]. The central factor in such processes in insulators is their large band gap. This allows localization of large units of energy in exciton type states (bound electron-hole pairs or electron-ion pairs). The band gap energy is in general more than enough to produce atomic displacements and some fraction of excitons give up their energy in this way. Sputtering of alkali halides by electronic excitation is another manifestation of energy conversion to atomic motion and interpretation of this phenomenon is based on repulsive interactions among electronically excited or ionized atoms [24].

The condensed gas solids (either rare gases or molecular gases) are a special class of insulators with big band gaps (as in the alkali halides) and very weak interatomic or intermolecular bonds. For these materials the localizable units of electronic energy are very much larger than the energies required to form atomic displacements either in the bulk or at the surface. For H_2O ice, for example, the electronic band gap is ~ 10 eV and the intermolecular bond energy is ~ 0.5 eV. For solid argon, the exciton energy is ~ 11 eV and the sublimation energy is only 0.08 eV. As a result, these materials provide particularly sensitive opportunities for observation of electronic to motional kinetic energy transfer.

The sputtering yield of H_2O ice for protons is shown in Figure 11 [25]. The collision cascade prediction is indicated. According to linear collision theory, the sputtering yield should peak at the maximum of the "nuclear" stopping power. This occurs at ~ 500 eV, so in the region above 1 keV of Figure 11 the expected yield

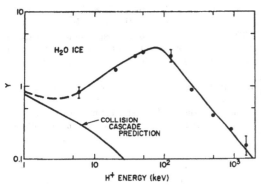

Fig. 11 The sputtering yield Y, molecules per ion, of H_2O ice by protons as a function of proton energy. The collision cascade prediction is also shown. (after Brown, et al. [25])

decreases continuously. However, the experimental yield does not drop in this energy range. It increases and reaches a peak at ~ 100 keV, approximately the maximum of the electronic stopping power in H_2O ice. This is very direct evidence for electronically stimulated sputtering which has now been observed in a variety of molecular and rare gas solids [26].

Figure 12 schematically illustrates the type of electronic to kinetic energy transfer that is involved in many of the cases studied, at least at low excitation densities. The $B^{**} + B$ state is typical. The energy of this state, which involves a highly excited atom B^{**} in combination with a ground state atom, is minimized in a bound state at a particular interatomic spacing as shown. Starting from a normal interatomic spacing in a solid when B^{**} is produced by a passing ion, the pair relaxes (moves together) toward this minimum, giving up phonons as it does so. However, before its relaxation is complete there is an energy level crossing to the antibonding state of the $B^* + B$ pair. At this crossing the partially relaxed $B^{**} + B$ system may transfer to the antibonding $B^* + B$ state and repulsively release the energy difference as shown by the arrow pointed down and to the right. This type of dissociative deexcitation is well known in gas phase reactions and results in kinetically energizing each partner with

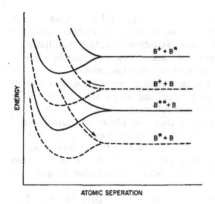

ENERGY

B⁺ + B*

B⁺ + B

B** + B

B* + B

ATOMIC SEPERATION

Fig. 12 A schematic diagram illustrating the relaxation of ion-excited state and excited state-ground state atom pairs. The repulsive energy release associated with a level crossing of $B^{**}+B$ to B^*+B is shown by the arrow down to the right. The arrow up to the left is a path for possible collisional excitation of B^++B^* from B^++B.

half the energy released (in the equal mass case illustrated).

The arrow upward to the left from the $B^+ + B$ state is a reminder that these processes are in principle reversible. If a B^+ ion is in close collision with a B atom the pair can travel up the antibonding repulsive curve. If it proceeds high enough on the repulsive curve before it separates again it may return to the $B^+ + B^*$ state rather than to the original $B^+ + B$ state. This is the type of process responsible for transfer of kinetic energy of motion to electronic excitation at collisional energies too low for ordinary coulomb excitation by the transient electric field of a moving ion to be effective.

Fig. 13 is a simplified energy diagram for solid argon, similar to Fig. 12 [27]. It identifies two repulsive deexcitation steps and a radiative transition which give rise respectively to sputtering and to luminescence. The sputtering yield and the luminescence yield of argon at 12 K are shown in Figure 14 as a function of the

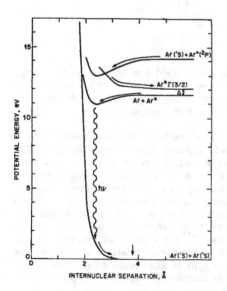

POTENTIAL ENERGY, eV

15

10

5

0

Ar('S)+Ar⁺(²P)

Ar*Γ(3/2)

Δ‡

Ar + Ar*

hν

Ar('S)+Ar('S)

INTERNUCLEAR SEPARATION, Å

Fig. 13 A simplified energy diagram for solid argon including the ionic exciton, Ar^+, the atom exciton Ar^* and their pairing with ground state atoms to form Ar_2^+ and Ar_2^*. The lowest energy of Ar_2^* decays by 9.8 eV photon emission. Repulsive energy releases are shown by arrows down and to the right, the first due to a level crossing, the second to repulsive separation following luminescence. Arrows down and to the left are multiphonon relaxations. (adapted from Johnson and Inokuti [27])

electronic energy loss, $(dE/dx)_e$ of H^+ and He^+ ions of different energies [28]: 2.5 MeV protons are at the lower left point, 1 MeV He^+ the upper right point. The rare gas solids are extremely efficient luminescent materials, a 9.8 eV radiative transition being the dominant deexcitation path for the fully relaxed $Ar^* + Ar = Ar_2^*$ excimer. The

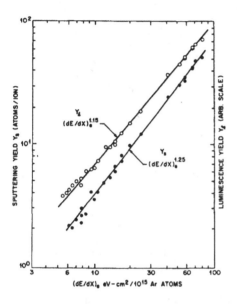

Fig. 14 Luminescence yield at 9.8 eV and sputtering yield for solid argon as a function of the electronic stopping power of protons and helium ions at different energies. (after Brown, et al. [28])

luminescence and sputtering yields are nearly proportional to the electronic stopping power. The small departure of their slopes from unity is not understood. In the stopping power range of Fig. 14 there is no indication of a break in the yield vs stopping power curve above some critical stopping power as there was in the collision cascade regime illustrated in Figure 5 for metals and has also been found in the collision cascade regime for solid xenon. The absence of a break may be due to the fact that argon ionic excitons (Ar^+) are mobile before they are self-trapped as they relax toward Ar_2^+ (Fig. 13) [29]. Even though the stopping power of 1 MeV He^+ is relatively high (resulting in ~ 3 ionization events per monolayer of the solid along the path of the ion) the exciton volume density may not be very high after diffusion, at which time the repulsive kinetic energy impulses are produced by decays such as those shown in Figures 12 and 13. Extensions of this data to higher electronic stopping power is clearly needed.

Sputtering of solid N_2 and O_2 have been compared using MeV He ions [30] and using keV electrons [31]. The electron results are shown Figure 15. The solid curves are drawn with the indicated power law dependence on electronic stopping power. As in the case of argon in Figure 14 the yields are nearly linear in the density of electronic excitation produced by individual electrons in this stopping power range. The ratio of the N_2 and O_2 sputtering yields is interesting in its own right but will not be discussed in detail here. It reflects very different efficiency in conversion of electronic energy to translational kinetic energy in these two molecular solids [30]. For H^+ and He^+ ion bombardment of solid N_2 there is a change in the slope of the sputtering yield as a function of $(dE/dx)_e$ [32]. Figure 16 shows this transition. The linear dependence of

66

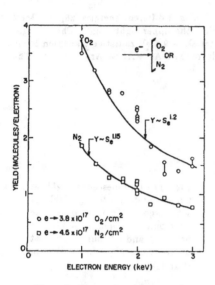

Fig. 15 Sputtering yields of O_2 and N_2 for electrons of different energies. (after Ellegaard et al. [31])

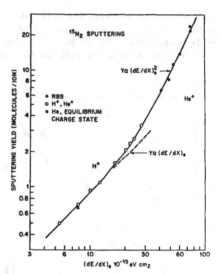

Fig. 16 Sputtering yield of N_2 as a function of electronic stopping power for H^+ and He^+ ions of different energies. (after Brown, et al. [32])

Figure 15 occurs in the low $(dE/dx)_e$ range where the ion bombardment yield is also low. But at high $(dE/dx)_e$ the yield increases nonlinearly to a slope of approximately 2 at the highest values on the curve. The transition is not nearly as sharp as that found in the collisional regime illustrated in Figure 5.

The origin of the change in slope of the yield vs stopping power curve in N_2 is still not certain. It clearly represents the onset of some collective behavior. The simplest suggestion is illustrated schematically in Figure 17. Each randomly occurring

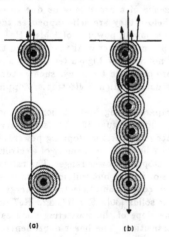

Fig. 17 Schematic illustration of (a) isolated and (b) overlapping mini cascades or "hot-spots" due to energy releases associated with individual ionization events along an ion path. (after Johnson and Brown [33])

ionization event along an ion track is envisioned as a source of a mini collision-cascade or a mini "hot-spot" as electronic energy temporarily stored in excited states is converted to kinetic energy of motion by repulsive deexcitation as in Figure 12 [33]. At low $(dE/dx)_e$ the sputtering yield is proportional to $(dE/dx)_e$ because the probability of a mini collision cascade occurring close enough to the surface to cause sputtering is proportional to $(dE/dx)_e$. As the electronic energy loss increases, however, the average spacing between these mini cascades or hot spots decreases until one expanding mini cascade overlaps with those adjacent to it. In the limit of very high ionization density along an ion track these overlapping hot spots look like an energized cylinder. The cylindrical geometry of this limit and a simple diffusion behavior for the increase in the radius of the cylindrically excited region with time leads to an expected sputtering yield proportional to $(dE/dx)_e^2$ [33]. However, a preliminary statistical treatment [35] of the approach to this limit as the average electronic stopping power increases yields a much more gradual transition than the data in Figure 16 shows. An alternative explanation considers the collective effect of high stopping powers as due not to overlap of individual isolated events but to creation of different electronically excited states. These might originate by interaction of two near neighbor excited or ionized atoms or molecules to yield a new state whose electron to kinetic energy transformation is larger than the sum of the individual states alone would have been. A particular electronic state for this behavior has not yet been identified, but current ideas about two-hole states make attractive possibilities [36].

The ultimate collective state is represented schematically in Figure 18. It leads to electronic to kinetic energy transfer via coulomb repulsion. At high excitation density it might be termed an ionization spike by analogy with the collision spikes discussed in section 2. There is coulomb repulsion among the ions of the central core of the spike that is surrounded by a shell of the electrons that had been ejected from the core in the ionization process [37]. The effectiveness of the coulomb "explosion" depends

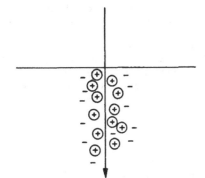

Fig. 18 Schematic illustration of an ionization spike with a positive ion core along the track of an incident ion surrounded by a shell of electrons expelled from the core by the ionizing events.

critically on the times required for relaxation of the energized electrons so that they can assume a screening role in reducing and finally eliminating repulsive fields between the ions. Until this occurs the net effect of the positive core of the track is repulsive. At the surface the repulsion will result in ejection of atoms from the surface in proportion to the square of the linear ionization density along the track. The quantitative validity of this idea for sputtering is still not established although this mechanism has been identified as responsible for latent track formation in minerals due to passage of a high energy heavy ion [38]. For solid H_2O [25] and CO [39] no linear dependence of sputtering yield on stopping power has been found over the stopping power range measured for argon and N_2 and shown in Figures 14 and 16. The dependences are approximately quadratic. At low enough electronic excitation density it seems inconceivable that a linear region doesn't exist, but the yield for linear processes may be too small to be observed in comparison with naturally quadratic processes, such as two-hole states.

Measurements of electronically stimulated sputtering of some insulating materials have been made with 10-100 MeV heavy ions that have much higher electronic stopping powers than those of MeV helium ions. For H_2O ice the sputtering yields tend to fall above the extrapolation of the approximately quadratic dependence observed for protons and helium ions [40]. For UF_4 the results favor an approximately fourth power dependence which has been suggested to arise from a coulomb repulsive energy transfer to produce a "hot" cylinder and then thermal evaporation where the hot cylinder intersects the surface [41]. Nonlinear collective effects that can be associated with electronic "spikes" are clearly present in many experimental measurements but their qualitative interpretation is still uncertain.

Summary of the Electronic Regime

Linear responses to electronic excitation of insulating solids by fast ions (or electrons) are observed at low excitation energy densities through measurements of luminescence, hole-electron pair collection, secondary electron emission and in some cases even sputtering of these materials. At high excitation energy density (ionization spikes) the responses deviate from linear. In cases of luminescence and hole-electron pair collection the deviations are to lower responses and they arise from the competition of other deexcitation pathways that are available at high energy densities. In the case of sputtering the deviations are to higher responses as new deexcitation pathways apparently serve to transfer electronic excitation more efficiently directly to kinetic energy of atoms. A number of different suggestions have been made for how this may arise but no definitive conclusion has yet been reached. At low excitation densities energy transfer from electronic excitation to kinetic energy are clearly intimate properties of each particular solid. It seems very likely that the effectiveness of energy transfer even at high energy densities, is still strongly material dependent. Even though a concept as general as a coulomb explosion may be the principal energy transfer mechanism, the cooling time for the electrons to shield the ion core of the track is expected to be material sensitive.

REFERENCES

[1] J. Lindhard, M. Scharff and H. E. Schiott, Kgl. Danske Vid. Selsk. Mat. Fys. Med. *33* No. 14 (1963).

[2] R. L. Fork, C. V. Shank, and R. Yen, Appl. Phys. Lett. *41*, 273 (1982).

[3] R. S. Walker and D. A. Thompson, Rad. Effects *37*, 113 (1978).

[4] S. Ahmad, B. W. Farmery, M. W. Thompson, Phil. Mag. A *44*, 1387 (1981).

[5] M. W. Thompson, Phil. Mag. *18*, 377 (1968).

[6] D. A. Thompson, Rad. Effects *56*, 105 (1981).

[7] G. H. Vineyard, Rad. Effects *29*, 245 (1976).

[8]   R. E. Johnson and R. Evatt, Rad. Effects *52*, 187 (1980).

[9]   D. A. Thompson, and R. S. Walker, Rad. Effects *36*, 91 (1978).

[10] G. H. Kinchin and R. S. Pease, Rep. Prog. Phys. *18*, 1 (1955).

[11] P. Sigmund, Appl. Phys. Lett. *25*, 169 (1974) (and an erratum in volume *27*, 52).

[12] J. M. Liu, H. Kurz and N. Bloembergen in "Laser-Solid Interactions and Transient Thermal Processing of Materials", eds. J. Narayan, W. L. Brown, R. A. Lemons (North-Holland, New York, 1983), p.3.

[13] M. O. Thompson and G. J. Galvin in "Laser-Solid Interactions and Transient Thermal Processing of Materials", eds. J. Narayan, W. L. Brown, R. A. Lemons (North-Holland, New York, 1983), p. 57.

[14] L. Csepregi, E. F. Kennedy, J. W. Mayer, and T. W. Sigmon, J. Appl. Phys. *49*, 3906 (1978).

[15] J. M. Poate in "Laser and Electron Beam Interactions with Solids", eds. B. R. Appleton and C. K. Celler (North-Holland, New York, 1982) p. 121.

[16] G. L. Olson, S. A. Kokorowski, J. A. Roth and L. D. Hess, in "Laser-Solid Interactions and Transient Thermal Processing of Materials", eds. J. Narayan, W. L. Brown, R. A. Lemons (Noth-Holland, New York, 1983), p. 141. New York 141 (1983).

[17] B. Schweer and H. L. Bay, Appl. Phys. A *29*, 53 (1982).

[18] H. D. Hagstrum in "Inelastic Ion Surface Collisions", eds. N. H. Tolk, J. C. Tully, W. Heiland and C. W. White (Academic Press, New York, 1977), p. 1; H. D. Hagstrum, Phys. Rev. *96*, 336 (1954).

[19] C. W. White, E. W. Thomas, W. F. Van der Weg and N. H. Tolk in "Inelastic Ion Surface Collisions", eds. N. H. Tolk, J. C. Tully, W. Heiland and C. W. White (Academic Press, New York, 1977), p. 201.

[20] J. B. Birks, Proc. Roy. Soc. *A64* 874 (1951).

[21] P. A. Tove and W. Seibt, Nucl. Inst. Meth. *51*, 261 (1967).

[22] "Desorption Induced by Electronic Transitions, DIET I", eds. N. H. Tolk, M. M. Traum, J. C. Tully, T. E. Madey (Springer Verlag, New York, 1983); "Desorption Induced by Electronic Transition DIET II", eds. W. Brenig and D. Menzel (Springer Verlag, Berlin, 1985).

[23] N. Itoh in "Desorption Induced by Electronic Transitions, DIET I", eds. N. H. Tolk, M. M. Traum, J. C. Tully and T. E. Madey (Springer Verlag, New York, 1983), p. 224.

[24] N. H. Tolk, W. E. Collins, J. S. Kraus, R. J. Morris, T. R. Pian and M. M. Traum, N. G. Stoffel and G. Margaritondo in "Desorption Induced by Electronic Transitions, DIET I", eds. N. H. Tolk, M. M. Traum, J. C. Tully and T. E. Madey (Springer Verlag, New York, 1983), p. 156.

[25] W. L. Brown, L. J. Lanzerotti, and R. E. Johnson, Science *218*, 525 (1982).

[26] W. L. Brown and R. E. Johnson, Proc. Atomic Collisions in Solids (1985) to be published.

[27] R. E. Johnson and M. Inokuti, Nucl. Inst. & Meth. *206*, 289 (1983).

70

[28] W. L. Brown, C. T. Reimann, R. E. Johnson, in "Desorption Induced by Electronic Transitions, DIET II" eds. W. Brenig and D. Menzel (Springer Verlag, Berlin, 1985), p. 199.

[29] C. T. Reimann, R. E. Johnson and W. L. Brown, Phys. Rev. Lett. *53*, 600 (1984).

[30] F. L. Rook, R. E. Johnson and W. L. Brown, Surf. Sci. (1985) in press.

[31] O. Ellegaard, J. Schou, H. Sorensen and P. Borgesen, Surf. Sci. (1985) submitted.

[32] W. L. Brown, L. J. Lanzerotti, K. J. Marcantonio, R. E. Johnson and C. T. Reimann, Proc. of Inelastic Ion Surface Collisions-V, Tempe AZ, (1985), Nucl. Inst. & Meth. to be published.

[33] R. E. Johnson and W. L. Brown, Nucl. Inst. & Meth. *198*, 103 (1982).

[34] W. L. Brown, W. M. Augustyniak, K. J. Marcantonio and E. H. Simmons, J. W. Boring, R. E. Johnson and C. T. Reimann, Nucl. Inst. & Meth *B1* 307 (1984).

[35] R. E. Johnson, Private Communication.

[36] T. Nakayama, M. Ikogawa and N. Itoh, Nucl. Inst. & Meth. *B1* 301 (1984).

[37] P. K. Haff, Appl. Phys. Lett. *29*, 245 (1976).

[38] R. L. Fleischer, P. B. Price and R. M. Walker, "Nuclear Tracks in Solids, Principles and Applications" (University of California Press, Berkeley, 1975).

[39] W. L. Brown, W. M. Augustyniak, K. J. Marcantonio and E. H. Simmons, J. W. Boring, R. E. Johnson and C. T. Reimann, Nucl. Inst. & Meth. *B1* 304 (1984).

[40] B. H. Cooper, thesis, California Institute of Technology (1982); L. E. Seiberling, C. K. Meins, B.H. Cooper, J. E. Griffith, M. H. Mendenhall, T. A. Tombrello, Nucl. Inst. & Meth. *198*, 17 (1982).

[41] L. E. Seiberling, J. E. Griffith, T. A. Tombrello, Rad. Effects *51*, 201 (1980).

TRANSITION BETWEEN CONDENSED PHASES IN Si AND Ge

DAVID TURNBULL*
Division of Applied Sciences, Harvard University, Cambridge, MA 02138, USA

ABSTRACT

The thermodynamic interrelation between the amorphous semiconducting (a-sc), the diamond cubic (c-sc) and the liquid metal (ℓm) states of Ge and Si is reviewed with especial emphasis on the question of the thermodynamic uniqueness of the a-sc state following its structural relaxation. The experience on the occurrence of the direct ℓm→a-sc transition and its reverse is surveyed and interpreted. This experience, in conjunction with the ℓm undercooling studies of Devaud and the author, indicates that the formation, in the metastable regime, of a-sc from ℓm results from preferential growth rather than preferential nucleation.

INTRODUCTION

The interphase transitions discussed here may be depicted as follows:

where c-sc, ℓm and a-sc denote, respectively, the diamond cubic, liquid metal, and amorphous semiconducting phase. What is known about the thermodynamics of these transitions has been reviewed and discussed in earlier publications [1-3], most recently by Donovan et al. [4]. There have been a number of studies of the kinetics and morphology of the a-sc→c-sc transition, which may occur via the normal solid state mode [5-8] or via "explosive" crystallization [9,10], and of the growth of c-sc into ℓm [11-14]. Rather little is known about the superheating of the semiconducting phases or the kinetics of their nucleation in ℓm.

In the author's judgment, the major issues concerning these transitions are as follows:

(1) structural relaxation of the a-sc state, its characterization and end state; a corollary issue is the thermodynamic nature of the a-sc ↔ ℓm transition.

(2) kinetics of nucleation and growth of c-sc and a-sc in ℓm.

(3) microscopic mechanism of a-sc→c-sc epitaxial regrowth.

(4) superheating of c-sc and a-sc.

The main emphasis of this paper is on these issues.

*Currently: B.T. Matthias Visiting Scholar, Center for Materials Science, Los Alamos National Laboratory, Los Alamos, NM 87545

STRUCTURAL RELAXATION OF a-sc

As prepared, a-sc specimens are configurationally frozen. The thermal data indicate that a-sc is thermodynamically less stable than c-sc at all temperatures from 0°K up to and well above the thermodynamic melting temperature, $T_{c\ell}$, of c-sc. Thus it is also less stable than ℓm at $T_{c\ell}$. However, extrapolation of the available thermal data of ℓm into the undercooled regime indicates that a-sc should become more stable than ℓm at some temperature, $T_{a\ell} > 0°K$. According to the thermal calculations $T_{a\ell} \approx 0.8\ T_{c\ell}$ for variously prepared a-sc Ge specimens [15,16] and ≈ 0.85 to 0.88 for a-sc Si formed by ion implantation [4].

When heated, a-sc specimens may exhibit substantial changes in certain properties--e.g., enthalpy, Raman spectra, diffraction-prior to detectable crystallization. In contrast with the nucleation and growth mode of crystallization, these changes occur homogeneously and indicate that *as prepared*, a-sc specimens are not in unique and thermodynamically well-defined states.

We wish to know the microscopic mechanisms of relaxation and the state toward which the relaxation is directed. The resolution of these questions will depend strongly on the structure of the a-sc specimens. If this structure were microcrystalline relaxation would occur by normal grain growth and a-sc would evolve continuously into the c-sc state. Actually, the transition from a-sc to c-sc always occurs by nucleation and growth and microcrystallite models are in poor accord with the diffraction patterns of a-sc. These patterns are in much better agreement [17,18] with continuous random network (CRN) models in which the average distortions $\Delta\bar\theta$, of the bond angles from their ideal c-sc values are in the range ~7 to $11°$.

We will suppose that the as-formed structure of a-sc is of the CRN type in which are dispersed various point defects -- voids, dangling bonds, interstitial atoms -- in non-equilibrium distributions and concentrations. Structural relaxation might then occur by the migration, redistribution and annihilation of these defects with attendant changes in $\Delta\bar\theta$.

Donovan et al. [4] measured the enthalpy changes of ion amorphized a-sc Ge specimens when heated at 40°/min from 400°K through their crystallization temperature. Irreversible enthalpy changes associated with relaxation occurred over the temperature range 400-660°K. Crystallization began at $\sim 660°K$ and was virtually complete at 745°K. The enthalpy release in the crystallization stage, $\Delta H_{ac} = 0.12$ ev, was in close agreement with that measured in several earlier studies [19-21] for a-sc Ge prepared by a variety of deposition processes. This agreement suggests that by crystallization onset, at least, specimens prepared by widely differing methods are in closely similar states of relaxation. The enthalpy release in the relaxation stage was quite large, ~ 0.06 ev, about $1/2\ \Delta H_{ac}$.

Lannin and coworkers [22] found that the Raman scattering spectrum of a-sc Ge sharpens considerably as a result of relaxation annealing. From this sharpening, in conjunction with X-ray scattering results, they inferred that $\Delta\theta$ decreased from ~11 to $9°$ from their most unrelaxed to their well-annealed specimens. We expect that the energy of an ideal CRN should scale as $\Delta\theta^2$. Thus, ignoring possible direct contributions from point defect annihilation, the Lannin et al. [22] results are consistent with an energy loss in relaxation near that found by Donovan et al. [4]. However, the uncertainties in this comparison are very large, considering the assumption of linear elasticity and the sensitivity of the energy to $\Delta\theta$. It is possible that $\Delta\theta$ in the vicinity of point defects is larger than in the fully connected part of the network. Thus, the presence of point defects might skew $\Delta\bar\theta$ to larger values than that of the ideal CRN so that the decrease in $\Delta\bar{\theta}$ with annealing would be coupled closely with point defect annihilation.

In contrast with Ge, ion amorphized a-sc Si specimens exhibited no irreversible enthalpy release when heated at 40°/min from 400°K to the temperature at crystallization onset [4]. Waddell et al. [23] had reported a factor of 2 drop in dangling bond density (at a level $< 10^{-3}$/atom) and a marked decrease in index of refraction in ion amorphized Si after anneals in the range 500-600°C. However, the atomic volume of a-sc ($\sim$1.015 to 1.02 that of c-sc) was not changed measurably in this treatment and the energy change may have been negligible as well; indeed, it seems likely that the energy decrease accompanying annihilation of no more than 10^{-3} dangling bonds/atom would be quite small relative to the measured ΔH_{ac}. Also, in accord with the energy measurements, Maley, Lannin and Cullis [24] found that the Raman spectrum of a-sc Si formed by melt quenching, presumably at temperatures well in excess of 1000°C, is in close agreement with that of CVD a-sc Si deposited on a substrate at 525°C, i.e., $\Delta\bar{\theta}$ apparently was the same for specimens formed very differently at temperatures more than 500° apart. These results indicate that at homologous temperatures the rates of structural relaxation in Si are much more rapid than in Ge, possibly reflecting that relaxation is effected mainly by the migration and annihilation of point defects present in the amorphous films, as formed, and that these defects are relatively more mobile in Si than in Ge.

Prokes and Spaepen's recent measurements [25] indicate that the Ge/Si interdiffusivity in compositionally modulated a-Ge/a-Si films is more than 10 orders of magnitude higher than either of the pure crystal self-diffusivities. At the average of the temperature of crystallization onset of a-Ge and a-Si, 800-850°K, their results give a mean time for Ge-Si diffusive exchange of order 1 sec which is 2 orders of magnitude smaller than that required to heat the specimens from 800-850°K in the Donovan et al. experiments. Thus, structural relaxation, if limited by the same atom movements as those controlling interdiffusion, would have been virtually complete at crystallization onset.

The experience and analysis just described supports the view that the a-sc phase does relax toward a thermodynamically well-defined state, perhaps characterized in part by unique $\Delta\bar{\theta}(T)$ and concentration, $x_i(T)$, of point defects. It follows that there would be a temperature, $T_{a\ell}$, at which a-sc and ℓm would be in metastable equilibrium or around which one phase would evolve continuously into the other.

While the CRN $\Delta\bar{\theta}$'s which give the best fit to the radial distribution functions of a-sc Ge and Si have been calculated, there is no microscopic theory for what determines $\Delta\bar{\theta}(T)$ and the $\Delta\theta$ distribution at metastable equilibrium [26]. Apparently they should be dictated by some compromise of the configurational entropy with the energies of bond distortion and point defects. That at equilibrium $\Delta\bar{\theta}(T) \neq 0$ for a-sc exists should, however, be hardly more surprising than its existence for the topologically similar structure for fused SiO_2.

UNDERCOOLING OF THE LIQUID METAL (ℓm) PHASE

Were the transition $\ell m \rightarrow$ a-sc structurally and thermodynamically continuous, it should occur readily and homogeneously when ℓm is undercooled to and below the calculated $T_{a\ell}$. If discontinuous, it would be effected by nucleation and growth. According to the simple nucleation theory, the thermodynamic barrier to homogeneous nucleation of a-sc is $b\sigma_{a\ell}^3/(\Delta G_{a\ell})^2$ and that for c-sc is $b(\sigma_{c\ell}^3)/(\Delta G_{a\ell})^2$ where: $\sigma_{a\ell}$ and $\sigma_{c\ell}$ are the tensions of, respectively, the ℓm/a-sc and ℓm/c-sc interfaces, $\Delta G_{a\ell}$ and $\Delta G_{c\ell}$ are the free energy changes/volume of, respectively, the $\ell m \rightarrow$ a-sc and $\ell m \rightarrow$ c-sc transitions, b is a constant.

If, as Spaepen and the author [15] suggested, $\sigma_{a\ell}$ and $\sigma_{c\ell}$ were determined primarily by the change in coordination accompanying the transitions, their magnitudes would be similar. Then since $|\Delta G_{c\ell}|$ always greatly exceeds $|\Delta G_{a\ell}|$, it is likely that nucleation of c-sc would become copious at undercoolings, relative to $T_{c\ell}$, much less deep than that required for measurable a-sc nucleation. With imposed thermal gradients of the usual magnitude recalescence would then cut off the possibility of a-sc nucleation.

Early determinations by Cech and the author [27], using the isolated droplet technique, and by Powell [28], using a fluxing technique indicated that the scaled undercoolings, ΔT_r°, at measurable nucleation onset of various specimens of ℓm Ge were ~0.2; ΔT_r° is defined as $(T_{c\ell} - T^{\circ})/T_{c\ell}$ where T° is the temperature (in °K) at nucleation. In these experiments T_r° ($= T^{\circ}/T_{c\ell}$) was near the calculated $T_{a\ell}$ (~0.8 $T_{c\ell}$) for Ge but the ℓm phase crystallized by nucleation and growth with evident recalescence.

Recently Devaud and the author [29] determined ΔT_r° of numerous ℓm Ge specimens immersed in fluxes of dehydrated B_2O_3. Some specimens exhibited ΔT_r° as large as 0.35, corresponding to a T_r° about 0.15 below $T_{a\ell}$. At these large undercoolings, the initial solidification occurred dendritically and was marked by sharp recalescence. When examined following solidification, these specimens exhibited the c-sc structure with an average grain size, δ_c; ~10 microns. In sharp contrast δ_c in specimens which began to solidify at $\Delta T_r < 0.2$ was of the order of the specimen dimension (~1 mm). There was no evidence that a $\ell m \rightarrow$ a-sc transition occurred either continuously or by nucleation and growth, at temperatures ranging as low as 0.84 $T_{a\ell}$ (~0.65 $T_{c\ell}$).

Devaud and the author [30] also investigated the undercooling of ℓm Si droplets contained in fused silica capsules. Some droplets exhibited scaled undercoolings as large as 0.19, corresponding to T_r° about 0.05 to 0.08 below the scaled $T_{a\ell}$. Solidification was accompanied by sharp recalescence and the as-solidified droplets had the c-sc structure. Here also there was no indication of a $\ell m \rightarrow$ a-sc transition to the lowest T_r° observed. The thermal gradients which develop during droplet crystallization are not large and must be several orders of magnitude below those reached in the laser pulsing experiments such as those of Cullis et al. [31], in which a-sc formed from ℓm by movement of an a-sc/ℓm interface originating at the c-sc/ℓm interface.

GROWTH OF SOLID PHASES

Formal theory of plane front growth

The formal theory for the rate of plane front growth, u, without significant composition or density change, has been reviewed in a number of earlier publications [3]; it leads to the following relation:

$$u = fk_i \lambda \left(1 - e^{\Delta G/RT_i}\right) \tag{1}$$

where f = the fraction of interfacial sites at which growth can occur

k_i = frequency of interfacial rearrangement

λ = interface displacement/rearrangement

T_i = *interfacial* temperature

ΔG = free energy change/mole accompanying the transition.

In the linear kinetic regime, which results when $RT_i \gg |\Delta G|$ and where $\Delta G \simeq -\Delta S \Delta T_i$, the relation reduces to:

$$u \simeq fk_i \lambda \frac{|\Delta S||\Delta T_i|}{RT_i} \tag{2}$$

here ΔS is the entropy change/mole in the transition. As growth occurs, the heat of transition, ΔH/mole, must be conducted away from the interface. At steady state the interfacial undercooling, ΔT_i may be related to the thermal gradient, $(\text{grad } T)_i$, imposed at the interface as follows:

$$\Delta T_i \simeq \left[\frac{\kappa (\text{grad } T)_i}{fk_i} \frac{\bar{V} RT_i}{\lambda |\Delta H||\Delta S|} \right] \tag{3}$$

where $\bar{V}$ = the molar volume and κ is the thermal conductivity of the phase through which the heat is conducted away.

Growth into ℓm phase

Specializing equation (2) to the rates of growth of c-sc or a-sc into ℓm gives

$$u^c \simeq f^c k_i^c \lambda \frac{|\Delta T_i^c||\Delta S^c|}{RT_i} \tag{4}$$

and

$$u^a \simeq f^a k_i^a \lambda \frac{|\Delta T_i^a||\Delta S^a|}{RT_i} \tag{5}$$

where c and a refer, respectively, to the c-sc and a-sc phases. The condition that the a-sc phase, *if nucleated*, would "outrun" the c-sc phase, i.e., $u^a > u^c$, is then:

$$f^a k_i^a |\Delta T_i^a||\Delta S^a| > f^c k_i^c |\Delta T_i^c||\Delta S^c| \tag{6}$$

the f's and k_i's are generally T_i dependent. This condition, which was derived earlier by Bucksbaum and Bokor [11], for the emergence of a-sc in ℓm by competitive growth is rather less stringent than that for its formation by preferential nucleation. The appearance, under certain experimental conditions, of a-sc layers on c-sc surfaces following irradiation by high energy laser pulses [32,33,11] indicates, as interpreted by Spaepen and the author [3], that a-sc growth was initiated at the c-sc/ℓm interface and proceeded more rapidly than that of c-sc. It was presumed that the thermal gradients imposed must have been high enough to depress T_i to levels well below $T_{a\ell}$. Cullis et al. [11] inferred the speeds of the solidification fronts in ℓm following various laser irradiation treatments and found that the regrowth product changed from epitaxial c-sc to a-sc when a critical speed, u^*, dependent on surface orientation, was exceeded. In regrowth normal to (100) $u^* \sim 15$ meters/sec. The essence of these conclusions have been confirmed by

Thompson and coworkers [13] who monitored the interface speeds more directly by means of their resistometric method. Expressing k_i^c as $(u_s/\lambda)e^{-\Delta G_c'/RT_i}$, where u_s is the sound speed in ℓm and $\Delta G_c'$ is the free energy of activation for interfacial rearrangement, the author showed [1] that if T_i at u^* is e.g. 50° below $T_{a\ell}$ (i.e. $T_i \sim 1450°K$) then $f^c e^{-\Delta G_c'/RT_i} \approx 10^{-2}$; this result implies that $f \gg 0.01$ and $G_c' \leqslant 0.57$. Under these conditions $u^a = u^c$ if $f^a k_i^a/f^c k_i^c \approx 10$. It is noteworthy that Ohdomari et al. [34] found that Si deposits from the vapor on clean c-sc surfaces at 20°C in a-sc rather than c-sc form indicating that the attachment rate in this process is also more rapid on a-sc than on c-sc.

At small undercooling, Si and Ge crystals grow faceted from their melts, in accord with the Jackson [35] correlation. This morphology implies that there is a thermodynamic barrier, due to a substantial ledge tension, 6_e, to 2-dimensional nucleation of crystal layers on the (111) faces, at least. The magnitude of σ_e is not known but it is likely that at $T_{a\ell}$, the large undercooling, relative to $T_{c\ell}$, is sufficient to insure copious nucleation of crystal layers and thereby a substantial fraction, f, of growth sites.

Recently, several investigators have attempted the extremely difficult experimental evaluation of ΔT_i during the melting in of ℓm and subsequent regrowth of c-sc Si in laser pulsing. The results were expressed in terms of the derivative $\beta = dT_i/du$. Depending on their assumption on the thermal conductivities, Thompson et al. [13] obtained $\beta \sim 10$-17 deg-sec/meter for regrowth normal to (100). A *linear* extrapolation of ΔT_i with u then leads to $\Delta T_i^* \sim 150$-250° at the critical speed, u^*, of transition from c-sc to a-sc regrowth normal to (100) c-sc. Given the large experimental and calculational uncertainties these results are consistent with the Donovan et al. [4] estimate that $T_{a\ell} \sim (0.84 \pm 0.03) \, T_{c\ell}$.

Larson et al. [36] have estimated T_i during melt-in and regrowth from their measurements of the attendant thermal strain in c-sc. They calculate that at a melt-in rate estimated to be 10 meter/sec, normal to (100), T_i is $110 \pm 40°$ higher than T_i during regrowth at 6 meter/sec. It appeared that these results could be reconciled with those of Thompson et al. [13] only by supposing that the kinetic coefficients for melt-in are much higher than those for regrowth. However, in these kinetic analyses, account must be taken that, considering the high melt-in and regrowth rates and extreme thermal gradients in the experiments, one or both of the phases in the interfacial region may be somewhat out of their internal equilibrium states at T_i. Consequently, the temperature, $T_{c\ell}'$, at which the free energies of the actual ℓm and c-sc phases would be equivalent may be considerably displaced from $T_{c\ell}$ corresponding to zero u and (grad T)$_i$. Also, where the ℓm/c-sc interface moves in at an initial rapid rate, slows, turns around and then moves out again, the interfacial configuration, which will define the site factor, f, may never reach a steady state value characteristic of constant (grad T)$_i$ and T_i. Such deviations from steady state f values could lead to an apparent asymmetry between melt-in and regrowth kinetics at $T_{c\ell}$ of the nature suggested by the results of Larson et al.

Thompson et al. [12] measured the melt-in rate of a-sc on c-sc Si during pulsed laser irradiation and observed that the melt-in was delayed for several nanoseconds when the ℓm layer reached the initial c-sc/a-sc interface. This delay was associated with the time required to heat the undercooled ℓm from the melting point, $T_{a\ell}$ of a-sc to the melting point of the crystal and from it and a thermal analysis they estimated that $T_{a\ell} = 0.88 \pm 0.03$ in reasonable agreement with the value indicated by the measurements of Donovan et al. [4].

Solid state regrowth

There have been several studies [5-8] of the rate of epitaxial regrowth (SPE), and its temperature dependence, of a-sc→c-sc for Ge and Si at 1 atm. pressure. All of the results are in fair agreement with the earliest ones, which were obtained by Czepregi et al. [5,6] The reported activation energies, q_u, are near the covalent bond energies, ε_b, taken to be 1/2 the cohesive energies of the c-sc phases. Recently Williams et al. [37] reported that the regrowth rate of c-sc Si is sharply increased by a fluence of Ne^+ ions and deduced that, under these conditions q_u was reduced to a level ~ 0.1 ε_b. These results indicate that at least 90% of the thermal activation energy for growth must be spent in forming a defect which can then migrate with little thermal activation to effect interfacial reconstruction. They are consistent with the Spaepen mechanism [15,3] in which interfacial rearrangement is effected by the formation and migration of dangling half-bonds, which subsequently recombine. It had also been suggested [38] that regrowth in SPE is effected by migrating vacancies. However, it appears that, at least in c-sc Si, the energy of vacancy formation greatly exceeds the formation energy of the defect responsible for regrowth. It is also established [39] that in Si the regrowth rate is enhanced by the separate additions of n- or p-type impurities but is not significantly affected by compensating n and p additions. Uncompensated dopant additions can increase the concentration of vacancies by altering their charge state. However, Mosley and Paessler [40,41] have shown that the probability of dangling bond formation also can be increased by parallel effects.

The possible role of vacancy formation in a-sc regrowth should be partly accessible, at least from the pressure dependence of u. In transition state rate theory this dependence is expressed approximately as:

$$\left(\frac{\partial \ln u}{\partial P}\right)_T \cong -V^* \tag{7}$$

where V^* is the activation volume. In regrowth occurring by the migration of thermally-created point defects, the activation volume would be: $V^* = V_f^* + V_m^*$, where V_f^* and V_m^* are, respectively, the volume of defect formation and migration. V^* would reduce to V_m^* if growth were effected entirely by defects present in the films as formed. However, the results of Williams et al. [37] strongly suggest that in Si the defects must be created thermally or by irradiation. The volume of formation of vacancies, V_v^*, and of interstitials, V_i^*, may be expressed [42]:

$$V_v^* = V_a(1 + \alpha_v) \tag{8}$$

and

$$V_i^* = -V_a(1 + \alpha_i) \tag{9}$$

where V_a is the atomic volume of the solid and α_v and α_i are coefficients which correct for the volume changes due to relaxation around the defect. Thus, provided the relaxation volumes are each less than the atomic volume, pressure application should promote interstitial and suppress vacancy formation.

Actually Nygren et al. [43] found that the SPE regrowth rate in Si increases markedly with pressure. Over the range 0-20 kilobar at 500°C the effect was described satisfactorily by a negative, roughly constant $V^* \sim -0.7 \, V_a$, corresponding to a lowering of 0.01 ev/kilobar in the free energy of activation opposing regrowth. These results suggest that the predominant mechanism in regrowth might require interstitial formation and migration. However, pressure application might also promote dangling bond formation. The sharp increase of the rate of growth of quartz crystals into fused silica with pressure, described by an activation volume $V^* \sim -1.0$ times the molecular volume of the fused SiO_2, found by Fratello et al. [44] was satisfactorily interpreted by adapting [45] Spaepen's dangling bond model to SiO_2. A parallel mechanism of pressure enhancement may be operative in c-sc Si regrowth.

"Explosive" crystallization

The explosive mode of a-sc$\rightarrow$c-sc seems reasonably accounted for by the mechanism of Gilmer and Leamy [10] whereby T_i at the a-sc/c-sc interface reaches $T_{a\ell}$ when the rate of transport of ΔH_{ac} from the interface is sufficiently slow. At $T_i > T_{a\ell}$ a-sc melts and crystallization then proceeds via the rapid $\ell m \rightarrow$ c-sc mode with attendant rapid melting of a-sc.

SUPERHEATING OF SOLIDS s-c

The problem of *internal* stability of the c-sc and a-sc phases has been reviewed by the author in earlier papers [1,2]. According to simple nucleation theory, the thermodynamic barrier to homogeneous nucleation of ℓm *within* c-sc or a-sc should be of the same form, with the same interfacial tension, $\sigma_{c\ell}$ or $\sigma_{a\ell}$, as that opposing nucleation of the reverse process. Assuming that $\sigma_{c\ell}$ and ΔS^c are T independent, and neglecting the strain energy contribution to nucleation resistance, the scaled superheating $\Delta T_r^{+\circ}$ at the onset of measurable ℓm nucleation in c-sc may be related to the scaled undercooling, ΔT_r°, at onset of measurable c-sc nucleation in ℓm, as follows:

$$\Delta T_r^{+\circ} \cong \Delta T_r^\circ (1 - \Delta T_h^\circ) \quad . \tag{10}$$

For Ge, with $\Delta T_r^\circ > 0.35$, $\Delta T_r^{+\circ}$ is estimated to be > 0.23.

As is well-known, melting generally begins and proceeds *heterogeneously* from the external surfaces or extended internal imperfections of crystals. Such melting proves nothing more than that $\Delta T_r^{+\circ}$ must exceed the interfacial superheating, ΔT_{ri}^+, driving the movement of the ℓm/c-sc interface. ΔT_{ri}^+ and the consequent melt-in speed, $\overset{+}{u}$, increase with the imposed thermal gradient. From equation (4) and supposing that $u_s > k_i^c \lambda$ we infer that

$$\Delta T_{ri}^+ \gg \frac{u^+}{u_s} \frac{R}{|\Delta S^c|} \tag{11}$$

Thus for Si with $u_s \sim 5000$ m/sec, $dT_i^+/du^+ > 0.1$ deg-sec/m and a melt-in speed of e.g. 200 m/sec would imply that $\Delta T_i^+ \gg 20°$.

We have noted that in laser pulsing experiments, speeds of melt-in of c-sc$\rightarrow \ell m$ of order 10 meters/sec or more are often indicated [see, e.g., experiments of Larson et al. [36]. Similar melt-in speeds of a-sc$\rightarrow \ell m$ are indicated by the experiments of Thompson et al. [12]. The apparent persistence of a-sc Si to $T \rightarrow T_{c\ell}$ for periods of several millisecs reported by Olson et al. [7] would indicate that the a-sc became substantially superheated. In view of the results of Thompson et al. [12] such superheating

could hardly be due to slowness of the motion of the a-sc/ℓm interface. It
appears that it could only result [2] from a high resistance of the a-sc film
to ℓm nucleation.

ACKNOWLEDGMENT

This article was prepared while the author was B.T. Matthias Visiting
Scholar at the Center for Materials Science, Los Alamos National Laboratory.
The author is grateful to the Center for its provision of support and a
highly stimulating ambience for thought and analysis. The author's research
at Harvard in the subject area of this paper was supported in part by Grants
from NASA (NAS-8-35416) and ONR (N00014-85-K-0023). Also he is indebted to
M.J. Aziz for critically reviewing the manuscript.

REFERENCES

1. D. Turnbull, J. de Physique 43, C-1, 259 (1982).

2. D. Turnbull, a) Mats. Res. Soc. Symp. Proc. 7, 103 (1983); b) ibid,
   13, 131 (1983).

3. F. Spaepen and D. Turnbull, "Laser Annealing of Semiconductors," (ed.
   J.M. Poate and J.W. Mayer), pp. 15-42, Academic Press, NY (1982).

4. E.P. Donovan, F. Spaepen, D. Turnbull, J.M. Poate, and D.C. Jacobson, J.
   Appl. Phys. 57, 1795 (1985).

5. L. Czepregi, R.P. Kullen, J.W. Mayer and T.W. Sigmon, Sol. State Commun.
   21, 1019 (1977).

6. L. Czepregi, E.F. Kennedy, J.W. Mayer, and T.W. Sigmon, J. Appl. Phys.
   49, 3906 (1978).

7. G.L. Olson, S.A. Kokorowski, J.A. Ross, and L.D. Hess, Mats. Res. Soc.
   Proc. 13, 141 (1983).

8. A. Lietoila, A. Wakita, T.W. Sigmon, and J.F. Gibbons, J. Appl. Phys. 53,
   4399 (1982).

9. a) G. Gore, Phil. Mag. 9, 73 (1855); b) T.T. Takamori, R. Messier, and
   R. Roy, J. Mat. Sci. 8, 1809 (1973); c) G. Auvert, D. Bensahel, A. Perio,
   V.T. Nguyen, G.A. Rozgoni, Appl. Phys. Lett. 39, 724 (1981).

10. G.H. Gilmer and H.J. Leamy, "Laser and Electron Beam Processing of
    Materials," (eds. C.W. White and P.S. Peercy), pp. 227-233, Academic
    Press, NY (1980).

11. A.G. Cullis, H.C. Webber, N.G. Chew, J.M. Poate, and P. Baeri, Phys. Rev.
    Lett. 49, 219 (1982).

12. M.O. Thompson, G.J. Galvin, J.W. Mayer, P.S. Peercy, J.M. Poate, D.C.
    Jacobson, A.G. Cullis, and N.G. Chew, Phys. Rev. Lett. 52, 2360 (1984).

13. M.O. Thompson, P.H. Bucksbaum, and J. Bokor, Mat. Res. Soc. Symp. Proc.
    35, 181 (1985).

14. P.H. Bucksbaum and J. Bokor, Phys. Rev. Lett. 53, 182 (1984).

15. F. Spaepen and D. Turnbull, Am. Inst. Phys. Conf. Proc. 50, 50 (1979).

16. B.G. Bagley and H.S. Chen, Am. Inst. Phys. Conf. Proc. 50, 97 (1979).

17. W. Paul, G.A.N. Connell, and R.J. Temkin, Adv. Phys. 22, 529 (1973).

18. a) G. Etherington, A.C. Wright, J.T. Wenzel, J.C. Dore, and J.H. Clarke, J. Non-Cryst. Solids 48, 265 (1982); b) A.C. Wright, J. Non-Cryst. Solids 75, 15 (1985).

19. H.S. Chen and D. Turnbull, J. App. Phys. 40, 4214 (1969).

20. F.W. Lytle, D.E. Sayers, and A.K. Eikum, J. Non-Cryst. Solids 13, 68 (1973).

21. R.J. Temkin and W. Paul, Proc. V-th Int. Conf. on Amorphous Semiconductors (ed. J. Stuke and W. Brenig), pp. 1193-1200, Taylor and Francis (1974).

22. J.S. Lannin, N. Maley, and S.T. Khirsagar, Sol. State Comm. 53, 939 (1985).

23. C.N. Waddell, W.G. Spitzer, J.E. Frederickson, G.K. Hubler, and T.A. Kennedy, J. Appl. Phys. 55, 4361 (1984).

24. N. Maley, J.S. Lannin, and A.G. Cullis, Phys. Rev. Lett. 53, 1571 (1984).

25. S.M. Prokes and F. Spaepen, Appl. Phys. Lett. 47, 234 (1985).

26. F. Wooten and D. Weaire, J. Non-Cryst. Solids 64, 325 (1984).

27. D. Turnbull and R.E. Cech, J. Appl. Phys. 21, 804 (1950).

28. G.L.F. Powell, Trans. Met. Soc. A.I.M.E. 239, 1662 (1967).

29. G. Devaud and D. Turnbull, Mats. Res. Soc. Proc., in press.

30. G. Devaud and D. Turnbull, Appl. Phys. Lett. 46, 844 (1985).

31. A.G. Cullis, H.C. Webber and N.G. Chew, Appl. Phys. Letters 40, 998 (1982).

32. a) P.L. Liu, R. Yen, N. Bloembergen, and R.T. Hodgson, Appl. Phys. Lett. 34, 864 (1979); b) J.M. Liu, R. Yen, H. Kurz, and N. Bloembergen, Appl. Phys. Lett. 39, 755 (1981).

33. R. Tsu, R.T. Hodgson, T.Y. Tan, and J.E.E. Baglin, Phys. Rev. Lett. 42, 1356 (1979).

34. I. Ohdomari, M. Kakumu, H. Sugahara, M. Hori, and T. Saito, J. Appl. Phys. 52, 6617 (1981).

35. K.A. Jackson, "Growth and Perfection of Crystals" (ed. R.H. Doremus, B.W. Roberts and D. Turnbull), pp. 319-325, Wiley, NY (1958).

36. B.C. Larson, J.Z. Tischler, and D.M. Mills, "Nanosecond Resolution Time-Resolved X-Ray Study of Si During Pulsed-Laser Irradiation," to be published.

37. J.S. Williams, R.G. Elliman, W.L. Brown, and T.E. Seidel, Phys. Rev. Lett. 55, 1482 (1985).

38. a) J.A. Van Vechten, "Lattice Defects in Semiconductors," 212, Inst. of Phys., London (1975); b) J.A. Van Vechten and C.D. Thurmond, Phys. Rev. 14B, 3539 (1976).

39. I. Suni, G. Goltz, M.G. Grinaldi, M.A. Nicolet, and S.S. Lau, Appl. Phys. Lett. 40, 269 (1982).

40. L. Mosley, P.J. Germain, M.A. Paessler, and K. Zellema, J. Non-Cryst. Solids 59-60, 273 (1983).

41. L.E. Mosley and M.A. Paessler, J. Appl. Phys. 57, 2328 (1985).

42. R.O. Simmons and R.W. Balluffi, Phys. Rev. 117, 52, 62 1960; ibid. 119, 600 (1960).

43. E. Nygren, M.J. Aziz, D. Turnbull, J.M. Poate, D.C. Jacobson, and R. Hull, Appl. Phys. Lett. 47, 232 (1985).

44. V.J. Fratello, J.F. Hays, and D. Turnbull, J. Appl. Phys. 51, 4718 (1980).

45. V.J. Fratello, J.F. Hays, F. Spaepen, and D. Turnbull, J. Appl. Phys. 51, 6160 (1980).

AMORPHISATION, CRYSTALLISATION AND RELATED PHENOMENA IN SILICON

J.S. WILLIAMS
Microelectronics Technology Centre, RMIT, Melbourne 3000, Australia.

ABSTRACT

This review examines recently observed phenomena associated with amorphisation and crystallisation of silicon under ion bombardment and furnace annealing. Ideally, heavy ion damage should completely amorphise the silicon surface layers so that the underlying crystal can provide a perfect template for subsequent epitaxial growth. However, in practise the ion bombardment and annealing behaviour can be decidedly more complex. During ion bombardment of silicon, several correlated processes can take place depending on the target temperature and the precise bombardment conditions. These processes include: defect production; amorphisation; diffusion and segregation of defects and impurities; and ion-beam-induced (epitaxial) crystallisation. During subsequent heat treatment, amorphous layers can exhibit anomalous impurity diffusion and precipitation effects, nucleation of random crystallites, and solid phase epitaxial growth. In addition, the kinetics of the epitaxial growth process are sensitive to the type and state of implanted impurities present in the silicon. The competition between random nucleation and epitaxy is also dominated by impurity effects. Finally, correlations between all of these phenomena provide i) considerable insight into impurity and defect behaviour in amorphous and crystalline silicon, and ii) a better understanding of the amorphous to crystalline phase transition, including mechanisms of solid phase epitaxial growth.

INTRODUCTION

Ion implantation and subsequent thermal annealing are key processes in the fabrication of integrated circuits and, consequently, they have been the subject of much scientific interest over the past two decades. The primary technological requirement of ion implantation processing is the controlled introduction of electrically active dopants into perfect single crystal silicon wafers. Thus, the processes by which damage is generated in silicon by ion bombardment and subsequently removed by thermal treatment have been most extensively studied. The immediately preceeding papers in this volume by Brown [1] and Turnbull [2] have introduced some of the fundamental concepts associated, respectively, with ion damaging and subsequent crystallisation processes. This review further develops both these areas, concentrating on recently observed phenomena which provide new insights into amorphisation and crystallisation processes in silicon.

IDEAL BEHAVIOUR

When an energetic ion penetrates a silicon target, sufficient energy can be imparted to the lattice atoms during nuclear collisions to cause a cascade of atomic displacements [1]. Such violent displacement processes can directly produce an amorphous zone about the individual ion track in the case of heavy ion impact, as illustrated in Fig.1a. The accumulation and overlap of disorder formed by individual ions can lead to the evolution of a continuous amorphous layer with increasing ion dose as shown schematically in Figs.1 b) - d) for Sb implantation into silicon. The cross-sectional TEM micrograph of Fig.2a illustrates this ideal behaviour. In this case, a continuous amorphous layer (A) is formed on the surface of silicon, above a buried transition region (T) of partially amorphised silicon which, in turn, overlays perfect crystalline silicon (C). This damage structure is only formed when the defects produced by the ion beam are essentially immobile during bombardment at the irradiation temperature. Usually, implantation at liquid nitrogen temperatures or below is necessary to avoid in-situ annealing and hence a more complex disorder structure, consisting of crystalline defect clusters and extended defects. Non-ideal damage structures which originate from dynamic annealing during bombardment are discussed later in this review.

Having produced a well defined amorphous silicon layer on perfect single crystal silicon, it is possible to remove this ion damage by solid phase epitaxial growth (SPEG) at temperatures of 500-600°C [3]. This annealing process ideally gives rise to perfect epitaxy in which the regrown layer is free of extended-defects and essentially similar in quality to the unbombarded silicon wafer. We illustrate such a situation in Fig.2b, where SPEG has been employed to partially remove the amorphous layer on the (100) silicon sample of Fig. 2a. The annealing process was arrested before complete growth had taken place, in this case to illustrate the epitaxial nature of crystallisation. Note that the regrown layer is free of extended defects, a situation which is most desirable for applications of ion implantation and SPEG in the semiconductor industry. When dopants are implanted into silicon for controllably altering the electrical behaviour, the further ideal requirement is that all dopant atoms are incorporated onto substitutional, electrically active sites within the silicon lattice during SPEG [4]. Although it is possible, in practise, to closely approach this ideal behaviour, i.e. formation of well-defined amorphous layers during ion bombardment and perfect SPEG and dopant incorporation during thermal annealing, the most general ion bombardment and annealing conditions do not give such optimum results. Situations which are less ideal but which often provide more insight into the operative damaging and annealing processes are illustrated in subsequent sections of this review.

Before reviewing more general ion bombardment processes, it is appropriate to summarise some important aspects of SPEG of silicon which have emerged during the past decade. Pioneering work at Caltech [3,5-7] established that the process was characterised by an activation energy of 2.5 eV. Furthermore, the SPEG rate exhibits a marked dependence on substrate orientation, whereby amorphous silicon layers on (100) substrates regrow 3 times faster than those on (110) and 25 times faster than on (111). SPEG can also be dramatically enhanced (by as much as a factor of 50 in rate) by the introduction of very small concentrations (0.1 atomic percent) of n- or p-type dopants. Fig.3 shows the more recent and extensive data of Olson et al [8], who measured the growth kinetics of Si-implanted amorphous silicon layers on (100) silicon over

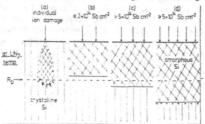

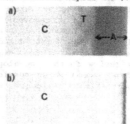

Fig. 1 Schematic representation of a continuous surface amorphous layer during heavy ion (Sb) irradiation.

Fig. 2 TEM cross-section micrographs showing a) the amorphous layer (A) of the order of 1000Å and partly amorphous transition region (T) resulting from 50keV Si⁺ implant at 77K to a dose of 2×10^{14} cm⁻² into (100) silicon, and b) the same layer after SPEG, leaving a surface amorphous layer (A) of ~50Å. The surface is denoted by S.

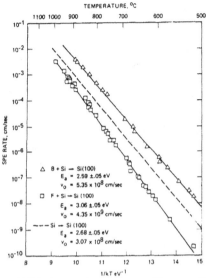

Fig.3 SPEG kinetics for (100) silicon preamorphized with Si⁺ ions. Dashed curve, impurity-free growth and for boron doped (▲) and fluorine doped (□) layers. After Olson et al [8].

a 600°C temperature span. These authors established an activation energy of 2.7eV for the process. Fig.3 also shows the effect of impurities, boron and fluorine, on the growth kinetics. Boron enhances the growth rate at a given temperature whereas fluorine retards growth. Various models for treating SPEG and explaining its orientation and doping dependencies have emerged over recent years [6,9-12]: these invoke bond breaking and/or migrating point defects as the primary initiation mechanism for SPEG. Some of the newer results presented in this review provide further insight into SPEG and its dependencies.

DYNAMIC ANNEALING PROCESSES DURING ION BOMBARDMENT

We illustrate the influence of dynamic annealing on the final disorder structure by the cross-sectional TEM micrograph in Fig.4. This particular structure (consisting of a buried amorphous layer (BA) of ~ 1500Å in thickness, sandwiched between regions of gross crystalline disorder both near the surface (SC) and deeper (DC)) originated from a 200keV Si⁺ implant into (100) silicon at a temperature slightly above room temperature. In contrast, if the implantation temperature is lowered to ~ -180°C, a continuous amorphous layer of ~ 3300Å thick is formed, similar to the structure shown in Fig.2a. In this section, the nature of ion-beam-induced disorder is examined and the various dependencies of dynamic annealing are illustrated. In particular, the final damage structure is shown to result from competition between damage production, arising from energy deposited by the ion beam into atomic displacement processes, and dynamic annealing, arising primarily from migrating defects. When the former process dominates, ion bombardment leads to a crystalline-to-amorphous phase transition, and when the latter process dominates, the originally perfect crystal can be converted to a highly defective one. Furthermore, it is possible to induce an amorphous-to-crystalline transition during ion bombardment under conditions which strongly favour dynamic annealing processes.

Dynamic Annealing Dependencies.

The nature and extent of dynamic annealing during ion bombardment of silicon depends strongly on the various implantation parameters. These are

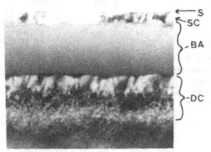

Fig.4 A cross-section TEM micrograph of a buried amorphous layer (BA) of ~1500Å in thickness, generated by 200keV Si^+ implantation into (100) silicon. Note the defective surface crystalline layer (SC) and the deep crystal damage (DC). The sample surface is denoted by S.

summarised in Table I, together with typical dependencies and key references. Implantation temperature is clearly an important variable since it controls the mobility of bombardment-induced defects. The data in Fig.5 [22] illustrate, for the case of heavy ion irradiation, the temperature-dependent observations which are summarised in Table I. As the implantation temperature increases from -200°C to +200°C, the amount of damage (as measured by ion channeling) decreases slowly for the higher doses of Sb^+, indicating that considerable dynamic annealing is present in this temperature range. The remnant disorder is primarily amorphous in this regime but, as the temperature is increased further (above ~ 230°C), only crystalline damage accompanies implantation. (Note the sharp decrease in measured damage at ~230°C for the two lowest Sb^+ doses in Fig.5).

At low implantation temperatures (e.g. room temperature and below) damage essentially accumulates by the overlap of amorphous zones (heavy ion case) or the build up of discrete damage clusters (light ion case) to produce an amorphous layer by one of two proposed mechanisms: the spike model [18] whereby an amorphous layer is formed when all lattice atoms have been displaced at least once by the overlap of ion cascades, or the critical damage density model [16,17] in which only 10-14% of lattice atoms need to be displaced before the lattice will undergo a crystalline to amorphous transition. The former model is particularly appropriate for heavy ion damage whereas the latter model is often employed to treat light-ion-damage built up. Both models give rise to a near-linear build up of disorder with increasing fluence until saturation (amorphisation) is reached.

Disorder accummulation at higher implantation temperatures is more difficult to model as a result of the more important role of dynamic annealing. Much higher fluences are needed to produce amorphous layers at the higher temperatures and Morehead and Crowder [13] modelled this in terms of annealing via vacancy migration at the periphery of direct impact-induced amorphous zones. This model and variations of it [16] have been reasonably successfully employed in the higher temperature regime. However, recent studies [22,23] have shown that such models may not be well suited to modelling of amorphous layer formation at temperatures of the order of 200°C. For example, Fig.6 shows that damage (N_d), as measured by ion channeling, does not build up linearly with fluence for Sb^+ implantation at 230°C, but, rather, exhibits a sharp increase at a fluence of $3 \times 10^{14} cm^{-2}$. TEM indicates gross crystalline damage (but no observable amorphicity) at doses $< 3 \times 10^{14} cm^{-2}$ and continuous amorphous layers at and above this dose. It has been suggested that this behaviour is best described by "nucleation" of an amorphous phase at a critical dose at either regions of gross crystalline damage or high implanted impurity concentrations.

With regard to ion mass and energy dependence of damage accumulation, the nuclear energy deposition density appears to control the process[14,16,18]. This is not only the case for the production of amorphous zones and layers but is also appropriate for the build up of ion-induced crystal damage at higher temperatures, as we illustrate in Fig.7. Fig.7a gives the implant range distribution and the nuclear energy deposition density for 1.5 MeV Ne^+

Table I. Implantation variables which influence dynamic annealing.

IMPLANTATION VARIABLES	OBSERVATIONS AND DEPENDENCIES	ILLUSTRATIVE REFERENCES
TEMPERATURE	. $\lesssim$ -180°C:essentially amorphous damage only . - 180°C to 250°C: amorphous and crystalline damage . > 250°C:only crystalline damage	[4,13,14]
FLUENCE	. Near linear damage build up at low temperatures. . Decidedly non-linear build up at higher temperatures.	[13-16]
ION MASS AND ENERGY	. Low mass ions: discrete defect production and significant dynamic annealing. . High mass ions: direct amorphisation (spike effects) and reduced dynamic annealing. . Disorder controlled by nuclear energy deposition density.	[13-18]
FLUX	. Damage can increase with increasing flux at low flux (e.g. < 10μA cm^{-2} for keV ions) . Damage can decrease with flux at high flux (e.g. >> 10 A cm^{-2} for keV ions) due to beam heating.	[14,19] [14,20,21]

irradiated silicon, whereas Fig.7b gives a TEM cross-section micrograph of the defect structure following irradiation at 318°C [24]. Clearly, the disorder, in the form of a sharp, dense band of dislocation loops, is confined to depths at which the energy deposition density exceeds ~150eV/nm. The surface region down to ~ 1.4µm is free of extended defects, suggesting that this lower density of discrete defects formed during irradiation can completely anihilate. This particular regime has considerable significance for ion-beam-induced crystallisation processes discussed in a following section.

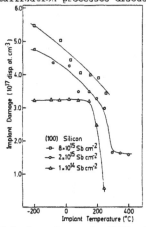

Fig.5 Implant damage in (100) silicon, as measured by He$^+$ ion channeling, plotted as a function of implant temperature for three doses of 80 keV Sb$^+$. From [22].

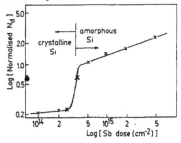

Fig.6 Damage (N_D) build up for 80keV Sb$^+$ implantation into (100) silicon at 230°C. Data from He$^+$ ion channeling. After [23].

There are two major consequences of varying ion flux on damage accumulation. In temperature regions where dynamic annealing is important and at low values of ion flux (e.g. ≤ 1 µA cm^{-2}), increases in ion flux can give rise to increased damage [14,19]. This behaviour can be attributed to a competition between the rate of damage production and the rate of dynamic annealing at a given irradiation temperature. In contrast, at high ion fluxes the opposite dependence can be exhibited whereby increased flux leads to reduced damage. This latter effect is a consequence of localised heating of the silicon substrate by the ion beam [20,21] and is illustrated in Fig.8 for the case of 1.7 MeV Ar$^+$ irradiation of silicon at a nominal substrate temperature of 77K. The local substrate temperature depends on the local power deposited with the ion beam and on the degree of heat sinking of the substrate [20].

Ion Beam Induced Crystallisation

The data presented in the previous section clearly illustrate that dynamic annealing during ion bombardment is an extremely efficient process for repairing the displacement damage generated by the ion beam. Indeed, ion irradiation at temperatures as low as 200°C can induce crystallisation in previously amorphised layers, as illustrated in Fig.9. This figure shows RBS/channeling spectra for 80keV Ar$^+$ bombardment (at 300°C) of (100) silicon containing a surface amorphous layer produced by a previous (low temperature) Sb$^+$ bombardment [23]. The Ar$^+$ irradiation clearly induces epitaxial growth of the amorphous layer at 300°C, as indicated by the single arrow, but produces significant crystalline disorder (double arrow) at a depth corresponding to the peak in the Ar$^+$ energy deposition density. Very recently this process of ion beam induced epitaxial crystallisation (IBIEC) has been studied in considerable detail [26-28] and an understanding of the process is now beginning to emerge [28]. In particular, under the appropriate energy deposition density and irradiation temperature conditions (e.g. those appropriate to the data presented in Fig.7) it is possible to produce complete epitaxial crystallisation (without the formation of extended defects) of surface amorphous layers at temperatures well below those at which such layers

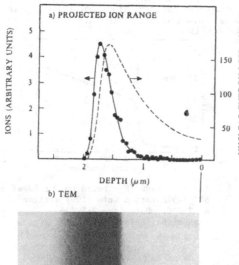

a) PROJECTED ION RANGE

IONS (ARBITRARY UNITS)

NUCLEAR ENERGY LOSS (eV/nm)

DEPTH (µm)

b) TEM

Fig.7:
a)   Plots of ion range and nuclear energy density distributions for 1.5 MeV Ne$^+$ irradiation of (100) silicon. Data from Monte Carlo simulations.

b)   TEM cross-section micrograph showing defect structure for the irradiation in a) at 318°C to a dose of 5x10^{16}cm^{-2}. After [24].

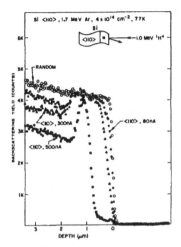

Fig.8 H⁺ ion channeling spectra
indicating flux dependence of damage
induced by 1.7MeVAr⁺ ions at 4x10¹⁴cm⁻². After [25].

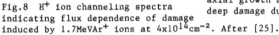

Fig.9 He⁺ ion channeling spectra
indicating ion beam induced annealing of
amorphous silicon on (100) silicon. The
layer was previously amorphised with
80keV Sb⁺ at 77K and the beam annealing
was carried out at 300°C using 80keV Ar⁺
ions. The single arrow shows the epitaxial growth and the double arrow shows
deep damage due to Ar⁺. After [23].

recrystallise thermally. The irradiating beam does, however, produce visible
crystalline defects at the end of the ion range where the nuclear energy
deposition is highest (see Figs. 7 and 9).

Recent investigations [26,29] have clearly shown that IBIEC is not simply a
beam heating effect but, rather, strongly dependent on the energy deposited
into nuclear collisions, at the amorphous-crystalline interface by the fast
ions. This is illustrated by the RBS/channeling spectra in Fig. 10, which show
IBIEC for various energy He⁺ irradiations at 400°C of an amorphous layer
(initially 60nm thick) on (100) silicon. The He⁺ dose was 5x10¹⁶cm⁻² for all
four irradiation energies. The data indicate that the extent of epitaxy
increases with decreasing energy, consistent with an annealing mechanism based
on the production of defects (by nuclear energy deposition). Details of IBIEC
are reviewed elsewhere in these proceedings [28], suffice it to say here that
the process is only very weakly dependent on irradiation temperature (0.24eV

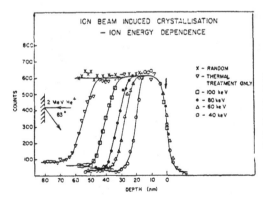

Fig.10 He⁺ ion channeling spectra
indicating the epitaxial growth
of a 60nm amorphous layer on
(100) silicon induced by 80keV
He⁺ irradiation at 400°C to a
dose of 5x10¹⁶cm⁻² and at various
energies. After [29].

activation energy) and that the ion beam appears to be generating the nucleation sites for epitaxy (the overwhelming fraction of the 2.7eV activation energy of thermal epitaxy). Finally, a more complete understanding of IBIEC may give important insights into the mechanisms involved in conventional thermally-induced epitaxy.

PROCESSES DURING LOW TEMPERATURE THERMAL PROCESSING

This section gives a selective overview of low temperature thermal processing subsequent to ion implantation. It concentrates on recently observed phenomena which have important scientific and technological implications for silicon processing.

Non-Ideal Solid Phase Epitaxial Growth (SPEG)

When ion implantation does not generate a completely amorphous layer or when the substrate orientation is other than (100), perfect SPEG, such as that depicted in Fig.2b, does not usually take place. Non-ideal SPEG is illustrated in Fig.11. Fig.11a shows partial regrowth of the buried amorphous layer shown previously in Fig.4. The poor quality seed crystal above and below the buried amorphous layer (original thickness IA) provide nucleation sites for the propagation of spanning defects (dislocations and twins) during epitaxy. Note that the amorphous layer (final thickness, A) has regrown from both sides and that the thermal treatment has resulted in coarsening of the end of range defects (ED). Fig.11b shows twinning (T) during regrowth of an amorphous layer (IA) on (111) silicon.

Additional factors which contribute to non-ideal SPEG pertain to the state of the implanted impurity prior to and during epitaxy. Impurities with low solubility in silicon and/or those which can segregate and precipitate in amorphous silicon can inhibit epitaxy (see, for example, [14]). In particular, non dopants such as the noble gases [30] or impurities which retard epitaxy (e.g. fluorine in Fig.3) can cause nucleation of crystalline defects during SPEG, and, in some cases, the formation of polycrystalline silicon [14]. Dopants in silicon can also interfere with epitaxy at concentrations well above their (equilibrium) solubility limits [14]. In such cases they are invariably

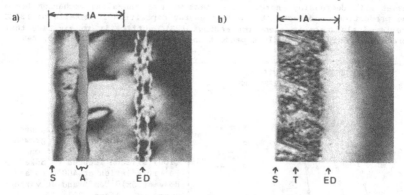

Fig.11 TEM cross-sectional micrographs indicating non-ideal SPEG for:
a) a buried amorphous layer in (100) silicon (that corresponding to Fig.4). IA indicates the original amorphous layer of 1500Å and A the final amorphous layer after SPEG. The surface is indicated by S and the end of range damage by ED: the arrows denote the directions of epitaxial growth from both interfaces. b) a continuous amorphous on (111) silicon of original thickness IA of 3300Å. End of range damage (ED) and twins (T) are indicated.

non-substitutional following crystallisation and hence exhibit low electrical activity.

Intriguing redistribution processes of impurities in silicon can also be observed during SPEG: one such case [31] is illustrated in Fig.12 for indium implanted (100) silicon. The RBS/channeling spectra show epitaxial regrowth (Fig.12a) for isothermal anneals at 550°C and the indium profiles in Fig.12b indicate segregation of a fraction of indium at the amorphous-crystalline interface during SPEG. The hatched area in Fig.12b indicates that most of the indium which remainds in crystalline silicon following SPEG is substitutional. It is interesting to note that appreciably lower doses of indium than that shown in Fig.12 do not redistribute during SPEG and the indium is totally substitutional. Recent observations outlined in the following section have indicated possible explanations for both redistribution phenomena and non-ideal growth during SPEG.

Diffusion in Amorphous Silicon

Although annealing of ion implanted silicon has been studied extensively, few studies have examined the possibility of impurity diffusion in amorphous silicon at temperatures below that at which SPEG is stimulated. However, a recent investigation by Elliman et al [32-34] has identified some intriguing processes which can take place in amorphous silicon prior to SPEG. Figs.13-15 illustrate diffusion of both gold (expected fast diffuser) and indium (expected slow diffuser) in amorphous silicon [32]. In Fig.13, RBS/channeling spectra show that the annealing of a dose of 10^{16} Au cm^{-2} in silicon at 450°C results in the complete redistribution of gold within the amorphous silicon layer. The diffusing gold, quite remarkably, does not cross the phase boundary from amorphous to crystalline silicon. In Fig.14, the diffusion of gold in a thick (2μm) amorphous layer is shown. Measurements of profile broadening for

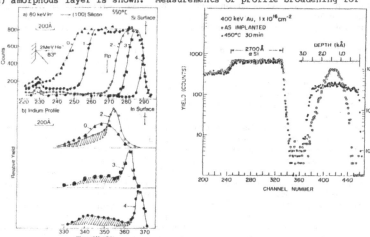

Fig.12 Ion channeling spectra showing SPEG for 80keV in implanted (100) silicon (1x10^{15} cm^{-2}) annealed iso-thermally at 550°C for 0 mins (0); 3 mins (1); 10 mins (2); 15 mins (3); 30 mins (4).

a) shows silicon prottion of spectra.
b) shows indium profiles (hatched areas indicating substitutional indium).
After [31].

Fig.13 Ion Channeling spectra indicating gold diffusion in ion implanted amorphous silicon. After [32].

92

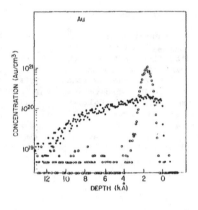

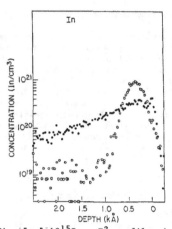

Fig.14 10^{16}Au cm^{-2} profiles in 2 μm
amorphous silicon before (open symbols)
and after (closed symbols) ramp annealing to 600°C in 10 mins. After [32].

Fig.15 5×10^{15}In cm^{-2} profiles in 2μm
amorphous silicon before and after ramp
annealing to 600°C in 10 mins. After[32].

isothermal annealing give a √time dependence, consistent with a diffusion-
limited process. Furthermore, extraction of diffusion coefficients from the
data gives values close to those expected from extrapolation to lower
temperatures of trace gold diffusion measurements in crystalline silicon.
However, the diffusion observations in Figs.13 and 14 are made at gold
concentrations at least 5 orders of magnitude higher than the solubility limit
of gold in crystalline silicon [35]. This may well explain the lack of
observable gold diffusion into crystalline silicon in Fig.13 and suggest very
high solubility of gold in amorphous silicon at the annealing temperatures
examined.
    The illustration of indium diffusion in amorphous silicon shown in Fig.15
is considerably more surprising than the gold diffusion results since the
measured diffusion coefficient is at least 10 orders of magnitude larger than
that expected for indium diffusion in crystalline silicon at temperatures of
the order of 500-600°C. More extensive investigations [33,34] of this
intriguing result have indicated that only very high indium concentrations
(well above the equilibrium solubility limit of indium in crystalline silicon
[35]) are observed to diffuse rapidly in amorphous silicon. Although further
details of the process are given elsewhere [33,34], it is significant to note
here that the strong concentration/temperature dependence of indium diffusion
in amorphous silicon may provide a possible explanation for the indium
segregation phenomenon illustrated in Fig.12. It is suggested that the
annealing of a 1x10^{15}In cm^{-2} dose at 550°C will initiate diffusion of indium in
amorphous silicon and, if indium has a higher solubility limit in amorphous
compared with crystalline silicon then segregation will result during
epitaxy. Inhibition of SPEG at higher impurity concentrations may involve
interactions between a moving amorphous-crystalline interface or to the
concomitant processes of random crystallisation and phase separation. Indeed,
as reported elsewhere in these proceedings [33], both of these latter processes
have been observed to take place at temperatures below those at which SPEG
occurs. For example, high indium concentrations ($\geq$ 5 atom percent) can result
in the nucleation of random crystallisation of silicon at temperatures as low
as 350°C. Furthermore, the processes of indium diffusion in amorphous silicon,
suppression of the crystallisation temperature and phase separation are closely
correlated, whereby diffusion is observed to precede the latter processes.
    These related phenomena, although not yet understood, obviously have
important implications for low temperature processing of silicon and, in

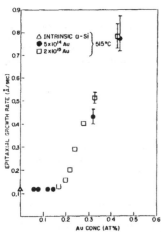

Fig.16 SPEG growth rate plotted as a function of gold concentration. Measurements from ion channeling from annealed Au ion implanted (100) silicon. After [36].

particular, SPEG. For example, the non-ideal SPEG behaviour observed for various implanted impurities is most likely attributable to such competing processes. In this regard, the effect of gold on the SPEG growth kinetics is shown in Fig.16 [36]. Low concentrations of gold are observed to dramatically enhance the growth velocity at 515°C, by up to a factor of 8 at 0.4 atomic percent gold. Since, in other studies [37], gold has been observed to markedly lower the crystallisation temperature of amorphous silicon, it could be speculated that this SPEG effect is attributable to a lowering of the crystallisation temperature. Indeed, gold concentrations in excess of 0.7 atomic percent inhibit SPEG [36] and one might further suggest, in such concentration regimes, that gold diffusion, phase separation and/or silicon crystallisation have taken place ahead of the advancing SPEG interface.

CONCLUSIONS

Although amorphisation and crystallisation processes associated with ion implantation have been the subjects of intense study for many years, the field remains a fruitful one from both the standpoint of scientific endeavour and that of technological importance. Processes which occur during bombardment, particularly the competition between damage production and dynamic annealing, are not yet fully understood nor are they fully appreciated and utilized to advantage by technologists. In particular, the process of ion-beam-induced epitaxial crystallisation has only recently been investigated in any detail, and results to date are providing much insight into the mechanisms of solid phase epitaxial growth. The process may also have technological significance, especially for low temperature processing and selective (lateral) epitaxy of silicon.

The understanding of low temperature thermal annealing, particularly of crystallisation processes, has also been significantly advanced by recent data. In particular, the characterisation of impurity diffusion in amorphous silicon, suppression of silicon crystallisation temperature and phase separation phenomena are already providing insight into non-ideal crystallisation and impurity redistribution processes during solid phase epitaxy.

94

ACKNOWLEDGEMENTS

Dennis Maher, John Poate, Walter Brown, Robert Elliman and Ken Short are particularly acknowledged for providing data and figures prior to publication and for stimulating discussions on the topics embodied in this review. The Australian Special Research Centres Scheme is acknowledged for financial support.

REFERENCES

[1]   W.L. Brown, these proceedings.
[2]   D. Turnbull, these proceedings.
[3]   L. Csepregi, J.W. Mayer and T.W. Sigmon Phys. Lett. 54A, 157 (1975).
[4]   J.W. Mayer, L. Eriksson and J.A. Davies, "Ion Implantation in Semiconductors", Academic Press, N.Y. (1970).
[5]   L. Csepregi, E.F. Kennedy, T.J. Gallagher, J.W. Mayer and T.W. Sigmon. J. Appl. Phys. 48, 4234 (1977).
[6]   L. Csepregi, E.F. Kennedy, J.W. Mayer and T.W. Sigmon, J. Appl. Phys. 49, 3906 (1978).
[7]   E.F. Kennedy, L. Csepregi, J.W. Mayer and T.W. Sigmon, J. Appl. Phys. 48, 4241 (1977).
[8]   G.L. Olson, S.A. Kokorowski, J.A. Roth and L.D. Hess, Mat. Res. Soc. Symp. Proc. 13, 141 (1983).
[9]   F. Spaepen and D. Turnbull. In "Laser Annealing of Semiconductors", J.M. Poate and J.W. Mayer, Eds. Academic Press, N.Y. (1982) p.15.
[10]   J. Narayan, J. Appl. Phys. 53, 8607 (1982).
[11]   I. Suni, G. Goltz, M.A. Nicolet and S.S. Lau, Thin Solid Films 93, 171 (1982).
[12]   J.S. Williams and R.G. Elliman, Phys. Rev. Lett. 51, 1069 (1983).
[13]   F. Morehead Jr. and B.L. Crowder, Rad. Eff. 25, 49 (1970).
[14]   J.S. Williams and J.M. Poate "Ion Implantation and Beam Processing", Academic Press, Sydney (1984) Ch.2.
[15]   J.W. Corbett, J.P. Karins and T.Y. Tan, Nucl. Inst. Meth. 182/183, 457 (1981).
[16]   L.A. Christel, J.F. Gibbons and T.W. Sigmon, J. Appl. Phys. 52, 7143 (1981); and Ch.3 in [14].
[17]   M.L. Swanson, J.R. Parsons and C.W. Hoelke, Rad. Eff. 9, 249 (1971).
[18]   J.A. Davies, Ch.4 in [14].
[19]   F.H. Eisen and B. Welch, Rad. Eff. 7, 143.
[20]   D.G. Beanland, Ch.8 in [14].
[21]   O.W. Holland, J. Narayan and D. Fathy, Nucl. Instr. Meth.7/8, 243 (1985).
[22]   I.F. Bubb et al. Nucl. Instr. Meth. B2, 761 (1984).
[23]   K.T. Short et al. Mat. Res. Soc. Symp. Proc. 27, 247 (1984).
[24]   R.G. Elliman, J.S. Williams, S.T. Johnson and A.P. Pogany, Nucl. Instr. Meth. (in press).
[25]   E.P. Donovan, Ph.D. thesis, Harvard University (1982).
[26]   J. Linnros, B. Svensson and G. Holmen, Phys. Rev. B30, 3629 (1984).
[27]   J.S. Williams, W.L. Brown, R.G. Elliman and T.E. Seidel, Mat. Res. Soc. Symp. Proc. 35, (1985).
[28]   R.G. Elliman, J.S. Williams, N.L. Brown and D.M. Maher, these proceedings.
[29]   R.G. Elliman, S.T. Johnson, A.P. Pogany and J.S. Williams, Nucl. Inst. Meth. B7/8, 310 (1985).
[30]   A.G. Cullis, T.E. Seidel and R.L. Meek, J. Appl. Phys. 49, 5188 (1978).
[31]   J.S. Williams and R.G. Elliman, Appl. Phys. Lett. 40, 226 (1982).
[32]   R.G. Elliman, J.M. Gibson, J.M. Poate, D.C. Jacobson and J.S. Williams, Appl. Phys. Lett.46, 478 (1985)
[33]   R.G. Elliman, J.S. Williams, J.M. Poate, D.C. Jacobson, J.M. Gibson and D.K. Sood, these proceedings.

[34] R.G. Elliman, J.M. Poate, K.T. Short and J.S. Williams to be published.
[35] F. Trumbore, Bell. Syst. Tech. Journal 35, 205 (1960).
[36] D.C. Jacobson, J.M. Poate and G.L. Olson, Appl. Phys. Lett. (in press).
[37] S.R. Herd, P. Chaudhari and M.H. Brodsky, J. Non. Cryst. Sol. 7, 309 (1972).

PART II

Laser-Induced Phase Transitions
in Semiconductors

ANOMALOUS BEHAVIOR DURING THE SOLIDIFICATION OF SILICON IN THE PRESENCE OF IMPURITIES

MICHAEL O. THOMPSON and P. S. PEERCY*
Dept. of Materials Science, Cornell University, Ithaca, NY 14853
*Sandia National Laboratories, P.O. Box 5800, Albuquerque, NM 87185

ABSTRACT

Nanosecond pulsed laser irradiation of silicon results in the melting of the surface with subsequent solidification from a crystalline substrate. In the presence of impurities or alloys, the solidification dynamics are greatly affected by the nature of the impurities. Five classes of materials have been investigated: amorphous Si, mutually soluble alloys, high solubility impurities, low solubility impurities, and compound forming materials. Solidification of each of these classes of materials is discussed. Several anomalous kinetic regimes are observed, including explosive crystallization, surface nucleation of solid, and internal nucleation of melt. These results are interpreted in terms of thermodynamic modifications of the melting temperature in the alloy regions.

INTRODUCTION

Pulsed laser annealing of silicon has recently been the subject of extensive investigations [1-2]. Under a wide range of conditions, nanosecond laser irradiation has been shown to melt the near surface region of Si to depths of ≤ 1 μm, with subsequent liquid phase epitaxial growth as heat is conducted into the substrate. Numerous experiments have elucidated the solidification kinetics and the resulting microstructures, leading to our current detailed understanding in the case of the pure crystalline material. However, it has been suggested, also from studies of the microstructures, that impurities may play a major role in both the melting and the liquid phase solidification. Impurities themselves have also been the subject of many other experiments, especially studies of the effects of rapid solidification velocities on the interface segregation coefficient [3,4]. In contrast to the previous suggestion that impurities affect the solidification behavior, these studies have assumed, primarily due to the lack of direct experimental evidence to the contrary, that the interface motion is not perturbed by the incorporation of impurities. In order to clarify some of these questions, we have performed real-time measurements of the melt and solidification dynamics, using the transient conductance and reflectance techniques, to investigate the effects of several classes of impurities and alloys on the melt kinetics.

Almost any impurity, at concentrations greatly in excess of the solid solubility, can be introduced into silicon by ion implantation. For a wide range of implanted species studied to date, we have observed that the behavior under laser irradiation can be generally characterized by five classes of materials. The first of these catagories (Class I) is, by nature of the implantation process, amorphous Si (a-Si) formed by the self implantation of Si into Si. Ion implantation of most elements results in the formation of an a-Si layer which has distinctly different melt and solidification characteristics than the crystalline phase (c-Si). Melt characteristics observed for amorphous layers produced by such self implantation are thus also important in the behavior of the other four classes of materials. These characteristics include a reduced melting temperature for the amorphous region and the possibility of explosive crystallization of unmelted material [5,6].

We have categorized a second class of materials by the ability of the elements to form continuous solid solutions with silicon, with germanium as the major example. Within this class, the melting temperature of an alloy region is a smoothly varying function of the alloy concentration and can be easily controlled. Laser melting in these alloys can exhibit behavior similar to the self implants, class I, and has been used to verify measurements of the melting point of a-Si [7].

Relatively soluble impurities comprise the third category of materials, with solubilities up to several atomic percent (at.%). Included in this class are the electrically active dopants such as As, Sb and P. Because these systems do not form solid solutions over the entire composition range, there is a greater reduction in the melting temperature for a given impurity concentration than for class II materials. Low concentrations of these impurities exhibit behavior similar to that observed for the class II alloys due to this reduction in the melting temperature. At high concentrations, however, the melting temperature can be reduced sufficiently to allow surface solidification, and, in some cases, internal nucleation of the melt [8,9].

Finally, completing the solution impurities are the highly insoluble elements such as In and Sn, which we classify as the fourth category of materials. In these cases, the melting temperature is an even stronger function of impurity concentration, and behavior such as internal melting and surface solidification are observed under a wide variety of conditions. Under short pulse irradiation, one of the unusual solidification behaviors is the formation of a buried layer of the impurity when an interface propagating from the surface collides with the interface propagating from the bulk [10,11].

One further class of interesting impurities were studied which do not, in general, form solid solutions but instead tend to form compounds. Oxygen and nitrogen are two of the most technologically important members of this class since they form insulating compounds and can be used to produce buried insulating layers. At concentrations below the stoichiometry required for compound formation, these impurities behave similar to class II and exhibit behavior consistent with a modified melting temperature. As the concentration is increased, effects of the resulting thermally insulating layers are observed. Solidification is initially retarded by the presence of pockets of impurities and, at a sufficiently high concentration, the surface layer of pure Si becomes fully insulated from the substrate thermally and crystallographically. Solidification under this condition requires the nucleation of solid either at the liquid/insulator interface, or at the liquid/air interface [12].

EXPERIMENTAL

Samples, containing impurities of Si, As, P, Zn, Sn, In, O, and N, were produced by implantation of the elements into crystalline Si and silicon-on-sapphire (SOS) at fluences between 2×10^{14} and 6×10^{17} and energies between 80 and 300 keV. The range of implant fluences studied for each class of materials are summarized in table I. In all cases, these implantations resulted in the formation of an amorphous layer, ranging in thickness from 80 nm to 600 nm. Impurity concentration and depth distribution were characterized by Rutherford backscattering and ion channelling techniques.

Table 1: Definition of the 5 classes of materials to be discussed and the range of implant fluences examined in each case.

Class	Elements	Fluences Studied	Type of Impurity
I	Si	2×10^{15} - 5×10^{15} /cm^2	a-Si
II	Ge	10 at.% - 75 at.%	Fully miscible
III	As,P	2×10^{15} - 2×10^{17} /cm^2	Slightly soluble
IV	Zn,Sn,In	2×10^{14} - 5×10^{15} /cm^2	Insoluble
V	O,N	1×10^{16} - 6×10^{17} /cm^2	Compound Forming

For Ge, the desired impurity concentrations exceeded the range which can be achieved by implantation; hence these samples were produced by electron beam evaporation of $Si_x Ge_y$ onto chemically cleaned SOS substrates. These samples were subsequently annealed at 800°C to produce large grain poly-crystalline Si:Ge alloys on a crystalline Si substrate. In contrast to the impurities described above, which consist of amorphous layers with an almost Gaussian impurity profile, the Si:Ge alloys consist of poly-crystalline layers with almost uniform impurity profiles.

Laser irradiation was provided by a pulsed ruby laser operating at a wavelength of 694 nm with pulse durations between 3 and 30 ns FWHM. Samples were irradiated through a quartz homogenizer to assure uniform illumination [13]. Melt depth versus time was monitored using the transient conductance technique [14-16] which utilizes the 30 fold increase in the electrical conductivity of Si [17] upon transformation to the liquid phase. Knowledge of the sample geometry and the liquid phase conductivity allows precise determination of the total liquid thickness during irradiation. In addition to the electrical conductance measurements, the surface reflectance $R(t)$ was monitored using a CW argon laser operating at 488 nm [18]. Concurrent with the jump in electrical conductivity is an increase from 35% to 70% in the optical reflectivity at this wavelength. The optical probe is sensitive to the near surface 25 nm and allows accurate determination of the melt duration in this region. Combined measurement of the transient conductance and $R(t)$ allow unambiguous distinction between melts extending to the surface and molten layers which are entirely contained within a sample.

RESULTS AND DISCUSSION

Pure Crystalline Si

An example of data obtained from transient conductance and transient reflectance measurements are shown for 30 ns irradiation of virgin SOS in fig. 1. Data from the transient conductance measurements are converted to the equivalent molten layer thickness required to produce the observed electrical conductance. Under conditions in which electrical conduction in the solid can be neglected, and only a single molten layer exists at the surface, this thickness is equivalent to a melt depth. The simultaneous increase of the transient conductance and $R(t)$ indicates that, within the experimental timing accuracy of 1 ns, melting initiated at the surface at -3 ns, measured relative to the peak of the incident laser pulse. The conductance indicates that the liquid/solid interface propagated to a maximum depth of 305 nm, whereupon it changed direction and began epitaxial solidification toward the surface. At 115 ns, $R(t)$ shows the surface to be fully solidified indicating termination of the molten layer. The transient conductance confirms the collapse of the molten

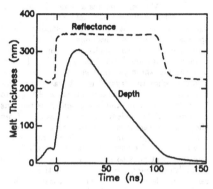

Fig. 1 Typical surface reflectance $R(t)$ and melt depth $d(t)$ determined from transient conductance for 0.5 μm SOS. The reflectance and conductance data indicate melt initiating and terminating at the surface.

layer at this time, with only a slowly decaying thermal tail remaining. This thermal tail, an artifact of the measurement which does not indicate molten Si, is caused by finite electrical conduction of thermally generated carriers in the resulting hot solid Si.

The interface solidification velocity as a function of time $v(t)$ can be determined from a numerical derivative of the data shown in fig. 1. These data show that a maximum solidification $v(t)$ of ~3.5 m/s is achieved shortly after the maximum melt, followed by a slow, monotonic decrease in $v(t)$ as the interface propagates toward the surface. Under normal conditions, $v(t)$ can be estimated from simple heat flow; in steady-state the enthalpy (ΔH) release at a solidification front is balanced by thermal conduction into the underlying substrate. Since thermal gradients in the substrate decrease with time because of thermal diffusion, $v(t)$ is expected to decrease as the inverse square root of time once steady state conditions have been established.

$$v\Delta H = \kappa_{Al_2O_3} \left(\frac{\partial T}{\partial x} \right) = \kappa_{Al_2O_3} \cdot \frac{(T^m - T^{RT})}{\sqrt{4Dt}} \qquad (1)$$

$$v(m/s) = \frac{27.5}{\sqrt{t(ns)}} \qquad (2)$$

For these equations, κ is the thermal conductivity of the substrate, and time (t) is measured from approximately the center of the incident laser pulse.

Class I, Self Implant - Amorphous Si

Following implantation of impurities into Si, an a-Si layer with a thickness between 50 nm and ~600 nm is usually formed. Since this amorphous layer is formed during implantation of all the impurities studied, melt and solidification characteristics of the pure a-Si layer are exhibited, to some degree, in all of the melting scenarios to be presented. In the case of Si self implants, this amorphous layer has the composition of the crystalline substrate but a structure corresponding to the metastable phase a-Si. It is now well established that a-Si is a unique phase in the Si system with a melting temperature ~220 K below the melting point of the crystalline phase [19,5]. During pulsed laser irradiation, the effect of this melting temperature reduction is to produce a ledge in the melt depth versus time at the interface between the amorphous layer and the crystalline substrate, as illustrated in fig. 2.

The formation of the ledge can be understood as follows. Melting initiates at the surface when the laser has delivered sufficient energy to raise the surface temperature to the melting point of the metastable phase, designated T^{al}. Since the a-Si layer is uniform, melting continues at this reduced temperature until the liquid/solid interface reaches the crystalline substrate. Once the liquid reaches the crystalline interface, the temperature is insufficient to continue melting into the substrate. Thus the liquid thickness remains constant until sufficient energy is absorbed from the incident radiation to raise the entire liquid layer to the crystalline melting temperature T^{cl}. After the interface temperature exceeds T^{cl}, the interface continues to propagate into the sample as additional energy is deposited by the incident radiation. Measurements of the total width of this ledge were used to obtain the first direct experimental determination of the melting temperature reduction for a-Si relative to the crystalline phase [5,6].

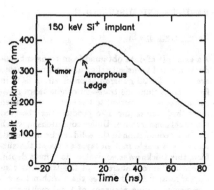

Fig. 2 Melt depth versus time for ~340 nm a-Si films on SOS illustrating the plateau in $d(t)$ at the a-Si/c-Si interface. The thickness of the a-Si layer was measured by ion backscattering.

For incident energy densities insufficient to melt completely through the overlying a-Si layer, it is now known that explosive crystallization can occur in unmelted a-Si layers and results in an unusual microstructure [5-6,20-22]. An example of the kinetics which occur during explosive crystallization is shown in fig. 3a for irradiation of a 120 nm a-Si layer with a 30 ns laser pulse at 0.23 J/cm². As in fig. 1, $R(t)$ and the electrical conductance increase at the same time indicating surface nucleation of melt. The reflectance, however, returns to the solid level while the conductance still shows the existence of 30 nm of melt, thus indicating the formation of a buried molten layer. Transmission electron microscopy of similar samples show the transformation of the amorphous layer to a fine-grained polycrystalline (FG poly) state to a depth much greater than the maximum molten thickness. These results have been interpreted in terms of a moving buried molten layer which transforms the material into the FG poly-Si as it propagates.

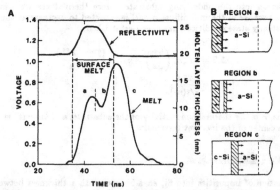

Fig. 3 a) Transient conductance measurements of melt thickness and simultaneous surface reflectance for a ~0.23 J/cm pulse incident on a ~210 nm a-Si layer on SOS. b) Schematic illustration described in the text for the explosive crystallization process.

The currently accepted scenario for melting of a-Si follows essentially the one proposed by Leamy et. al [23] for explosive crystallization of thick Ge films, illustrated diagrammatically in fig. 3b. During the first phase of the irradiation, sufficient laser energy is supplied to melt the surface of the a-Si to a depth of ~12 nm. This molten layer is produced by direct irradiation from the laser and is referred to as the primary melt. As this primary melt solidifies forming poly-Si near T^{cl}, the enthalpy released is sufficient to heat the underlying a-Si above its melting temperature and create a second, buried, molten layer as illustrated in step b, referred to as the secondary melt. Propagation of this secondary melt layer, with poly-Si at one interface and a-Si at the other interface, will be self sustaining if the liquid temperature is between the T^{al} and T^{cl}. Under these conditions, the underlying a-Si continues to melt, while the near surface interface solidifies as FG poly-Si. Because of the high thermal conductivity of the liquid phase compared to that of a-Si, the thin buried layer can be assumed to be almost isothermal. If the overheating required to melt the a-Si is significantly less than the undercooling required to solidify the FG poly-Si, the temperature of the liquid layer will be near T^{al} [24]. At this temperature, solidification of the FG poly-Si interface occurs near 15 m/s, the maximum crystal growth velocity before the onset of amorphous formation [25]. The observation of FG poly-Si at this solidifying interface is probably related to either the interface motion at this velocity in the absence of a well defined planar crystalline seed, or to the volume change which accompanies solidification of Si (~10%) [17].

For a period of time after the secondary melt is established, these two primary and secondary melt layers exist concurrently. Like the secondary melt, the primary melt interface is also highly undercooled and continues solidifying toward the surface as the secondary melt propagates into the a-Si layer. Since the primary melt interface is somewhat higher in temperature, solidification occurs as a coarse grained poly-Si as observed in TEM. Eventually, the primary melt solidifies to the surface, leaving only the buried secondary melt which continues to propagate. This regime can be seen in fig. 3a when $R(t)$ has returned to the crystalline value, and the conductance shows the thickness of the secondary melt only. The secondary melt is finally quenched by either thermal conduction of heat away from the interface or by reaching the a-Si/c-Si interface.

Propagation of this secondary melt front releases a net enthalpy of $(v^c \Delta H^{cl} - v^a \Delta H^{al})$ per unit time, where v^c is the velocity of the crystalline interface, and v^a is the velocity at the a-Si interface. The interface will continue to propagate as long as this heat release is greater than the thermal conduction into the substrate. Due to the low thermal conductivity of a-Si [26] and the large value of ΔH^{ac} [19], the secondary melt can extend to almost the a-Si/c-Si interface at very low energy densities. TEM observations confirm that at energy densities only slightly above the threshold for melting, the secondary melt does indeed propagate almost the entire distance to the original a-Si/c-Si interface.

The above results apply to irradiation of a-Si samples with a 30 ns laser pulse. While it might be expected that short laser pulses would not result in the establishment of the secondary melt due to the steeper temperature gradients, we have observed explosive crystallization with pulses as short as 2 ns. These results suggest that explosive crystallization may play a major role in determining the melt behavior and microstructure of amorphous silicon irradiated in the picosecond regime where numerous unusual results have been observed [27].

Class II, Fully Miscible Systems - Si-Ge alloys

The Si:Ge system is unusual in that it forms a continuous solid solution at all compositions, as indicated by the phase diagram in the inset of fig. 4. For studies of laser melting and solidification, equilibrium conditions are seldom reached, especially with respect to the segregation of impurities. Indeed, the so called T_0 curve, shown by the dashed curve in the inset, is more applicable in the laser melting regime. This T_0 curve is the locus of points where the liquid and solid are in thermodynamic equilibrium at the same composition, i.e., under conditions of no phase separation. During pulsed laser melting, the low diffusion coefficient in the solid phase makes the T_0 approximation extremely accurate. For solidification, the use of T_0 is justified only as the effective segregation coefficient approaches unity.

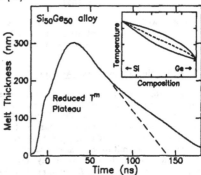

Fig. 4 Melt depth versus time for a SiGe alloy illustrating the plateau due to T^m differences between the alloy and c-Si. The inset shows the phase diagram and calculated T_0 curve. [28]

Pulsed laser melting of surface layers of Si:Ge alloys can be accurately approximated as melting at the T_0 value for the appropriate composition. Thus alloy compositions can be selected to provide melting temperatures from 1210 K to 1685 K. An example of the melting of a 50/50 atomic percent alloy is shown in fig. 4. The melting temperature of this alloy is approximately 1460 K, or 225 K below the melting temperature of c-Si. This layer is equivalent, in terms of the melting behavior at high velocities, to an a-Si layer of the comparable thickness. Indeed, the thermal conductivity of these Si:Ge alloys [29] is almost an order of magnitude lower than the pure Si values, which is also in correspondence with a-Si. A ledge occurs during the melting of these Si:Ge surface layers, similar to the case of pure a-Si, and is interpreted in an identical fashion. The liquid-solid interface is stopped as it reaches the c-Si substrate until the liquid temperature can be raised to the pure Si melting temperature. Comparisons of the width of the ledge in these Si:Ge alloys with those formed in a-Si samples confirms the melting temperature depression of ~220 K for amorphous Si.

Unlike a-Si, however, the Si:Ge alloys have no enthalpy of crystallization to release and thus do not exhibit explosive crystallization. Furthermore, during solidification these alloys retain their lowered melting temperature and thermal conductivity. Thus, while a-Si solidifies as c-Si with crystalline thermal properties, the Si:Ge alloys exhibit behavior during solidification consistent with thermal properties which remain distinct from pure Si. In fig. 4, the movement of the solidifying interface into the Si:Ge alloy is readily apparent as $v(t)$ decreases from 3.5 m/s to less than 2 m/s.

These Si:Ge alloys are an ideal model system for the study of effects of melting temperature differences during rapid solidification. Complex temperature profiles can be "constructed" in a sample by the simple selection of composition. Thus, conditions of superheating or supercooling can be established, and the effects studied, by the appropriate selection of multiple alloy layers.

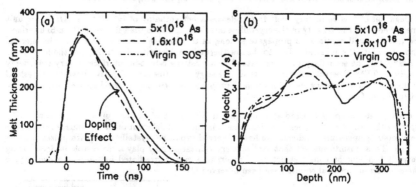

Fig. 5 a) Melt depth versus time for an unimplanted sample and samples implanted to peak As concentrations of 2 and 7 at.%. b) Velocity versus depth obtained from the numerical derivative of the data in a).

Class III, Soluble Impurities - As doped

Class III impurities includes most of the electrically active impurities, including As, P, and B. These impurities are characterized by their relatively high equilibrium solid solubilities, up to several atomic percent, in the crystalline lattice. Under metastable conditions induced by either laser quenching or solid phase recrystallization, these impurities have been shown to have substitutional solubilities in excess of 10 at.% [3]. It has been pointed out [30], however, that there is a fundamental thermodynamic limit to the concentrations of these impurities, given by the concentration for which the liquid becomes thermodynamically stable to the solid even at 0 K. This is, by definition, the concentration at which the T_0 curve reaches absolute zero temperature and has been estimated at 15 at.% for As in Si [30]. Below this concentration, these substitutional impurities are expected to act similar to the Si:Ge alloy due to a reduction of the melting temperature, although the melting temperature now decreases much more rapidly with concentration than in the previous classes.

Melting behaviors of an undoped and two As doped samples, implanted to peak concentrations of 2 and 7 at.%, are illustrated in fig. 5. Since As is electrically active, the data of fig. 5a involves deconvolution of the electrical conduction of the As impurities from the transient conductance [8-9]. During melting, the ledge due to the amorphous layer is again observed. Although As affects the thermodynamics of a-Si, the ledge is due primarily to the normal melting temperature reduction of a-Si with only slight perturbations from the As. In all cases, the melt extends into the crystalline substrate and solidification occurs epitaxially. Since As is known to have a unity segregation coefficient at these velocities [3], solidification occurs at temperatures near T_0, reduced only by the undercooling necessary to drive the interface at the observed velocity. During solidification, the effect of the impurities can be seen as the interface passes through the peak of the As profile at 190 nm. As the peak As concentration increases, the effect becomes increasingly apparent. Figure 5b shows these data as $v(t)$ versus depth, and illustrates the decrease in $v(t)$ as the liquid-solid interface approaches the impurity layer and the increase in $v(t)$ as the interface propagates back into the undoped material near the surface.

Solidification of these samples where the As concentration varies spatially is controlled by melting temperature changes induced by the impurities. An idealized melting temperature (i.e. T_0) profile with depth would show a symmetric depression which follows the impurity concentration. Solidification thus occurs when the interface temperature (T_i) is undercooled below this melting temperature. Since the interface velocity depends on the driving force for solidification, $v(t)$ can be linearized about the melting temperature T_m to obtain $v(t) = \beta(T - T_m)$, where β varies slowly with temperature. The β^{-1} value for pure silicon is estimated to be between 4 and 15 K/(m/s) [31-33]. Following the peak melt depth, steady state solidification in pure Si is quickly established with the required undercooling. As the interface propagates into the As region, T_i must decrease below the new T_0 value. However, the thermal inertia of the overlying liquid prevents the liquid from instantaneously changing its temperature. A portion of the heat being conducted into the substrate must be used to cool the entire liquid layer and reestablish the interfacial undercooling. Thus, as the liquid-solid interface propagates into a region with a lower melting temperature, $v(t)$ is reduced. Conversely, $v(t)$ increases as the interface returns to regions of higher T_m.

These qualitative conclusions can be made quantitative under the assumptions that the heat conducted into the substrate is constant and that the velocity/undercooling constant β^{-1} is independent of the concentration and velocity. Under these constraints, if v_0 is the steady state velocity in pure Si, and T^{liq} is the liquid temperature, $v(t)$ can be determined from energy constraints as:

$$\kappa_{Al_2O_3}\left(\frac{\partial T}{\partial x}\right) = v_0\Delta H = v(t)\Delta H + C_p^{liq}d(t)\left(\frac{\partial T^{liq}(t)}{\partial t}\right) \tag{3}$$

$$= v(t)\Delta H + C_p^{liq}d(t)\left(\frac{\partial(T_m(x) - \beta^{-1}v(t))}{\partial t}\right) \tag{4}$$

$$= v(t)\Delta H - \beta^{-1}C_p^{liq}d(t)\left(\frac{\partial v(t)}{\partial t}\right) + C_p^{liq}d(t)v(t)\left(\frac{\partial T_m}{\partial x}\right) \tag{5}$$

Furthermore, neglecting the second order correction represented by the β^{-1} term, $v(t)$ at any position can be expressed in terms of the spatial derivative of the melting temperature as:

$$v(t) = v_0 - \left(\frac{C_p^{liq}d(t)v(t)}{\Delta H}\right)\left(\frac{\partial T_m}{\partial x}\right) \tag{6}$$

As shown by eq. 6, $v(t)$ should follow the derivative of the melting temperature, and consequently the derivative of the impurity profile, as is observed. If the T_0 curve is linear with composition, measurements of changes in $v(t)$ can be directly translated into measurements of the T_0 value. These measurements allow some of the structure of non-equilibrium transformations to be studied [9].

As the concentration of As is increased above 7 at.%, the normal melt and solidification scenario breaks down. At 11 at.%, the melting temperature is reduced sufficiently at the peak of the As concentration that nucleation of solid at the surface, in relatively pure Si, occurs. This behavior is illustrated in fig. 6a which shows the molten thickness and $R(t)$ for a sample irradiated at 0.46 J/cm². As in the case of a-Si explosive crystallization, $R(t)$ decreases while substantial liquid remains buried in the sample. However, the surface does not solidify due to an interface propagating toward the surface, but rather by nucleation of solid at the surface to generate a new moving liquid/solid interface. Nucleation occurs at the highly undercooled surface as the normal liquid-solid interface passes through the high As concentration region. This nucleation forms a second interface which

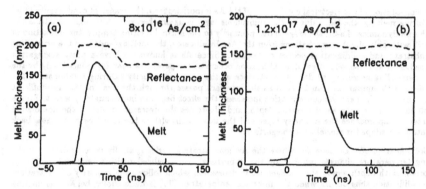

Fig. 6 a) Transient conductance and reflectance measurements for ~11 at.% peak concentrations of As in Si illustrating surface nucleation of solid. b) Data similar to that in a) for a peak As concentration of ~15 at.% illustrating internal melt.

propagates into the sample, and the two interfaces eventually collide within the sample when melt terminates.

As the concentration is increased further, the melting temperature near the As peak continues to decrease with reductions expected for As in both the c-Si lattice and in the a-Si phase. While the thermodynamics of the a-Si:As system are unknown, we observe that the melting point is reduced sufficiently to allow internal nucleation of melt to occur near the peak of the As profile, before the melting temperature of the pure a-Si is reached at the surface. This behavior is illustrated in fig. 6b for an As doped sample with a peak concentration near 15 at.%. In this case, $R(t)$ indicates that the surface remained in the solid state throughout irradiation. The entire melt indicated by the transient conductance is confined to the region near the impurity peak. These observations are in agreement with the prediction that T_0 decreases drastically at 15 at.% for the crystalline phase if comparable reductions also occur for the amorphous alloy.

Class IV, Insoluble Impurities - Zn, In, Sn

The class IV impurities are a natural extension of class III to the elements for which the maximum equilibrium solubilities are well below 1 at.%. The corresponding T_0 curves for these impurities in Si simply decrease more rapidly with concentration than for more soluble impurities. Thus, it might be expected that much of the behavior seen in the As case above would also occur in the class IV impurities at correspondingly lower concentrations. For example, the steep drop of T_0 with composition causes the "high" concentration regime to be reached with relatively low implantation fluences, typically requiring only 1-2x10^{14} ions/cm^2, corresponding to 0.05 at.%. For the concentrations of class IV impurities studied to date ($\geq 10^{14}$), the melting temperature is always dramatically reduced from the pure Si equilibrium values and consequently the T_0 curves for these systems are extremely important. Unfortunately, due to the very low solid solubilities, the T_0 curve cannot be

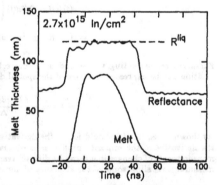

Fig. 7 Transient conductance and reflectance for SOS implanted with 2.7x10^{15} In/cm^2 and irradiated at 0.57 J/cm^2 at 30 ns, illustrating internal nucleation of melt.

accurately determined for even the crystalline systems. Furthermore, during solidification the segregation coefficient in these systems is not unity [4], and a velocity dependent equation for the solidus and liquidus is required. Despite these difficulties in obtaining quantitative descriptions of the melt and solidification process, numerous unusual solidification scenarios are understood qualitatively.

An example of the transient conductance and $R(t)$ behavior of In implanted into Si at a fluence of 2.7×10^{15} is shown in fig. 7. The implant energy of 100-120 keV places the impurity peak only 40-80 nm below the surface. As for explosive crystallization, interpretation of the behavior is only possible with the combination of the reflectance and conductance data. The transient conductance shows behavior consistent with melting of a surface amorphous layer, with a ledge at the a-Si/c-Si interface (~80 nm). The reflectance increases simultaneously with the transient conductance, indicating a molten layer near the surface. However, until the melt has extended through the amorphous layer, $R(t)$ does not rise to the full level expected for a liquid layer shown by the dashed R^{liq} line. This behavior is consistent with internal nucleation of melt near the peak of the impurity distribution. Since the probe depth of the reflectance laser is ~100 nm in the solid phase, comparable to the depth of the impurity peak, the internal melt exhibits a reduced $R(t)$ level rather than the complete suppression observed in the As case. The eventual increase from this low level to the full liquid $R(t)$ occurs when the melt finally reaches the a-Si/c-Si interface. After melt initiates at the peak of the In distribution, the molten layer increases its thickness both toward the surface and into the bulk of the sample. At all times, the thickness is approximately determined by the points on the impurity distribution where the T_0 curve equals the liquid temperature. This internal nucleation of melt has been observed under 30 ns irradiation for all energy densities sufficient to produce melting for Sn, Zn and In with implant fluences $\geq 5 \times 10^{14}$ ions/cm². Under certain conditions, it is also possible to observe an explosive-like crystallization propagating from the buried melt toward the surface [11].

At lower implant fluences, rather than approaching the behavior of class III, the situation actually becomes increasingly complex. Figure 8 shows the melt and $R(t)$ behavior for a sample implanted with 2×10^{14} Sn/cm² and irradiated with 0.66 J/cm². Again, the transient conductance indicates an almost normal a-Si melt profile and ledge. The reflectance, however, shows a rapid increase shortly after melt initiation, followed by a decrease to almost the crystalline level before again increasing to the full liquid $R(t)$. While this behavior is not fully understood, our interpretation is as follows. Melt initiates at or near the surface. The steep T_0 gradient causes the molten layer to "break away" from the surface and propagate toward the lowest melting point region. This initial thin melt reaches the impurity peak, stops and slowly spreads both directions until the entire amorphous layer is consumed and melting continues into the substrate. The surface eventually again melts and $R(t)$ regains the fully molten level. It is very interesting to note that while the impurities have a great impact on the melting behavior,

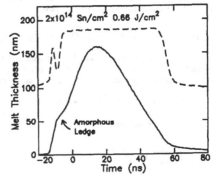

Fig. 8 Transient conductance and reflectance for 2×10^{14} Sn/cm² implants, illustrating internal nucleation of melt followed immediately by transient growth of the surface solid layer.

they do not have an observable effect during the solidification. A full quantitative analysis of these melt and solidification dynamics will require detailed knowledge of the thermodynamics of both the amorphous and crystalline alloys.

Further evidence for the strange behavior of these impurities has been observed during short pulse irradiation of In doped samples [34,11]. Samples implanted to a fluence of 2×10^{15} In/cm² and irradiated at 694 nm with 2 ns laser pulses results in the formation of an In layer buried 40 nm below the surface. Figure 9a shows the melt kinetics and $R(t)$ behavior for a sample irradiated at 0.31 J/cm². The corresponding RBS profile for this sample is shown in fig. 9b. Normally, melt and solidification of In doped samples irradiated with sufficient energy to melt to the underlying crystalline substrate results in the segregation of most of the In to the surface. At lower energy densities, however, the $R(t)$ and conductance measurements demonstrate nucleation of solid at the surface during the melt and thus the formation of two moving interfaces. The buried In layer results from impurity segregation at both interfaces, with the final layer occurring at the position where the interfaces collide. Although the material solidifies in the amorphous phase, the sum of the velocities for the two interfaces ranges from 2 m/s for shallow melts to 8 m/s for melts

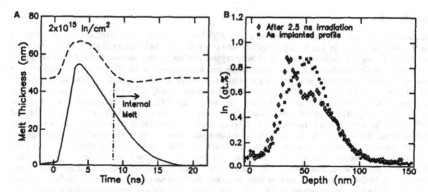

Fig. 9 a) Transient conductance and reflectance for In implanted SOS irradiated with
a 2.5 nsec pulse at 0.31 J/cm², illustrating surface solidification on molten Si.
b) Comparison of the In depth distribution measured after irradiation with the
as-implanted profile.

that approach a-Si/c-Si interface. These velocities are well below velocities necessary for direct
amorphous phase formation on c-Si [25].

Class V, Compound Forming - Oxygen and Nitrogen

The class V impurities, typified by nitrogen and oxygen, exhibit by far the most unusual behavior
of materials studied to date. Because little effect is observed on the melting behavior until the
concentrations approach stoichiometry for either the SiO_x or the Si_xN_y compounds, these materials
are characterized only by their compound forming ability. While both N and O are common
impurities in silicon, their solid solubilities are relatively low [35] and higher concentrations, above
$2-3 \times 10^{15}$, are expected to precipitate in the solid as compounds. At concentrations in excess of
3×10^{17}, buried insulating layers are known to form near the peak of the implanted concentration [36].

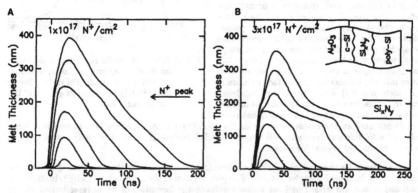

Fig. 10 a) Melt depth versus time for SOS implanted with 1×10^{17} N/cm². b) Similar data
for 3×10^{17}.

Laser melting of O and N implanted samples shows no discernible deviation from undoped melt
behavior for concentrations $\leq 3 \times 10^{16}$ (~5 at.%). The melt depth versus time is shown in fig. 10,
for N implanted to fluences of 1×10^{17} and 3×10^{17} ions/cm². Oxygen shows similar behavior at
approximately two times greater ion fluences. RBS studies show that the O and N impurities do
not diffuse significantly during laser irradiation. At the low implant fluences, $v(t)$ follows the $\sqrt{t}$

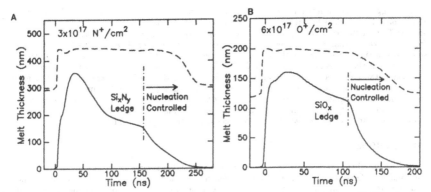

Fig. 11 a) Time dependent reflectance and transient conductance for 3×10^{17} N/cm² irradiated with 0.87 J/cm² at 30 ns. b) Data similar to that in a) for 6×10^{17} O/cm². Note the difference in the $R(t)$ behavior for the O and N implants.

behavior expected for normal solidification, but modified solidification behavior can be observed in the 1×10^{17} implant. For cases where solidification begins in pure Si beyond the range of the N implant, epitaxial growth is initially observed. As the interface reaches the region of the N, the apparent velocity is reduced and exhibits a derivative type behavior similar to that discussed above for As. Once past the peak of the N, $v(t)$ again reaches a steady state value lower than the initial v_0 because of the reduced thermal conductivity of the Si_xN_y layer formed. It is not understood why these concentrations of insoluble impurities can be incorporated without a greater effect on either the melting temperature or the interfacial kinetics.

As the implant fluence is increased to 3×10^{17}/cm², a second dramatic change is observed in the solidification behavior; the velocity $v(t)$ essentially stops when the interface reaches the N peak. With the outer layer of Si thermally and crystallographically isolated from the underlying substrate, the surface liquid cools through the Si_xN_y layer for 80 ns until nucleation of solid occurs. Under these conditions, the behavior of O and N implants diverge as illustrated by the melt depth and $R(t)$ traces in fig. 11 for the highest implant fluences studied. Consider first the N implant. During the period the liquid-solid interface is stopped at the buried insulating layer, $R(t)$ drops slightly which is interpreted as indicating severe cooling of the surface liquid layer. Once the interface begins to move at 155 ns, $R(t)$ returns to the full liquid value, and ultimately decreases again when the conductance indicates termination of the melt. The resulting microstructure for a similar sample after irradiation is shown in fig. 12. The amorphous Si_xN_y layer is visible starting at a depth of 150 nm, as is also observed by the transient conductance. Nucleation of poly-Si is observed in the

Fig. 12 Cross-sectional micrograph of a 3×10^{17} N/cm² implanted sample irradiated at 0.87 J/cm². Region A shows the single crystal Si material, region B the amorphous Si_xN_y layer, and C is the poly-Si near the surface.

surface layer and confirms the breakdown of epitaxy as the interface propagates through the Si_xN_y layer [12]. In the case of O, however, $R(t)$ does not rise back to the liquid value after the interface begins to move as for N. Instead, $R(t)$ drops monotonically to the solid level over a period of ~80 ns, with the transient conductance exhibiting non-uniform solidification velocity.

We interpret the differences in the $R(t)$ behavior for O and N in terms of the nucleation sites which are available above the buried insulating layers. The increase of $R(t)$ at 155 ns for the N-implanted sample suggests that nucleation of the solid in the undercooled liquid occurs at the Si_xN_y interface. Hence, once the interface begins to solidify, the latent heat released increases the liquid temperature to near T^{cl} and $R(t)$ rises correspondingly. For oxygen, random nucleation appears to occur at the free surface rather than at the buried SiO_x interface. The reflectance thus averages the molten and solid area of the surface and exhibits a slow decrease as the entire layer solidifies. The extremely rapid drop of the transient conductance is further evidence for this random surface nucleation. The conductance technique provides an accurate estimate of the molten thickness only for uniform melt layers; for a spatially varying melt thickness, the technique biases the measurment toward the shallow melt regions. Thus the rapid drop and slow "exponential" type decay suggest small pockets which solidify rapidly and are followed by a slow "filling in the gaps" in the sample. Further microscopic studies of the surface layers are necessary to fully understand these processes.

CONCLUSIONS

Impurities in silicon exhibit a wide and varied range of behavior during pulsed melting. The melt and solidification dynamics are found to depend critically on such factors as the solid phase solubility and the compound forming abilities. We have catagorized our observations to date into five classes of materials. These classes consist of: (I) self implants of Si creating a-Si, (II) fully miscible alloy systems such as Si:Ge, (III) relatively soluble systems such as As:Si, (IV) insoluble systems such as In:Si, and (V) compound forming impurities such as N and O.

ACKNOWLEDGEMENTS

We would like to acknowledge the large number of collaborators on this work, including J. Y. Tsao, S. Stiffler, J. M. Poate, D. C. Jacobson, A. G. Cullis, and M. A. Aziz. Work at Cornell has been supported by NSF (L. Toth), and by NSF through the Materials Science Center. Samples were prepared in the National Research and Resource Facility for Submicron Structures (NSF). Work performed at Sandia National Laboratories was supported by U.S. Department of Energy under contract DE-AC04-76DP00789.

REFERENCES

1 *Laser Annealing of Semiconductors*, edited by J. M. Poate and J. W. Mayer, (Academic Press, New York, 1982)

2 For a review of the current literature, see other volumes in this series, Mat. Res. Soc. Symp. Proc. **1, 4, 13, 23, 35** (1981-85)

3 C. W. White, B. R. Appleton and S. R. Wilson, in ref. 2, Chap. 5

4 P. Baeri, J. M. Poate, S. U. Campisano, G. Foti, E. Rimimi and A. G. Cullis, Appl. Phys. Lett. **37**, 912 (1980)

5 Michael O. Thompson, G. J. Galvin, J. W. Mayer, P. S. Peercy, J. M. Poate, D. C. Jacobson, A. G. Cullis and N. G. Chew, Phys. Rev. Lett. **52**, 2360 (1984)

6 Michael O. Thompson, P. S. Peercy and J. Y. Tsao, submitted J. Mat. Res.

7 Michael O. Thompson, P. S. Peercy and J. Y. Tsao, submitted Appl. Phys. Lett.

8 Michael O. Thompson, P. S. Peercy and J. Y. Tsao, Mat. Res. Soc. Symp. Proc. **35**, 169 (1985)

9 P. S. Peercy, Michael O. Thompson, and J. Y. Tsao, Appl. Phys. Lett. **47**, 244 (1985)

10 P. S. Peercy, J. M. Poate, Michael O. Thompson and J. Y. Tsao, these proceedings

11 P. S. Peercy, J. M. Poate, Michael O. Thompson and J. Y. Tsao, submitted Appl. Phys. Lett.

12 Michael O. Thompson, P. S. Peercy, J. Y. Tsao and S. Stiffler, submitted Appl. Phys. Lett.

13 A. G. Cullis, H. C. Webber and P. Bailey, J. Phys. E **12**, 688 (1979)

14 G. J. Galvin, Michael O. Thompson, J. W. Mayer, R. B. Hammond, N. Paulter and P. S. Peercy, Phys. Rev. Lett. **48**, 33 (1982)

15 Michael O. Thompson and G. J. Galvin, Mat. Res. Soc. Symp. Proc. **13**, 57 (1983)

16 Michael O. Thompson, G. J. Galvin, J. W. Mayer, P. S. Peercy and R. B. Hammond, Appl. Phys. Lett. **42**, 455 (1983)

17 *Liquid Semiconductor*, V. M. Glazov, S. N. Chizhevskaya and N. N. Glagoleva, Eds. (Plenum Press, New York, 1969)

18 D. H. Auston, C. M. Surko, T. N. C. Venkatesan, R. E. Slusher and J. A. Golovchenko, Appl. Phys. Lett. **33**, 437 (1978)

19 E. P. Donovan, F. Spaepen, D. Turnbull, J. M. Poate and D. C. Jacobson, Appl. Phys. Lett. **42**, 698 (1983)

20 J. Narayan and C. W. White, Appl. Phys. Lett. **44**, 35 (1984)

21 J. Narayan, C. W. White, O. W. Holland and M. J. Aziz, J. Appl. Phys. **56**, 1821 (1984)

22 W. Sinke and F. W. Saris, Phys. Rev. Lett. **53**, 2121 (1984)

23 H. J. Leamy, W. L. Brown, G. K. Cellar, G. Foti, G. H. Gilmer and J. C. C. Fan, Mat. Res. Soc. Symp. Proc. **1**, 89 (1981)

24 J. Y. Tsao, private communication and work in progress.

25 Michael O. Thompson, J. W. Mayer, A. G. Cullis, A. G. Webber, N. G. Chew, J. M. Poate and D. C. Jacobson, Phys. Rev. Lett. **50**, 896 (1983).

26 H. J. Goldsmid, M. M. Kaila and G. L. Paul, Phys. Stat. Sol., **76**, K31 (1983).

27 For example, Y. Kanemitsu, I. Nakada and H. Kuroda, Appl. Phys. Lett. **47**, 939 (1985).

28 R. Olesinski and G. J. Abbaschian, Bull. Alloy Phase Diag. **5**, 180 (1984)

29 B. Abeles, P. S. Beers, D. G. Cody and J. P. Dismukes, Phys. Rev. **125**, 44 (1962).

30 J. W. Cahn, S. R. Coriell and W. J. Boettinger, in *Laser and Electron Beam Processing of Materials*, C. W. White and P. S. Peercy, Ed., (Academic Press, New York 1980), p. 89

31 G. J. Galvin, J. W. Mayer and P. S. Peercy, Appl. Phys. Lett. **46**, 644 (1985).

32 M. O. Thompson, P. H. Bucksbaum and J. Bokor, Mat. Res. Soc. Symp. Proc. **35**, 181 (1985).

33 B. C. Larson and J. Z. Tischler, these proceedings

34 S. U. Campisano, D. C. Jacobson, J. M. Poate, A. G. Cullis and N. G. Chew, Appl. Phys. Lett. **46**, 846 (1985)

35 W. Kaiser and P. H. Keck, J. Appl. Phys. **28**, 882 (1957)

36 R. Pinizzotto, Mat. Res. Soc. Symp. Proc **27**, 265 (1984)

OVERHEATING IN SILICON DURING PULSED-LASER IRRADIATION?*

B. C. Larson, J. Z. Tischler, and D. M. Mills[†]
Solid State Division, Oak Ridge National Laboratory, Oak Ridge, TN 37831
[†]CHESS and Applied and Engineering Physics, Cornell Univ., Ithaca, NY 14853

ABSTRACT

Nanosecond resolution time-resolved x-ray diffraction measurements of thermal strain have been used to measure the interface temperatures in silicon during pulsed-laser irradiation. The pulsed-time-structure of the Cornell High Energy Synchrotron Source (CHESS) was used to measure the temperature of the liquid-solid interface of <111> silicon during melting with an interface velocity of 11 m/s, at a time of near zero velocity, and at a regrowth velocity of 6 m/s. The results of these measurements indicate 110 K difference between the temperature of the interface during melting and regrowth, and the measurement at zero velocity shows that most of the difference is associated with undercooling during the regrowth phase.

INTRODUCTION

The relationship between interfacial undercooling and crystal growth in silicon has been of interest both from the standpoint of fundamental crystal growth thermodynamics[1] and with regard to the amorphization of silicon for crystal growth rates of more than 15 m/s. It has been suggested that >200 K undercooling is required to drive 15 m/s crystal growth and that such undercooling favors a liquid-amorphous transition rather than a liquid-crystal transition.[2] Likewise, the relationship between overheating and melting velocity is of interest.

We have previously reported[3] time-resolved x-ray diffraction measurements of the overheating and undercooling of silicon during pulsed laser irradiation; a greater than 100 K difference was measured between the interface temperature during 11 m/s melting and the interface temperature during 6 m/s regrowth along the <111> direction. However, because of uncertainties in the absolute temperature scale brought about by uncertainties in the temperature dependence of the elastic constants and the thermal expansion coefficients, the proportion of overheating and undercooling could not determined. We have now analyzed additional time-resolved x-ray scattering measurements made at a time of near-zero interface velocity, and in this paper we present results that separate the overheating and under-cooling contributions.

EXPERIMENTAL

We have used nanosecond resolution time-resolved x-ray diffraction measurements of thermal expansion-induced strain to determine the liquid-solid interface temperatures in Si during pulsed-laser irradiation. The time-resolved nature of the measurements and the analysis procedures have been discussed elsewhere[3,4] and only a brief description of the experiment will be presented here. Figure 1 shows a schematic picture of x-ray scattering from a crystal with a depth dependent thermal gradient.

*Research sponsored by the Division of Materials Sciences, U.S. Department of Energy under contract DE-AC05-840R21400 with Martin Marietta Energy Systems, Inc.

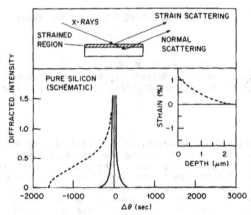

Fig. 1. Schematic representation of the scattering geometry depicting
the thermally strained region of the laser irradiated sample
and a schematic view of the x-ray scattering pattern resulting
from the thermal strain distribution.

Through an analysis of Bragg scattering profiles, thermal strain profiles
have been determined and lattice temperature distributions have, in turn,
been extracted from the strain using the thermal expansion coefficients.
Laser pulses of 248 nm wavelength and 25 ns time duration (as shown in Fig.
2) were synchronized with x-ray pulses from the Cornell High Energy
Synchrotron Source (CHESS) to achieve nanosecond resolution x-ray diffraction measurements. The laser pulses used had energy densities of 1 J/cm^2
and the silicon samples were <111> oriented wafers of 0.5 mm thickness.
The x-ray measurements were made in a step scanning mode in the region of
the Bragg reflection using 18 laser shots at each angular setting of the
crystal; the jitter in the laser firing was ±2 ns.
    Figure 2 shows heat flow simulations (using a computer code supplied
by M. O. Thompson)[5] of the laser melting and regrowth process pertaining
to the measurements in this experiment. The laser power, the interface
velocity, and the melt-front penetration depth are shown as a function of
time. As can be seen in the figure, measurements made at 15 and 45 ns
correspond to interface velocities of 11 m/s melting and 6 m/s regrowth,
respectively. Measurements made at these times have been reported previously;[3] of particular importance for this work is the interface velocity
indicated for measurements made at 27 ns. This time corresponds to that of
maximum melt-front penetration and hence represents the time at which the
interface velocity vanishes as the interface temperature passes through the
equilibrium melting temperature.

RESULTS

    Figures 3 and 4 show time-resolved x-ray measurements on <111> Si for
delay times of 16, 27, and 45 ns relative to the arrival time of the laser
pulses. The error bars on the measured data represent the statistical
uncertainty in the measured data and the solid lines correspond to least-
squares fits of the data with x-ray scattering calculations for silicon
containing a strained surface layer. In the fitting procedure the lattice
strains as a function of depth were used as the adjustable parameters. The
full lattice temperature distributions corresponding to Figs. 3 and 4 have
been reported elsewhere[6] and will not be presented here; however, the

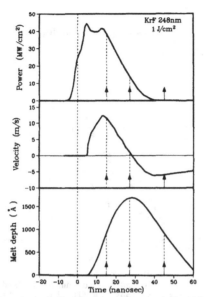

Fig. 2. A composite view of the laser pulses and the timing used in this experiment together with calculations of the liquid-solid interface velocity and melt depth for 1 J/cm^2, 248 nm laser pulses on silicon.

temperature distributions to depths of 1500 Å are shown in Fig. 5. The hatched areas indicate the estimated uncertainty in the determination of the lattice temperatures, and it can be seen that the measured difference in the interface temperatures during melt-in (16 ns) and regrowth (45 ns) is 110 ± 30 K. Of most significance, though, is the temperature profile for the 27 ns case. We see that the interface temperature at 27 ns (which corresponds to a time of near zero velocity) falls in the same range as that measured at 16 ns (11 m/s velocity). That is, the 110 K temperature difference seems to be associated mainly with undercooling during the 6 m/s regrowth while only a small fraction is associated with overheating during the 11 m/s melt-in. Under the assumption of prorating overheating and undercooling with velocity, an undercooling of ~45 K had been reported[3] earlier.

The estimated precision in the interface temperatures is ±25 K for the 16 and 27 ns measurements and ±10 K for the 45 ns measurements. Within these uncertainties we can assign velocity coefficients for undercooling and overheating (defined by $\Delta T_u = \beta^u V$ and $\Delta T_0 = \beta^0 V$) for <111> silicon. Using these values we get limiting values of $\Delta T_0 < 60$ K and $145 > \Delta T_u > 70$ K, which in turn lead to $\beta^0 < 6$ K/m/s and $11 < \beta^u < 24$ K/m/s.

DISCUSSION

The result to be emphasized here is that overheating is observed to be small so that most of the measured differences in the interface temperatures during melt-in and regrowth must be attributed to undercooling during regrowth. Recognizing that β^0 must be greater than zero, the midrange of the allowed β's, consistent with the measurements made here, are $\beta^0 \sim 3$ K/m/s for overheating and $\beta^u \sim 17$ K/m/s for undercooling. These values are to be compared with values of $\beta u = 15 \pm 5$ K/m/s reported by

116

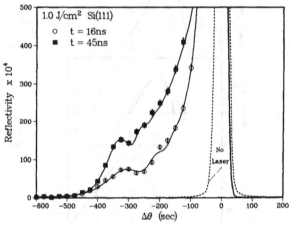

Fig. 3. Time-resolved Bragg profile measurements made at the (111)
reflection on <111> oriented silicon during pulsed-laser
irradiation. The solid lines are fits to the data measured
at 16 ns and 45 ns.

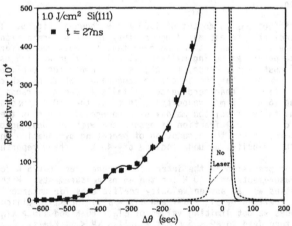

Fig. 4. Time-resolved Bragg profile measurements made under the same
conditions as those in Fig. 3, but at a time of 27 ns.

Galvin et al.[7] from transient conductivity measurements on 3.5 ns ruby-
laser irradiated <100> silicon-on-sapphire (SOS) and they can be compared
to $\beta^u = 17\pm3$ reported by Thompson et al.[8] from the analysis of transient
conductivity measurements on 28 ns ruby-laser irradiated SOS. Thompson et
al.[8] reported much lower values of $\beta^u = \beta^0 = 4\pm2$ K/m/s (for low interface velocities) obtained from the analysis of optical transmission
measurements on picosecond range UV laser pulses on silicon. Although
Thompson et al. expressed caution in interpreting the picosecond results,

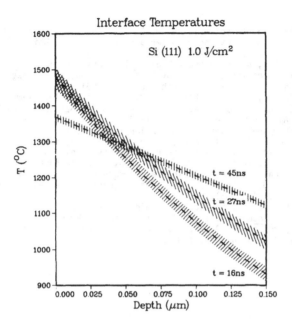

Interface Temperatures

Si (111) 1.0 J/cm²

t = 45ns

t = 27ns

t = 16ns

Depth (μm)

Fig. 5. A composite view of the temperature profiles at the liquid-solid interface corresponding to the fits to the measured data for 16, 27, and 45 ns in Figs. 3,4.

picosecond results, the data showed significant non-linearities at high melting and regrowth velocities and indicated the velocity coefficient for overheating to be smaller than that for undercooling.

The fact that the the indicated interface temperature in Fig. 5 for 27 ns is 1478 ± 25°C rather than 1412°C (as would be expected for the equilibrium melting temperature) is interpreted to be a result of inaccuracies in the temperature dependencies of the elastic constants and the thermal expansion coefficients. That is, because data on the temperature dependence of the elastic constants of silicon above 900°C is not available and because the temperature dependence of the thermal expansion coefficients are not believed to be accurate to better than ~2%, the absolute temperature scale inferred from the lattice strain measurements has relatively large uncertainties. The relative temperatures are expected to be accurate, however, because systematic errors present in the absolute magnitudes are largely eliminated in the subtraction.

The large range of values reported in the literature for the velocity coefficients for overheating and undercooling indicates the need for additional work in this area. The transient conductivity results were made on <100> SOS while the results presented here were for <111> silicon; x-ray measurements are not available for the 27 ns time on <100> silicon, but the temperature differences measured between melt-in and regrowth are about half that found for the <111> direction. Additional measurements with higher resolution and lower uncertainties would be helpful with each of the techniques, and additional theoretical work is needed in this area as well.

118

REFERENCES

[1] J. W. Cahn, S. R. Coriell, and W. L. Boettinger, p. 89 in Laser and Electron Beam Processing of Materials, ed. by C. W. White and P. S. Peercy, Academic Press, NY (1980).

[2] A. G. Cullis, p. 16 in Energy Beam-Solid Interactions and Transient Thermal Processing 1984, ed. by D. K. Biegelsen, G. A. Rozgonyi, and C. V. Shank (Materials Research Society, 1985).

[3] B. C. Larson, J. Z. Tischler, and D. M. Mills, p. 187 (same as Ref. 2).

[4] B. C. Larson, C. W. White, T. S. Noggle, J. F. Barhorst, and D. M. Mills, Appl. Phys. Lett. 42, 282 (1983).

[5] M. O. Thompson, private communication (see Ref. 8).

[6] B. C. Larson, J. Z. Tischler, and D. M. Mills, J. Mat. Res. (in press).

[7] G. J. Galvin, J. W. Mayer, and P. S. Peercy, Appl. Phys. Lett. 46, 644 (1985); P. S. Peercy and M. O. Thompson, p. 54 in Energy Beam-Solid Interactions and Transient Thermal Processing 1984, ed. by D. K. Biegelsen, G. A. Rozgonyi, C. V. Shank (Materials Research Society, Pittsburgh, PA, 1985).

[8] M. O. Thompson, P. H. Bucksbaum, and J. Bokor, p. 181 in Energy Beam-Solid Interactions and Transient Thermal Processing 1984, ed. by D. K. Biegelsen, G. A. Rozgonyi, and C. V. Shank (Materials Research Society, Pittsburgh, PA, 1985).

TIME-RESOLVED X-RAY ABSORPTION OF AN AMORPHOUS Si FOIL
DURING PULSED LASER IRRADIATION

KOUICHI MURAKAMI*, HANS C. GERRITSEN, HEDSER VAN BRUG, FRED BIJKERK,
FRANS W. SARIS AND MARNIX J. VAN DER WIEL
FOM-Institute for Atomic and Molecular Physics, Kruislaan 407,
1098 SJ Amsterdam, The Netherlands.

ABSTRACT

We report time-resolved X-ray absorption and extended X-ray absorption
fine structure (EXAFS) measurements on amorphous silicon under nanosecond
pulsed-laser irradiation. Each measurement was performed with one laser shot
in the X-ray energy range from 90 to 300 eV. An X-ray absorption spectrum
for induced liquid Si (liq*Si) was first observed above an energy density of
0.17 J/cm^2. It differs significantly from the spectrum for amorphous Si and
characteristically shows the disappearance of the Si-L(II,III) edge structure at around 100 eV. This phenomenon is interpreted in terms of a significant reduction in the 3s-like character of the unfilled part of the conduction band of liq*Si compared to that of amorphous Si. This is the first direct evidence that liq*Si has a metallic-like electronic structure. Time-
resolved EXAFS results are also discussed briefly.

INTRODUCTION

Many time-resolved (TR) measurements such as optical reflectivity [1-3],
electrical conductivity [4], X-ray diffraction [5], etc. [2,3] have been
performed for the study on extremely fast phase transition. They have indicated that intense picosecond (ps) and nanosecond (ns) pulsed-laser irradiation on Si (coordination number 4) induces an abrupt temperature rise in
the solid state and a phase transition to a short-lived liquid state (liq*Si).
Most of the time-resolved measurements give no direct information about the
long- and short-range geometric structures. However, TR electron diffraction
[6], TR low energy electron diffraction (LEED) [7] and TR X-ray diffraction
[5] techniques have been used to obtain information about the long range
structure of short-lived liq*Al, liq*Ge and liq*Si, respectively. They revealed that the long-range order disappears in the liquid states like in the
normal liquid state. However, there have been no experiments about the short-
range structure, or the local atomic arrangement, and electronic structure of
short-lived liq*Si. It is known that normal liquid Si has a coordination number 6.5 at the melting point [8]. It remains to be solved whether or not
short-lived liq*Si, produced on ps or ns time scales, is entirely equal to
normal liquid Si in these respects [2,3]. Studies on the local atomic arrangement or electronic structure will be an important key to clarify the nature
of short-lived liq*Si and undercooled liq*Si [4,9].
In order to study these subjects, we have performed TR X-ray absorption
and TR extended X-ray absorption fine structure (EXAFS) measurements on amorphous Si films under ns-pulsed laser irradiation [10]. A similar experiment
on short-lived liq*Al was reported earlier by Epstein et al. [11]. They showed
some short-range order remains in short-lived liq*Al. In this paper we report the first observation of the time-resolved X-ray absorption near and
above the Si-L(II,III) edge for liq*Si and describe also the electronic structure of liq*Si.

*on leave from: Inst. of Materials Science, Univ. of Tsukuba, Sakura,
Ibaraki 305, Japan.

120

EXPERIMENTAL

Amorphous Si (a-Si) films of 600 Å thickness were produced by electron-
beam evaporation of Si (purity 99.999%) and deposition on carbon films of
440 Å thickness (on thin NaCl film coated on glass) in a vacuum of 7×10^{-8}
Torr. During the deposition, substrate temperature was raised up to 230°C to
reduce the formation of voids and to prevent indiffusion of O_2 and H_2O. The
a-Si on C (a-Si/C) films were floated off in distilled water and picked up
by stainless steel sample holders which have holes with a diameter of 3 or 4
mm. Free standing a-Si foils of 600 Å thickness were also prepared for comparison.

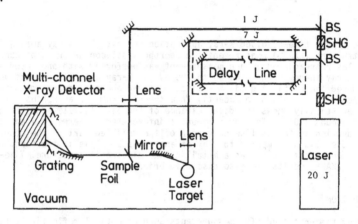

Fig. 1

Outline of the experimental system of time-resolved X-ray absorption measurements.

Figure 1 shows an outline of our experimental system. X-ray absorption
and EXAFS measurements were carried out by transmitting a broad continuum of
X-rays through the a-Si/C foil and dispersing the radiation afterwards. As an
X-ray source we used a plasma that was created by focusing the output of frequency doubled Nd-YAG/glass laser (7 J, 18 ns, 532 nm) on a Ta target. The
created hot Ta plasma emits an intense pulse of X-ray photons (10^{15} in one
pulse of 18 ns) with a smooth spectral distribution, so that we can obtain
the structural information averaged over a period of 18 ns. Our measuring
system allows us to record an absorption spectrum in one single laser shot
with good statistics. The energy resolution was approximately 4 eV. Part of
the experimental technique was described in an earlier paper [12]; more details will be given in a later publication [13].

As an annealing beam a fraction of the laser output was used to irradiate the a-Si/C foils in a vacuum of 10^{-7} Torr, as shown in Fig. 1. The
laser beam was focused to 4-5 mm, while the diameter of the X-ray probe beam
on the sample was approximately 0.2 mm. The laser beam has a spatial profile
close to TEM-00 and TEM-01, and remaining intensity variations are believed
to occur on a scale larger than the probe area. By performing some of the measurements using a diffuser in the annealing beam, it was also clear that possible intensity inhomogeneities has no significant effect on the X-ray absorption spectra. The irradiation energy density was varied from 0.1 to 3.6 J/cm²
with an uncertainty of 20%. To do time-resolved measurements, the irradiation
laser pulse on the foil was followed by the X-ray probe pulse with a variable
delay time τ_d of 12, 30 and 60 ns, as shown in Fig. 1.

RESULTS AND DISCUSSION

Typical X-ray absorption spectra at τ_d of 12 ns are shown in Fig. 2 for various annealing energy densities. For free standing a-Si foils without C backing, identical results were obtained. This indicates no fast reactions of alloying between Si and C occur. From the results of our TR X-ray absorption spectra and our additional TR optical (633 nm) transmission measurements, the laser energy density can be roughly divided into three ranges. Range I is below approximately 0.17 J/cm², at which energy density we observe the first significant changes in the absorption spectra. This value is therefore thought to correspond to the annealing threshold E_{th}. Range II is from 0.17 to approximately 1.0 J/cm², where the liq*Si phase is produced so that annealing takes place. Range III is above 1.0 J/cm² at which further density changes in the spectra are observed. This corresponds to the damaging threshold E_d, as we confirmed by TR optical transmission measurements.

A clear edge at 98 eV and a broad absorption peak at 125 eV are seen for the spectrum 2(a) without pulsed-laser irradiation. The edge comes from the Si-L(II,III) absorption corresponding to excitation of an electron from the 2p core level to the bottom of the conduction band [14-16] (see Fig. 3(a)).

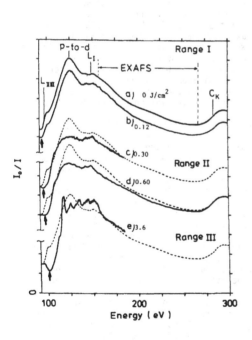

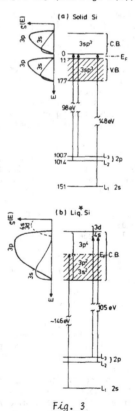

Fig. 2

Typical X-ray absorption spectra at τ_d of 12 ns for laser energy densities from 0 to 3.6 J/cm². The edge at 280 eV is due to the carbon foil (C_K absorption). The vertical axis indicates the ratio of the incident X-ray intensity I_0 to that I of the transmitted X-rays.

Fig. 3

Energy level diagram and rough sketch of the density of state $\rho(E)$ for solid Si (a) and liq*Si (b).

122

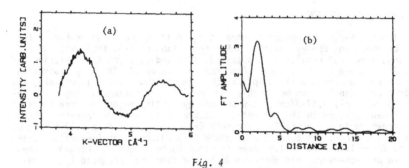

Fig. 4

(a) EXAFS oscillation as a function of photoelectron wavevector k and
(b) the Fourier transformation for a-Si. By considering phase shift
(+0.20 Å), a bond distance of 2.30 ± 0.10 Å is obtained.

The broad absorption is thought to be due to so-called "centripetal barrier"
for p-to-d transitions [15]. Hereafter we call this the "p-to-d" maximum. The
Si-L(I) edge is also observed at about 150 eV, which corresponds to excitation
from the 2s core level to the bottom of the conduction band [14,15] (see Fig.
3(a)). It should also be stressed that in the energy range from 155 to 270 eV
EXAFS oscillations can be observed. We can obtain clearer EXAFS oscillation
as a function of photoelectron wavevector k, as shown in Fig. 4(a), by subtracting an absorption background. Figure 4(b) shows the Fourier transformation of this oscillation, indicating a Si-Si atomic distance of 2.30 ± 0.10 Å,
which shows good agreement with literature values (2.35 Å). At an energy density of 0.12 J/cm^2 in Range I, there is no significant change in the spectrum
(see Fig. 2(b)).

At energy densities of 0.30 and 0.60 J/cm^2 in Range II (Fig. 2(c) and
(d)), the first observation is a decrease in total area. This will be discussed below. The structure due to the Si-L(II,III) absorption at 100 eV nearly
disappears and the observed spectra are much different from the original one
for a-Si. Although the actual onset of the Si-L(II,III) absorption shown by
the arrows in Fig. 2 is not changed considerably, there is a distinctly slower
rise towards the p-to-d maximum. If we define simply the point of maximum
slope as an "effective edge" of the Si-L(II,III) in Ranges II and III, there
is a clear edge shift of about 7 eV to higher energy. It should also be noticed that there is a decrease of the p-to-d maximum height and a small shift
of the peak position to lower energy. The Si-L(I) edge seems to be unchanged
or to move slightly to lower energy, in contrast to the behavior of the Si-L
(II,III) edge. This behavior of opposite shifts has not been observed in Si
compounds such as SiO$_2$, Si$_3$N$_4$, SiH$_4$, etc. [14]. This fact indicates strongly
that the observed edge shifts cannot be attributed to a chemical shift of the
core levels.

The EXAFS amplitude becomes so weak that the inter-atomic distances in
liq*Si can no longer be determined, in contrast to the result of short-lived
liq*Al [11]. The weakness of the EXAFS amplitude could be ascribed to a temperature rise to at least 1400 K, which is compared to the Si melting point
of 1690 K.

At an energy density of 3.6 J/cm^2 in Range III, a complicated structure
near the absorption peak and a large shift in the Si-L(II,III) edge can be
seen in a typical spectrum (e) of Fig. 2. The complicated peaks may be connected with the formation of a Si plasma or fine cluster [17]. A detailed analysis will be published elsewhere. The new structure disappears with time and
the observed spectrum at τ_d of 60 ns becomes very similar to those shown in
Fig. 2 (c) and (d) for liq*Si. This suggests that the produced Si plasma or

fine clusters change again into a neutral liq*Si-like state with time.
We interpret the Si-L(II,III) and Si-L(I) edge shifts as reflecting the
change in final states, i.e. electronic structure, between solid and liq*Si.
The cross section for X-ray absorption by each initial core state ψ_i is described in the electric dipole approximation as follows:

$$\sigma_{if}(E) \propto \Sigma_f |<\psi_f|P|\psi_i> \cdot A|^2 , \tag{1}$$

where p is the momentum operator, A the vector potential, ψ_f the final states
in the conduction band above the Fermi level E_F and ψ_i the 2s or 2p core
levels from which the transition takes place. According to the dipole selection rule, the final states which can be reached depend on the symmetry of
ψ_i, i.e. from 2s, only p-like states can be reached, while from 2p both s-
and d-like states can be reached, as shown in Fig. 3(a). Consequently, the observed large change in the spectra or large shift of the Si-L(II,III) edge can
be attributed to a reduction of the density of 3s-like states just above E_F
in liq*Si; i.e. the main transition strength occurs from 2p to 4s-like and
3d-like states (see Fig. 3(b)). However, the first onset of Si-L(II,III) absorption is changed little. This indicates that a small amount of 3s-like
states remains just above E_F, which is consistent with recent band calculations [18] for normal liquid Si. On the other hand, the small change in the
Si-L(I) edge indicates little or no change in the density of 3p-like states
just above E_F for liq*Si. In general, solid Si is thought to have nearly equal
amounts of s-like and p-like states in both the conduction ((sp^3)* antibonding
states) and valence (sp^3 bonding states) bands [19,20] (see Fig. 3(a)). On
the other hand, in metallic solids s-like states tend to lie in the part of
the band below E_F, while p-like states occur both below and above E_F [16,30]
(see Fig. 3(b)). Therefore, we conclude that liq*Si produced has a metallic-
like electronic structure.

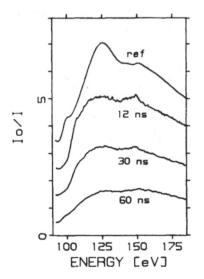

Figure 5 gives the change in the
X-ray absorption spectrum at an energy
density of 0.30 J/cm^2 with various
delay times. The most striking feature
is that the integrated absorption decreases with time. We interpret this
as indicative of the formation of
liq*Si droplets on C-foil surface with
time: the droplet formation leads to
an increase in X-ray transmission.
Therefore the liq*Si state resembles
normal liquid in this respect. After
60 ns, weak EXAFS giving correct bond
distance, becomes visible again [13].

It should also be noticed that, at
the 12 ns delay, some averaging may
take place only near the threshold
energy density E_{th} of 0.17 J/cm^2, because the X-ray pulse width is 18 ns.

← Fig. 5

Time evolution of the X-ray absorption
spectra of short-lived liq*Si produced
by irradiation at 0.30 ± 0.04 J/cm^2.

CONCLUSION

In summary, we have reported a large change in the X-ray absorption spectra of Si under intense pulsed-laser irradiation. At the energy range from

124

0.17 to 1.0 J/cm^2 corresponding to the annealing range, characteristic changes in the onset and edge of the Si-L(II,III) absorption and in the Si-L(I) edge for the liq*Si are interpreted in terms of a significant reduction of the 3s-like character of states just above the Fermi level. This is consistent with a metallic electronic structure. The time evolution of the spectra for liq*Si also shows the formation of droplets, which indicates fluid properties for liq*Si.

Unfortunately, TR-EXAFS measurements have given no clear information about local atomic arrangement of liq*Si. An experiment is in progress on the X-ray absorption of normal liquid Si, in order to establish whether EXAFS can be observed. This would allow a conclusion as to whether liq*Si and normal liquid Si are equal with respect to the local structure and electronic structure. In addition, we believe it is of interest to enhance the time-resolution of the time-resolved X-ray absorption measurements up to the picosecond scale in order to clarify the dynamics and structure of the induced, short-lived liq*Si.

ACKNOWLEDGEMENT

We are grateful to J.F. van der Veen for helpful discussions on X-ray absorption data. This work is sponsored by FOM with financial support from ZWO.

REFERENCES

1. For example; C.V. Shank, R. Yen and C. Hirliman, Phys.Rev.Lett. 50, 454 (1983); L.-A. Lompre, J.M. Liu, H. Kurz and N. Bloembergen, Appl.Phys.Lett. 44, 3 (1983); K. Murakami, K. Takita and K. Masuda, Jpn.J. Appl.Phys. 20, L867 (1981).
2. H.M. van Driel, Semiconductors Probed by Ultrafast Laser Spectroscopy, vol. II (ed. by R.R. Alfano) (Academic Press, 1984) pp.57-94.
3. K. Murakami and K. Masuda, ibid, pp.171-195.
4. M.O. Thompson, G.J. Galvin, J.W. Mayer, P.S. Peercy, J.M. Poate, D.C. Jacobson, A.G. Cullis and N.G. Chew, Phys.Rev.Lett. 52, 2360 (1984).
5. B.C. Larson, C.W. White, T.S. Noggle, J.F. Barhorst and D.M. Mills, Appl.Phys.Lett. 42, 282 (1983).
6. S. Williamson, C. Mourou and J.C.M. Li, Phys.Rev.Lett. 52, 2364 (1984).
7. R.S. Becker, G.S. Higashi and J.A. Golovchenko, Phys.Rev.Lett. 52, 307 (1984).
8. Y. Waseda and K. Suzuki, Z.Physik B 20, 339 (1975).
9. W. Sinke and F.W. Saris, Phys.Rev.Lett. 53, 2121 (1984).
10. K. Murakami, H.C. Gerritsen, H. van Brug, F. Bijkerk, F.W. Saris and M.J. van der Wiel (to be published).
11. H.M. Epstein, R.E. Schwerzal, P.J. Mallozzi and B.E. Campbell, J.Am.Chem.Soc. 105, 1466 (1983).
12. H.C. Gerritsen, H. van Brug, M. Beerlage and M.J. van der Wiel, Nucl.Instr.and Meth. A 238, 546 (1985).
13. H.C. Gerritsen, H. van Brug, F. Bijkerk, K. Murakami, F.S. Saris and M.J. van der Wiel (to be published).
14. F.C. Brown, R.Z. Bachrach and M. Shibowski, Phys.Rev. B 15, 4781 (1977).
15. J.J. Risko, S.E. Schnatterly and P.C. Gibbons, Phys.Rev.Lett. 32, 671 (1974).
16. G. Wiech and E. Zopf, Band Structure Spectroscopy of Metals and Alloys, ed. D.J. Fabian and L.M. Watsonn (Academic Press, N.Y., 1973) pp.629-640.
17. L.A. Bloomfield, R.R. Freeman and W.L. Brown, Phys.Rev.Lett. 54, 2246 (1985).
18. J.P. Gaspard, Ph. Lambin, C. Mouttet and J.P. Vigneron, Phil.Mag. B 50, 103 (1984).
19. J.C. Phillips, Bonds and Bands in Semiconductors (Academic Press, N.Y., 1973) pp.98-125.
20. C.F. Hague, C. Sénémand and H. Ostrowiecki, J.Phys. F 10, L267 (1980).

NUCLEATION OF INTERNAL MELT DURING PULSED LASER IRRADIATION

P. S. PEERCY, MICHAEL O. THOMPSON,† and J. Y. TSAO
Sandia National Laboratories, Albuquerque, NM, 87185

J. M. POATE
AT&T Bell Laboratories, Murray Hill, NJ, 07974

ABSTRACT

Real-time measurements of the molten layer thickness and simultaneous measurements of the melt duration at the surface reveal that melt nucleates internally when Si implanted with low solid-solubility impurities such as In is melted with a 30 nsec laser pulse at 694 nm. Internal nucleation of melt was observed for all energy densities examined. Furthermore, at energy densities insufficient to melt the entire amorphous layer, internal nucleation of melt is followed by an explosive-like process in which a buried molten layer propagates toward the irradiated surface.

INTRODUCTION

When a laser pulse of sufficient energy density is incident on a solid, melt is induced in the near-surface region to produce a molten layer which resolidifies after the laser pulse. The normal scenario is for melt to initiate at the irradiated surface and propagate in to some maximum depth, whereupon the motion of the liquid-solid interface reverses and the melt front returns to the surface as the molten layer solidifies [1].

While such surface-initiated melt is the accepted scenario and has been assumed in the analysis of numerous impurity redistribution experiments, it was recently shown that, under certain conditions, buried molten layers can be produced inside samples. Examples include explosive crystallization of amorphous Si to produce fine-grained polycrystalline Si, where the explosive event is mediated by a thin molten layer [2], and pulsed laser-induced melting of Si containing very high concentrations (>12%) of As, which results in both surface nucleation of solid and internal nucleation of melt due to the reduced melting point of the Si-As alloy compared to that of pure Si [3,4].

In the present paper, recent studies are described in which it is shown that melt initiates inside the sample when Si, implanted with In, is irradiated with a 30 nsec pulse at 694 nm from a ruby laser. Furthermore, for low energy densities, the internal nucleation of melt is followed by explosive-like crystallization that propagates toward the irradiated surface producing significant redistribution and surface segregation of impurities.

Although the present paper will be limited to In in Si, these phenomena appear to be general. They were observed for Si implanted with a variety of dopants, including Sn, Zn and In, at fluences ≥5 x 10¹⁴ ions/cm², over a wide range in energy density. These impurities are characterized by a low solid solubility in equilibrium and are expected to reduce the melting temperature dramatically.

EXPERIMENTAL ARRANGEMENT

The melt and solidification dynamics were studied in real time using the transient conductance technique [5,6] to measure the molten layer thickness versus time and glancing angle reflectance [7] at 488 nm to measure the melt duration at the surface. The transient conductance was measured using a charge line configuration [8]. Because the resistivity of Si changes by a factor of ~30 upon melting [9], measurement of the resistance permits direct determination of the molten layer thickness.

The measurements were made on thin (0.5 μm) silicon on sapphire (SOS); SOS was used to avoid difficulties in the analysis caused by contributions from photogenerated carriers to the electrical conductance. Transient conductance samples were patterned photolithographically to yield resistors with a length-to-width ratio of 55:1. After patterning, the samples were implanted with In at fluences from 5×10^{14} to 3×10^{15} ions/cm^2 at 100 or 120 keV, producing amorphous layers ~100 nm thick. These In concentrations have previously been shown to be insufficient to cause interfacial breakdown [10] for the present experimental conditions.

Samples were irradiated with 30 nsec pulses at 694 nm through a fused silica beam homogenizer [11]. Changes in the depth distribution after irradiation were measured using ion backscattering techniques. Because the irradiation crystallized part or all of the amorphous material and altered the impurity profile, each sample was irradiated only once.

RESULTS AND DISCUSSION

A. Transient Conductance and Reflectance Measurements

Transient conductance and reflectance measurements are shown in Fig. 1 for a sample implanted with 2.7×10^{15} In/cm^2 at 120 keV and irradiated at 0.67 J/cm^2. This energy density is sufficient to melt through the amorphous layer; data for all higher energy densities are similar. Consider first the transient conductance trace. The molten layer thickness d(t) increases rapidly until the entire amorphous Si (a-Si) layer is molten; d(t) then decreases slightly as the liquid Si (ℓ-Si) starts to solidify as crystalline Si (c-Si). This reversal occurs because the a-Si melting temperature is ~225 K below that of c-Si [2, 12]. The melt front propagating at the a-Si melting temperature cannot propagate into c-Si until the laser deposits sufficient energy to increase the

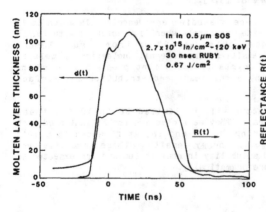

Fig. 1. a) Molten layer thickness d(t) and time-dependent surface reflectance R(t) for In-implanted Si melted with a 30 nsec pulse at 0.67 J/cm^2.

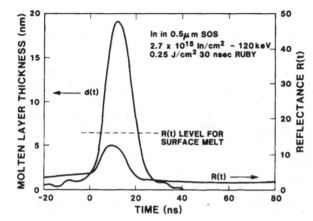

Fig. 2. Molten layer thickness d(t) and time-dependent reflectance R(t) for In-implanted Si irradiated with a 30 nsec pulse at 0.25 J/cm².

temperature of the molten layer to the melting temperature of c-Si. After sufficient energy has been absorbed, the interface again reverses direction and melts into the underlying c-Si.

The transient reflectance R(t) displays additional anomalous behavior. R(t) increases rapidly and reaches an initial plateau; however, this plateau occurs at a level below that for reflection from the molten phase. This plateau, together with the fact that the melt is thicker than the ~15-20 nm absorption depth of the probe laser, indicates that the melt nucleated inside the sample, producing a buried molten layer. Approximately 15 nsec later, the reflectivity increases to the value corresponding to a molten surface, indicating that the molten layer has reached the surface. The reflectivity remains high throughout the melt, indicating that the surface remains molten throughout this time. As the sample solidifies, R(t) decreases to a value consistent with that of c-Si.

Because the time-dependent reflectance indicates that melt nucleated internally, the transient conductance data should be interpreted as a molten layer thickness rather than a melt depth. One implication of this difference is that, because there are two liquid-solid interfaces, the derivative of the molten layer thickness versus time during melt-in cannot be interpreted simply as an interface velocity. This distinction is necessary until the melt reaches the surface and only the deeper liquid-solid interface remains. After the surface melts, $\partial d(t)/\partial t$ yields the true solidification velocity.

Internal nucleation of melt is observed for all energy densities sufficient for complete melt of the amorphous layer; it is also observed for irradiation at lower energy densities as illustrated by the 0.25 J/cm² data in Fig. 2. For reference, the full liquid reflectance level observed when the surface is molten is indicated. The surface reflectance never attains the metallic value for this laser energy density, which implies that either the surface never melted or that the thickness of the molten layer was always less than the absorption depth (~15-20 nm) of the probe laser. Since d(t) measurements show the maximum thickness of the molten layer to be 20 nm, these data imply that melt nucleated internally to produce a buried molten layer 20 nm thick. As will be discussed below, the redistribution of impurities under these irradiation conditions indicates that the buried molten layer actually propagates from the point of nucleation toward the surface.

B. Impurity Redistribution Measurements

Previous measurements of interfacial distribution coefficients and surface segregation have shown that irradiation of In-implanted Si at energy densities sufficient to melt to the underlying single crystalline material results in epitaxial solidification and partial zone refining of In to the surface. This surface segregation is illustrated by the data in Fig. 3 in which the In depth distribution after irradiation at an energy density of 1.50 J/cm² is compared to the as-implanted distribution. Although the solidification velocity for this particular sample was not directly measured, other samples irradiated under similar conditions show solidification velocities of 3.2 m/sec.

The impurity redistribution produced by a lower energy density irradiation of 0.25 J/cm², in which the melt front does not penetrate to the underlying crystalline interface, is shown by the spectra in Fig. 4. The spectra indicate that this melt caused In to migrate from the peak of the In distribution to the surface. This migration is qualitatively that expected for zone-refining under melt and solidification. Detailed examination of this spectrum, however, reveals that In has been removed from depths in excess of 60 nm, whereas the transient conductance data show that the molten layer thickness never exceeded 20 nm -- i.e., less than one-third of the depth over which In has been redistributed.

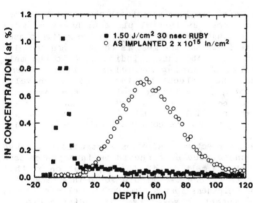

Fig. 3. Comparison of the In depth distribution after irradiation with a 30 nsec pulse at 1.5 J/cm² with as-implanted profile.

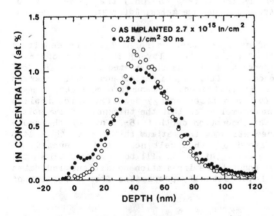

Fig. 4. Comparison of the In depth distribution after a single pulse at 0.25 J/cm² with the as-implanted profile.

Two important conclusions can be drawn from these data, in conjunction with the transient reflectance measurements. First, melt initiated near the peak of the In distribution, presumably because of the In-induced depression of the melting temperature. This internal nucleation of melt is similar to that observed for As in Si [3,4], which was also interpreted in terms of a reduced melting temperature. Second, after nucleation, the buried molten layer propagated toward the surface in a manner analogous to that observed in explosive crystallization of amorphous Si [2], although, in the present case, the buried layer propagates in the opposite direction. This difference can be understood as follows. In explosive crystallization, the melting temperature of a-Si is uniformly below that of c-Si, and the transformation is driven by the free energy difference between a- and c-Si. Since the surface nucleates the melt, the explosive crystallization propagates into the uniform a-Si layer. In the present case, however, the melting temperature profile is symmetric about the In peak concentration. The asymmetry in the propagation direction is provided by the laser heating, resulting in the melt front, and the corresponding transformation, proceeding toward the hotter part (surface) of the sample.

The maximum time rate of change of d(t) is ~3 m/sec. As noted above, because there are two moving liquid-solid interfaces, this derivative is not simply the interface velocity. The actual velocity of this molten layer can be estimated by comparing the melt duration from the transient conductance with the penetration depth estimated from changes in the In depth distribution. Specifically, the In has been redistributed over a depth of ~70 nm in a total solidification time of ~10 nsec. Under these assumptions, the velocity of the liquid-solid interface is ~70/10 = ~7 m/sec, which is more than twice the 3 m/sec obtained from d(t)/t. This value is still well below the maximum solidification velocity of 15 m/sec observed for crystalline Si [13] or the steady-state value of 13 m/sec for explosive crystallization of a-Si [14].

SUMMARY AND CONCLUSIONS

In summary, combined transient conductance measurements of the melt depth versus time and transient reflectance measurements of the melt duration at the surface reveal that melt nucleates internally in Si implanted with certain low solubility impurities. This phenomenon was studied in detail in In-implanted Si, and melt was found to nucleate internally under 30 nsec laser irradiation at 694 nm for all laser energy densities examined. Furthermore, at low laser energy densities, internal nucleation of melt was followed by an explosive-like crystallization process, mediated by a thin molten layer propagating toward the irradiated surface, that zone-refined impurities to the surface.

The discovery of these previously unexpected phenomena raises questions concerning the accuracy of interfacial distribution coefficients determined previously in pulsed laser melting studies. For example, data such as those in Fig. 3 are used to deduce interfacial segregation coefficients; however, because of internal nucleation of melt, these analyses are conceptually incorrect. At the very least, the melt duration of the peak of the impurity profile and in the near-surface region are incorrectly evaluated from numerical simulations in which melt is assumed to initiate at the surface. In addition, at lower laser energy densities, the impurity redistribution produced by the explosive-like crystallization has not previously been considered. (It should be noted that Bi is an exception to this behavior in that it does not appear to exhibit internal melt under similar irradiation conditions [15]). Not only might the distribution coefficients be in error, but the velocity dependence might

130

also be qualitatively incorrect if different laser conditions were used to vary the solidification velocity. Experiments are currently underway in In-implanted Si to evaluate the magnitude of these effects by studying the distribution coefficient as a function of velocity.

ACKNOWLEDGEMENTS

The authors would like to acknowledge the excellent technical assistance of Greg Brue in the transient conductance and reflectance measurements and Dan Buller in the ion backscattering analysis. This work was performed at Sandia National Laboratories supported by the U.S. Department of Energy under contract number DE-AC04-76DP00789.

†Permanent address: Department of Materials Science, Cornell University, Ithaca, New York.

REFERENCES

1. See, e.g., J. M. Poate and J. W. Mayer, in Laser Annealing of Semiconductors (Academic Press, New York, 1982).
2. Michael O. Thompson, G. J. Galvin, J. W. Mayer, P. S. Peercy, J. M. Poate, D. C. Jacobson, A. G. Cullis and N. G. Chew, Phys. Rev. Lett. 52, 2360 (1984).
3. P. S. Peercy, Michael O. Thompson and J. Y. Tsao, Appl. Phys. Lett. 47, 244 (1985).
4. P. S. Peercy and Michael O. Thompson, in Energy Beam-Solid Interactions and Transient Thermal Processing 1984 - MRS Proceedings 35, ed. by D. K. Biegelsen, G. A. Rozgonyi and C. V. Shank, (MRS, Pittsburgh, 1985), p. 53.
5. G. J. Galvin, Michael O. Thompson, J. W. Mayer, R. B. Hammond, N. Paulter and P. S. Peercy, Phys. Rev. Lett. 48, 33 (1982).
6. Michael O. Thompson, G. J. Galvin, J. W. Mayer, P. S. Peercy and R. B. Hammond, Appl. Phys. Lett. 42, 445 (1983).
7. D. H. Auston, C. M. Surko, T. N. C. Venkatesan, R. E. Slusher and J. A. Golovchenko, Appl. Phys. Lett. 33, 437 (1978).
8. Michael O. Thompson, PhD Thesis, Cornell University, 1984.
9. Liquid Semiconductors, ed. by V. M. Glazov, S. N. Chizkovskaga and N. N. Glagoleva (Plenum Press, New York, 1969).
10. See, e.g., C. W. White, B. R. Appleton and S. R. Wilson in Ref. 1, Chapter 5.
11. A. G. Cullis, H. C. Weber and P. Bailey, J. Phys. E12, 688 (1979).
12. E. P. Donovan, F. Spaepen, D. Turnbull, J. M. Poate and D. C. Jacobson, Appl. Phys. Lett. 42, 698 (1983).
13. Michael O. Thompson, J. W. Mayer, A. G. Cullis, H. C. Webber, N. G. Chew, J. M. Poate and D. C. Jacobson, Phys. Rev. Lett. 50, 896 (1983).
14. Michael O. Thompson, P. S. Peercy, Phys. Rev. B, to be published.
15. M. J. Aziz, J. Y. Tsao, Michael O. Thompson, P. S. Peercy and C. W. White, Appl. Phys. Lett., to be published.

DIRECT IMAGING OF "EXPLOSIVELY" PROPAGATING BURIED MOLTEN LAYERS IN AMORPHOUS SILICON USING OPTICAL, TEM AND ION BACKSCATTERING MEASUREMENTS*

D. H. Lowndes, G. E. Jellison, Jr., S. J. Pennycook, S. P. Withrow,
D. N. Mashburn, and R. F. Wood
Solid State Division, Oak Ridge National Laboratory, Oak Ridge, TN 37831

ABSTRACT

The behavior of pulsed laser-induced "explosively" propagating buried molten layers (BL) in ion implantation-amorphized silicon has been studied in a time- and spatially-resolved way, using nanosecond time-resolved reflectivity measurements, "Z-contrast" scanning transmission electron microscope (STEM) imaging of implanted Cu ions transported by the BL, and helium ion backscattering measurements. Infrared (1152 nm) reflectivity measurements allow the initial formation and subsequent motion of the BL to be followed continuously in time. The BL velocity is found to be a function of both its depth below the surface and of the incident KrF laser energy density (E_ℓ); a maximum velocity of about 14 m/s is observed, implying an undercooling-velocity relationship of about 14 K/(m/s). Z-contrast STEM measurements show that the final BL thickness is less than 15 nm. Time-resolved optical, TEM and ion backscattering measurements of the final BL depth, as a function of E_ℓ, are also found to be in excellent agreement with one another.

INTRODUCTION

During the past several years, techniques for transient electrical conductance (TEC) [1-4] and optical reflectivity and transmission [5-8] measurements have been used to study the dynamics of rapid melting and solidification in both crystalline (c) and amorphous (a) silicon, following pulsed laser melting of a shallow (<1 μm) near-surface region. For crystalline semiconductors, TEC measurements provide a powerful, relatively simple means for determining the position and velocity of the liquid-solid interface and the depth of melting, since the melt is continuously-connected to the surface and its electrical conductivity can be assumed to be known.

However, the interpretation of TEC measurements on amorphous semiconductors is complicated by the fact that a thin, propagating "buried molten layer" (BL), isolated from the surface, may be formed. In such a case, TEC measures the total thickness of molten material present at any time, but not its depth or velocity. For example, TEC measurements during pulsed ruby laser (694 nm, 28 ns FWHM) melting of a-Si layers at low laser energy densities (E_ℓ = 0.2–0.3 J/cm²) [1] reveal a double peak in the electrical conductance vs time, the second peak occurring after the surface has solidified (as determined by simultaneous transient reflectance measurements). Thompson et al. [1] interpret the second conductance peak as indicating ℓ-Si in the interior of the sample, a BL [7]. From the magnitudes of the two conductance peaks they estimate the initial melt depth as 12 nm (at 0.2 J/cm²) and the thickness of the buried layer (BL) as 18 nm. Cross-sectional TEM micrographs following low E_ℓ irradiation of a-Si layers (insufficient to melt entirely through the a-Si layer) typically show a region of large-grained polycrystalline Si (LG p-Si) near the surface and fine-grained (FG) p-Si at greater depth [9]. Thompson et al. hypothesize that recrystallization of

*Research sponsored by the Division of Materials Sciences, U.S. Department of Energy, under Contract No. DE-AC05-84OR21400 with Martin Marietta Energy Systems, Inc.

the original a-Si layer occurs in two steps: The LG p-Si region results from solidification of a thin, "primary" melt. The latent heat released during this process raises the temperature of the LG p-Si above the melting point (T_a) of a-Si and the a-Si beneath begins to melt, resulting in formation of a thin, nearly self-propagating, molten layer whose forward motion is driven by the difference in the heats of fusion of the a- and c-phases ($L_a \approx 0.75$ L_c [10]). This process of "explosive crystallization" of amorphous materials has been known for a long time [11], though the suggestion that its mechanism sometimes involves a thin, molten layer is recent [12,13]. TEC measurements also show [1] that T_a lies about 200 (± 50) K below the c-Si thermodynamic melting point (T_c), in agreement with earlier free energy measurements that indicated a T_a depression of several hundred degrees [10].

The buried molten layer produced in laser-irradiated a-Si is expected to be highly undercooled relative to T_c; Thompson et al. postulate that the FG p-Si observed via TEM is formed by recrystallization at the back interface of this propagating undercooled molten layer. From the duration of the second conductance peak and the FG p-Si layer thickness observed in TEM they inferred that the BL velocity lies in the 10—20 m/s range, which is consistent with crystal growth occurring at an undercooled liquid temperature near the depressed melting point, T_a. (The undercooling vs velocity relationship has been estimated to lie in the range of 17 K per m/s, or less, for Si [14].)

As a complement to TEC, time-resolved optical reflectivity (R) and transmission (T) measurements have several advantages for dynamical studies of the formation and propagation of the BL in a-Si [7]. Because the 1152 nm HeNe wavelength is only weakly attenuated by either c- or a-Si, interference maxima and minima appear in R(1152 nm) due to reflections by the sample surface and by the BL moving beneath the surface; from these the BL position can be directly determined as a function of time, so long as the depth of any additional surface melt is not too great.

In this paper we present the first direct measurements of BL velocity in a-Si and of its variation with depth. We also present direct, high resolution measurements of the final BL thickness: These thickness measurements were obtained by chemical "Z-contrast" STEM imaging, following low-E_ℓ irradiations of Cu-implanted a-Si specimens. The low interface segregation coefficient of Cu in Si results in substantial liquid phase transport of Cu both back to the surface and deep into the specimen, via the BL. Thus, the Z-contrast image of the Cu deposited when the BL comes to rest provides an upper limit on the final thickness of the BL. Finally, we also present He ion backscattering measurements of the E_ℓ-dependence of Cu redistribution back to the surface and forward via the BL. The combination of time-resolved optical, TEM and RBS measurements is shown to provide a consistent picture of the motion of pulsed laser-induced BLs in a-Si at low E_ℓ. Thus, there seems to be no doubt that formation of FG p-Si is mediated by a thin, rapidly moving BL and not by a volume nucleation process [5,6].

MEASUREMENTS OF THE VELOCITY AND THICKNESS OF BURIED MOLTEN LAYERS

A KrF excimer laser (248 nm, 45 nsec FWHM pulse duration) was used to irradiate a-Si layers that were produced by 80- and 180-keV Si ion implantation (at 77 K) at a dose of $10^{16}/cm^2$ into (100) c-Si substrates. Some of the 180-keV Si-implanted specimens were pre-implanted with $5 \times 10^{15}/cm^2$ of Cu ions, using five different implantation energies to produce a "flat" Cu implantation profile that was confined to the top 300 nm. The a-layer thicknesses determined by TEM were 195 nm and 410 nm ($\pm 5\%$), for the 80- and 180-keV (both with and without Cu) implants, respectively. Time-resolved reflectivity measurements were carried out simultaneously at 633 and 1152 nm using low-power HeNe probe lasers that were aligned to be accurately colinear and centered on the same spot; the $1/e^2$ diameter for both probe beams was measured to be about 60 μm in the plane of the sample's front surface.

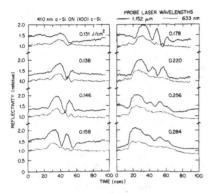

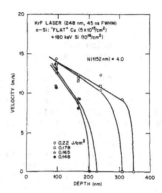

Figure 1. Reflectivity (in arbitrary units) as a function of time and KrF laser E_ℓ at visible (633 nm) and infrared (1152 nm) probe laser wavelengths (see text).

Figure 2. Buried layer velocity vs depth and laser E_ℓ, calculated as described in the text. The data points are plotted at depths corresponding to the calculated mean depth of occurrence of successive pairs of 1152 nm reflectivity minima and maxima, in the absence of any surface melt. The points on the abscissa correspond to the final depth of the buried layer.

Figure 1 shows the 633 nm and 1152 nm R signals following irradiation of the (Cu + 180 keV Si)-implanted specimens at various E_ℓ. For $E_\ell < 0.16$ J/cm^2, strong interference maxima and minima are seen to modulate both R signals, due to interfering reflections from the sample surface and from the BL moving away from the surface. At higher E_ℓ (Fig. 1), the duration of high reflectivity is much longer at 1152 nm than for 633 nm, and the 1152 nm signal still exhibits deep interference maxima and minima when the BL is at depths for which the 633 nm beam is strongly attenuated. Its attenuation results in part from a thin molten region at the surface (particularly at earlier times) but is also due to the hot p-Si that crystallizes behind the moving BL (especially at later times). The depth of the BL beneath the surface actually varies slightly (see below) within the probe laser beam diameter; this causes a further "phase smearing" reduction of the oscillatory R signals that is greatest at 633 nm [15]. The rapid decay of the 633 nm R signal, after its initial peak, shows that at low E_ℓ the surface has already solidified at times when the BL is still moving and is easily observable at 1152 nm (e.g., $E_\ell = 0.158$ or 0.178 J/cm^2 in Fig. 1). In this case the BL is separated from the sample's surface only by a layer of recrystallized FG p-Si, so that the velocity of the BL, and its variation with depth, can be obtained directly from the measured time intervals between successive 1152 nm interference maxima and minima in Fig. 1, which correspond to optical path differences of ¼-wave in the FG p-Si. Figure 2 is a plot of BL velocity vs depth obtained in this way, using the index of refraction N(1152 nm) = 4.0 for FG p-Si [15,16] at $T \leq T_c$ (¼-wavelength $\approx$ 72 nm at 1152 nm wavelength). We have observed a similar velocity variation using these same techniques with implantation-amorphized silicon-on-sapphire specimens [15].

Figure 2 shows that the maximum BL velocity (13–14 m/s) observed in these experiments is just slightly below the limiting velocity of 15 m/s, above which direct amorphous regrowth from an undercooled Si melt occurs in the (100) direction [2]. A maximum BL velocity of 14 m/s also implies an undercooling-velocity dependence < (200 K)/(14 m/s) $\approx$14 K/(m/s), if the BL temperature is ~ T_a. Figure 2 also shows that the BL velocity is a function both of depth and (weakly) of the incident laser E_ℓ.

It was recently shown [17] that heavy dopants in Si can be directly imaged by detecting Rutherford-scattered transmitted electrons in a scanning transmission electron microscope (STEM), using an annular detector placed around the transmitted beam. By collecting electrons scattered at large angles ($\gtrsim$ 0.1 rad) one obtains an image intensity that is proportional to $\Sigma \, n_i Z_i^2$, where n_i is the number of atoms of atomic number Z_i under the electron probe. Very strong Z-contrast imaging is obtained, and the contrast in the image accurately reflects dopant concentration variations at peak doping levels of order 1%, provided that care is taken to avoid channeling of the incident beam which occurs in crystalline samples oriented near low order Bragg reflections.

Because Cu has very low solid solubility in Si (and a low interface segregation coefficient), pulsed laser melting of Cu-implanted a-Si is expected to result in substantial surface segregation of Cu. For the same reason, if a BL is generated in Cu-implanted a-Si, then segregation of Cu to a point deep inside the original a-layer should also occur, as the Cu is transported by the propagating BL. Two other groups have recently used Rutherford backscattering spectrometry (RBS) measurements to observe such a deep Cu segregation peak and have pointed out that it provides primary evidence that explosive crystallization is mediated by a thin, propagating molten layer [18,19]. However, the depth resolution of RBS measurements is insufficient to provide precise information about the thickness of the BL (see below), so we have instead obtained direct measurements of the BL thickness from Z-contrast STEM.

Figure 3 shows Z-contrast STEM images of Cu that was transported by the BL, following irradiation of (Cu + 180 keV Si)-implanted specimens at 0.143 and 0.167 J/cm^2; the corresponding conventional cross-sectional TEM images are also shown. Comparison of the pairs of images in Fig. 3 shows that the Cu image of the BL does lie at the boundary between the FG p-Si and the remaining, unmelted a-Si, consistent with the model of Thompson et al. [1]. The width of the image of the deposited Cu provides a direct measure of the thickness of the BL at the time that it solidifies. A line scan perpendicular to the Cu image in Fig. 3 gave a final BL thickness of 15 nm, slightly thinner than the initial 18 nm maximum thickness estimated from TEC measurements on Si-implanted SOS [1]. Our Z-contrast measurements actually provide an upper limit on the final BL thickness, because of averaging over the non-planar BL interface (Fig. 3) through the finite specimen thickness and the ~2.5 nm electron probe beam diameter. Quantitative Z-contrast analysis revealed a Cu concentration $\geq$0.4 at. % in the BL.

ION BACKSCATTERING MEASUREMENTS; BURIED LAYER DEPTH VS E_ℓ

Figure 4 shows results of 2.0 MeV He ion backscattering measurements of Cu segregation, both toward the surface and toward the a-c interface, following pulsed laser irradiation of Cu-implanted specimens. In order to accentuate the net transport of Cu by the BL, we have plotted in Fig. 4 only the differences of the Cu distributions, as a function of depth, for the as-implanted and laser-irradiated specimens. The net transport of Cu to greater depth with increasing E_ℓ, throughout the low E_ℓ range, and the beginning of substantial surface segregation at ~0.2 J/cm^2 are both clear. However, the width of the region in which backscattering shows a net increase in Cu is not a measure of the BL thickness, because of the nonplanar BL interface; backscattering necessarily averages over the different BL depths found within the ~1 mm ion beam diameter. Taking into account the backscattering instrumental depth resolution of ~21 nm at the depth of the BL, and assuming the 15 nm BL thickness measured by TEM, the backscattering measurements imply a ±24 nm (RMS) variation in the mean BL depth at 0.146

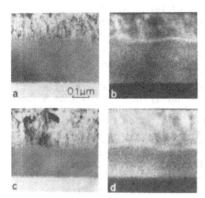

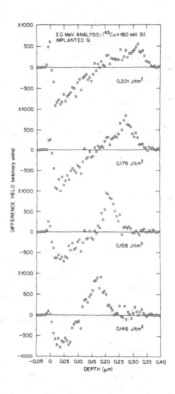

Figure 3. (Left) Conventional cross-sectional TEM images and (right) Z-contrast STEM images of (180 keV Si + Cu)-implanted Si, following KrF laser irradiation at 0.143 (a and b) and 0.167 (c and d) J/cm^2.

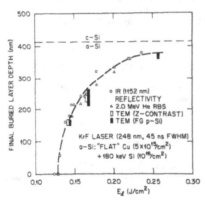

Figure 4. Change in Cu distribution vs depth, for various E_ℓ. (The difference yield is defined as Cu[laser-irradiated specimen] - Cu[as-implanted specimen].)

Figure 5. Terminal depth of the back (deepest) side of the buried layer as a function of laser E_ℓ, determined by 1152 nm optical reflectivity, Z-contrast STEM, cross-sectional TEM and RBS imaging (see text); the dashed curve is only a guide for the eye.

J/cm^2, in excellent agreement with cross-section TEM measurements of the variation in the FG p-Si depth at different locations. The total Cu backscattering yield was constant for E_ℓ < 0.20 J/cm^2 but decreased for E_ℓ > 0.25 J/cm^2, indicating the onset of Cu loss by vaporization from the surface in the same E_ℓ range for which surface segregation of Cu becomes significant.

Figure 5 summarizes our measurements of the terminal depth of the BL in the (180 keV Si + Cu)-implanted specimens, as a function of laser E_ℓ, using a variety of experimental probes. The optical measurements of BL depth were made by counting the number of interference oscillations (or fraction thereof) in the 1152 nm R data (Fig. 1), taking 1/4-wave = 72 nm and with the first interference minimum occurring when the top of the BL is 59 nm below the surface. (This was determined from model calculations [15].) Two types of TEM data are also included in Fig. 5: direct Z-contrast imaging of the Cu transported by the BL and TEM cross-section (structural) imaging of the boundary between the FG p-Si and the remaining unmelted a-Si. RBS measurements of the redistributed Cu profiles (Fig. 4) are also shown. The agreement between the different probes of BL depth in Fig. 5 is remarkably good, providing reassurance that our modeling of the optical properties of l-Si, a-Si and FG p-Si (at least) is not seriously in error. Figure 5 nicely illustrates the ease with which the BL penetrates deep into the original a-Si layer even at low E_ℓ, and the plateau in BL penetration that occurs with increasing E_ℓ as the a-c crystalline interface is approached [1]. Model calculations show that these phenomena result from the very low thermal conductivity [5] of the thick layer of a-Si that remains unmelted at low E_ℓ and from the quenching of explosive crystallization by thermal conduction to the c-Si substrate, as the a-c interface is approached at higher E_ℓ.

REFERENCES

1. M. O. Thompson, G. J. Galvin, J. W. Mayer, P. S. Peercy, J. M. Poate, D. C. Jacobson, A. G. Cullis, and N. G. Chew, Phys. Rev. Lett. 52, 2360 (1984).
2. M. O. Thompson, J. W. Mayer, A. G. Cullis, H. C. Webber, N. G. Chew, J. M. Poate, and D. C. Jacobson, Phys. Rev. Lett. 50, 896 (1983).
3. G. J. Galvin, M. O. Thompson, J. W. Mayer, R. B. Hammond, N. Paulter, and P. S. Peercy, Phys. Rev. Lett. 48, 33 (1982).
4. P. S. Peercy and M. O. Thompson, Mat. Res. Soc. Symp. Proc. 35, 53 (1985).
5. D. H. Lowndes, R. F. Wood and J. Narayan, Phys. Rev. Lett. 52, 561 (1984).
6. R. F. Wood, D. H. Lowndes and J. Narayan, Appl. Phys. Lett. 44, 770 (1984).
7. D. H. Lowndes, G. E. Jellison, Jr., R. F. Wood, and S. J. Pennycook, and R. F. Carpenter, Mat. Res. Soc. Symp. Proc. 35, 101 (1985). Note: The 190 nm a-layer thickness mentioned in the caption to Fig. 4 of this reference (and in its discussion in the text) is incorrect; the correct thickness was 440 nm.
8. D. H. Lowndes, G. E. Jellison, Jr. and R. F. Wood, Phys. Rev. 26B, 6747 (1982).
9. J. Narayan and C. W. White, Appl. Phys. Lett. 44, 35 (1984).
10. E. P. Donovan, F. Spaepen, D. Turnbull, J. M. Poate, and D. C. Jacobson, Appl. Phys. Lett. 42, 698 (1983).
11. G. Gore, Phil. Mag. 9, 73 (1855).
12. G. H. Gilmer and H. J. Leamy, in Laser and Electron Beam Processing of Materials (ed. by C. W. White and P. S. Peercy, Academic Press, New York, 1980), p. 227.
13. H. J. Leamy, W. L. Brown, G. K. Celler, G. Foti, G. H. Gilmer, and J. C. C. Fan, Appl. Phys. Lett. 38, 137 (1981).
14. M. O. Thompson, P. H. Bucksbaum and J. Bokor, Mat. Res. Soc. Symp. Proc. 35, 181 (1985).
15. D. H. Lowndes, G. E. Jellison, Jr., S. J. Pennycook, R. F. Wood, S. P. Withrow, and D. N. Mashburn, in preparation for The Physical Review.
16. G. E. Jellison, Jr. and H. H. Burke, manuscript in preparation.
17. S. J. Pennycook and J. Narayan, Appl. Phys. Lett. 45, 385 (1984).
18. J. Narayan, C. W. White, O. W. Holland, and M. J. Aziz, J. Appl. Phys. 56, 1821 (1984).
19. W. Sinke and F. W. Saris, Phys. Rev. Lett 53, 2121 (1984).

PHASE DIAGRAM OF LASER INDUCED MELT MORPHOLOGIES ON SILICON

JOHN S. PRESTON, JOHN E. SIPE AND HENRY M. VAN DRIEL
Dept. of Physics and Erindale College, University of Toronto, Toronto,
Canada, M5S 1A7

ABSTRACT

During cw laser induced melting, over a large range of intensities, silicon films phase separate into patterns of coexisting solid and molten regions.
We have identified several distinct and reproduceable morphologies of this
inhomogeneous coexistence region ranging from random lamellae ("amorphous")
structures to periodically ordered ("crystalline") strips. The type of morphology formed is a function of laser intensity and spot size and these parameters can be viewed as constituting the two axes for a steady state "phase
diagram" of the structures. "Phase transitions" between these structures occur
for small changes of the experimental parameters with the fraction of liquid
being an order parameter.

INTRODUCTION

The linear optical properties of many materials change discontinuously
upon melting. As a result, such materials exhibit manifestly nonlinear behavior during laser induced melting. Instabilities can occur [1,2,3] which
lead to a breaking of the surface symmetry through a coupling of coherent,
polarized electromagnetic radiation to the surface morphology of the material.
For silicon, in particular, a simple argument for the source of these instabilities can be given for cw laser irradiation. There exists a range of intensities for which the solid surface will be significantly overheated and hence
melt, while a uniformly molten surface would become undercooled and hence resolidify, due to the significantly higher reflectivity of the melt. Since a
uniform solution is not allowed, a mixed state of coexisting solid and molten
regions must form. This argument, based on the reduced coupling of the molten
surface to the laser radiation, neglects the continuous transition of the reflectivity from the solid to the molten value due to skin depth effects.
However, we have previously shown [4] that although a thin uniform melt can
satisfy the relevant coupled electrodynamic and thermal transport equations,
such a solution is unstable with respect to infinitesimal periodic perturbations in the melt depth. Hence such perturbations grow and evolve, breaking
the translational symmetry of the surface. Bosch and Lemons [5] and Biegelsen
and his coworkers [6,7] have demonstrated such mixed states experimentally for
silicon films.

We have studied the evolution of the steady state of silicon films on
sapphire substrates as a function of the incident laser intensity and spot
size. It became evident during this work that within the mixed state regime,
there are several subregions, each of which is characterized by a significantly different steady state morphology. The existence of these subregions was
highlighted by the relatively sharp boundaries separating them. These observations have led us to borrow concepts from thermodynamics and describe the
subregions of the phase diagram and the boundaries between them as nonequilibrium phases and phase transitions, respectively, of silicon films during
laser irradiation.

EXPERIMENTAL DETAILS

The experiments were conducted by laser irradiating 0.5 μm thick films of silicon on sapphire substrates. The laser was a linearly polarized 20W CO_2 laser operating at $\lambda = 10.6$ μm. The laser's output was in a TEM_{00} transverse mode and single longitudinal mode. The samples were set on a movable resistive heating stage with the temperature being adjustable up to 700°C. The laser's output was focussed onto the silicon film using a 10 cm focal length ZnSe lens with the position of the lens being variable so that the location and size of the irradiated spot could be manipulated. The mixed states of the silicon film were observed through the back of the sapphire substrate using a 10x or 20x microscope objective. The illumination could be provided either by reflected light or the sample's own blackbody emission which is quite intense in the vicinity of the melting point of 1415°C. The microscope was configured such that the resulting image could be viewed directly or stored either as a still photograph or on tape via a video camera.

It is clear that for cw laser irradiation the energy balance equations for the liquid and solid regions place important constraints on any theoretical treatment of this system. The magnitude of the source term is determined by the intensity of the incident radiation and its coupling to the surface morphology of the sample. The loss term will be due to blackbody emission, thermal conduction through the film and thermal conduction into the sapphire. Simple estimates indicate that for our geometry and experimental conditions the latter mechanism dominates. The conditions for inhomogeneous melting of semiconductors are greatly different for cw and pulsed laser irradiation. In the pulsed case the inhomogeneous energy deposition is almost completely determined by the surface electrodynamics and heat flow plays little role in influencing optical feedback and hence pattern formation. Indeed, for nanosecond pulse illumination, we have shown [3] that the patterns are nearly always in the form of ripples whose morphology is determined entirely by the laser characteristics. While we defer to a later publication the detailed discussion of the manner in which the cw energy balance can be incorporated into a complete theory, we note here that the two most salient experimental parameters to vary are the incident intensity and the efficiency of thermal conduction in removing heat. The rate of heat loss due to conduction into the sapphire will not be particularly sensitive as to whether the local state of the system is solid or liquid. The heat loss efficiency can clearly be controlled by adjusting the temperature of the sample holding stage. A second, less obvious technique involves changing the size of the focussed laser spot. A simple dimensionality argument indicates that for a given intensity, the heat generated will be conducted away more efficiently from a smaller spot. Both techniques were found to produce similar results; however, we will report on experiments in which spot size was altered since a larger variation in the thermal loss could be affected in this manner.

EXPERIMENTAL RESULTS

The steady states that were observed as a function of the incident intensity and spot size are shown in Fig. 1. Within region II the melt is confined to randomly distributed regions of approximately 3 μm diameter. Since this behavior is observed only for large spot sizes, where the temperature gradients and heat flow are low, it is possible that region II is not a phase of the system but rather appears because the system cannot, within laboratory times, reach its steady state. In region III a grating-like structure exists with a periodicity equal to the incident wavelength of 10.6 μm to within experimental error. In region IV a grating with a periodicity of 2λ is formed. A sequence of still higher harmonics of the original gratings occurs within region IV. The blackbody emission from a mixed state which exhibits two of the grating-like structures is shown in Fig. 2. The dark regions in the photograph correspond to molten silicon due to its lower emissivity. Gratings with a spacing of λ (region III) cover most of the spot. However, near the center the onset of the doubled grating structure (region IV) can be seen.

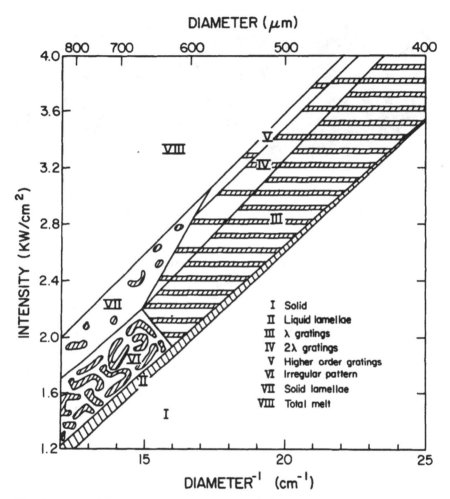

Fig. 1. Nonequilibrium phase diagram for silicon under laser irradiation.

Both phases exist in the same steady state due to the laser intensity being higher at the center of the Gaussian spot. Irregular patterns of liquid and solid with an average spacing of approximately λ occur in region VI. The morphology of these patterns is indicated in Fig. 2b. Within region VI, as the intensity is increased, the topology of the structure undergoes a rather subtle transition from isolated liquid regions to isolated solid regions. In region VII the isolated solid regions are sufficiently separated to move independently, interacting only if the separation is less than λ.

From Fig. 1, we see that the grating-like structures are associated with a relatively large heat loss efficiency. Experimentally a large heat loss can be expected to result in a rapid evolution of the structures since the time scale of the system is determined by the ratio of the total energy of the

140

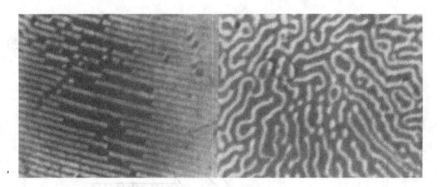

Fig. 2. Photographs of black-body radiation from solid (white regions) and liquid morphology depicting a) the ripple period doubling in going from region III to region IV and b) the random nature of region VI; spatial variations are on a scale of λ.

system to the energy flux. We find that the grating structure can be produced by manipulating the laser spot, whereas the disordered structures observed in region VI and VII may not be truly steady states, but rather the metastable "amorphous" phases associated with the "crystalline" grating phase. The fraction of liquid increases with increasing laser intensity as expected. However, this increase is not smooth but rather occurs in sporadic jumps as each new pattern emerges. We can thus identify the fraction of liquid as an order parameter whose change marks the onset of a transition between morphological phases. In Fig. 4, the fraction of liquid is plotted as a function of laser intensity for two different spot sizes. The discrete jumps in Fig. 3a occur as a result of the grating period changing. It is worth noting that the change in the fraction of liquid that can be attributed to changes in the width of the solid regions was found to be negligible in these experiments, resulting in the striking plateaus seen in the figure. In the corresponding plot for the "amorphous" structures, Fig. 3b, the fraction of liquid increases smoothly although some remnant of the plateaus remain.

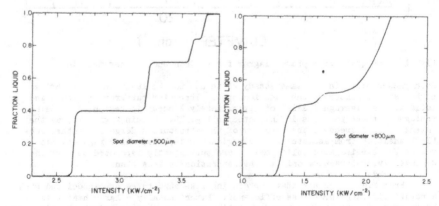

Fig. 3. The fraction of liquid versus the laser intensity for a spot size of a) 500 μm and b) 800 μm.

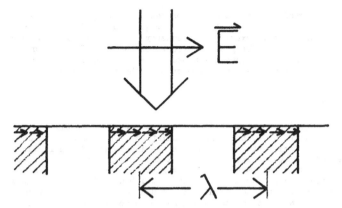

Fig. 4. Diagram used to explain interference shielding.

DISCUSSION

Any fundamental understanding of this system's evolution will require a description of the coupling of the incident fields to the surface structures. It is clear that geometrical optics cannot be used to describe the coupling since the sizes of the structures formed are on the order of λ. At 10.6 μm the optical properties of the crystalline and molten silicon correspond to a dielectric ($\hat{\varepsilon} = 7 + i10$) and a metal ($\hat{\varepsilon} = -30 + i800$) respectively. Thus our original perturbation approach [4], while capable of describing the initial symmetry breaking instabilities, is inappropriate for the description of the final states of the system for cw illumination. Instead we have developed a scheme for calculating the fields that accounts for the large differences in the solid and melt optical properties. To understand the concepts involved consider radiation incident on a metal at normal incidence. The applied field induces a uniform current density within the skin depth of the metal. The absence of any fields penetrating to the interior of the metal can be considered to result from the complete destructive interference of the incident fields and the fields generated by the induced currents. Fig. 4 depicts the situation if the radiation is incident on metallic strips separated by λ and oriented perpendicular to the polarization. Again a current density is induced within the strips which will result in radiated fields. Now consider the Fourier transform of the induced current density and retain the two leading terms. The second term, which has the periodicity of the incident radiation, is clearly phase matched to radiate along the surface, perpendicular to the strips. It is, however, unable to radiate since that direction is parallel to the current density. Thus the uniform component of the current density must be responsible for cancelling the incident field in the interior of the metal, and since it is uniform it must also cancel the incident radiation in the gaps between the strips. Thus within this picture the metallic strips shield the intermediate regions. In our calculation scheme, we approximate the response of the surface to the incident radiation by a uniform polarization within the solid regions, a current density at the top of the molten regions and a current density at the melt substrate interface. We are then able to calculate the "best" values of these parameters via the variational principle generalized from that described in Ref. 3. Once the polarization and current densities are known it is straightforward to calculate the fields produced. Although the inclusion of the substrate and the solid's dielectric constant add complications, the presence of the metallic strips separated by λ still significantly reduce the fields within, and hence the energy deposited in, the solid regions. We refer to this mechanism as interference shielding. As might be expected from the argument presented above, interference shielding is most effective for an array of metallic regions aligned perpendicular to the polarization, which is exactly the geometry of the grating structures. The formation of higher order grating structures can be understood qualitatively since the Fourier components discussed above will still be present if we consider more than the first two components of the higher order grating.

CONCLUSIONS

In conclusion we have studied the steady states associated with the melting of silicon films with CO_2 laser radiation. Our results indicate that the nature of the steady states depends on the intensity of the incident radiation and the efficiency of heat removal from the irradiated spot. By varying these two parameters we have mapped the steady states onto a nonequilibrium, but steady state, phase diagram. Parameters which describe the steady states, in particular the fraction of liquid, are found to change discontinuously upon crossing between two such phases and hence represent the order parameters of the system. The stability of the most prevalent phase, the grating structure, can be understood as resulting from the metallic molten regions shielding the solid silicon, within the framework of a simplified electrodynamic theory.

We would like to acknowledge the generosity of D. Biegelsen and D. Guidotti for providing the silicon on sapphire samples used in this work. This work is supported by the Natural Sciences and Engineering Research Council of Canada, including a postgraduate scholarship to J.P.

REFERENCES

1. J. F. Young, J. E. Sipe, M. I. Gallant, J. S. Preston and H. M. van Driel in Laser and Electron-Beam Interactions with Solids, B. R. Appleton and G. K. Celler eds. (North Holland, New York, 1982) pp. 233.

2. P. M. Fauchet, Z. Guosheng and A. E. Siegman in Laser-Solid Interactions and Transient Thermal Processing of Materials, J. Narayan, W. L. Brown and R. A. Lemons eds. (North Holland, New York, 1983) pp. 205.

3. J. E. Sipe, J. F. Young, J. S. Preston and H. M. van Driel, Phys. Rev. B 27, 1411 (1983).

4. J. E. Sipe, J. F. Young and H. M. van Driel in Laser-controlled Chemical Processing of Surface, A. W. Johnson, D. J. Ehrlich and H. R. Schlossberg eds. (North Holland, New York, 1984) pp. 415.

5. M. A. Bosch and R. A. Lemons, Phys. Rev. Lett. 47, 1151 (1981).

6. D. K. Biegelsen, N. M. Johnston, W. G. Hawkins, L. E. Fennell and D. M. Moyers in Laser and Electron-Beam Interactions with Solids, B. R. Appleton and G. K. Celler eds. (North Holland, New York, 1982) pp. 537.

7. R. J. Nemanich, D. K. Biegelsen and W. G. Hawkins in Laser and Electron-Beam Interactions with Solids, B. R. Appleton and G. K. Celler eds. (North Holland, New York, 1982) pp. 211.

TIME-RESOLVED REFLECTIVITY MEASUREMENTS OF SILICON AND GERMANIUM USING A PULSED EXCIMER LASER*

G. E. Jellison, Jr., D. H. Lowndes, D. N. Mashburn, and R. F. Wood
Solid State Division, Oak Ridge National Laboratory, Oak Ridge, TN 37831

ABSTRACT

Time-resolved reflectivity measurements of silicon and germanium have been made during pulsed KrF excimer laser irradiation. The reflectivity was measured simultaneously at both 1152 and 632.8 nm wavelengths, and the energy density of each laser pulse was monitored. The melt duration and the time of the onset of melting were measured and compared with the results of melting model calculations. For energy densities just above the melting threshold, it was found that the melt duration was never less than 20 ns for Si and 25 ns for Ge, while the maximum reflectivity increased from the value of the hot solid to that of the liquid over a finite energy range. These results, along with a reinterpretation of earlier time-resolved ellipsometry measurements, indicate that, during the melt-in process, the near-surface region does not melt homogeneously, but rather consists of a mixture of solid and liquid phases. The reflectivity at the onset of melting and in the liquid phase have been measured at both 632.8 and 1152 nm, and are compared with the results found in the literature.

INTRODUCTION

Pulsed laser melting of semiconductors has been the object of an extensive study for the last 7–8 years. Because of this work, the nature of the pulsed laser annealing process is well-known: if a light pulse of sufficient energy density (E_ℓ) is incident upon the surface of a semiconductor, the near-surface of the sample will melt [1]. The time of onset of melting and the melt duration will depend in a complicated fashion on the parameters of the laser pulse, including its wavelength, pulse duration and shape, and E_ℓ. Detailed melting model calculations have been successfully carried out to describe the results of time-resolved reflectivity measurements using a ruby laser [2], and a Nd:YAG frequency-doubled laser [3], as well as the results of time-resolved electrical conductivity measurements using a ruby laser [4].

In this paper, we present the results of a series of time-resolved reflectivity (TRR) measurements during pulsed KrF (248 nm) excimer laser irradiation of both silicon and germanium. The TRR measurements were performed at two different wavelengths (632.8 nm and 1152 nm), using cw laser beams focused on the same spot. Detailed melting model simulations were also carried out to compare with the experiments. This work represents the first detailed and extensive TRR measurements of germanium and silicon irradiated with an excimer laser, which provides an independent check on both the generality and the details of melting model calculations. Since these measurements have been carried out using an excimer laser, the pulse energy and shape are reasonably reproducible (unlike ruby or doubled Nd:YAG) and the absorption length in both silicon and germanium is very short (< 10 nm). Perhaps the strongest motivation for carrying out these measurements is that excimer lasers are potentially very useful for semiconductor processing, and the data presented here should be useful for these applications.

---

*Research sponsored by the Division of Materials Sciences, U.S. Department of Energy, under Contract No. DE-AC05-84OR21400 with Martin Marietta Energy Systems, Inc.

EXPERIMENT

The Lambda-Physik 210 KrF excimer laser (248 nm = 5 eV) used for the experiments has a pulse width of ~38 ns full width at half maximum, and can produce up to 1.3 J/pulse. Energy densities up to 1.8 J/cm^2, uniform to ~5% over an area ~3 mm in diameter were obtained by focusing the incident laser beam. A typical time profile of the pulse from this laser is shown in the bottom panel of Fig. 1. A beam splitter (Supracil polished on both sides) was placed in front of the sample to monitor the energy of each laser pulse with a microjoule meter.

The TRR measurements were made using two probe laser beams of 632.8 nm and 1152 nm. The two beams were coincident on the sample surface (within ~30 μm) and were focused to a spot size <50 μm diameter. The reflected light was collected simultaneously using a polished Si beam splitter and two Si avalanche photodiodes (APD's), connected to two Tektronix 7912A/D waveform digitizers. The digitized data was transferred to a laboratory computer for convenient data acquisition and analysis. The rise time of the entire system was ~1 ns.

During the course of these experiments, it was found that it was necessary to clean all samples with appropriate solvents to remove any organic surface contaminants just before the experiment. In addition, it was necessary to remove the surface oxide from the Ge samples within a few hours of the experiment. This is necessary because Ge tends to oxidize quickly, and the Ge oxide (unlike SiO_2) is not transparent to the 248 nm light. In addition, excimer laser irradiation of Ge with energy densities large enough to melt the front surface region results in a significant growth of oxide. (~100 shots results in ~80 nm of oxide.) Therefore, we had to be careful to irradiate each spot on the sample only once. Although this problem is much less important for the case of silicon, the same precaution was taken for the silicon experiments.

RESULTS AND DISCUSSION

Figure 1 shows the reflectivity of both germanium (top) and silicon (bottom) as a function of time for several different E_ℓ's near the melting threshold. The reflectivity of silicon increases monotonically with time during the heating by the laser pulse, until the surface melts, which results in a large increase of the reflectivity. On the other hand, the reflectivity of germanium at 632.8 nm first increases and then decreases with time during laser heating (this is not clear from Fig. 1, but becomes obvious on an expanded scale), until the surface becomes molten, which again results in a large increase in the reflectivity. The different behavior for germanium in the solid phase is due to the close proximity of the E_1 transition (~2.10 eV at 300 K) to the probe wavelength used (632.8 nm = 1.96 eV). As the temperature of the sample is increased, the E_1 transition will decrease in energy, passing through the cw laser photon energy at ~550 K [5].

Two observations can be made from the data presented in Fig. 1 for energy densities just above the threshold for melting: 1) the maximum reflectivity increases with increasing energy density from the value expected for the hot solid to that expected for the liquid over a finite energy range (from 0.32 to 0.40 J/cm^2 for Ge, and from 0.70 to 0.85 J/cm^2 for Si). 2) The duration of the partial melt (defined as the time the reflectivity spends above the reflectivity of the hot solid, ~0.42 for Si and ~0.50 for Ge), undergoes a discontinuous jump with increasing energy density: no melt durations were observed less than ~25 ns for Ge or ~20 ns for Si.

A model of the melt-in phenomenon where there is a well-defined melt front propagating into the material for energy densities near the melt

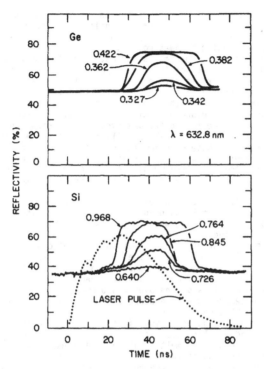

Fig. 1. Time-resolved reflectivity of Ge (top) and Si (bottom) at 632.8 nm during pulsed laser irradiation. The time profile of the excimer laser pulse is shown in the bottom panel. The numbers in the figure refer to the energy density in J/cm² of the laser pulse.

threshold is inconsistent with observation 2 listed above, since it would predict melt durations smaller than 20 ns for Si and 25 ns for Ge (see Ref. 6), for energy densities just above the melt threshold. In addition, the results of time-resolved ellipsometry (TRE) [7] performed on Si irradiated with 0.8 J/cm² excimer laser pulses cannot be fit with the molten layer thickness as the only free parameter.

A more realistic model is one in which the near-surface region consists of a layer of liquid and solid coexisting just after the melting begins. If the near-surface region is modeled as an effective medium [8], with liquid and solid coexisting, two free parameters (the thickness of the effective medium layer and the fraction of liquid) can be determined from the ellipsometric parameters Psi and Delta, as functions of time. If Si is irradiated with a 0.8 J/cm² pulse, the thickness of the effective medium layer is ~19 nm thick and is constant while the front surface is partially molten, but the fraction of liquid in the effective medium layer increases continuously to 0.80 at the maximum of the reflectivity and then decreases continuously to 0 at the end of the laser pulse. The reflectivity can be calculated separately from this model and is found to agree with the experimentally determined TRR at 0.8 J/cm².

When TRE measurements of Psi and Delta were examined for a more energetic laser pulse (1.4 J/cm²), it was found that the effective medium theory was again needed to explain the observations. At this energy density, the thickness of the effective medium layer increases to 13 nm in 1.5 ns, and stays at this thickness for 2.5 ns before it continues to increase. The fraction of liquid increases monotonically with time, taking ~7 ns before the entire near-surface region becomes molten.

Both the TRR and the TRE experiments indicate that the melting process occurs inhomogeneously. The use of the effective medium approximation assumes that the average particle size of one material in another is small compared with the wavelength of light (632.8 nm in this case). If the average particle size were larger than the wavelength of light, then another averaging scheme based upon incoherent reflections should be used; this averaging scheme was not able to fit the observed data, indicating that the average liquid particle size in crystalline silicon was less than the wavelength of light. Although these experiments indicate that the near-surface region melts inhomogeneously, no indication is given as to the origin of the inhomogeniety; it may come from inhomogenieties of the heating laser beam (on the μm scale), or it may be a fundamental property of the melting process.

It was also possible to determine accurately the reflectivities of the hot solid Si and Ge at their melting points and the reflectivities of ℓ-Si and ℓ-Ge at both 632.8 and 1152 nm, which are shown in Table I. Also shown in Table I are values of the reflectivities of ℓ-Si and ℓ-Ge determined from the literature and extrapolations to the melting point of constant temperature reflectivity and ellipsometry measurements of the hot solid [9]. As can be seen, the reflectivities of ℓ-Si and ℓ-Ge agree well with the literature values. The results from the TRR measurements are only slightly greater than the extrapolations of the oven measurements, indicating either a nonlinearity in the temperature dependence of the reflectivity or possibly a superheating up to ~200°C on melt-in.

Two parameters that can be determined from TRR mearurements for comparison with melting model calculations are the melt duration and the time of onset of melting. The surface melt duration is defined here as the time the reflectivity is above that of the hot solid, while the time of onset of melting is defined as the time after the beginning of the laser pulse at which the shape of the TRR signal changes suddenly, indicating surface melting. The surface melt duration for Si and Ge is shown in the top panel of Fig. 2, while the time of onset of melting is shown in the bottom panel of Fig. 2. As can be seen from Fig. 2, the melting threshold for Si is significantly larger than it is for Ge. For energy densities well above the melt threshold, the melt duration increases monotonically with energy density for both materials, with the increase being much larger for Ge than for Si. The lines represent the results of melting model calculations (described below), which indicate good agreement with experiment. The same

Table I: The reflectivities of Si and Ge at 1152 and 632.8 nm for the hot solid and liquid states determined from this work and literature (lit.) values.

| Material | | Reflectivities (±0.005) | | | |
| | | 632.8 nm | | 1152 nm | |
		this work	lit.	this work	lit.
Si	liquid	0.715	0.715(a)	0.750	-----
	hot solid	0.415	0.408(b)	0.355	0.346(b)
Ge	liquid	0.755	0.750(c)	0.795	0.790(c)
	hot solid	0.505	0.500(b)	0.455	0.438(b)

(a) Ref. 7
(b) Extrapolations of the data presented in ref. 9.
(c) Interpolated from the data of ref. 10.

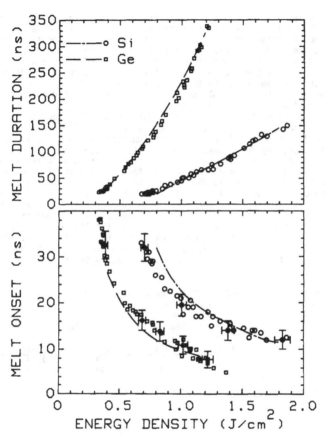

Fig. 2. The measured (open symbols) and calculated (lines) surface
melt durations (top) and times of the onset of melting bottom) for Si and Ge
as a function of versus KrF laser E_ℓ.

good agreement is obtained between the measured and calculated time of onset
of melting for both Si and Ge.

CALCULATIONS

Melting model calculations were carried out using both the HEATING5
code [11] and a new code LASER8 (described in detail in ref. 12). The major
problem in implementing melting model calculations is that temperature-
dependent thermal and optical properties must be included in the computer
code. The thermal conductivity of Si and Ge in the solid and liquid states
are discussed in Chapter 4 of ref. 1 and plotted in Fig. 2 of that
reference. The specific heat of solid Si and Ge are discussed in Section
8.b of Ch. 4 of ref. 1, while the specific heat of the liquid was taken to
be the same as the solid at the melting point. (The melt duration is not
very sensitive to the specific heat of the liquid.) No undercooling was
assumed at the solid liquid interface. However, the optical properties of
Si and Ge in both the solid and liquid phases have not been sufficiently

studied at 248 nm. The absorption coefficient of liquid and solid Si and Ge is large (>10^6 cm^{-1}), and therefore is not a very important parameter in the melting model calculations. The reflectivity, on the other hand, is large (>60%), can vary with temperature in the solid phase, and is very important in determining the amount of energy deposited in the sample. In this paper we have treated the reflectivities in the hot solid and in the liquid as fitting parameters; the best fits were obtained using the reflectivities shown in Table II, resulting in the calculated melt durations and times of onset of melting shown in Fig. 2.

The reflectivities of the solid and liquid phases of Si and Ge can be compared with other results in this laboratory, and with literature values. Viña et al. [5] have measured the dielecric functions of solid Ge at elevated temperatures, and found that the reflectivity decreased from ~63% at 300 K to ~57% at 825 K, the average of which is consistent with the value of R given in Table II. We have also performed some melting model calculations with a reflectivity in the solid phase which goes linearly from 65% at 20°C to 55% at 940°C, and found only minor differences with these results. The reflectivity of Si in the solid phase has also been measured by Francois et al. [13], where they found R ~65% and nearly independent of temperature up to 700°C. In order to measure the reflectivity of the liquid phase, we have performed self-reflectivity measurements of Si and Ge irradiated with 248 nm radiation; we found that the reflectivity increased ~10—15% for both Si and Ge upon melting. All these results are in agreement with the results listed in Table II.

Table II. The optical properties of Si and Ge at 248 nm.

Material	Absorption Coefficient (X10^6 cm^{-1}) solid	liquid	Reflectivity solid	liquid
Si	1.8	1.5	0.63	0.70
Ge	1.6	1.3	0.60	0.65

REFERENCES

1. Pulsed Laser Processing of Semiconductors, (Semiconductors and Semimetals, Vol. 23, R. F. Wood, C. W. White, and R. T. Young, eds., Academic Press, New York, 1984).
2. D. H. Lowndes, G. E. Jellison, Jr., and R. F. Wood, Phys. Rev. B 26, 6747 (1982); D. H. Lowndes, Phys. Rev. Lett. 48, 267 (1982).
3. D. H. Lowndes, R. F. Wood, and R. Westbrook, Appl. Phys. Lett. 43, 258 (1983).
4. G. J. Galvin, M. O. Thompson, J. W. Mayer, P. S. Peercy, R. B. Hammond, and N. Paulter, Phys. Rev. B 27, 1079 (1983); M. O. Thompson, J. W. Mayer, A. G. Cullis, H. C. Weber, N. G. Chew, J. M. Poate, and D. C. Jacobson, Phys. Rev. Lett. 50, 896 (1983).
5. L. Viña, S. Logothetidis, and M. Cardona, Phys. Rev. B 30, 1979 (1984).
6. G. E. Jellison, Jr. and D. H. Lowndes, Proc. Mat. Res. Soc. Symp. 35, 113 (1985).
7. G. E. Jellison, Jr. and D. H. Lowndes, Appl. Phys. Lett. 47, 718 (1985).
8. D. A. G. Bruggeman, Ann. Phys. (Leipzig) 24, 636 (1935).
9. H. H. Burke and G. E. Jellison, Jr., to be published.
10. J. N. Hodgson, Phil. Mag. 64, 509 (1961).
11. R. F. Wood and G. Giles, Phys. Rev. B 23, 2923 (1981).
12. G. A. Geist and R. F. Wood, ORNL-6242 (1985).
13. J. C. Francois, G. Chassaing, L. Argeme, and R. Pierrisnard, J. Optics 16, 47 (1985).

RAMAN MICROPROBE ANALYSIS OF
LASER-INDUCED MICROSTRUCTURES

P. M. FAUCHET
Department of Electrical Engineering
Princeton University
Princeton, NJ 08544

ABSTRACT

We study the composition, stress and structure variations across periodic surface undulations produced by pulsed laser illumination of semiconductors, by explosive crystallization
of amorphous films, and by laser-assisted CVD. These variations are mapped out with a one
micron spatial resolution using a Raman microprobe. Similarities and differences between the
three cases are pointed out. These results are also compared to those obtained by deli-
berately exposing the sample to interfering beams.

INTRODUCTION

In laser processing of solids, the energy deposition from the laser is often not uniform.
An extreme case is that of the formation of periodic structures on solid surfaces during laser
processing of semiconductors or during laser-assisted deposition [1]. Calculations suggest
that the modulation of the power flow across the surface can reach 100 percent. If the power
flow is so inhomogeneous, we expect that during and after processing, various properties of
the surface will also be inhomogeneous. For example, one could obtain stripes of laser
annealed material alternating with stripes of unannealed material. In explosive recrystallization, the surface of the film is corrugated in a quasi-periodic fashion [2]. Theory predicts this
behavior, including the alternance of amorphous and crystalline material [3]. Finally,
periodic surface structures have been observed during c.w. u.v. laser-assisted photodeposition of metal films [4] and during CO_2 laser-assisted chemical vapor deposition of Si and SiC
[5]. The large surface undulations in these two cases can be traced to a significant modulation of the absorbed power, on which the growth rate is dependent. Calculations very similar to those performed to describe ripple formation upon laser annealing indicate that in
addition to the film thickness, the film composition or structure could also be affected.

A study of these expected variations had not been performed so far. We have now
recorded the variations of the Raman lines across these laser-induced microstructures with a
microprobe having a spatial resolution of one micron. For a given material, the Raman
Stokes line is sensitive to crystal structure, to composition and to stress or strain. Stress
shifts the line peak in a linear fashion and does not alter the linewidth or shape. The spectrum of amorphous solids is easily distinguished from that of the same solids in the crystalline phase. It is also possible to measure the grain size in polycrystalline films since, as size
decreases, the line peak shifts, the linewidth broadens, and the line shape becomes asym-
metric. However, grains above 20 nm have a Raman spectrum undistinguishable from that
of single crystal. In compound or mixed solids, the composition or nature of the alloy is
measurable by observing the appearance and the strength of vibrational lines typical of
specific bonds. For example, a Si-Si bond vibrates at a frequency that is quite different from
that of a Si-C bond.

LASER-INDUCED PERIODIC STRUCTURES AFTER PULSED ILLUMINATION

Laser-induced surface ripples have generated a large body of work over the past five years. The major features of the electromagnetic interactions producing them are now understood [1,6], although some observations remain unexplained [7]. Comparatively little attention has been paid to a proper description of the material's response during ripple formation. Clearly, the material structure should be strongly modulated during illumination since theory predicts that power flow could be zero on the hills (or valleys) of the corrugation. All or part of the modulation could be frozen-in after illumination and may be observable with the Raman microprobe.

To verify the theoretical calculations and learn something about the material's transformations, we produced approximately 4 microns spacing ripples by laser melting a Ge wafer. The laser was a Nd:YAG operating at 1064 nm and was obliquely incident on the wafer. The Raman microprobe was focussed alternatively where the absorbed power was maximum (as evidenced by the relative roughness of the surface) and where it was minimum. The results are shown in Figure 1. The two spectra are nearly identical, given the ± 0.1 wavenumber repeatability of the instrument. This indicates that any stress present is below $10^9 \mathrm{dyne/cm^2}$, and that the amount of amorphous or fine grain microcrystalline Ge, if present, is below our detection limit.

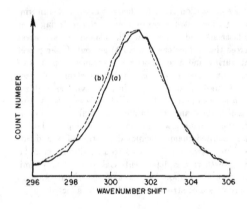

Figure 1

Stokes line of Ge (a) on top of a ripple and (b) away from a ripple.

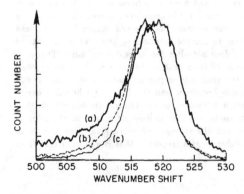

Figure 2

Stokes line of Si in a polycrystalline film on insulator. (a) virgin material; (b) region of destructive interference; (c) region of constructive interference.

This somewhat surprising result was confirmed by the result of another experiment also performed on the Ge wafers. Two equal intensity beams were overlapped at the sample so as to create an interference pattern. The period of the grating was chosen nearly equal to that of the ripples. The writing beams came from the same Nd:YAG laser. The absorbed power modulation was 100 percent across the fringes, corresponding to a temperature gradient of the order of 300 K/micron. Where the interference was constructive, the melting threshold was exceeded. Raman spectra taken on melted and unmelted regions were also nearly identical. We therefore conclude that the lack of variation of the line across ripples is not an indication of a small modulation of the absorbed power.

The situation appears somewhat different in Si thin films. We repeated the last experiment in a 1 micron thick polycrystalline film of Si deposited on fused quartz. The result is shown in Figure 2. Several facts are worth noting. First, the Raman line in the entire illuminated region (whether at the position of destructive or constructive interference) is altered. A transformation has taken place which leaves the material under larger tensile stress (increase by $510^9 dyne/cm^2$). The average grain size is also increased as can be seen from the symmetrization and narrowing of the line. Second, there is a small but observable difference between regions where interference was constructive and regions where it was destructive. The latter show a small wavenumber tail which we interpret as due to smaller average grain size. These observations are not surprising: we have reported earlier that melting of polycrystalline silicon on insulator (SOI) films produced larger Si grains and that, for the case of highly localized melting, the material's properties relaxed slowly to their initial values only several microns away from the illuminated area [8]. The reason for the different results in thin films and in wafers is not completely certain presently. The constraints imposed on the heat transport by the film/substrate boundary may result in a lateral heat flow that will carry enough energy to melt the destructive interference regions.

EXPLOSIVE CRYSTALLIZATION OF AMORPHOUS FILMS

Explosive crystallization of amorphous films has been known for some time. It is triggered by a thermal or mechanical impulse which produces crystallization of a domain in the amorphous layer. The process can be repeatedly triggered by scanning the impulse in order to transform the entire film. Large quasi-periodic surface undulations are often observed in the transformed film. Self-sustained explosive crystallization leading to a thermal instability has been predicted by theory for a relatively narrow range of temperature [3]. Alternating regions of amorphous and recrystallized material are expected.

We have probed the structural properties of a 6 micron thick film of explosively crystallized amorphous Ge deposited on glass [9]. The sample was doped with Ga. Undulations with many different periods were observed. Here, we report results obtained in a region where the period was 1 to 2 microns. Typical Raman spectra are shown in Figure 3, together with that of single crystal Ge. Note first the large difference between the film and the wafer. Second, the spectra taken at the top and at the bottom of a corrugation are very similar (the small differences are within the uncertainty of the measurement).

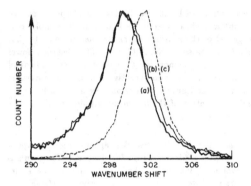

Figure 3

Stokes line of explosively crystallized a-Ge on glass. (a) on top of a ripple; (b) away from a ripple; (c) is the line of crystalline Ge.

The crystallized film displays a Raman line typical of fine grain polycrystalline material. A quantitative interpretation of the data is complicated by the presence of Ga. Large concentrations of acceptors soften the phonon frequency. Following the calculations of Compaan et al. [10], we find that if the peak shift is attributed to the presence of Ga acceptors only, the concentration of electrically active acceptors is 410^{19}cm^{-3} in rough agreement with the nominal 0.1 atomic percent Ga concentration. We believe this analysis is not entirely correct here, because the film is not well-ordered, as evidenced by the large low wavenumber tail indicative of disorder. If the lineshape is analyzed in terms of microcrystals, the average size is smaller than 10 nm. The influence of stress is probably negligible, because we are probing a 100 nm thick surface layer, whereas the total film thickness is 6 microns. We conclude that although the surface is rather strongly corrugated, the top layer is essentially homogeneous, with small grains and amorphous-like grain boundaries. The lack of variation between hills and valleys in the surface layer is in contradition with the theory for explosive crystallization [3], since we do not observe the expected alternating regions of amorphous and crystalline Ge. Depth profiling, either by etching of successive layers or by using longer wavelength probe lasers, could tell us whether such alternating regions exist deeper in the film.

LASER-ASSISTED CHEMICAL VAPOR DEPOSITION

Photodeposition of metal films has been demonstrated as a useful technique to write fine patterns. In previous studies, involving the deposition of metallic compounds (Cd [4], Cu/C [11]), a very rich surface structure has been observed. The most dominant surface corrugations are produced in a manner analogous to the surface ripples observed after laser annealing of wafers [12]. Very recently, Si and SiC films have been produced by CO_2 laser-assisted CVD on various substrates. We have examined the variations of structure and composition of Si and C films deposited on a SiC substrate [5]. The surface showed large periodic undulations, across which the ratio of Si and C seemed to vary.

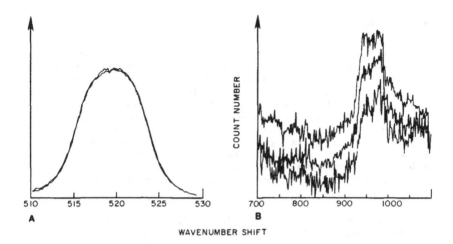

WAVENUMBER SHIFT

Figure 4

Partial Raman spectrum of laser-assisted CVD of Si and C. (a) The strong crystalline Si line taken at the top and the bottom of a corrugation is identical (b) from top to bottom, spectra taken at the bottom and the top of a corrugation, and spectrum of crystalline Si. The count number is smaller by a factor of 100. The three spectra are rather similar but not identical and are produced mostly by second-order Raman scattering of Si.

The results obtained with the Raman microprobe are shown in Figure 4. In order to maximize the count rate, we opened the slit to 1 mm (instead of 300 microns), thus trading spectral resolution for signal to noise. All across the surface, in the hills and the valleys, we observe a strong crystalline Si line. Its peak frequency and shape show that the grain size is rather large (above 20 nm). Furthermore, the constant line intensity across the ripples shows that the percentage of the film that is in the form of Si microcrystals remains constant. We have been unable to detect a clear SiC line. For example, the strong 790 wavenumbers line of SiC that is easily detected in the substrate is absent from the film. The broad but weak peak around 950 wavenumbers is clearly attributed to a second order Raman signal (2 TO phonons). We have detected C with graphite structure (line at 1580 wavenumbers), which appears as a rather weak line at the peak of the corrugation, but seems absent at the valleys.

Auger sputtering experiments have shown [13] that the stoichiometry varied across the ripples, with in general an excess Si. Our Raman microprobe results further indicate that the SiC phase is most likely absent everywhere, that pure Si grains are most likely present everywhere, and that C is present at the peaks in the graphite form. These measurements have been performed on a very limited set of samples so far. In particular, the deposits we have looked at so far have been produced in a Si-rich atmosphere and have undergone Auger sputtering prior to the Raman microprobe measurements. It is not clear whether films deposited under different conditions (laser power, gas composition, exposure time) will show similar results. Further tests are underway. A possible interpretation of these results is that Si tends to form pure grains and that C surrounds them. In addition, we have observed significant variations in composition between hills and valleys.

CONCLUSIONS

We have presented Raman microprobe measurements across surface undulations produced by various methods. The laser annealing results show that despite local melting, the material is essentially unchanged and uniform across the undulations. Other results suggest that even when the treated material is very different from the initial one, variations across periodic structures can be quite small (fringes in SOI, explosively crystallized films). Finally, the results of the previous section indicate that the composition can be altered across ripples (laser CVD). Rather ironically, it is in the case where the expected temperature modulation is estimated to be only 50 K [13] that we obtain the most obvious example of compositional modulation. These results clearly indicate that one must carefully analyze the processes involved before predicting large material modulations. They also suggest several other experiments, some of which have been indicated in the text. These experiments are in progress or will be performed in the near future.

I acknowledge the expert help of Fran Adar and Ian Campbell in the experiments. I would like to thank Chuck Wickersham (Varian) who provided the explosively crystallized Ge film, and Susan Allen and her student Farhad Shaapur (USC) who provided the SiC films. This work was supported in part by NSF grant ECS-8504202.

REFERENCES

1. Z. Guosheng, P.M. Fauchet, and A.E. Siegman, *Phys. Rev. B26*, 5366 (1982).

2. C.E. Wickersham, G. Bajor, and J.E. Greene, *J. Vac. Sci. Technol. A3*, 336 (1985).

3. W. van Saarloos and J.D. Weeks, *Phys. Rev. Lett. 51*, 1046 (1983).

4. R.M. Osgood, Jr. and D.J. Ehrlich, *Opt. Lett. 7*, 402 (1982).

5. F. Shaapur, S.M. Copley, and S.D. Allen, CLEO 1985, paper TUJ2.

6. J.E. Sipe, J.F. Young, J.S. Preston, and H.M. van Driel, *Phys. Rev. B27*, 1141 (1983).

7. P.M. Fauchet and A.E. Siegman, in *Energy-Beam Solid Interactions and Transient Thermal Processing 1984*, Biegelsen et al. editors, Materials Research Society, 1985, pp. 199-204.

8. P.M. Fauchet, I.H. Campbell, and F. Adar, *Appl. Phys. Lett. 47*, 479 (1985).

9. provided by Dr. C. Wickersham, Varian Specialty Metals Division.

10. A. Compaan, G. Contreras, and M. Cardona, in *Energy-Beam Solid Interactions and Transient Thermal Processing*, Fan and Johnson editors, North-Holland 1984, pp. 117-122.

11. R.J. Wilson and F.A. Houle, *Phys. Rev. Lett. 55*, 2184 (1985).

12. S.R.J. Brueck and D.J. Ehrlich, *Phys. Rev. Lett. 48*, 1678 (1982).

13. S.D. Allen, private communication.

CHARACTERISTICS OF LASER/ENERGY BEAM-MELTED
SILICON MECHANICAL DAMAGES

EL-HANG LEE
Monsanto Electronic Materials Co., St. Peters, MO 63376
Present Address: AT&T, Engineering Research Center, P. O. Box 900,
Princeton, NJ 08540

ABSTRACT

We describe what appears to be a first attempt to melt and recrystallize macroscopic (10-20 μm deep) silicon mechanical damage that is induced from wafer modification such as slicing and lapping. Recrystallized surfaces appear mirror shiny, with significantly improved surface smoothness, as compared to the coarse texture of damaged surfaces. The crystallinity also appears good in general. Through the depth of melt were observed indications of impurity migration, probably caused by accumulated segregation at the advancing solid-liquid interface. Recrystallized surfaces, despite their smoothness, remain topologically uneven as a result of lateral mass transport. In addition, the extensive heat required to melt thick layers of silicon causes slip dislocations.

INTRODUCTION

Energy beam-assisted annealing and recrystallization techniques have been extensively used in recent years to remove structural damage and restore electrical activity in ion-implanted silicon layers [1]. As beam energy is absorbed by damaged layers, atoms are activated to restore their order, either in the solid phase or in the liquid phase. Energy beams have included pulsed or cw electron beams, lasers, high-power lamps, ion beams, and strip heaters, to name a few. However, this technique has not been much used for other forms of damage, such as that caused by wafer slicing or lapping. As-sliced or as-lapped wafers contain severe mechanical damage that extends tens of microns into the depth of wafers in the form of microcracks, indentations, scratches, dislocations, microplastic deformations, and the like [2-4]. In the current practice of wafer modification, such damage is removed by polishing and etching until the undamaged sub-layer emerges to form a smooth surface. This process, however, inevitably involves the removal of costly single-crystal material. We have thus employed the energy-beam melting and recrystallization technique in order to investigate the possibility of restoring and annealing process-induced mechanical damage in silicon. The quality of recrystallized layers was then examined using various characterization techniques. To the best of the author's knowledge, this appears to be the first reported attempt to apply the energy-beam melting technique to the annealing of macroscopic silicon damage.

EXPERIMENTAL

Figure 1 shows a schematic illustration of the recrystallization system used in the present study. The system consisted of a large area graphite base plate, upon which silicon wafers were placed with damaged surfaces facing up. A thin graphite strip heater (8 cm long and 2 mm diameter), heated to 2000°C at a distance of 1 mm above the wafer, or a line-shaped cw argon-ion laser beam, was scanned along the sample surface

156

at a constant speed between 1-5 mm/sec. Sub-atmospheric pressures of argon were used as ambient. During melting, the base plate was heated to 1250°C in order to provide the required energy. Silicon wafers were p-type or n-type of (100) orientation.

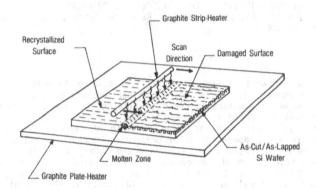

Fig. 1. A schematic illustration of the laser- or strip-heater recrystallization system.

RESULTS AND DISCUSSION

Upon melting and solidification, the dull, gray texture of as-sliced wafers changed into a mirror-smooth, shiny one. Figure 2 shows optical microscopic views of wafer surfaces before and after strip-heater melting. The contrast between the coarse texture of an as-sliced surface and the smooth, refined one is quite noticeable. The microscopic formations of voids and bumps on the recrystallized surface, mostly in sizes of 1 μm or less, have been observed on laser-melted surfaces, too, and are under continued investigation in search of their origin. At present they are speculated to be SiO-evaporated crater pits or oxygen-related dislocations.

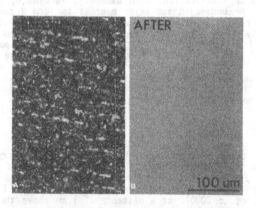

Fig. 2. Optical microscopic views of a sliced wafer surface (a) before and (b) after melting/recrystallization.

Figure 3 shows an angle-lapped (ca. 11° with the surface), cross-sectional view of an as-sliced wafer, both before and after laser melting. Before melting, the surface damage ranged to a 20-μm depth, but after melting the damage was confined to 2 μm or less. Also, the highly corrugated, original surface smoothed out to a uniformly even profile.

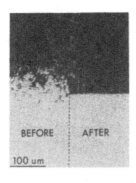

Fig. 3.
Angle-lapped, cross-sectional view of an as-sliced wafer, one portion of which was laser-melted. Note the remarkable improvement in microscopic surface topography of the recrystallized layer.

One of the intriguing characteristics of the recrystallized layer is that it usually contains far fewer defects than the bulk, sometimes evolving to a denuded-zone-like layer. Figure 4 shows a representative view of such a denuded-zone layer induced by surface melting. It is now well established that silicon wafers under a controlled thermal treatment yield denuded-zone layers (20-30 μm depth) as a result of oxygen out-diffusion, with the bulk silicon underneath containing extensive oxygen precipitate defects [5]. The effect of oxygen out-diffusion (or venting) has been observed in other processes too, such as the recrystallization of thick-film (20-30 μm) silicon on oxide layers [6].

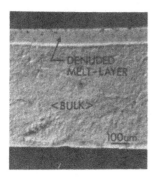

Fig. 4.
An angle-lapped (11°) cross-sectional view of a recrystallized wafer reveals a denuded zone, after etching, in clear demarcation with the dislocation-laden bulk underneath.

In these studies, high-quality, defect-free layers were obtained in the areas where oxide capping was removed, whereas in the areas still covered with oxide caps, the layers were laden with a high level of oxygen precipitate defects. In our studies no oxide capping was employed.

158

Selected area TEM diffraction has shown that recrystallized layers
consistently replicated the crystallinity and orientation of the mother
substrates, as shown in Fig. 5. Similarly, a Raman spectroscopic

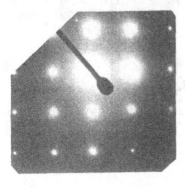

Fig. 5.
A TEM diffraction pattern of (100)
recrystallized surface revealing a
high degree of crystallinity.

analysis also revealed good crystallinity of melted layers. Single
crystalline silicon is known to emit Raman radiation with a shift of
520 cm^{-1}, with polycrystalline silicon at about 505 and 511 cm^{-1} and
amorphous silicon broadly at about 480 cm^{-1} [7]. Figure 6 shows Raman
spectra of the recrystallized silicon layer, where the peak occurred at
521 cm^{-1}, suggesting a good quality crystallinity.

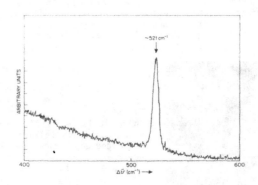

Fig. 6. Raman spectra of a recrystallized surface.

Angle-lapped surfaces of recrystallized layers were also examined
using spreading resistance measurements. Figure 7 shows a typical
spreading resistance profile, which, like many other samples, reveals
(1) a broad peak at a depth of 5-10 μm from the surface, and (2) a
surface value different from that of the bulk. Since the crystalline
quality has been found to be good throughout the layer (as mentioned
above), the resistance variation is attributed to the effect of impurity
migration within the melted layer. As the melted layer solidified toward
the surface, the impurities may have been segregated out to the surface

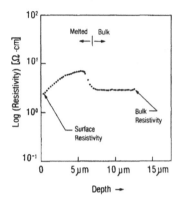

Fig. 7. A spreading resistance profile across an angle-lapped, recrystallized layer. A broad peak is seen at the melt junction, possibly indicating an impurity depletion, and also a lower surface resistance value than that of the bulk, suggesting a higher impurity concentration at the surface than in the bulk.

along with the advancing solid-liquid interface. Frequently, the types of recrystallized layer were observed to be opposite to those of bulk substrates (n-type vs. p-type, or vice versa), indicating a reversal in the relative concentration levels of doping and compensating layers between bulk and melt layer. A SIMS study [8] indeed supported this observation by showing, for example, a higher arsenic (n-type) concentration than boron (p-type) in the layer melted on a p-type substrate (having higher boron concentration than arsenic). The impurity segregation phenomena have been a common occurrence in laser-annealed implantation damages, whether in the liquid phase or in the solid phase [1]. Despite the good crystallinity of melted layers, several undesirable aspects were observed. First, the melted layer showed considerable mass transport along the direction of scan, resulting in a mound formation toward the beam exit end of the wafer. The mass transport phenomenon has been consistently observed in energy beam-recrystallized layers, such as silicon-on-insulator, and is regarded as an important issue. Another pronounced result is that the substrate developed, without fail, heavy slip dislocations caused by severe heating. Figure 8 shows an X-ray topographic image of a surface-melted wafer which reveals a high level of strain coupled to slip dislocations.

SUMMARY AND CONCLUSION

An attempt has been made to utilize the energy-beam-recrystallization technique to remove silicon mechanical damage caused by slicing and lapping. Various characterization results show that the restored crystalline quality is good, but the wafer contains surface unevenness, impurity migration, and a high degree of slip dislocations induced by mass transport and excessive heating. These undesirable results pose a somewhat discouraging prospect for the practical use of this technique.

160

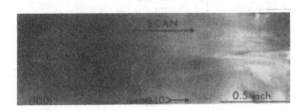

Fig. 8. An X-ray topographic image of a surface-melted wafer, showing a high level of strain with slip dislocations.

ACKNOWLEDGMENTS

The author acknowledges the following individuals: R. Sandfort, R. Craven, R. Hockett for support of this research; P. Doerhoff' for technical assistance; D. Ruprecht for spreading resistance measurement; John Freeman for Raman measurement; and G. Fraundorf for TEM studies, all at Monsanto.

References

1. See, for example, Mater. Res. Soc. Symp. Proc., Vol. 1 (1981), Vol. 4 (1982), Vol. 13 (1983), Vol. 35 (1984) (North Holland, New York).
2. T. M. Buck and R. L. Meek, in Silicon Device Processing, edited by C. P. Marsden, NBS Special Publication 337 (NBS Washington, DC, 1970) p. 419.
3. A. W. Fisher and J. A. Amick, J. Electrochem. Soc. 113, 1054 (1966).
4. A. Mayer, RCA Rev., June 1970, p. 414.
5. See, for example, K. Kugimiya, S. Akiyama, and S. Nakamura, in Semiconductor Silicon/1981, edited by H. R. Huff, R. J. Kriegler, and Y. Takeishi (The Electrochemical Society, Pennington, NJ, 1981) p. 294.
6. L. Pfeiffer, K. W. West, S. Paine, and D. C. Joy, Mater. Res. Soc. Proc. 35, 583 (1984).
7. M. H. Brodsky, M. Cardona and J. J. Cuomo, Phys. Rev. B 16, 3556 (1977); E. Anastassokis, A. Pinczuk, E. Burstein, F. H. Pollack, and M. Cardona, Solid State Commun. 8, 133 (1970).
8. E. H. Lee (to be published).

PULSED CO₂-LASER INDUCED MELTING AND NONLINEAR OPTICAL STUDIES OF GaAs

R. B. James*, W. H. Christie**, B. E. Mills*, and H. L. Burcham, Jr.*
*Sandia National Laboratories, Livermore, CA 94550
** Analytical Chemistry Division, Oak Ridge National Laboratory, Oak Ridge, TN 37831

Abstract

We report new optical and structural properties of p-type GaAs that result from the absorption of high-intensity 10.6 μm radiation. Prior to the onset of surface melting, we find that the absorption coefficient decreases with increasing intensity in a manner predicted by an inhomogeneously broadened two-level model. As the energy density of the CO_2 laser radiation is increased further, the surface topography shows signs of melting, formation of ripple patterns, and vaporization. Auger spectroscopy and electron-induced x-ray emission show that there is loss of As, compared to Ga, caused by the melting of the surface. Using plain-view TEM we find that Ga-rich islands are formed near the surface during the rapid solidification of the molten layer. Auger and SIMS measurements are used to study the incorporation of oxygen in the near-surface region, and the results show that oxygen incorporation can occur for GaAs samples that have been irradiated in air.

Introduction

There exists an extensive list of publications on the processing of GaAs using a pulsed ruby, excimer, or Nd:YAG laser [1]. The primary concern of this work is to study changes in the optical and structural properties of GaAs that result from pulsed CO_2 laser irradiation. These studies were stimulated by two important observations recently reported for the CO_2-laser processing of Si [2]. First, it was found that by controlling the absorption coefficient via control of the free-carrier density, one can achieve surface melting and subsequent defect-free recrystallization to depths in excess of 1 μm. Second, one can use the differential absorption between layers with different free-carrier concentrations to melt near-surface regions which are embedded in Si, without melting the material that encapsulates it. Using the preferential energy deposition of the CO_2 laser radiation, new applications for the processing of heterostructures with varied or modulated doping in the near-surface region are made possible. (Furthermore, by avoiding the melting of the surface, the loss of arsenic by evaporation would be greatly reduced, which might solve some of the previous problems associated with deviations from stoichiometry.)

In this paper we report measurements of the nonlinear absorption coefficient of p-type GaAs for energy densities less than the melt threshold. For higher laser energy densities, we have obtained direct evidence of the CO_2-laser induced melting of the near-surface region. Using Auger spectroscopy, secondary ion mass spectrometry, and transmission electron microscopy, we report results for the deviations from stoichiometry, incorporation of oxygen, and the microdefects in the near-surface region of samples that have been melted by a laser pulse.

Experiment

A gain-switched, TEA CO_2 laser is used to generate the pulses at a wavelength of 10.6 μm. The laser is operated with a low nitrogen mix so that the amplitude of the long tail on the pulses can be greatly suppressed. About 80% of the energy in each pulse is contained in the form of a Gaussian-like peak with a duration of 70 ns (FWHM), and the remainder of the energy is in a long tail following the spike which lasts for several hundred nanoseconds. The output beam impinges on a CO_2 laser beam integrator which spatially homogenizes the pulse to within $\pm 10\%$. The laser pulses have a size of 12x12 mm in the target plane of the integrator and can be adjusted by using additional lenses and linear attenuators. A photon-drag detector and volume absorbing calorimeter are used to measure the intensity and energy of the laser pulses, respectively.

Results and Discussion

For light with a wavelength near 10 μm, the absorption in p-type GaAs is dominated by direct free-hole transitions between states in the heavy- and light-hole bands. For laser energy densities less than the melt threshold, the results of our transmission measurements show that the intervalence band absorption decreases with increasing intensity. The intensity-dependence of the absorption coefficient is found to be well described by an inhomogeneously broadened two-level model [3-4]. For a GaAs:Zn sample with a free-hole concentration of 1×10^{17} cm^{-3}, the saturation intensity has a value of 20 MW/cm^2.

We attribute the observed nonlinearity to a redistribution of hole states in the resonant region of the heavy- and light-hole bands. At low laser intensities (≤ 1 MW/cm^2), the distribution of hole states is maintained close to the equilibrium value by hole-phonon scattering. As the excitation rate becomes sufficiently large, hole-phonon scattering cannot maintain the equilibrium population of heavy-hole states in the resonant region of k-space, and they become depleted. Since the absorption is governed by the population difference of the resonantly coupled states, the depletion of these relevant states leads to a decrease in the absorption coefficient.

For more heavily doped samples, there is an increase in the intervalence band absorption coefficient and a decrease in the melt threshold. For GaAs:Zn crystals with a free-hole density of greater than about 10^{18} cm^{-3}, we find that large areas of the surface are melted by the absorption of the laser radiation. This is in contrast to the lightly doped samples where small sites of damage would initially appear.

For a sample with a hole concentration of 5.1×10^{18} cm^{-3}, we find that the reflectance of the surface remains almost constant up to incident energy densities of 2.3 J/cm^2. For higher laser energy densities, the reflectance increases rapidly from near its linear value up to a value of 80-90%. Using the photon-drag detector to temporally analyze the pulse, we find that the reflectivity of the trailing edge of the laser pulse increases much more than the reflectivity of the leading edge. Since the reflectance of molten GaAs to CO_2 laser light should be about 90-100%, we attribute the increased reflectance at energy densities greater than 2.3 J/cm^2 to the presence of a thin layer of liquid GaAs at the surface of the sample.

The surface topography of samples with $N_h = 5.1 \times 10^{18}$ cm^{-3} was studied with scanning electron and Nomarski interference microscopes. For energy densities slightly above the melt threshold, the surface of the (100) GaAs becomes discolored, although a high degree of planarity is maintained. As the incident energy density is further increased, smooth periodic ripples are visible on the surface, which have a spacing equal to the wavelength of the light. For energy densities greater than about 3.4 J/cm^2, we find that fissures develop on the surface. Since the linear expansion of GaAs at the melting point is approximately one percent of the room temperature lattice spacing, the formation of the fissures is most likely due to the relief of mechanical stress during the time that the hot resolidified material cools and contracts. The depths of the fissures are several microns, which are somewhat deeper than the melt depths, and they are oriented along the (011) and (0$\bar{1}$1) planes. At slightly higher energy density ($\gtrsim$ 8 J/cm^2), spall fragments are ejected from the surface at the interaction region. The thickness of the fragments varies from about 5 to 8 μm. The exfoliation of surface fragments facing the laser source is likely due to a shock wave which produces a rarefaction that propagates into the material. The presence of a laser-induced shock wave could result from the recoil of the GaAs substrate due to the rapid vaporization of material from the surface.

For the GaAs samples that had been melted by a CO_2 laser pulse, we used Auger spectroscopy and electron-induced x-ray emission to study the deviations from stoichiometry by measuring the relative signals for the Ga and As atoms. We find that As loss does occur, and the degree of the loss depends on both the laser energy density and the number of shots at a fixed laser density. This preferential loss of As is apparent in the Auger profiles in fig. 1 for an unirradiated sample and a sample irradiated in air with ten shots at $E_L = 5.4$ J/cm^2.

The Auger profiles also show that oxygen is incorporated into the near-surface region of the material. Furthermore, we find that a decreased amount of arsenic at the surface is directly correlated to the presence of an increased amount of oxygen. This correlation could result from the reaction of oxygen with the available gallium to liberate arsenic, which has a relatively high vapor pressure.

Using secondary ion mass spectrometry with a N_2^+ primary beam, we

164

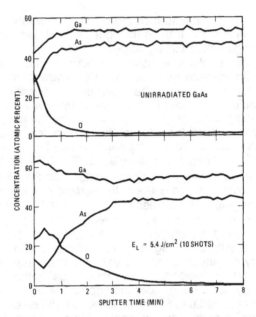

Fig. 1 Auger profiles for gallium, arsenic, and oxygen in an unirradiated and a CO_2 laser-irradiated GaAs sample. Sputter rate ≈ 140 angstroms/min.

measured the depth profiles of ^{16}O for several different laser energy densities. The four curves shown in fig. 2 are for an unirradiated GaAs crystal, a sample irradiated at an energy density of 4.2 J/cm^2, a sample irradiated at an energy density of 5.7 J/cm^2, and a sample irradiated with five shots at an energy density of 7.3 J/cm^2. We find that for energy densities greater than about 4 J/cm^2, there is a noticeable diffusion of oxygen atoms into the near-surface region. (A small amount of diffusion of oxygen from the native oxide or ambient may occur at lower energy densities, but this is not clearly discernible in the experiment.) For energy densities larger than about 6 J/cm^2, the oxygen incorporation greatly increases. Since the melting of the surface occurs for $E_L \geq 2.3$ J/cm^2, we see that surface melting is not sufficient for significant incorporation of oxygen. These measurements are consistent with the work by Bentini et al. [5], in which they proposed that the presence of the native oxide hinders the oxygen incorporation, so that energy densities somewhat greater than the melt threshold are required for the uptake of oxygen from the ambient.

In order to obtain microstructural information, we performed plainview transmission electron microscopy on the laser-irradiated GaAs samples. Figure 3 shows a TEM micrograph of a sample that was irradiated by a single shot at an energy density of 4.2 J/cm^2. The clear regions in the figure are single crystal GaAs, and the dark regions are Ga-rich islands as determined by electron diffraction and x-ray fluorescence. The Ga-rich islands extend throughout the irradiated region and contain little arsenic.

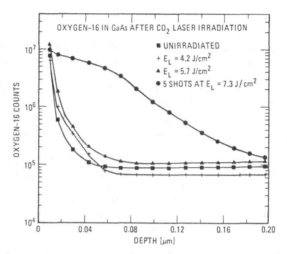

Fig. 2 SIMS measurements of oxygen-16 counts as a function of depth for samples irradiated in air at different CO_2 laser energy densities.

The size of the islands are found to increase with increasing laser-energy density. (Similar observations of a "network" of Ga-rich regions have been observed by Fletcher et al. [6] using plain-view TEM on pulsed ruby laser-irradiated GaAs.) For energy densities at which exfoliation of the solid GaAs fragments occurs, we observe no Ga-rich islands at the new surface, although there exists some damage that results from the spallation of the original surface.

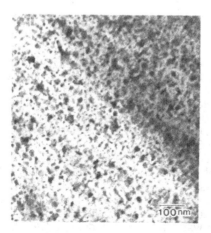

Fig. 3 Plain-view TEM micrograph showing Ga-rich islands formed after laser-induced melting of the near-surface region.

Summary and Conclusions

We have measured the intensity dependence of the free-hole absorption of CO_2 laser radiation in p-type GaAs crystals. We find the intensity dependence of the nonlinear absorption to be closely approximated by an inhomogeneously broadened two-level model. For more heavily Zn-doped GaAs, we show that a pulsed CO_2 laser can be used to melt the near surface region of the material. For samples that have been melted by the laser radiation, we find that Ga-rich islands are formed within the first few hundred angstroms from the surface. These islands are surrounded by stoichiometric or nearly stoichiometric GaAs. Auger and SIMS measurements are performed to study the Ga-rich layer and the incorporation of oxygen from the native oxide and/or ambient. The results of our measurements show that oxygen incorporation occurs for laser-energy densities that exceed a threshold value of about 4 J/cm^2. For samples irradiated in air, the oxygen incorporation is greatest in areas that are most deficient of arsenic.

In conclusion, we have established that melting and crystallization of GaAs can be accomplished by pulsed CO_2 laser radiation. The occurrence of surface evaporation of arsenic greatly increases the difficulty in obtaining defect-free single-crystal material. Our results show that the "window" for optimal pulsed CO_2 laser annealing is fairly narrow, and possibly nonexistent, for samples in which surface melting occurs.

Acknowledgments

This research was sponsored by the U. S. Department of Energy. We would like to acknowledge M. I. Baskes, D. H. Lowdnes, F. Gruelich, R. F. Wood, A. J. Antolak, and J. R. Hogan for many useful discussions.

References

1. See, for example, D. H. Lowdnes, in Pulsed Laser Processing of Semiconductors, Vol. 23, edited by R. F. Wood, C. W. White, and R. T. Young (Academic Press, New York, 1984), p. 471.
2. R. B. James, J. Narayan, W. H. Christie, R. E. Eby, O. W. Holland, and R. F. Wood, in Energy Beam-Solid Interactions and Transient Thermal Processing, edited by D. K. Biegelson, G. A. Rozgonyi, and C. V. Shank (MRS, Pittsburgh, PA, 1985), p. 413.
3. R. B. James and D. L. Smith, Phys. Rev. Lett. 42, 1495 (1979).
4. R. B. James, W. H. Christie, R. E. Eby, B. E. Mills, and L. S. Darken, Jr., J. Appl. Phys., in press.
5. G. G. Bentini, M. Berti, C. Cohen, A. V. Drigo, E. Ianitti, D. Pribat, and J. Siejka, J. de Physique 43, C1-229 (1982).
6. J. Fletcher, J. Narayan, and D. H. Lowdnes, in Defects in Semiconductors, edited by J. Narayan and T. Tan (North Holland, New York, 1981), p. 421.

IN SITU TIME-RESOLVED REFLECTIVITY MEASUREMENTS OF GROWTH KINETICS DURING SOLID PHASE EPITAXY : A TOOL TO ESTIMATE INTERFACE NON PLANARITY DURING GROWTH

C. LICOPPE and Y.I. NISSIM
C.N.E.T. Laboratoire de Bagneux*- 196 rue de Paris - 92220 Bagneux - FRANCE

ABSTRACT

The time resolved reflectivity technique is shown to give informations on the amorphous-crystalline interface evolution during solid phase epitaxial (SPE) regrowth in semiconductors. Two specific cases have been treated here. The first case is encountered in laser annealing when the growth front exhibits a curvature due to the combination of an inhomogeneous temperature distribution and a steep dependence of SPE growth rates with temperature. A computer simulation is carried out from an analytical determination of the laser induced temperature profiles to shape up the resulting reflectivity signal. The second case is obtained when there is an evolution of interface roughness during regrowth. In order to simulate this effect a simple model is developed to treat the influence of diffusion of the reflected light at the interface, on the reflectivity modulation during SPE regrowth.

INTRODUCTION

Solid phase epitaxial (SPE) regrowth of semiconductors is usually studied by a posteriori techniques such as Rutherford backscattering or transmission electron microscopy, where one has to stop the regrowth to determine the relevant growth parameters. The time resolved reflectivity (TRR) technique has provided an optical means to study regrowth in situ [1]. It has extensively been used to study SPE kinetics. The aim of this work is to show that it can bring valuable information in two cases where interfaces are not perfect. In the first case, the effect of a curved interface on the reflectivity signal is examined. This is encountered during CW laser annealing of Si and GaAs. In the second case, that of thermal annealing of GaAs, interface roughening occurs during regrowth. In both cases, the TRR signal is perturbed, as this presentation will show and, information on interface structure can be gained from analysis of the TRR data.

THE SMOOTH CURVATURE OF THE LASER-INDUCED REGROWTH FRONT

A typical set-up for the use of the TRR technique in the case of CW laser annealing involves a reflectivity probe beam, usually provided by a focussed He-Ne laser beam with a waist of 15 microns [1]. This probe beam is directed toward the center of a spot irradiated by a CW laser beam which is used for annealing purposes. Energy deposition by a laser beam follows a gaussian distribution. Hence, laser annealing induces an inhomogeneous temperature distribution in the sample. This distribution has a radial symmetry and can be calculated analytically [2].

$$T(r) = \frac{\sqrt{\pi}}{2} \frac{wI_0}{K} \exp(-r^2/2w^2) \, I_0(r^2/w^2) \qquad (1)$$

---

*Laboratoire associé au CNRS (LA 250)

where r is the radial distance to the center of the beam, w is the gaussian beam waist of the laser, I_0 is the total absorbed intensity, K is the thermal conductivity of the semiconductor and $I_0(x)$ is the modified Bessel function of order zero.

In semiconductors such as Si or GaAs, SPE growth rates are thermally activated [3,4]. That is, the growth rate V is expressed in terms of the activation energy E_a as follows :

$$v = v_0 \exp(-E_A/k_B T) \tag{2}$$

In a typical laser annealing situation such as the one described above, the temperature variation scale is much larger than the interfacial fluctuations scale. Thus at a given point in the irradiated spot whose temperature is $T(r)$, the SPE growth rate is given by equation (2) where T stands for the local temperature $T(r)$. In the inhomogeneous laser induced temperature distribution, the growth front departs from planarity because of the steep dependance of SPE kinetics on temperature. Fig.1 gives a computer calculation of the shape of the growth front for a laser-induced SPE experiment in silicon.

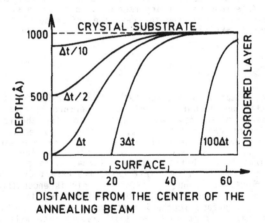

DISTANCE FROM THE CENTER OF THE ANNEALING BEAM

Fig. 1 : Computer calculation of the profile of the recrystallization front at different stages. Δt is the characteristic time needed by the regrowth front to reach the surface at the center of the annealing beam. The sample was held at 200°C and irradiated with an Ar$^+$ laser beam of diameter 150Å and power 8W. Amorphized layer width is 1000Å.

In the case of an interface between disordered and crystalline material located at depth z, the reflectivity is given by : [5]

$$\rho(z) = \left| \frac{\rho_{01} + \rho_{12} \, e^{2ikz} \, e^{-\alpha z}}{1 + \rho_{01}\rho_{12} \, e^{2ikz} \, e^{-\alpha z}} \right|^2 \tag{3}$$

where ρ_{01} is the reflectivity at the surface of the disordered material, ρ_{12} the reflectivity at the interface ; α is the light absorption coefficient in the disordered material and k its wavenumber. Equation (3) is valid for a planar interface. When the interface is not planar, but vary smoothly enough to be considered as plane on a local scale, the validity of formula (3) can be extended, with depth z now depending on the surface

coordinates (x,y).
The net reflected light intensity, I is given by :

$$I = \int_{-\infty}^{+\infty}\int_{-\infty}^{+\infty} \frac{I'_0}{\pi w'^2} [\exp(-(x^2+y^2)/2w'^2)] \rho(x,y)\, dx\, dy \qquad (4)$$

the parameters I'_0 and w'_0 are respectively the intensity and the diameter of the He-Ne probe beam. The reflectivity depends on the growth front profile. The consequence of the growth front curvature is a loss of coherence in the interference, and an attenuation of the oscillatory pattern in the TRR signal during regrowth.
Fig. 2 shows the results of a computer simulation for the time-resolved reflectivity signal during regrowth. Annealing parameters are identical to those used in fig. 1. In Fig. 2.a, the width of the probe beam has been increased, while it has been supposed to be perfectly centered in the annealing beam. It starts from a waist $w'=0$, where the probe beam strikes a flat interface at the center of the irradiated spot. It then increases toward positive values. In the latter situation, a wider beam probes curved regions of the interface. In Fig.2.b, w' is held constant at a typical experimental value. The center of the probe beam is increasingly displaced from the center of the irradiated spot, by a length Δr. Basically the effect is the same as in the situation of Fig.2.a since the He-Ne beam probes regions of increasing curvature at the interface.

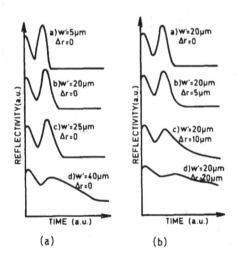

(a)   (b)

Fig.2 : Evolution of the TRR signal during regrowth when :
a) The waist of the probe beam is increased
b) The distance Δr from the center of the annealing beam is increased for a constant beam waist of the probe.

The deformation of the TRR signal due to the curvature of the recrystallizing front has been checked experimentally to follow the evolution of Fig.2. This property can be very useful experimentally to align precisely a probe beam and a test beam. It also permits us to precisely focus the probe.

THE NON-PLANAR GROWTH FRONT DURING SPE IN GaAs

In a compound semiconductor material such as GaAs, SPE recrystallization at low temperature (400°C) leaves many defects such as stacking
faults in the regrown layer. SPE in GaAs is not a layer by layer process
but involves a complex interfacial process occurring at a rough interface
[6]. The aim of this section is to show that the TRR technique is sensitive
to the structure of the interface during regrowth.
    The situation will be different from that described in section I.
Temperature will be taken as homogeneous in the whole sample. Light diffusion by defects at the interface will be accounted for by introduction of a
random phase factor $e^{i\phi(x,y)}$ depending on surface coordinates in equation
(3). This leads to an equation very similar to equation (4) where $\rho(x,y)$ at
a given point takes the value ρ with probability $p(\rho)$. If the characteristic fluctuation scale of the phase factor ϕ which stands for interface
roughness is much smaller than the visible light probe beam waist, it then
is possible to write equation (4) whose $\rho(x,y)$ has been replaced by its
mean value ρ .
    The consequence will be that the presence of the random phase factor
decreases the effect of the term e^{2ikz} in equation (3), which is responsible for the oscillatory behaviour of the TRR signal during regrowth [6].
Hence, the introduction of a defective non-planar interface structure will
influence on the TRR signal, which in turn can be taken as a source of
information on interface structure during regrowth.
    In Fig.3 it is shown how the TRR signal evolves with increasingly
rough interfaces. The change of random variable $\phi=2k\ell$ has been made where ℓ
is a new variable that has the dimension of a length. It allows comparison
of the problem outlined above, where the exact nature of interface defects
and their interaction with light is not known, with the case of a stepped
interface with random distribution and height of steps. Such steps introduce a random variation in optical paths in the reflectivity equation (3).

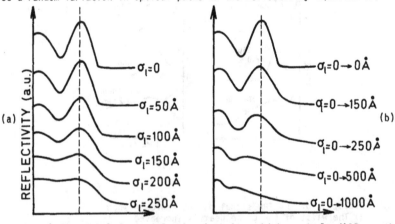

Fig.3 : a) TRR signal for a defective interface which randomly diffuses the
    reflected light. $\phi=2k\ell$ according to text notations and $\sigma_\ell = \ell^{1/2}$ is
    a parameter characterizing interfacial disorder. For all curves σ_ℓ
    is kept constant during recrystallization.
    b) TRR signal for a defective interface which evolves during re-
    growth. Disorder parameter σ_ℓ increases linearly from 0 at the
    start to the indicated value for each curve. The upper curve cor-
    responds to a perfectly planar interface.

This description of interface roughness, though physically irrelevant, gives some meaning to the order of magnitude of the variance in ϕ needed to affect the TRR signal.

The most meaningful result is that it is possible to discriminate between an interface with a static structure during regrowth, and an interface dynamically evolving during SPE. In the first case, oscillations in the reflectivity are simply attenuated, while in the second case the last oscillation tends to collapse as interfacial disorder increases. This was also observed in section I with laser annealing, as the interface curvature increased. Thus the collapse of the last oscillation seems to be a characteristic feature of interfacial evolution during regrowth.

It must also be noticed that in Fig. 3.b, the time location of the last reflectivity maximum is shifted toward shorter times as the rate is increased. In laser experiments, it was the contrary due to convex growth front. In both cases, the reflectivity peak is submitted to a shift compared to the ideal case of a plane interface. In the first case, interface position is defined through an averaging. Fluctuations in local interface position prevent full evolution of the peak by partially destroying interference buildup. In the second case interface position is defined at the center of the beam. Convexity makes this a slight over estimation since at the border of the probe beam, local interface position is slightly behind that at the center. Since the TRR technique allows determination of kinetics by measuring the time between extrema, caution must be exerted into data recording and analysis. Without these considerations, the TRR technique would introduce overestimates in laser annealing experiments in silicon and underestimates with degraded growth fronts in III-V materials. Similar loss of interference contrast can also be caused by polycristalline material formation [1]. In the exemples presented here it has been check by TEM analysis that the regrown layers had no trace of polycristals [4].

In Fig.4 a typical case of interface roughness evolution in GaAs has been shown. The material is (111) oriented GaAs substrate implanted with As+ ions at energies leading to a different initial amorphized depth in each case. Regrowth took place at T=320°C. As depth is increased the interface has more time to degrade during regrowth. It also shows reflectivity oscillation shortening with time.

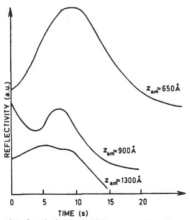

Fig.4 : TRR signals obtained during SPE regrowth of amorphized (111) GaAs samples with various initial amorphous thickness.

There is good agreement between the probabilistic description of the TRR result for a rough regrowth front and the experimental data.

CONCLUSION

It has been shown that the time-resolved reflectivity technique provides useful information on interface structure evolution during SPE in two cases. In a laser annealing configuration, it is sentitive to the curvature of the growth front induced by the inhomogeneous temperature distribution, while in III-V materials it detects roughness evolution at the interface. These examples show the versatility of the TRR technique in studies of solid phase epitaxial growth phenomena.

ACKNOWLEDGMENTS

The authors would like to thank Drs N. Duhamel, B. Descoux and P. Krauz for performing the implantation schemes and F. Bonnouvrier for computational assistance.

REFERENCES

1. G.L. Olson, J.A. Roth, Y. Rytz-Froidevaux and J. Narayan, Mat. Res. Soc. Symp. Proc., 35, 211 (1985)

2. A. Maruani, Y.I. Nissim, F. Bonnouvrier and D. Paquet in Laser-solid interactions and transient thermal processing of materials, edited by J. Narayan, W.L. Brown and R.A. Lemons, (North Holland, New York, 1982) p.123

3. L. Csepregi, J.W. Mayer and T.W. Sigmon, Phys. Lett. 54A, 157 (1978)

4. C. Licoppe, Y.I. Nissim, J. of Appl. Phys. 58, 3094 (1985)

5. G.L. Olson, S.A. Kokorowski, J.A. Roth and L.D. Hess in Laser and electron beam solid interactions and materials processing, edited by J.F. Gibbons L.D. Hess and T.W. Sigmon (North Holland, New York, 1981) p.125

6. C. Licoppe, Y.I. Nissim and C. Meriadec to be published.

LASER INDUCED OXIDATION OF HEAVILY DOPED SILICON*

E. FOGARASSY**, C.W. WHITE, D.H. LOWNDES, AND J. NARAYAN
Solid State Div., Oak Ridge National Laboratory, Oak Ridge, TN 37831, USA
**Centre de Recherches Nucléaires, Laboratoire Phase, 23, rue du Loess,
F-67037 Strasbourg Cedex, FRANCE

ABSTRACT

We have investigated the incorporation of oxygen into heavily doped silicon during UV excimer laser irradiation. For the case of repetitive laser irradiations in air, we observe that the amount of oxygen incorporated into Si depends markedly on the dopant. For As and Sb implanted silicon, there is no anomalous oxygen incorporation. By contrast, increasing amounts of O are incorporated into In implanted silicon as a function of number of laser shots. The incorporation of O is associated with degradation of the optical and structural properties of the surface, and a deep diffusion of the dopant. This behavior is believed to be partly related to specific chemical reactions between oxygen and indium present in the surface at high concentrations as the result of dopant segregation during solidification.

INTRODUCTION

In the past few years, several investigations have been performed to examine the incorporation of oxygen into silicon during laser annealing (in an oxygen or air environment) using continuous wave [1–4] or pulsed lasers [5], working in solid and liquid phase regimes respectively. Recent work performed on virgin crystalline silicon using ultra-violet laser irradiation resulted in a high rate of oxidation of the sample [6,7]. The oxide formation was attributed to adsorption of oxygen by the molten silicon when the near surface was melted by the laser treatment. In this study, we have investigated the incorporation of oxygen into heavily doped silicon during repetitive UV laser irradiation in air or in a controlled atmosphere. We observe that the rate of oxygen incorporation depends markedly on the dopant. In addition, O incorporation is associated with important modifications in the properties of the doped layer; these include degradation of optical and structural properties of the surface, a reduction in the dopant solubility, and deep diffusion of the dopant.

EXPERIMENTAL CONDITIONS

Different dopants such as arsenic (100 keV), antimony (150 keV) and indium (125 keV) were implanted into FZ silicon substrates of <100> orientation, to doses in the range 5×10^{15} cm^{-2} to 5×10^{16} cm^{-2}. Laser annealing was carried out in air using repetitive UV pulses (1 Hz, 35 nsec FWHM), provided by a KrF excimer laser (λ = 248 nm). The laser energy density could be varied between 0.5 and 1.3 J/cm^2. The amount of oxygen incorporated and the total and substitutional dopant distribution profiles were determined by Rutherford backscattering and ion channeling measurements performed along the <110> direction, using a 2.5 MeV ^{4}He$^+$ ion beam (diameter ≈ 1 mm). The energy of the backscattered particles was measured by a cooled detector, permitting a depth resolution of about 150Å. The structural quality of selected samples was analyzed by transmission electron microscopy.

*Research sponsored by the Division of Materials Sciences, USDOE under contract DE-AC05-840R21400 with Martin Marietta Energy Systems, Inc.

RESULTS

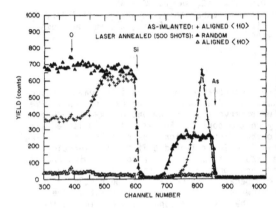

Figure 1: Backscattering spectra for ^{75}As (100 keV, 5×10^{16} cm^{-2}) implanted into <100> Si after excimer laser annealing (KrF, 1.2 J/cm^2, 500 shots).

Channeling spectra (along the <110> direction), reported in Fig. 1 show the recrystallization caused by repetitive laser irradiation (500 shots) of a 2000Å thick amorphous Si layer produced by As implantation (100 keV, 5×10^{16} cm^{-2}). Laser annealing was carried out at E = 1.2 J/cm^2. At this energy density the melt front penetrates both the amorphous and damaged region into the underlying substrate. Under these conditions eptiaxial regrowth occurs as shown by the low value of X_{min} (~3.5%) in the implanted region. The as-implanted As distribution profile is approximately gaussian with a peak near 650Å. Laser annealing causes a broadening of As both towards the surface and deeper in depth, which is characteristic of diffusion of As in liquid Si ($D_L^{As} \approx 1.0 \times 10^{-4}$ cm^2/sec [8]). After the repetitive laser treatment, we obtain a very uniform arsenic distribution, bounded by a sharp cut-off. This behavior reflects the fact that the effective diffusion distance of As in the liquid silicon after 500 shots is limited by the maximum melt depth (2500Å), a depth which is in reasonable agreement with the predictions of heat flow calculations. The As atoms are almost 100% substitutional in the silicon lattice, with a maximum substitutional concentration of 5×10^{21} cm^{-3}, as determined from channeling experiments; this value, which exceeds the retrograde maximum equilibrium solubility limit (1.5×10^{21} cm^{-3} at T = 1200°C [9]) does not depend on the number of laser shots (between 1 and 500 shots). In this case, oxygen incorporation is limited (as shown in RBS spectra of Fig. 1) to the formation of a thin surface oxide layer not exceeding 50—100Å thick. This is in good agreement with the thickness deduced from ellipsometry measurements. This thickness is comparable to that found on pulsed laser treated virgin Si ($\approx$ 60Å of SiO$_2$) after 500 KrF laser shots.

Similar experiments have been performed on a 2000Å thick amorphous Si layer produced by Sb implantation (150 keV, 5×10^{16} cm^{-2}). In this case, multipulse laser annealing (500 shots) was carried out at E = 1.0 J/cm^2. As shown in Fig. 2, epitaxial recrystallization occurs. In contrast to the results for As, the channeling yield stays high in the near surface region due to nonsubstitutional Sb atoms which are accumulated in the walls of a well defined cell structure which results from lateral segregation during solidification.

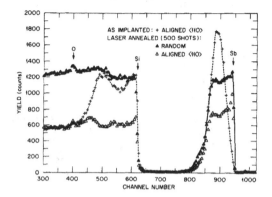

Figure 2: Backscattering spectra for [122]Sb (150 keV, 5×10^{16} cm^{-2}) implanted into <100> Si after excimer laser annealing (KrF, 1 J/cm^2, 500 shots).

The formation of a cell structure (which is observed by transmission electron microscopy) results from interfacial instability which develops during regrowth of the molten region and limits the maximum substitutional concentration of Sb to a value of 2×10^{21} cm^{-3} (the equilibrium solubility limit is 6×10^{19} cm^{-3} at T = 1200°C [9]). The distribution of Sb after 500 shots does not differ significantly from the distribution obtained after 1 pulse and is limited in depth by the thickness of the molten region ($\approx$ 2100A). The amount of oxygen incorporated is limited (after 500 shots) to a value corresponding to a surface oxide layer which does not exceed 100A in thickness.

By contrast, a large amount of oxygen is incorporated into In implanted (125 keV, 1.5×10^{16} cm^{-2}) silicon after multiple pulse laser irradiation as shown on channeling spectra obtained after 1 (Fig. 3) and 500 shots (Fig. 4). After 500 shots the amount of O incorporated into In doped Si is more than one order of magnitude higher (up to 5×10^{17}/cm^2) compared to As and Sb doped samples ($\approx$ 3×10^{16} O/cm^2). The compound formed is a complex mixture of silicon, oxygen, and indium. The proportion of In in the near surface does not exceed 1%. The atomic composition of this oxide layer has been estimated from the ratio of Si and O signal heights. The top surface consists of a thin stochiometric SiO$_2$ layer of about 200A thickness, covering a non-stochiometric SiO$_x$ (x < 2) compound. Due to the variation in composition with depth, we can obtain only a rough estimate of its thickness, which is of the order of 4000–5000A. TEM experiments revealed a very high level of structural damage in the surface region of this oxidized layer.

Several important modifications of the irradiated Si surface are associated with oxygen incorporation. There is a noticeable degradation.of surface optical properties. A dull surface develops which is believed to be induced by an increase of surface roughness. In addition, there is a considerable modification of the distribution profile of In following repetitive laser irradiation. After 1 shot the In profile is limited to a depth of ~1000A as shown in Fig. 3. After 500 shots In is clearly measured to depths exceeding 5000A, suggesting that melting occurred at least to this depth (Fig. 4). This result means that there is a strong modification of optical and thermodynamic properties of the surface oxidized layer as compared to pure silicon. Finally, we measure a substitutional In concentration of 3×10^{20} cm^{-3} after one shot (compared to the equilibrium solubility limit of ~ 8×10^{17}/ cm^3 [9]). The substitutional In concentration is limited by the formation of a cellular structure in the surface region. By contrast, after 500 shots the cell structure disappears, as confirmed by TEM, and channeling experiments show no preferred lattice site for In atoms in the oxidized layer.

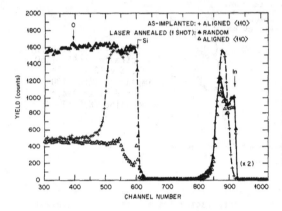

Figure 3: Backscattering spectra for ^{115}In (125 keV, 1.5×10^{16} cm^{-2}) implanted into <100> Si after excimer laser annealing (KrF, 1 J/cm^2, 1 shot).

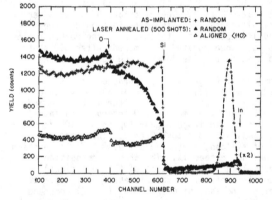

Figure 4: Backscattering spectra for ^{115}In (125 keV, 1.5×10^{16} cm^{-2}) implanted into <100> Si after excimer laser annealing (KrF, 1 J/cm^2, 500 shots).

DISCUSSION

During irradiation of In implanted (125 keV, 1.5×10^{16} cm^{-2}) Si in an argon atmosphere, we do not observe significant oxygen incorporation. This result indicates clearly that surface oxidation is mainly due to adsorption of oxygen from the atmosphere into silicon during the time the surface is molten. During regrowth of the molten layer, the adsorbed oxygen molecules can be trapped into the solid due to the very high regrowth velocity (V ≈ 5-6 m/sec as deduced from heat flow calculations [11]). However, oxygen trapping is observed to be strongly dependent on the nature of the dopant, since substantial incorporation is only observed with indium. The substitutional solubility of In in laser annealed Si is significantly lower than for As and Sb. Moreover, significant differences have been found between interfacial segregation coefficients for these elements during rapid solidification of silicon: $K_{As} \approx 1$, $K_{Sb} \approx 0.8$, and $K_{In} \approx 0.2$ for regrowth velocities near 5 m/sec[10,12].

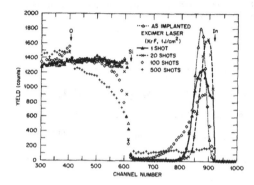

Figure 5: Backscattering spectra for [115] In (125 keV, 1.5×10^{16} cm^{-2}) implanted into <100> Si after excimer laser annealing as a function of the number of laser shots.

Random spectra plotted in Fig. 5 which show the evolution of the In distribution profiles as a function of the number of laser shots, suggest a possible relation between the segregation coefficient and O incorporation. Before diffusing deeply into laser treated Si layer, In atoms segregate strongly towards the surface during the first 20 laser shots. This effect is due to the value of segregation coefficient of In which is significantly lower than 1 during solidification. It is possible that a chemical reaction occurs between oxygen (in air) and indium located at high concentrations in the surface or in cell walls. (We note that the large amount of In atoms accumulated by segregation into the walls of cellular structure following the first laser pulse cannot be responsible alone for the mechanism of O incorporation since cells are formed in the Sb implanted case but no anomalous O incorporation is observed.) After 20 laser shots only the surface of sample is oxidized as confirmed by the presence of a small peak of oxygen ($\approx 5 \times 10^{16}$ cm^{-2}). Upon increasing the number of laser pulses, we observe a sharp increase of O incorporated ($\approx 5 \times 10^{17}$ cm^{-2} after 500 shots) into the irradiated layer, resulting in a large modification of Si random spectrum between 20 and 500 shots. Deep O incorporation is closely related to the extent of the indium distribution profile which diffuses deeper and deeper into the treated layer as a function of the number of shots. A large amount of incorporated O is also found following multipulse laser treatment of bismuth implanted Si. Bi also has an interfacial segregation coefficient which is significantly lower than 1 (≈ 0.3 for V ≈ 5 m/sec [12]). However, further experiments are needed to determine the exact nature of the chemical reactions responsible for oxygen incorporation into In and Bi doped Si. It is clear, however, that the photon wavelength is not important for the mechanism of oxygen incorporation. We have observed similar results using multiple pulse Ruby laser irradiation ($\lambda = 0.694$ nm) of silicon crystals implanted to high doses by In and Bi.

CONCLUSION

We have demonstrated that incorporation of oxygen into heavily doped silicon during multiple pulse laser irradiation is closely related to the nature of dopant. No anomalous oxygen incorporation is observed for the case of As and Sb implanted silicon. By contrast, massive oxygen incorporation is observed into the near surface of In and Bi implanted silicon during repetitive laser irradiations. This behavior which may be related to surface segregation phenomena during laser annealing, is associated with degradation of the optical and structural properties of the surface and deep diffusion of the dopant.

REFERENCES

1) J.F. Gibbons, Jpn J. Appl. Phys. Suppl. 19, 121 (1980).

2) I.W. Boyd and J.I.B. Wilson, Thin Solid Film, 83, L173 (1981).

3) I.W. Boyd and J.I.B. Wilson, Appl. Phys. Lett. 41, (2), 162 (1982).

4) S.A. Schafer and S.A. Lyon, J. Vac. Sci. Technol., 19, 494 (1981).

5) K. Hoh, H. Koyama, K. Uda, and Y. Miura, Jpn J. Appl. Phys. 19, n°7, L375 (1980).

6) Y.S. Liu, S.W. Chiang, and F. Bacon, Appl. Phys. Lett. 38 (12) 1005 (1981).

7) T.E. Orlowski and H. Richter, Appl. Phys. Lett. 45 (3), 241 (1984).

8) P. Baeri, S.U. Campisano, G. Foti, and E. Rimini, Appl. Phys. Lett. 33 (2), 137 (1978).

9) F. Trumbore, Bell Syst. Techn. J. 39, 205 (1960).

10) C.W. White, Laser and Electron Beam Interactions with Solids, edited by B.R. Appleton and G.K. Celler (North Holland, 1982) p. 109.

11) M. Aziz (private communication).

12) J.M. Poate, Laser and Electron Beam Interactions with Solids, edited by B.R. Appleton and G.K. Celler (North Holland, 1982) p. 121.

RAMAN STUDIES OF PHASE TRANSFORMATIONS IN PULSE LASER IRRADIATED DIELECTRIC FILMS

GREGORY J. EXARHOS
Pacific Northwest Laboratory,* P. O. Box 999, Richland, WA 99352

ABSTRACT

Sputter deposited TiO_2 and ZrO_2 films on silica have been characterized with regard to thickness, phase purity, and residual film stress using nondestructive Raman spectroscopic techniques. Irradiated anatase coatings exhibit vibrational features that irreversibly shift to higher frequencies, suggesting an increase in localized stress. Catastrophic damage is manifested by the appearance of a higher density rutile phase. Rutile coatings examined after irradiation show only intensity changes, while amorphous coatings were found to rapidly crystallize under high-pulse energy irradiation. Results of these studies are compared with equilibrium Raman measurements of thermally-induced transformations in these materials.

INTRODUCTION

Refractory oxides are widely used in the fabrication of multilayer structures designed to have specific wavelength-dependent optical properties. It is desirable to characterize the response of such dielectric coatings to high-energy pulsed laser irradiation on a molecular level to enable mechanistic interpretation of the interaction. Rapid energy transfer to the coating can produce catastrophic failure by thermal processes resulting in phase transformations (chemical bond rearrangement) or by multiphoton excitation/ionization processes (bond breaking).

Vibrational Raman spectroscopy has been successfully used as a localized structural probe of submicron dielectric films on silica [1,2]. Band intensities were shown to vary linearly with thickness over two orders of magnitude, and phase composition could be readily determined from measured spectra. For coatings thinner than .5μ, substrate scattering becomes appreciable. Methods have been developed to minimize this effect based upon the scattering geometry and polarization properties of the scattered radiation [3,4]. Characterization of very thin films is difficult owing to the inherent low Raman scattering efficiency; however, intensity enhancement using the interference properties of the dielectric film has been observed, making Raman measurements tractable [1,5,6]. Macroscopic sample properties, such as inherent film stress and surface temperature during irradiation, can also be obtained from measured Raman spectra and are manifested in band frequency shifts and intensity variations [7-9].

The application of Raman spectroscopy to characterize thermally-induced phase transformations in thin-film dielectrics has been previously reported [10]. Polycrystalline anatase films sputter-deposited on silica were found to irreversibly crystallize into the rutile phase at temperatures exceeding 800°C accompanied by marked contraction and catastrophic cracking of the coating due to the density change. Time-resolved Raman measurements during

*Pacific Northwest Laboratory is operated by Battelle Memorial Institute for the U. S. Department of Energy under contract DE-AC06-76RLO 1830.

heating of an amorphous TiO_2 coating showed that recrystallization to a rutile phase initiated at temperatures below 200°C. Rutile coatings, on the other hand, were stable up to the melting point of the silica substrates, verifying that rutile is the thermodynamically stable phase.

The response of these and similar oxidic dielectric films on silica to high-energy pulsed-laser irradiation has been characterized in this work using Raman spectroscopy as a molecular structural probe. Results indicate that dielectric films recrystallize to the thermodynamically stable phase at a sufficiently high pulse energy.

EXPERIMENTAL

Single-layer dielectric films (ca. 1μ thick) were prepared by reactively sputtering Ti or Zr in Ar/O_2 atmospheres onto fused silica substrates in an rf diode system [11]. Amorphous or polycrystalline films (150 nm grain size) could be prepared by varying the deposition conditions. Pure phase TiO_2 (anatase or rutile) or mixed phase films were also prepared. Polycrystalline grain orientation was achieved for several of the rutile films. HfO_2 was sputter-deposited in a similar manner to demonstrate the utility of Raman measurements for film characterization.

Raman spectra from samples mounted in a 180° backscattering geometry, $Z(X\bar{Y})\bar{Z}$, were excited by 488 nm CW emission from an Argon ion laser or low-energy 532 nm excitation from a pulsed Nd:YAG laser operating at a repetition frequency of 30 Hz. Scattered Raman radiation was collected at f/1 and imaged onto the slits of a 0.85 m SPEX Model 1403 double monochromator. Dispersed radiation was detected by a conventional photomultiplier arrangement or in a time-resolved manner using an intensified gated array detector (TN Model 1633). Coarse ruled gratings (150 g/mm) were used with the array detector giving an effective spectral range of ca. 1000 cm^{-1}. This arrangement also required that a narrow band rejection filter centered at the Raman probe wavelength be inserted in front of the entrance slits to minimize stray light in the monochromator. In order to study the effect of pulsed irradiation on dielectric films, CW excitation from the ion laser was combined with pulsed radiation (532 or 355 nm) from a Quantel Model 580 Nd:YAG laser through a dichroic mirror. The pulsed excitation was brought to a 1 mm focus at the sample, while the CW probe was focused to a 100μ spot size. In this manner, the sample could be characterized before and after a single high-energy pulse was delivered to the surface.

RESULTS

Sample Characterization

Raman spectra for the two tetragonal phases of TiO_2, anatase and rutile, are well documented [12]. Anatase exhibits vibrational features at 143, 199, 398, 515, and 636 cm^{-1} while the rutile bands occur at 244, 440, and 608 cm^{-1}. The stoichiometry of mixed phase coatings may be deduced from measured spectra [2]. Raman features from .9μ films on silica agree with bulk frequencies; however, a slight shift of the anatase 143 cm^{-1} mode to higher frequencies is indicative of compressive stress in the film [7]. Evidence for polycrystalline grain orientation in .6μ rutile films is observed in the polarized Raman scattering. When grain disorder is present, the A_{1g} mode at 608 cm^{-1} is observed in the $Z(XY)\bar{Z}$ geometry; this feature is absent in coatings that have a high degree of order. Glassy TiO_2 films (.6μ thick) show very weak broad Raman lines which form an envelope around the crystalline rutile features.

Raman spectra of polycrystalline ZrO_2 films (.6μ thick) are characteristic of the monoclinic phase with principal lines at 751, 636, 612, 535, 556, 500, 474, 379, 344, 332, 303, 220, 189, 177, and 99 cm^{-1}. Raman scattering from an amorphous film (1.1μ) yields a relatively featureless and weak spectrum. Polycrystalline HfO_2 films (.5μ thick) exhibit spectra characteristic of the monoclinic phase with features quite similar to those seen in ZrO_2 films. The strong 493 cm^{-1} band in bulk crystals, which corresponds to the 474 cm^{-1} ZrO_2 mode, is often observed at higher frequencies (496 cm^{-1}) in thin films, which suggests that it correlates with residual film stress produced during sputter deposition.

Equilibrium Damage Measurements

A .9μ anatase film sputter deposited on silica was subjected to .532 nm pulsed radiation from a Nd:YAG laser. Spectra before and after irradiation are displayed in Figure 1. Relative band intensities decrease, bands broaden by 50%, and exhibit frequency shifts. An increase in Rayleigh scattering is also observed. When the same film is exposed to multiple pulse irradiation at a higher pulse energy, new vibrational lines appear as shown in Figure 2. The bands at 440 and 607 cm^{-1} are indicative of the rutile phase induced by the pulsed excitation.

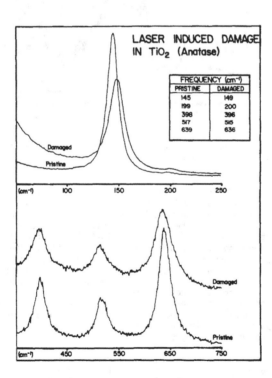

LASER INDUCED DAMAGE IN TiO_2 (Anatase)

FREQUENCY (cm⁻¹)	
PRISTINE	DAMAGED
145	149
199	200
398	396
517	515
639	636

FIG. 1. Raman spectra of a .9μ anatase film on silica before and after exposure to a single 532 nm high-energy pulse (488 nm CW excitation).

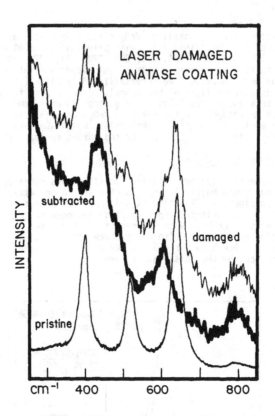

FIG. 2.   Difference Raman spectrum of multiple pulse damage to a .9μ single-layer anatase film.

Raman spectra of single-phase rutile coatings (.9μ thick) were investigated under similar conditions. Little change in the Raman spectrum following irradiation was observed; however, the relative band intensities were observed to decrease. The same measurement was performed on a poly-crystalline rutile film characterized by a high degree of grain orientation. In this case, Raman spectra acquired before irradiation in the $Z(XY)\bar{Z}$ scattering geometry showed an absence of the 607 cm^{-1} A_{1g} mode. Following irradiation, the 607 cm^{-1} mode was observed. These results suggest that the pulsed excitation has induced disorder in the grains and that the film has recrystallized. Raman spectra of high-energy pulse-irradiated mixed-phase TiO_2 coatings indicate preferential removal of the anatase phase with respect to the rutile phase. All irradiated coatings show catastrophic surface damage characterized by cratering and cracks propagating into undamaged regions.

Amorphous coatings also exhibit irreversible phase transformation phenomena following pulsed laser irradiation. Figures 3 and 4 depict changes in measured Raman spectra for amorphous TiO_2 and ZrO_2. These films are presumed to be amorphous based on the absence of crystalline X-ray diffraction features. The diffuse Raman spectrum for the ZrO_2 film supports this contention; however, the TiO_2 film (while exhibiting a relatively weak spectrum) exhibits shoulders at frequencies characteristic of an anatase phase. "Amorphous TiO_2," therefore, may be a conglomerate of polycrystalline anatase and rutile in an amorphous matrix. Following 532 nm irradiation, TiO_2 has fully recrystallized into the rutile phase. Upon irradiation at 355 nm, the amorphous ZrO_2 film exhibits Raman features characteristic of the monoclinic phase.

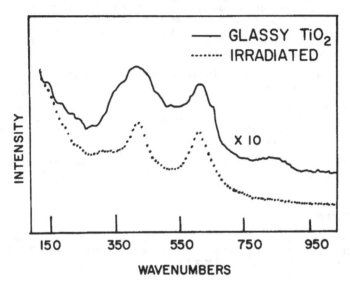

FIG. 3. Raman spectrum of pulsed laser irradiated glassy TiO_2 compared with that of the pristine coating (532 nm damage pulse).

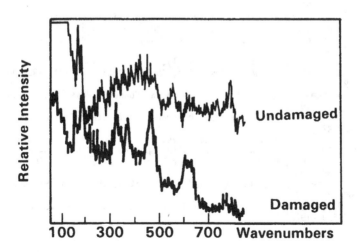

FIG. 4. Raman spectrum of pulsed laser irradiated glassy ZrO_2 compared with that of the pristine coating (355 nm damage pulse).

184

DISCUSSION

The interaction of high-energy laser pulses with dielectric films often results in catastrophic damage to the film accompanied by an irreversible transformation to the thermodynamically stable phase. This transformation occurs at relatively low temperatures (below 900°C) for amorphous TiO_2 and the anatase modification and results in the growth of a rutile phase for both materials. The rutile phase is stable at temperatures near the melting point of the substrate and does not transform to other phases under equilibrium conditions [2]. A similar effect must occur in glassy ZrO_2 albeit at much higher pulse energies (high surface temperature) resulting in the monoclinic phase at room temperature.

At laser pulse energies below those required to induce the irreversible phase transformation in polycrystalline anatase coatings, Raman measurements suggest the existence of induced film stress as manifested by a shift in frequency of the 143 cm^{-1} Eg mode. Based upon previous work regarding Raman band shifts as a function of applied pressure [7], the irradiated film yields a spectrum comparable to that from a bulk crystal under ca. 10 Kbar of pressure. This metastable state could result from partial transformation of the film to rutile with stress building up along grain boundaries or from substrate/film relaxations caused by a difference in thermal properties. The stressed state definitely accompanies catastrophic damage to the film and may be a precursor to such damage.

ACKNOWLEDGEMENT

This work was supported by the Air Force Weapons Laboratory under Contract PO-85-037. Sputter-deposited coatings were supplied by Dr. W. T. Pawlewicz at Pacific Northwest Laboratory.

REFERENCES

1. W. T. Pawlewicz, G. J. Exarhos, and W. E. Conaway, Applied Optics 22(12), 1837-1840 (1983).
2. G. J. Exarhos, in Applied Materials Characterization, Vol. 48. (Materials Research Society, Pittsburgh, 1985) pp. 461-469.
3. S. H. Long, C. Y. She, and G. J. Exarhos, Applied Optics 23(18), 3049-3051 (1984).
4. G. J. Exarhos, J. Chem. Phys. 81(11), 5211-5212 (1984).
5. G. J. Exarhos, and W. T. Pawlewicz, Applied Optics 23(12), 1986-1988 (1984).
6. R. J. Nemanich, C. C. Tsai, and G. A. N. Connell, Phys. Rev. Lett. 44, 273-276 (1980).
7. T. Ohsaka, S. Yamaoka, and O. Shimomura, Solid State Commun. 30, 345-347 (1979).
8. G. A. Samara, and P. S. Peercy, Phys. Rev. B. 7, 1131-1148 (1973).
9. M. Malyj and J. E. Griffiths, Applied Spectrosc. 37(4), 315-333 (1983).
10. G. J. Exarhos, Proc. Soc. Photo-Opt Instrumentation, Eng. 540, 460-466 (1985).
11. W. T. Pawlewicz, D. C. Hays, and P. M. Martin, Thin Solid Films 73, 169-175 (1980).
12. R. J. Capwell, K. Spagnolo, and M. A. DeSesa, Appl. Spectrosco. 26, 537-539 (1972).

FORMATION OF Al$_x$Ga$_{1-x}$As ALLOY ON THE SEMI-INSULATING GaAs SUBSTRATE
BY LASER BEAM INTERACTION

N. V. JOSHI* and J. LEHMAN
Honeywell Physical Sciences Center, 10701 Lyndale Avenue South,
Bloomington, MN 55420

ABSTRACT

Al$_x$Ga$_{1-x}$As alloy was obtained on semi-insulating GaAs by laser beam interaction. For this purpose, thin layers ($\sim$150 A) of AlAs and GaAs were deposited by MOCVD on the semi-insulating undoped GaAs substrate and irradiated with a high power laser beam. For a certain critical value of incident power (7.1 watts/mm^2), a layer of Al$_x$Ga$_{1-x}$As alloy was formed. The nature of the alloy was examined by Auger and Far infrared reflectance spectroscopy. The later technique reveals the characteristic mode of Al$_x$Ga$_{1-x}$As confirming its formation. Disorder induced LA mode was also observed.

INTRODUCTION

Recently, there has been a clear understanding of the importance of Al$_x$Ga$_{1-x}$As in the technology of electro-optical devices, particularly the formation of Al$_x$Ga$_{1-x}$As - GaAs heterojunctions[1-3]. The technological impact has come from the development of several types of diode lasers[1], optical buried structures[1,2], and other devices where the Al$_x$Ga$_{1-x}$As system plays a crucial role. Thin layers of Al$_x$Ga$_{1-x}$As on a GaAs substrate are generally obtained by well established techniques such as molecular beam epitaxy (MBE), liquid phase epitaxy (LPE), etc. There is no doubt that the above mentioned techniques are precise and reproducible, but for production purposes, they are rather expensive. The purpose of the present investigation, therefore, is to examine the possibility of the formation of the above system by a laser beam interaction technique which is generally computer-controlled, and hence, possesses a high degree of reproducibility. Moreover, it is possible to control the width of the alloy on the order of a micron, and therefore, circuit writing by laser beam could be practical.

In this paper we describe the formation of Al$_x$Ga$_{1-x}$As alloy on a semi-insulating GaAs substrate. The nature of the alloy is examined by Auger and Far infrared reflectance spectroscopy. The quality of the alloy from a device point of view is not analyzed in the present work. However, similar techniques have been employed for silicon devices[4].

EXPERIMENTAL

A thin layer of AlAs (150 A) and over it a layer of GaAs (150 A) were grown on undoped GaAs (100) by metal organic chemical vapor deposition (MOCVD) at 800°C. These layers were then irradiated by a high power laser beam obtained from a cw argon ion laser for a few seconds (Spectra Physics model 171-18). Power from all wavelengths was used and the beam was focused to a spot of 1 mm diameter. The incident power was varied from 5 to 8 watts/mm^2, and for this intensity, it was found that the interdiffusion has taken place between deposited layers (300 A). The thickness of AlAs and GaAs before irradiation was the same. Naturally, the percentage of aluminum and gallium in the alloy is also expected to be the same. In order to obtain direct experimental confirmation of the formation of

---

*On sabbatical leave from the University of Los Andes, Merida, Venezuela.

$Al_{0.5}Ga_{0.5}As$ on GaAs substrate, far infrared reflection spectroscopy was found to be adequate.

For reflectance measurements, Digilab FTIR spectrometer (model FIS-146) was employed. Spectra of several laser-treated samples were obtained at 300k. The angle of incidence was 13° from normal. The reference energy spectrum was obtained using the adjacent area which was not treated by laser. The reflectance spectrum of each sample, therefore, is a clear indication of the changes in the sample made by laser irradiation. In order to achieve a good signal/noise ratio, the signal was averaged over 5000 scans. The spectral resolution obtained in the present case was reasonably good ($4 cm^{-1}$).

RESULTS AND DISCUSSION

In order to get information about the homogeneity of the epitaxial layer (distribution of aluminum, gallium, and arsenic atoms), the sample was examined by using depth profile analysis. It has been found that 7.1 watts/mm^2 is an optimum power to obtain nearly homogeneous distribution of Al and Ga. A typical Auger electron spectrum is shown in Figure 1.

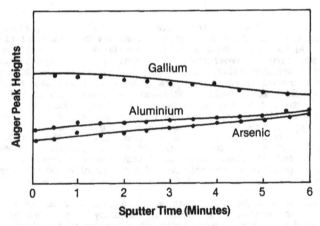

Figure 1. Auger electron spectrum of laser-grown $Al_xGa_{1-x}As$. (Sputter rate: 1 minute = 50 A)

It can be seen that gallium and aluminum maintain nearly the same concentration up to 300A within the sample (oxygen and carbon contamination was observed). It is true that distribution of aluminum and gallium is an indication for the formation of the alloy but does not guarantee the desirable role of aluminum and gallium in the bonding process of the alloy. Hence, it is necessary to investigate vibrational modes of the alloy formed by the laser-assisted technique. This will provide information about the site symmetry of the constituent elements if there is any misorder.

The alloy made in the present investigation is very thin (~300 A) and it is on the GaAs substrate, which has nearly the same value of the lattice constants, and therefore, X-ray diffraction method could not work to identify the formation of an alloy. Considering the change in the vibrational modes and thinness of the layer, Far infrared reflectance spectroscopy is a more adequate technique to identify the compounds and a typical reflectance spectrum recorded for the sample radiated with laser power of 7.5 watts/mm² is shown in Figure 2.

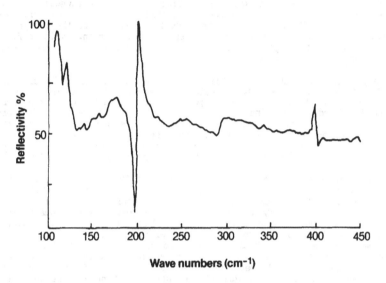

Figure 2. IR reflectance spectrum of $Al_xGa_{1-x}As$ – GaAs heterostructure after laser irradiation.

It is well known that the[5,6] vibrational spectrum of an alloy shows two phonon behavior, i.e., the prominent phonon structure of GaAs and AlAs should be revealed with the slight displacement of the peak in energy corresponding to the concentration of Al (or Ga).

Vibrational modes by Raman spectroscopy[7] and IR reflectance spectroscopy[8] of $Al_xGa_{1-x}As$ for various values of x have been extensively studied earlier. Even the epitaxial layer of $Al_xGa_{1-x}As$ on the substrate of GaAs has been investigated in detail by Durschlag, et. al.[8]. Because of the similarity in the structure [$Al_xGa_{1-x}As$ – GaAs], we compare the values of dominant transverse optical (TO) modes and longitudinal optical (LO) modes with the values reported earlier[8].

Two phonon behavior for $Al_xGa_{1-x}As$ alloy is well studied[7-9] and measurements for a particular value of x have been reported. For x=0.5, it is observed that TO and LO modes lie at 267 cm⁻¹ and 290 cm⁻¹, respectively. (See Table 1, Reference 8.)

We did observe weak but well marked peaks at 260 cm⁻¹ and at 290 cm⁻¹. A TO mode was reported[7,8] both by Raman and IR spectroscopy at about ⁻265 cm⁻¹ and the observed value in the present investigation is very close to the reported value. A value for the LO mode obtained in the present investigation lies slightly to higher energy than the earlier reported value[7] for a bulk sample. However, it has been found that in $Al_xGa_{1-x}As$ – GaAs submicron heterostructure, the LO peak is shifted to higher energy by 10 cm⁻¹. Durschlag has reported this peak at ⁻290 cm⁻¹ for a thin layer, which is close to the observed value in the present investigation, and therefore, the peak located at 290 cm^{-1} could be attributed to GaAs like LO mode.

A strong peak corresponding to AlAs like LO mode is expected [8] at 395 cm^{-1}. Our experimental results also confirm the same. Sharp structure is detected at ⁻395 cm^{-1}. In short, important features of the two phonon behavior are clearly revealed in the present investigation. This confirms that the alloy $Al_xGa_{1-x}As$ is really formed on the substrate of GaAs by high power laser beam interaction.

In $Al_xGa_{1-x}As$, the atoms of gallium (aluminum) are replaced by aluminum (gallium). This breaks the translation symmetry; k selection rule is violated, and therefore, one expects disorder activated transverse acoustical and longitudinal acoustical modes. This behavior is more obvious in the present case where the alloy is not formed with the conventional methods and density of the defects is expected to be high. Naturally, the intensity of the defect activated mode should also be higher. Experimentally observed spectrum was examined with this view.

Figure 2 shows a high intensity peak located at 200 cm^{-1}. This peak has been reported earlier by several investigators[7,9,10] and identified as defect activated LA mode (DALA). The value observed in the present investigation agrees very well with the value reported by previous workers (⁻200 cm^{-1}). Intensity of DALA mode observed in the present investigation is certainly high as compared to the earlier data[10]. The intensity for this mode has been theoretically calculated by Talwar, et. al.[12] on the basis of the parameters obtained from neutron scattering for GaAs and AlAs. They have used the simple perturbation model which takes into account only the change in the nearest neighboring coupling constant and the variation of mass at a particular site symmetry. This model is probably more adequate since long order is not expected in the alloy made by laser-assisted growth. According to this model, the intensity of DALA mode is much higher than reported earlier[10] and is very close to the intensity observed in the present investigation.

In the present paper, we do not intend to analyze all the details of the infrared reflectance spectra and defect-induced modes in the light of the variation of the force constants, but we wish to confirm the formation of an alloy $Al_xGa_{1-x}As$ with the laser-assisted technique.

From the information obtained in the present investigation, it seems that there exists a threshold power for laser interaction below which the alloy is not formed. Such laser-treated material when examined with IR reflectance shows the GaAs structure with slight modification. Details of the formation of an alloy with respect to structural changes will be discussed elsewhere.

CONCLUSION

In this paper, we have shown that it is possible to grow epitaxial layer of $Al_xGa_{1-x}As$ on the substrate of semi-insulating GaAs. The presence of interfacial impurities does not seem to hinder the formation of an alloy. Oxygen contamination is observed even up to 250 A, but it can be reduced by maintaining an inert atmosphere. A systematic study is needed to reduce the intensity of the defect-activated mode and to achieve device quality.

Acknowledgements: One of the authors (N. V. Joshi) is grateful to Prof. Laude, Mons, Belgium for introducing this exciting field. The authors are thankful to Mike Campbell for valuable discussion on experimental aspects.

The assistance of Lee Hallgren and Al Beck is appreciated for IR and Auger spectra.

REFERENCES

1) H. Kresel and J.K. Butter, "Semiconductor Lasers and Heterojunction LEDS," Academic Press, New York (1977).

2) H.C. Casey, Jr. and M.B. Panish, "Heterostructure Lasers," Academic Press, New York (1978).

3) R. Olshansky, C.B. Su, J. Manning, and W. Powazinik, IEE Journal of Quantum Electronics QE20, No. 8, 838 (1984).

4) M. Tamura, N. Natsuaki, M. Miyao, M. Ohkura, F. Murai, E. Takeda, S. Minagawa, and T. Tokuyama, "Laser and Electron Beam Interactions with Solids," Eds. B.R. Appleton and G.K. Celler, [Elsevier Science Publishing Company, New York (1982)], p. 567.

5) I.F. Chang and S.S. Mitra, Adv. in Phys. 20, 359 (1971).

6) A.S. Barker and A.J. Sievers, Rev. of Modern Phys. 47, 52 (1975) and references therein.

7) B. Jusserand and J. Sapriel, Phys. Rev. B, 24, 7194 (1981).

8) M.S. Durschlag and T.A. DeTemple, Solid State Communication, 40, p. 307 (1981).

9) O.K. Kim and W.G. Spitzer, J. Appl. Phys. 50, 4362 (1979).

10) N. Saiut-Cricq, R. Carles, J.B. Renucci, A. Zwick, and M.A. Renucci, Solid State Comm., 39, 1137 (1981).

11) J.L.T. Waugh and G. Dolling, Phys. Rev. B7, 3481 (1973).

12) D.N. Talwar, M. Vandevyver, and M. Zigone, Phy. Rev. B, 23, 1743 (1981).

OSCILLATORY MORPHOLOGICAL INSTABILITIES DURING RAPID SOLIDIFICATION — THE ROLE OF DIFFUSION IN THE SOLID

ATUL BANSAL AND ARIJIT BOSE
Department of Chemical Engineering
University of Rhode Island
Kingston, RI 02881

ABSTRACT

Recent results by Coriell and Sekerka [J. Crystal Growth, 61, 499 (1983)] on the oscillatory instability of a planar rapidly solidifying binary melt are extended to include diffusion in the solid phase. Under assumptions equivalent to those made by Coriell and Sekerka, it is shown that no matter how small the diffusion coefficient is in the solid, the system is stable to all oscillatory and non-oscillatory disturbance modes if the modified constitutional supercooling criterion is satisfied and if the non-equilibrium segregation coefficient is zero. Thus, a range of the non-equilibrium segregation parameter exists where these results allow the possibility of instability, whereas Coriell and Sekerka predict that the system will be stable.

System stability is increased for both oscillatory and non-oscillatory modes. It is necessary for the diffusivity ratio D_s/D_l to be nearly 0.1 before oscillatory modes are affected. Both the critical wavelength of the disturbance as well as the oscillation frequency are reduced slightly from the case where diffusion in the solid is ignored.

1. INTRODUCTION

Interest in rapid solidification has increased sharply in the past few years because of economical ways of achieving it through laser and electron beam annealing or splat cooling [1]. The key advantages of rapid solidification processing over conventional rates of solidification are the ability of the fast moving interface to trap solute atoms, thus producing solids of composition far different from those predicted from equilibrium phase diagrams. Several models [2,3] have been developed which predict the distribution coefficient as a function of the interface velocity, and this phenomenon has been demonstrated experimentally [4]. Additionally, the rapid motion of the interface allows little time for the growth of nucleation sites, thus producing unique microstructures. These advantages have been exploited extensively in producing superalloys.

In all these processes the performance of the finished product depends strongly on the uniformity of solute distribution in the solid phase. It is well known that microsegration free solids can only be produced by either partitionless solidification or planar growth. Since partitionless solidification is nearly impossible to achieve at practically realisable growth velocities, the importance of planar growth becomes apparent. Recent studies on morphological stability of a rapidly solidifying binary melt [5,6] have included interface kinetics [5] and non-equilibrium solute segregation [6]. This kind of study becomes especially important because classical theory predicts absolute stability beyond a critical growth velocity [7]. In [6], henceforth referred to as I, Coriell and Sekerka show that non-equilibrium solute segregation causes oscillatory instability to develop by a solute pumping mechanism at critical concentrations far below that required for non-oscillatory instabilities to occur. The effect

of this instability is a periodic segregation pattern in the growth
direction, which has recently been experimentally observed by Boettinger
et.al. [8].

We extend the work in I by including diffusion in the solid phase. The
motivation is severalfold: at temperatures near the melting point, the
ratio of the diffusion coefficient in the solid to that in the liquid can
be up to 10^{-2} so that it becomes unclear that ignoring solid diffusion is
realistic. Solid diffusion is necessary in describing melting problems
[9], and this has been the main restriction in applying solidification
results to melting. Diffusion in the solid phase is necessary when trying
to model the drop shedding phenomena in deep grooves [10]. From a
mathematical perspective, the problem is of interest because of the
singularity in the perturbed diffusion equation in the solid when the limit
$D_s \to 0$ is approached. The theory is outlined in section 2 – since the
formulation remains largely similar to I, only the additional governing
equations and boundary conditions necessary for accounting for diffusion in
the solid phase are shown. The key results along with a discussion are
presented in section 3.

2. THEORY

The physical system analysed, the choice of coordinate system and
notation remain the same as I. The subscript 'l' refers to the liquid
phase 's' to the solid phase. The diffusion equation in the solid is added
to the governing equations for solid and liquid temperature and liquid
composition (equations (1) in I)

$$\frac{\partial C_s}{\partial t} - v \frac{\partial C_s}{\partial z} = D_s \nabla^2 C_s, \tag{1}$$

and the far field boundary condition in the solid is added to those already
prescribed in I (equations (2))

$$C_s \text{ is constant} \quad , \quad z \to -\infty \tag{2}$$

The solute flux balance at the interface (equation (3c) in I) is modified
to include diffusion in the solid phase

$$v = (- D_l \frac{\partial C_l}{\partial z} + D_s \frac{\partial C_s}{dz}) / (C_l - C_s) \tag{3}$$

It is important to note that the steady state concentration profile in
the solid in still $C_s(z) = C_\infty$ where C_∞ is the far field concentration
in the liquid. The linear stability analysis is then carried out in the
standard way, where each of the temperature and concentration fields are
written as a sum of an unperturbed part, which is a function of 'z' alone
and a perturbed part of the form

$$F(z) = \exp[\sigma t + i(\omega_x x + \omega_y y)].$$

The position of the solid-liquid interface is given by

$$W(x,y,t) = d \exp[\sigma t + i(\omega_x x + \omega_y y)], \tag{4}$$

where d is the perturbation amplitude at t=0, and ω_x and ω_y are the spatial
frequencies in the x and y directions and σ the growth constant. Clearly,

the interface is unstable if the real part of σ is greater than zero for any perturbation.

Upon solving the differential equations and applying the associated boundary conditions, we find

$$\sigma = \frac{V\left\{[-k_1 G_1 (\alpha_1 - \frac{V}{\kappa_1}) - k_s G_s (\alpha_s + \frac{V}{\kappa_s})] \, U_A - 2\bar{k}T_m \Gamma \omega^2 \bar{\alpha} + \dfrac{2\bar{k}m' G_c \bar{\alpha}(\alpha - V/D_1)}{\alpha + \frac{D_s}{D_1}\beta - (\frac{D_s}{D_1}\beta + \frac{V}{D_1})p'}\right\}}{L_V V U_T + \dfrac{2\bar{k}m' G_c \bar{\alpha} \, [U_k + \frac{D_s \beta}{V}(U_k - 1)]}{\alpha + \frac{D_s}{D_1}\beta - (\frac{D_s}{D_1}\beta + \frac{V}{D_1})p'}} \tag{5}$$

The definitions of all the parameters in equation (5) are the same as those of equation (5) in I except

$$U_A = 1 - \mu_A/\mu_T + \frac{m' k_A C_o V \, (\frac{D_s \beta}{V} + 1) / D_1}{\alpha + \frac{D_s}{D_1}\beta - (\frac{D_s}{D_1}\beta + \frac{V}{D_1})p'}$$

and $\beta = -\dfrac{V}{2D_s} + \sqrt{(\dfrac{V}{2D_s})^2 + \omega^2 + \dfrac{\sigma}{D_s}}$

In the definition for β, we require that the real part of the square root be positive in order to satisfy the boundary condition at $z \to -\infty$.

3. RESULTS AND DISCUSSION

For non-oscillatory instability the real and imaginary parts of σ are set equal to zero, so that the critical conditions are determined by the zeros of the numerator of equation (5). If $(1-p')>0$, then it is apparent that increasing D_s will be stabilizing by reducing the magnitude of the destabilizing concentration gradient.

As in I, the thermal steady state assumption is invoked, so that $\omega \gg V/\kappa_1$ and $\omega \gg V/\kappa_s$. Additionally μ_A and k_A are taken to be zero, as in I. Equation (5) can then be rewritten as

$$\Sigma_1 = \frac{n_1 [-(Q+Pn_1^2)\{R_1 + R_s - p'(R_s + 1/2)\} + I(R_1 - 1/2)]}{(1 + Mn_1) \{R_1 + R_s - p'(R_s + 1/2)\} + In_1 [1 - H(R_s + 1/2)]} \tag{6}$$

where all quantities are the same as in I. Those with subscript 1 correspond to the unsubscripted variables in I, and

$$R_s = (\frac{1}{4} + \frac{\sigma D_s}{v^2} + \frac{D_s^2 \omega^2}{v^2})^{\frac{1}{2}}$$

The real parts of both R_1 and R_s must be positive to satisfy the far field boundary conditions. Equation (6) reduces to equation (6) in I with $D_s=0$ (since $R_s=1/2$ when $D_s=0$).

To investigate the possibility of oscillatory instability, equation (6) is rearranged to obtain a coupled third order equation in R_1 and R_s

$$R_1^3 a_3 + R_1^2 R_s b_2 + R_1^2 c_2 + R_1 a_1 + R_s a_0 + c_0 = 0 \qquad (7a)$$

where

$$a_3 = 1 + Mn_1$$

$$a_1 = n_1(Q + Pn_1^2 - I) - (\frac{1}{4} + n_1^2)a_3$$

$$b_2 = a_3(1-p') - In_1 H$$

$$c_2 = -a_3 \frac{p'}{2} + In_1(1-\frac{H}{2})$$

$$b_0 = n_1(1-p')(Q+Pn_1^2) - (\frac{1}{4} + n_1^2)b_2$$

$$c_0 = -(n_1^2 + \frac{1}{4}) c_2 - n_1(Q+Pn_1^2)\frac{p'}{2} + \frac{In_1}{2}$$

Additionally,

$$R_s^2 = R_1^2 R_m + (R_m-1)(n_1^2 R_m - 1/4) \qquad (7b)$$

where R_m is the ratio of diffusion coefficients D_s/D_1. Equation (7a) can be rearranged, and R_s eliminated to obtain a single sixth order equation in R_1:

$$A_6 R_1^6 + A_5 R_1^5 + A_4 R_1^4 + A_3 R_1^3 + A_2 R_1^2 + A_1 R_1 + A_0 = 0, \qquad (7c)$$

with

$$A_0 = c_0^2 + b_0^2 (R_m-1)(n_1^2 R_m - 1/4),$$

$$A_1 = 2a_1 c_0,$$

$$A_2 = a_1^2 + 2c_2 c_0 - R_m b_0^2 - 2b_2 b_0 (R_m-1)(n_1^2 R_m - 1/4),$$

$$A_3 = 2a_3c_0 + 2c_2a_1,$$

$$A_4 = c_2^2 + 2a_3a_1 - 2b_2b_0R_m - b_2^2(R_m-1)(n_1^2R_m-1/4),$$

$$A_5 = 2a_3c_2,$$

$$A_6 = a_3^2 - b_2R_m.$$

Clearly there are six roots of equation (7c) for R_1, which are obtained using the IMSL subroutine ZPOLR. Equation (7b) is used to obtain the twelve corresponding roots R_s. Some of the roots are eliminated upon using the condition that the real parts of R_1 and R_s must be positive. The surviving roots are required to satisfy equation (7a), cubic in R_1, which may exclude some of the remaining roots. Upon completing this procedure, only two roots for R_1 remain, either both real or a complex conjugate pair.

As in I, we set $1-p'>0$, $M>0$ and $I>0$, since violation of these inequalities is unlikely for a crystal growth process.

It is easily shown using a procedure similar to that outlined in I that if $I<Q$, which is the modified constitutional supercooling criterion and if $H<0$, then the system is stable to both non-oscillatory and oscillatory modes. Since H is physically restricted to being greater than or equal to zero, the condition on the non-equilibrium segregation parameter for no instability becomes H=0. This is in contrast to the result obtained in I with $D_s=0$ where the condition was $H<1$ for stability. No matter how small the solid diffusion coefficient, there is a region for H between 0 and 1 where there is a possibility for instability even if the modified constitutional supercooling criterion is satisfied. This region is entirely excluded in I.

The remaining results are now presented in graphical form. The curves for $R_m=0$ correspond to the results obtained by Coriell and Sekerka in I. Figure 1 illustrates the effect of increasing the diffusivity ratio R_m on the real and imaginary parts of the growth constants. Since $I>Q$, the modified constitutional supercooling criterion is not satisfied, but the system is stable under conditions of local equilibrium (M=0) and no non-equilibrium solute segregation (H=0). Increasing R_m decreases the range of wavelengths over which oscillatory roots exist and decreases the imaginary part of the growth constant σ_I. The growth constants differ numerically from those obtained in [6] for $R_m>0.1$. However, for the remaining figures R_m is chosen to be between 0.5 and 1.0.

In Figure 2, the non-equilibrium segregation parameter H is changed to 0.04 and all other parameters are retained the same as in Figure 1. Increasing the diffusivity ratio to 1.0 serves to stabilize the system slightly.

For H=0.1, shown in Figure 3, the oscillatory instability still is the critical mode even at $R_m=1.0$. The growth constant curve is shifted slightly towards lower wavenumbers as the diffusivity ratio increases.

The effect of the diffusivity ratio on the stability limit is synthesized in Figure 4. Since the region above the curve is unstable, it is clear that the region of stability is expanded as R_m increases. The modified constitutional supercooling criterion is plotted for reference as

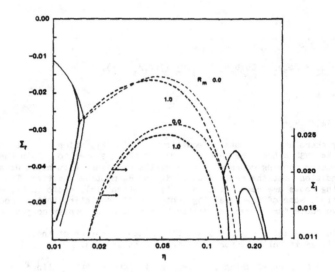

Figure 1. The real and imaginary parts of $\Sigma = \sigma D_l/V^2$ as a function of the spatial frequency $\eta = D_l \omega/V$ for $Q=1$, $P=10^3$, $k=0.5$, $M=H=0$, $I=500$. The thin lines are for $R_m=0$ and therefore correspond to the results of Coriell and Sekerka [6]. The dashed curves correspond to complex values of Σ.

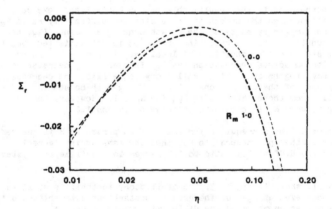

Figure 2. The real part of Σ as a function of the spatial frequency η for $H=0.04$. The other parameters are the same as in figure 1. The dashed curve indicates a non-vanishing value of the imaginary part of Σ. The system is stabilized slightly for $R_m=1.0$, although the effect is not major.

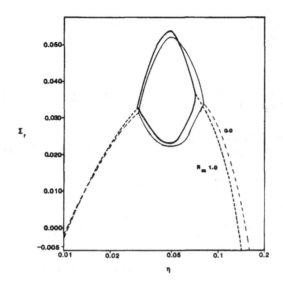

Figure 3. The real part of Σ as a function of the spatial frequency η
for H=0.1. The other parameters are the same as in figures 2
and 3. The solid and dashed curves indicate vanishing and non-
vanishing values of the imaginary part of Σ respectively.

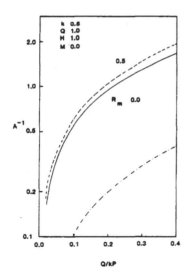

Figure 4. The dimensionless parameter I/kP plotted as a fuction of Q/kP
for Q=1, k=0.5, H=1 with M=0. The interface is unstable above
a given curve. The line -.-. corresponds to I=Q or the
modified constitutional supercooling criterion. The effect of
increasing R_α is to stabilize the system slightly.

the dashed line in Figure 4. The absolute stability criterion is the horizontal line $A^{-1}=1$.

4. REFERENCES

1. M. Cohen, B.H. Kear, R. Mehrabian, Rapid Solidification Processing Principles and Technologies II, edited by R. Mehrabian, B.H. Kear and M. Cohen (Claitor, Baton Rouge, 1980) p. 1.

2. K.A. Jackson, G.H. Gilmer, H.J. Leamy, Laser and Electron Beam Processing of Materials, edited by C.W. White, P.S. Percy (Academic Press, New York, 1980) p. 104.

3. R.F. Wood, Phys. Rev. B, 25, 2786 (1982).

4. J. Narayan, H. Naramoto, C.W. White, J. Appl. Phys., 53, 912 (1982).

5. S.R. Coriell, R.F. Sekerka, Rapid Solidification Processing Principles and Technologies II, edited by R. Mehrabian, B.H. Kear, M. Cohen (Claitor, Baton Rouge, 1980) p. 35.

6. S.R. Coriell, R.F. Sekerka, J. Crystal Growth, 61, 499 (1983).

7. W.W. Mullins, R.F. Sekerka, J. Appl. Phys., 35, 444 (1964).

8. Boettinger, W., D. Schechtman, R.J. Shaefer, F.S. Biancaniello, Met. Trans. A, 15A, 55 (1984).

9. D.P. Woodruff, The Solid-Liquid Interface (Cambridge University Press, Cambridge, 1973) p. 77.

10. L.H. Ungar, Ph.D. Thesis, Dept. of Chemical Engineering, MIT (1985).

PART III

Ultrafast Laser Excitation
of Semiconductors

TIME-RESOLVED SPECTROSCOPY OF PLASMA RESONANCES
IN HIGHLY EXCITED SILICON AND GERMANIUM

A. M. MALVEZZI, C. Y. HUANG[+], H. KURZ[*] and N. BLOEMBERGEN
Division of Applied Sciences, Harvard University, Cambridge, MA 02138
[+] Los Alamos National Laboratory, Los Alamos, NM 87545
[*] Technical University Aachen, D-5100 Aachen, Federal Republic of Germany

ABSTRACT

The dynamics of the electron-hole plasma in silicon and germanium samples irradiated by 20 ps, 532 nm laser pulses has been investigated in the near infrared by time-resolved picosecond optical spectroscopy. The experimental reflectivities and transmissions are compared with the predictions of the thermal model for degenerate carrier distributions through the Drude formalism. Above a certain fluence, a significant deviation between measured and calculated values indicates a strong increase of the recombination rate as soon as the plasma resonances become comparable with the band gaps. These new plasmon-aided recombination channels are particularly pronounced in germanium.

I. INTRODUCTION

The interaction of picosecond laser pulses with semiconductor surfaces has been extensively studied in recent times in view of assessing the temporal evolution of the carrier density and lattice heating prior to the melting of the semiconducting surface[1]. For these investigations, time-resolved spectroscopic techniques are a unique tool for the determination of the ratio N /m[*] of the carrier concentration to the optical reduced mass. By a proper selection of the probing wavelength, the complex index of refraction is more sensitive to the lattice temperature[2] or to the carrier density[3]. The optimum conditions to probe the latter occur near the plasma resonance where the real part of the dielectric constant approaches the zero value. A sharp rise of the reflectivity of the material occurs, which determines the carrier density.

The interpretation of the optical reflectivity and transmission measurements is usually carried on the basis of the Drude formalism which allows a direct determination of the ratio N/m[*]. This model has been successfully used in the past for the simulation of the optical response of laser irradiated semiconductors. Its validity in systems driven far out of the equilibrium has been more successful than anticipated. The main parameters which are critical in the interpretation of the experimental data are the optical reduced mass, the momentum relaxation time and their functional dependence on the carrier density and temperature. Experiments in the past have shown little or no dependence of the optical reduced mass on temperature in silicon[4]. Also, different experiments[5,6] have suggested extremely small momentum relaxation times during picosecond irradiation. The damping of the optical response, as described by the Drude model, occurs via carrier-phonon or carrier-carrier scattering. The short momentum relaxation times, however, point toward a predominance of intervalley scattering mechanisms.

The onset of a plasma resonance can be easily observed[7] by monitoring the reflectivity at a fixed delay after the excitation pulse as a function of the laser fluence (see e.g. fig. 4 in ref. 1) As soon as the surface carrier density exceeds the critical value $N = N_c(1-1/\epsilon'_\infty)$ with $N_c \approx m^*\omega \; \epsilon'_\infty/(4\pi e^2)$ the reflectivity rises sharply. The extent of its variation depends on the plasmon damping mechanism. For higher plasma densities the reflectivity remains at the same high level. Information on the further growth of the ratio N/m* is therefore not accessible. Using this technique with 2.8 μm probing it has been found[8] that the plasma density exceeds $N_c/m^* \approx 4 \times 10^{48}$ cm^{-3}g^{-1} in silicon and $\approx 3 \times 10^{48}$ cm^{-3}g^{-1} in germanium. These experimental values, however, refer to excitation fluences much below the critical value for surface melting. The plasma can in principle reach higher densities which can be monitored by using shorter probing wavelengths. At 1.9 μm, for example, the plasma resonance is reached at N/m* = 4.3×10^{48} and 4.8×10^{48} cm^{-3} g^{-1} for silicon and germanium, respectively.

This paper reports on experimental investigations specifically aimed to the study of the extreme excitation conditions that are reached in silicon and germanium just prior to the onset of the melting of the surface of the sample. A further motivation for investigations in this high excitation regime is the recent proposal of new plasmon-aided recombination channels in silicon[9] when the plasmon frequency becomes comparable with the energy gap. The case of germanium, with a relatively low band gap, is particularly favourable for such investigations, although different routes for plasmon-aided recombinations become possible in this case[9,10]. A careful comparison of numerical simulations with the time-resolved reflectivity and transmission data, taken at 1.06 μm, 1.55 μm and 1.9 μm under .532 μm, 20 ps laser irradiation, has been carried out.

For both elemental semiconductors the excitation at 532 nm ensures a linear generation rate of electron-hole pairs as long as picosecond laser pulses are used. In the case of germanium this high photon energy specifically avoids possible competing excitation processes such as intravalence band transitions between light and heavy holes. Also, the eventual filling of the central valley, where the primary excitation occurs, does not reduce the absorption coefficient at the highest excitation levels. These nonlinearities are detrimental for an exact analysis of the carrier dynamics when the excitation energy becomes comparable with the band-gap energy.

2. MODELING OF THE PLASMA DENSITY AND TEMPERATURE

The interaction of picosecond laser pulses with semiconductors can be modeled within the framework of the thermal model[11,12]. It is assumed that the photogenerated carriers and the lattice reach a common temperature T in times much shorter than the laser pulse duration. Ample experimental evidence that this is the case in the picosecond excitation regime has been provided by many independent tests on various semiconductors. The temporal evolution of the carrier density N(x,t) across the depth x of the material is described in a one-dimensional appproach by a diffusion equation where the laser driven source term is balanced by ambipolar diffusion and Auger recombination. This equation is coupled to a similar one for the evolution of the temperature profile T(x,t) in terms of a thermal diffusion and a source terms S(x,t). The latter has two contributions[13]:

$$S(x,t) = G(x,t)(h\nu - E_g(T) - HkT) + \gamma N^3(x,t)(E_g(T) + HkT) \qquad (1)$$

where hν is the exciting photon energy and k the Boltzmann constant. The first term describes the rate of direct heating provided to the phonons by the light absorption process. This quantitiy depends on the excess energy available to the carriers. The second term represents the heating rate of the lattice due to the Auger recombination process. The energy transferred to the lattice per single event is the sum of the band gap energy $E_g(T)$ and the thermal content of the carriers, which is described through the degeneracy parameter H. The latter is evaluated through the charge neutrality condition

$$N(x,t) = 2 \left[\frac{2\pi \, m_e \, kT}{h^2} \right]^{1/2} F_{1/2}(\eta_e) = 2 \left[\frac{2\pi \, m_h \, kT}{h^2} \right]^{1/2} F_{1/2}(\eta_h) . \qquad (2)$$

h being the Planck constant, $m_{e,h}$ the density-of-states reduced masses and $F_{1/2}(\eta)$ the Fermi integrals of order 1/2 which depend on the reduced quasi-Fermi levels for electrons and holes, respectively, $\eta_{e,h} = E^F_{e,h}/kT$. The degeneracy parameter H is then evaluated through

$$H = F_{3/2}(\eta_e)/ F_{1/2}(\eta_e) + F_{3/2}(\eta_h)/ F_{1/2}(\eta_h) \qquad (3)$$

The presence of the parameter H in eq. (3) accounts for the energy transfer modifications due to the smaller number of states available for scattering in degenerate systems. The effect is to increase the amount of Auger heating at expenses of the direct heating. It has been recently recognized[13] that the former contribution can substantially delay the heating of the lattice, the rate becoming comparable to the Auger time constant $(\delta N^2)^{-1}$. At the expected high carrier densities, however, Coulomb screening should provide a saturation to the Auger rate constant[14]. It has also been predicted that, when the plasmon energy becomes comparable to the band gap energy, strong carrier recombination can occur by plasmon emission[9,10]. These two effects, which are not included in the present calculation, give opposite contributions to the carrier density. Thus it is in principle possible to discriminate among them with suitable experiments.

Numerical calculations based on the above description have been performed for silicon and germanium excited by 20 ps, 532 nm radiation. The material parameters[15] used in the two cases are summarized in table I. Silicon has a higher melting temperature and has a thermal diffusivity almost four times bigger than germanium. The two materials also differ greatly in the the absorption coefficient α. In silicon it increases with temperature by almost two orders of magnitude at the melting point, reaching however a value only a fraction of the one of germanium. These different properties are mirrored in the resulting critical fluences for surface melting F^{th}, which agree well with the experimental results.

Figure 1 illustrates the temporal evolution of the calculated carrier density for silicon and germanium at half the critical fluence value F^{th}. The buildup of the carriers is driven by the laser pulse and the maximum density value is reached within few picoseconds of the maximum irradiation. The density decay is mainly determined by the Auger recombination in silicon. The higher ambipolar diffusivity in germanium also contributes to the process. The reduction in carrier density slows down the Auger recombination process and the smoothing of the spatial gradients decreases the carrier transport to the interior of the sample. Therefore at times longer than $\approx$ 70 ps the surface density exhibits a much slower decay. The excess energy available to the lattice per absorbed photon is much higher in germanium than in silicon, due to the differences in band-gap energies. Moreover, the energy required to melt the silicon surface is higher than the corresponding quantity in germanium. Therefore, the comparison at $F = F^{th}/2$ of the maximum calculated carrier densities shows that a higher value is reached for silicon.

The calculated behavior of the temperature at the surface, illustrated for the same excitation pulses in Fig. 2, shows a more pronounced material dependence. In the case of silicon, which exhibits a longer absorption depth, relatively low thermal gradients are generated which fully compensate for the higher thermal diffusivity. This results in an almost constant temperature profile after the laser irradiation. Germanium, instead, has an extremely short absorption depth. Correspondingly steeper thermal gradients are present which override the poor thermal diffusivity of the material. Therefore the surface

204

TABLE I

Material parameters for silicon and germanium used in the numerical
calculations

		Silicon	Germanium
a) <u>Thermodynamical parameters</u>:			
melting temperature, T_m(°K)		1680	1214
latent heat of fusion L_H (J/cm³)		4000	2644
density ρ (g/cm³)			
	300 °K	2.33	5.32
	T_m	2.37	5.2
specific heat c_p (J/g °K)			
	300 °K	.71	.32
	T_m	1.044	.377
thermal diffusivity (cm² /s)			
	300 °K	1.35	.35
	T_m	.196	.075
b) <u>electronic properties</u>:			
ambipolar diffusivity D_a	300 °K	20	65
(cm² /s)	T_m	3.5	7.98
Auger coefficient γ (cm⁶/s)		$2 - 4.10^{-31}$	2.10^{-31}
density of states mass (m_0)			
	e^-	.32 × 6	.22 × 4 (L)
			.32 × 6 (X)
	h	.59	.35
c) <u>optical properties at .53 μm</u>:			
absorption coefficient		$\alpha = \alpha_0 \exp(- T/\Theta)\ (cm^{-1})$	
	300 °K	$4.5\ 10^3$	$5.\ 10^5$
	T_m	$2.2\ 10^5$	$7.\ 10^5$
reflectivity R			
	300 °K	.37	.5
	T_m	.43	.5
critical fluence for melting F^{th} (mJ/cm²)		200	35

temperature decays more rapidly with time. In both materials the maximum
surface temperature is reached after the corresponding maximum surface
carrier density. This delayed heating is a direct consequence of the plasma
degeneracy, as explained earlier.

In figures 3 and 4 the calculated carrier surface density and temperature are
plotted as a function of the laser fluence, normalized to the critical value for
surface melting F^{th}, for three different time delays. These plots allow direct
comparison with the experimental data, which are obtained at fixed delay
times. Moreover, information of the interplay of the effects considered in the
calculations can be obtained. At 0 ps delay time, i.e. at the maximum of the
laser pulse, the buildup of the carriers in germanium is almost linear with
laser fluence since the absorption coefficient α has a weak temperature
dependence. This is further confirmed by the linear slope of the surface
temperature in fig. 4. At the same time delay, the bending of the carrier
density in silicon shows that the Auger recombination ($\gamma = 4 \times 10^{-31}$ cm⁶/s) is
the main limiting mechanism. At 20 ps the surface carrier density in

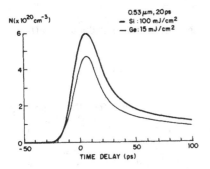

Fig. 1. calculated carrier density versus time for silicon and germanium under ps green illumination at the fluence levels indicated. The zero of the time scale is the maximum of the pulse.

Fig. 2. calculated temperature versus time for silicon and germanium The same parameters as in Fig 1 are used.

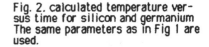

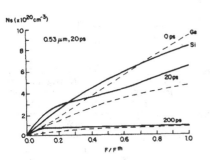

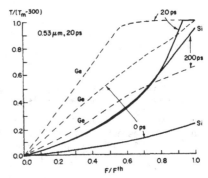

Fig. 3. Carrier density for silicon and germanium versus normalized laser fluence at the delay times shown.

Fig. 4. Normalized temperature for silicon and germanium versus normalized fluences at the times shown.

germanium is reduced by diffusion of the plasma into the bulk. In silicon, the Auger recombination dominates at low fluences. At $F/F^{th} > 0.5$ the generation rate of electron-hole pairs starts to become nonlinear due to thermal band-gap shrinkage during the picosecond excitation pulse. The simultaneous increase of the indirect absorption gives rise to a further increase of the plasma density at higher fluences. At long time delays between the pump and the probe pulses, the carrier density is reduced by the combined effects of ambipolar diffusion and Auger recombination to an asymptotic value which is independent from the laser fluence.

To illustrate the differences in thermal response between silicon and germanium, the temperature axis in fig. 4 is normalized to the difference between melting (T_m) and room temperature. The positive second derivative of the silicon curves indicates the nonlinear heating of the surface due to the strong temperature dependence of the absorption. The relatively low thermal gradients in silicon prevent a fast cooling of the surface at high irradiation levels. Little difference between the curves for 20 and 200 ps is therefore observed. In germanium a more linear dependence of the temperature versus laser fluence is calculated at all times. The weak temperature dependence of the light absorption is the primary cause for this behavior. The melting temperature is reached earlier in germanium than in silicon. The ratio F^*/F^{th} of the fluence necessary for reaching the melting temperature to the fluence for melting the surface can be estimated, in absence of thermal diffusion effects, as

$$F^*/F^{th} \approx \{1 + L_H \alpha^*/[\rho c_p \alpha(T_m) \Delta T]\}^{-1} \qquad (4)$$

with $\Delta T = T_m - 300°K$ and α^* a suitable average value for the absorption coefficient between the extreme temperatures. This expression gives F^*/F^{th} =0.87 for silicon and 0.45 for germanium, in fair agreement with the numerical results.

The calculated values of carrier density and temperature are used to predict the reflectivity and transmission changes induced by a 532 nm picosecond pulse striking the semiconductor surface, which are determined by the optical dielectric functions ϵ' and ϵ'':

$$n^2 - k^2 = \epsilon' = \epsilon'_\infty - 4\pi N e^2 \left[\frac{1}{m^*_e} \frac{\tau_e^2}{1 + \omega^2 \tau_e^2} + \frac{1}{m^*_h} \frac{\tau_h^2}{1 + \omega^2 \tau_h^2} \right] \qquad (5)$$

$$2nk = \epsilon'' = \epsilon''_\infty + \frac{4\pi N e^2}{\omega} \left[\frac{1}{m^*_e} \frac{\tau_e}{1 + \omega^2 \tau_e^2} + \frac{1}{m^*_h} \frac{\tau_h}{1 + \omega^2 \tau_h^2} \right] \qquad (6)$$

the first term in each expression represents the lattice contribution. n and k are the real and imaginary parts of the index of refraction, N the carrier density, e the electron charge and ω the frequency of the probing light beam. The Drude terms in eq. (5) and (6) are dependent on the values of the optical reduced masses $m^*_{e,h}$ and of the momentum relaxation times $\tau_{e,h}$. In eq. (5) and (6), terms related to possible intervalence-band and intraband transitions have been neglected. They play a minor role in the optical response at the wavelengths studied. The experimental results presented in the next section confirm this view by showing a temperature behavior opposite to the one expected from these contributions.

The fitting of the experimental data is critically dependent on the values of the optical reduced mass $m^* = (1/m^*_e + 1/m^*_h)^{-1}$, the Auger coefficient γ and the momentum relaxation time τ. A particularly useful relation can be derived for long delay times $t \gg (2\gamma N^2)^{-1}$. In this case the Auger effect has depressed the carrier density to a common value independent from the excitation laser fluence, $N \approx (2\gamma t)^{-1/2}$. The corresponding Drude terms in (5) and (6) then become proportional to $(m^*\sqrt{\gamma})^{-1}$. The same optical response is obtained at long delay times whenever the product of the reduced optical mass m^* and of the square root of the Auger coefficient γ is constant.

The use of the Drude formalism for the description of the dielectric functions suffers from the lack of reliable data of the optical reduced mass at high carrier densities. Thus, from the reflectivity and transmission changes, only the ratio N/m^* can be evaluated. By an extensive parametric study of the reflectivity and transmission changes as a function of time delays and laser fluences, however, the temperature dependence of the optical reduced mass can be estimated and the range of variation with the carrier density narrowed.

The final step for a meaningful comparison with the experimental data is the convolution of the actual reflectivity and transmission changes with the temporal profile of the probing pulse.

3. TIME RESOLVED OPTICAL REFLECTIVITY AND TRANSMISSION

The experiments were carried out by exciting the sample with 20 ps, 532 nm laser pulses from the frequency-doubled output of an active-passive mode locked Nd:YAG laser. The probe pulses, at 1.9 and 1.55 μm, were obtained by Raman shifting the fundamental frequency at 1.06 μm in high pressure H_2 and CH_4 gas cells, respectively. The output was spatially filtered and imaged in the center of the irradiated area of the sample. The ratio of the pump and probe spot diameters was larger than 10. Crystalline silicon wafers, polished on the front surface were used. In the case of germanium, the surfaces were also chemically etched. This procedure drastically reduces the plasma recombination on the surface. Indeed, experimental runs on unetched samples showed little or no plasma effects.

In the following figures 5 - 8, we present the measured and calculated reflectivity and transmission for both elemental semiconductors. From the full parametric time resolved study only two time delays between the pump and the probe are selected. At a time delay of 20 ps a high density regime is expected, while the temperature is not fully developed due to the delayed heating mechanism discussed in the preceding section. The reflectivity of silicon first decreases to a broad minimum located at F≈ 80 mJ/cm² and rises again at higher fluences. The high reflectivity phase associated with the transition to metallic, liquid, silicon is reached above the well known threshold value for surface melting, F^{th}= 200 mJ/cm² observed at long time delays. The energy absorbed from the laser pulse is not completely transferred to the phonon subsystem in 20 ps. The thermal energy stored at this time in the lattice for a 200 mJ/cm² laser pulse is not sufficient to melt the surface.

The lines in fig. 5 represent the result of numerical calculations using different optical reduced masses m* and Auger recombination coefficients.

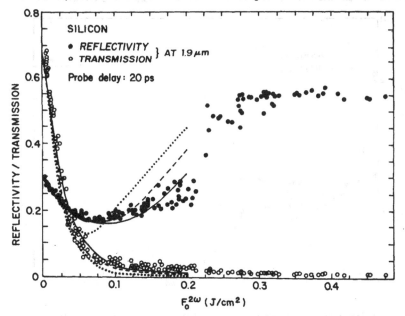

Fig. 5. Reflectivity and transmission at 1.9 μm of silicon samples versus incident 0.53 μm fluence for 20 ps delay. Lines: calculated values (see text).

Below 100 mJ/cm² a fair agreement between measured and calculated data confirm the validity of the model employed. The value of the reflectivity minimum and the transmission changes associated with free carrier absorption are consistent with a strong damping of the electron-hole plasma oscillations. Extremely short momentum relaxation times $\tau = 8 \times 10^{-15}$ (300/T), being T(°K) the lattice temperature, have to be introduced to fit the measured transmission changes. The reflectivity data rule out the possibility for a low value of the Auger coefficient ($\gamma = 2 \times 10^{-31}$ cm⁶/s, dotted lines). In this case the calculations show the reflectivity rising to a plasma resonance which is not observed. The agreement with the data can be extended somewhat above 100 mJ/cm² by using a constant Auger coefficient $\gamma = 4 \times 10^{-31}$ cm⁶/s. However, even using the high density reduced optical mass m* = 0.165 m₀ (fig.5, dashed lines) or a strong density dependent value m* = 0.123 m₀ (1 - βN)⁻¹, β = 5 × 10⁻²² cm³, (solid lines), a substantial discrepancy is noted in the high fluence regime which can be reconcilied only if the ratio N/m* is simulated to decrease.

In order to obtain more experimental information on the optical reduced mass m* we consider in the next figure 6 the data at long time delays (200 ps). At this time, the density has dropped to the asymptotic value $N \approx (2 \gamma t)^{-1/2}$ independent of the laser fluence, while the temperature is strongly increasing with F as illustrated in fig. 4. The incremental change of the real part of the dielectric constant is proportional to $(m^* \sqrt{2\gamma t})^{-1}$. Excellent agreement between measured and calculated values is found if a m*√γ product of 7.5 × 10⁻¹⁵ g cm³ s⁻¹/² is used in the calculations (solid lines of fig. 6). We know from figs 1 - 4 that at long time delays the density remains constant, while the lattice temperature increases strongly with fluence. This agreement certainly indicates that the m*√γ product is completely insensitive towards the drastic increase of the lattice temperature. Thus the optical reduced mass itself cannot exhibit a noticeable temperature dependance. The fitting value of

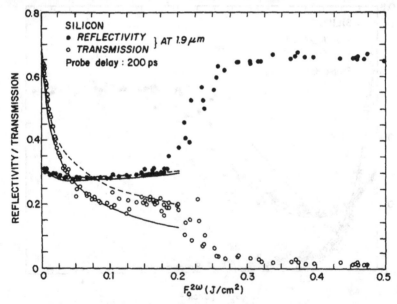

Fig. 6. Reflectivity and transmission at 1.9 μm of silicon samples versus incident 0.53 μm fluence for 200 ps delay. Lines: calculated values (see text).

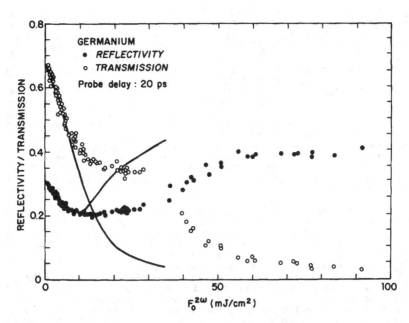

Fig. 7. Reflectivity and transmission at 1.9 μm of germanium samples versus incident 0.53 μm fluence for 20 ps delay. Lines: calculated values (see text).

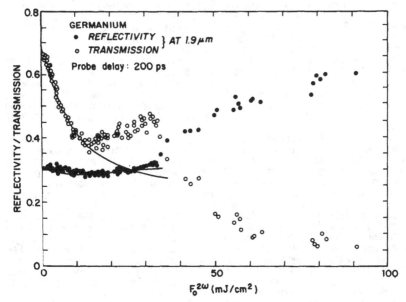

Fig. 8. Reflectivity and transmission at 1.9 μm of germanium samples versus incident 0.53 μm fluence for 200 ps delay. Lines: calculated values (see text).

the $m^*\sqrt{\gamma}$ product can be separated in a constant high density mass of $m^*=0.165\ m_0$ and $\gamma= 2 \times 10^{-31}\ cm^6/s$ or $m^*= .123\ m_0$ and $\gamma= 4 \times 10^{-31} cm^6/s$, respectively. At the highest fluences before melting, the transmission data are in better agreement with $m^*\sqrt{\gamma} = 1 \times 10^{-16}\ g\ cm^3 s^{-1/2}$ (dashed lines of fig. 6), corresponding to $m^* = 0.165\ m_0$ and $\gamma = 4 \times 10^{-31}\ cm^6/s$. Consistently with the observations at earlier times, the transmission indicates a lower density profile across the sample at the highest fluences.

From the comparison of the data obtained at 20 and 200 ps on silicon we can therefore deduce the following: 1) The insensitivity of the optical reduced mass towards temperature indicates degenerate carrier distributions, where the optical relevant parameter are only depending on the band structure and scattering process at the Fermi level. 2) Contrary to the expected drop of the Auger recombination by screening of the Coulomb interaction, the results indicate rather an increase of the recombination rate at high densities.

Thus, in silicon, the experimentally observed decrease of the N/m^* ratio is most likely due to the saturation of the obtainable density by the occurrence of new recombination channels as soon as the plasma resonance approaches the energy of the indirect band gap.

To prove this possibility of plasmon aided recombination in highly excited semiconductors, we have performed the same experiments in germanium, which has a lower band gap than silicon. Also, the maximum attainable carrier density is higher, as long as only Auger recombination is assumed to be the main density limiting process. The measurements on germanium at 20 and 200 ps delay times are shown in figures 7 and 8, respectively.

The plasma induced reflectivity drop is considerably less pronounced than in silicon, for comparable laser fluences. At a delay time of 20 ps the reflectivity drops from 30% to $\approx$ 20%. Again, a broad minimum is observed at $\approx F^{th}/2$. Contrary to the expected behavior, the reflectivity rises continuously with incresed fluence. A much slower and delayed rise to the value for molten germanium occurs than in silicon. In contrast with the observations in silicon, but in agreement with measurements on GaAs[16], the rise of the reflectivity towards the liquid value is always gradual. The transmission experiences a complementary behavior. Between $F^{th}/2$ and F^{th} a saturation of the absorption takes place. The fitting of the germanium data using $m^*=0.08\ m_0$ and $\gamma = 2\ 10^{-31}\ cm^6 s^{-1}$, indicates that at low fluences the calculations describe accurately the carriers. At higher fluences, however, the simulations are neglecting a basic mechanism in the evolution of the carrier plasma. Strong plasma resonance conditions, which would drastically increase the reflectivity of the surface, and reduce the transmission accordingly, are predicted above $\approx$ 10 mJ/cm^2. Instead, a saturation of the transmission is observed and a very slow rise of the reflectivity occurs. These results are strongly indicative of an increased recombination rate, which affects the surface and bulk carrier density.

The reflectivity measured at a delay time of 200 ps is qualitatively similar to the one at 20 ps. The initial decay slope of the transmission is comparable with the one observed at shorter times. The reflectivity minimum is less pronounced, due to the combined effects of Auger recombination and plasma diffusion into the bulk. Again, the slow recovery of the reflectivity versus fluence, when only an increase of the temperature occurs, supports the notion of a very weak temperature dependence of the optical reduced mass. The reflectivity data at 200 ps delay are well described by the calculations, indicating that at this long delay time the plasma evolution at the surface is again determined by the balance of ambipolar diffusion and slow Auger recombination. At $F > F^{th}/2$ the transmission, on the contrary, shows a marked increase with laser fluence which is abruptly interrupted by the melting of the surface. This behavior, which cannot be simulated by the present calculatons, indicates a progressive reduction of the bulk carrier density. The trend is suddenly changed by the onset of the melting of the surface.

The data on germanium, therefore, strongly suggest the that a new recombination channel opens up at a well-defined laser fluence, as soon as the

the plasma oscillations become resonant with the band gap. During the excitation laser pulse, these conditions are met at progressively lower density due to the reduction of the band gap energy with temperature. Once these recombination channels open up, the whole subsequent evolution of the plasma is affected. Lower densities occur at the surface of the sample, as the data at short time delays suggest. This in turns reduces the ambipolar diffusion of the carriers inside the material. At longer times, however, the resonance conditions cannot be sustained by the reduced plasma density and the evolution of the plasma on the surface can be again described in terms of diffusion and Auger recombination. In the bulk of the material, on the contrary, the initial reduced ambipolar diffusion still affects the overall carrier density and a transmitted beam will experience a smaller absorption at later times, as the transmission data at 200 ps show.

4. CONCLUSIONS

The interaction of picosecond laser pulses with silicon and germanium surfaces has been investigated by comparing the results of time-resolved spectroscopic measurements in the infrared region with numerical simulations. This analysis has revealed the importance of the plasma degeneracy in the evaluation of the carrier dynamics, which in turns affects the temporal evolution of the lattice heating even in the picosecond time-scale. The experimental results for both materials indicate extremely small momentum relaxation times, suggesting the dominance of intervalley scattering mechanisms over other competing processes.

Due to the high excess photon energy used for the excitation, the strong discrepancies at high laser fluence between the observed and the calculated optical response are not affected by either nonlinearities in the excitation process or by thermally induced interband transitions of the probing light. The predicted saturation of the Auger effect at high carrier density has not been observed. On the contrary, there are indications in silicon of an increase in the recombination rate of the carriers at the highest laser fluences. In germanium, due to the more favourable ratio $h\nu/E_g$, the experimental data provide strong evidence for a plasmon-assisted recombination which limits the carrier density as soon as the plasmons become resonant with the thermally reduced band gap.

ACKNOWLEDGEMENTS

This work is supported by the U.S. Office of Naval Research under contract no. N00014-83K-0030 and by the U. S. Department of Energy.

REFERENCES

1. N. Bloembergen, in *"Beam-Solid Interactions and Phase Transformations"*, edited by H. Kurz, G. L. Olson and J. M. Poate, Mat. Res. Soc. Symp. Proc., 1986 (to be published), and references quoted therein.
2. L.-A. Lompre', J. M. Liu, H. Kurz and N. Bloembergen, Appl. Phys. Lett. 43, 168 (1983)
3. L.-A. Lompre', J. M. Liu, H. Kurz and N. Bloembergen, Appl. Phys. Lett. 44, 3 (1984)
4. H. Kurz , A. M. Malvezzi , L.-A. Lompre', in *Proceedings of the 17th International Conference on the Physics of Semiconductors*, edited by J. D. Chadi, W. A. Harrison (Springer, Berlin, 1985)
5. A. M. Malvezzi , H. Kurz and N. Bloembergen, Appl. Phys. 36, 143 (1985)
6. L.-A. Lompre', H. Kurz and N. Bloembergen, in *"Ultrafast Phenomena IV"*, edited by D. H. Auston and K. B. Eisenthal (Springer, New York, 1984)
7. M. I. Gallant, H. M. van Driel, Phys. Rev. B 26, 2133 (1982)

8.   H. M. van Driel, L.-A. Lompre' and N. Bloembergen, Appl. Phys. Lett. $\underline{44}$, 285 (1984)

9.   M. Rasolt and H. Kurz , Phys. Rev. Lett. $\underline{54}$, 722 (1985)

10.   P. A. Wolff, Phys. Rev. Lett. $\underline{24}$, 266 (1970); A. Elci, M. O. Scully, A. L. Smirl, J. C. Matter, Phys. Rev. B $\underline{16}$, 191 (1977)

11.   P. Baeri, U. Campisano, G. Foti and E. Rimini, J. Appl. Phys. $\underline{50}$, 788 (1979)

12.   A.Lietola and J. F. Gibbons, Mat. Res. Soc. Symp. Proc. $\underline{4}$, 163 (1982)

13.   H. Kurz and N. Bloembergen, Mat. Res. Soc. Symp. Proc. $\underline{35}$, 3 (1985)

14.   E. Joffa, Phys. Rev. B $\underline{21}$, 2415 (1980)

14.   J. M. Meyer, M. R. Kruer, J. F. Bartoli, J. Appl. Phys. $\underline{51}$, 5513 (1980)

15.   J. M. Liu, A. M. Malvezzi and N. Bloembergen, in *Beam-Solid Interactions and Phase Transformations*, edited by H. Kurz, G. L. Olson and J. M. Poate, Mat. Res. Soc. Symp. Proc., 1986 (to be published)

CROSS-SECTIONAL TEM CHARACTERIZATION OF STRUCTURAL CHANGES PRODUCED
IN SILICON BY ONE MICRON PICOSECOND PULSES

ARTHUR L. SMIRL,* IAN W. BOYD,** THOMAS F. BOGGESS,** STEVEN C. MOSS,**
AND R.F. PINIZZOTTO[†]

* Hughes Research Laboratories, 3011 Malibu Canyon Road, Malibu, CA 90265
** Center for Applied Quantum Electronics, Department of Physics, North
Texas State University, Denton, TX 76203
† Ultrastructure, Inc., 1850 Greenville Ave., Richardson, TX 75081

ABSTRACT

Plan and cross-sectional transmission electron microscopy (TEM) has
been used to examine the various bulk and surface structural changes
observed in crystalline silicon following melting caused by the absorption
of 1-μm pulses that are 4 ps in duration. We show for the first time that
for picosecond excitation polycrystalline Si (p-Si), rather than single-
crystal Si, is always formed when the melt resolidifies. Specifically,
cross-sectional TEM analyses indicate that a thin layer of fine-grained
p-Si is formed at the interface of the melted region with the bulk. When
the incident fluence is more than 10% above the melting threshold, the
region between the surface and the fine-grained p-Si regrows as larger-
grain p-Si. This is the first reported observation of a large-grain p-Si
layer on a fine-grain p-Si base in crystalline material. If the incident
fluence is less than 10% above threshold, the region near the surface
resolidifies as amorphous Si, but the narrow layer of fine-grained p-Si
remains at the interface with the single crystal material. Present models
for resolidification must be modified to account for these features.

INTRODUCTION

The strong absorption of picosecond pulses can produce fast melting
and steep temperature gradients on a time scale too short to allow
significant thermal diffusion. Rapid thermal transport following
ultrarapid melting can then produce a highly undercooled liquid, which in
turn can result in an unusually high resolidification velocity. Such
conditions provide a unique opportunity to investigate the rapid kinetic
processes that control the formation of novel metastable surface patterns
and bulk resolidification structures. Even so, resolidification studies
that take advantage of the capabilities of ultrashort pulses have been
limited primarily to the qualitative identification of a crystalline-to-
amorphous phase transition in c-Si following melting by visible and
ultraviolet pulses[1]. To date, most information concerning the rapid
solidification of highly undercooled liquid Si (l-Si) has been extracted
from experiments where the undercooling was provided by melting layers of
a-Si on crystalline substrates with visible nanosecond optical pulses (see
Refs. 2 through 10).
By comparison with the visible (Refs. 11 through 16), investigations
of both the melt dynamics and resolidification morphologies of c-Si
following irradiation with subnanosecond pulses at 1 μm are rare. Studies
of the melt dynamics at the latter wavelength have been limited to isolated
temporally resolved reflectivity[17] and imaging measurements[18]. Similarly,
descriptions of the resolidification morphologies in c-Si at 1 μm consist
primarily of a report of recrystallization following irradiation with 30-ps
pulses[1] and studies of the surface damage morphologies of 1.5 to 2.5-μm-
thick free-standing films.[19] In fact, it was initially believed that the

absorption depth for 1 μm light in c-Si was too large to produce
substantial undercooling.[13]
    We have previously pointed out that, when a sample of single-crystal
silicon is melted with pulses at 1 μm as short as 4 ps, two distinct
surface morphologies are clearly visible upon resolidification.[20,21] For
regions where the incident fluence is more than 10% above threshold, the
melt apparently recrystallizes. By contrast, for regions where the fluence
is less than 10% above threshold, the resolidified material has been
tentatively identified as amorphous silicon; that is, a crystalline-to-
amorphous phase transition is observed. Typically, then, what is visible
in Nomarski micrographs of the surface following a single laser firing is a
recrystallized disk surrounded by an amorphous ring. Moreover, ripple
patterns are formed within and near the melted region. The objective of
our transmission electron microscopy study was to obtain microstructural
images and electron diffraction patterns to aid in the characterization of
the regrowth in each of these areas. The latter indicate that the
nonlinear absorption[22] and thermal runaway associated with the shorter
pulses produce a highly undercooled melt.

EXPERIMENT

    The silicon samples were optically polished, high-purity, undoped,
high-resistivity (10^3-10^4 Ω-cm), <111>-orientation single-crystal wafers.
The Si disks were approximately 25 mm in diameter and 1 mm thick and were
antireflection coated on the back surface to eliminate complications from
multiple reflections. Melting was achieved with pulses having a duration
of 4 ps (half-width at e^{-1} of maximum irradiance) at 1 μm provided by a
mode-locked Nd:glass laser. The sample was mechanically translated after
each laser firing. Following irradiation, the surface morphologies were
first studied using Nomarski interference microscopy and high contrast
imaging techniques. The bulk microstructure of the previously laser-melted
material was then examined using plan and cross-sectional TEM techniques.
    For use in the cross-sectional TEM studies, the silicon wafers were
prepared in the following way. First, a large number of laser melted spots
were produced in lines across the surface of a sample by 4-ps optical
pulses. Then the silicon wafer was cut in half through an array of laser-
melted spots using a diamond saw. The face of one piece with laser-melted
spots was then glued to the face of an unirradiated piece with epoxy. This
sandwich structure was then cut into approximately 250-μm-thick slices
using the diamond saw. The slices were mechanically polished to roughly
80 μm. Finally, the sample specimens were dimpled and thinned to electron
transparency using Argon ion milling. Bright- and dark-field imaging, as
well as selected area diffraction, were used to obtain crystallographic
information on the various resolidified regions.

RESULTS AND DISCUSSION

    A dark field TEM image of a typical cross section taken from the
central (or recrystallized) portion of a laser melted spot is shown in
Figure 1. The black arrows mark the interface between the epoxy resin,
which was used to bond the virgin c-Si reference sample to the irradiated
sample. The large slanted bands in the substrate region arise from
thickness variations in the sample. Examination of the laser treated
material reveals a fine-scale microstructure that is clearly
polycrystalline material. The surface of the melted region has an
irregular step-like structure. Also, when the cross-section shown in
Figure 1 is viewed under lower magnification, it is clear that the surface
undulates more gently and regularly. This long range variation is
characterized by a peak-to-trough amplitude of ~200 Å and by a period of

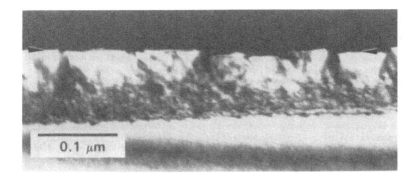

Figure 1. Cross sectional TEM dark-field image of an area near the
center of the previously laser-melted region.

1.07 ± 0.02 μm, in excellent agreement with the predicted spacing of
surface ripples. Although suggested, this periodic behavior is not
altogether apparent in Figure 1, since less than one period of the surface
oscillation is shown.
    It is clear under the higher magnification of Figure 2 that the
resolidified region is not homogeneous with depth, but rather consists of
two layers. The top zone (labeled A) consists of relatively large grains
(LG) whose boundaries are slightly askew of vertical. These grains are
about 30 nm in size and, when viewed under bright field conditions, clearly
have many twins. These grains start abruptly at an interface above a
second layer (labeled B) that appears to be fine grain (FG) material with
an average grain size of only 3 nm. The interface of this layer with the
unmelted substrate is somewhat ragged and is indicative of inhomogeneous
penetration of the melt front. The overall thickness of the melted region
shown in Figure 1 is only roughly 80 nm. In fact, we did not observe a
melt depth thicker than 100 nm. This is indicative of the extremely short
absorption depths that accompany the nonlinear absorption in Si at 1 μm.
The approximate depth of the LG p-Si region is 55 nm, while the FG p-Si
region is 25 nm thick.
    Much nearer the edge of the laser melted spot, the microstructural
properties are very different (see Figure 3). A dual-layered structure
still exists, but now the top layer is amorphous rather than
polycrystalline. The new layer is characteristically grey in bright-field
images of this area, and it exhibits no obvious structure down to a 2 to
3 Å level. This region is dark in dark-field images (Figure 3),
demonstrating that it does not contribute intensity to the diffracted
beams. A second layer of FG p-Si is still present below this layer. In
this case, the total extent of the melted region is 40 nm, and is composed
of 25 nm of amorphous material and 15 nm of fine-grained material.
    Figure 4 is a bright-field TEM image at the extreme edge of the melted
region. The layer of FG p-Si has reached the surface. There is no longer
a layer of a-Si or LG p-Si remaining between this layer and the surface.
It can be seen that the thickness of this zone is irregular, having an
average thickness of roughly 15 nm. It therefore appears that the fine-
grained zone forms independently of the incident fluence once threshold has
been reached.
    Recently, resolidification at high velocities in undercooled melts has
been investigated by melting a-Si layers produced on c-Si by ion implantation with ns laser pulses.[2-5,7,8] In these previous studies, LG and FG

Si layers were produced that are similar in appearance to those observed in our work. That is, FG p-Si was formed at the farthest extent of the melt front into the material, while LG p-Si was formed in the region extending from the surface to the boundary of the fine grain region. These are indicative of high resolidification velocities expected of an undercooled melt. An undercooled liquid is expected in the earlier studies if one believes the melting temperature of a-Si to be lower than that of c-Si. In our studies, however, the undercooling can be attributed to the nonlinear absorption associated with the short pulses. In previous experiments the LG p-Si was thought to have formed as a result of rapid resolidification, while it has been suggested that the FG p-Si is formed by explosive recrystallization[9] or by bulk nucleation of a highly undercooled melt.[10] It must be emphasized, however, that such layers were produced only when

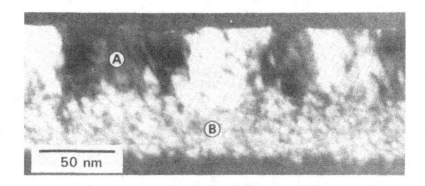

Figure 2. A magnified image of the same area as Figure 1, showing both LG p-Si (A) and FG p-Si (B) regions.

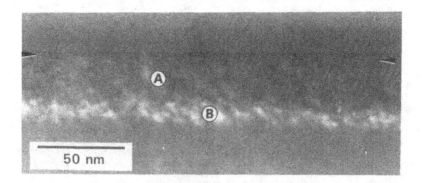

Figure 3. Cross sectional TEM of a region nearer the edge of the resolidified spot showing an a-Si layer (A) over a FG p-Si layer (B).

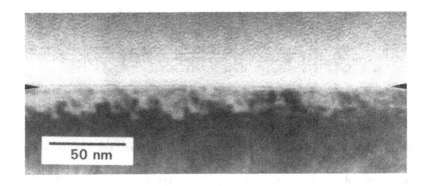

Figure 4. Cross sectional TEM of the extreme edge of the
melted region showing only a thin layer of
FG p-Si near the surface.

the incident laser fluence was not sufficient to melt completely through
the pre-existing amorphous layer. The existence of an unmelted amorphous
layer between the melt-front and the crystal is an essential feature of
previous work. We have shown, however, that the presence of a pre-existing
amorphous layer is not a necessary prerequisite for the formation of large-
or fine-grain p-Si. Furthermore, we have observed a fine-grain p-Si layer
below an amorphous layer. It is not clear to us how present models can be
modified to account for the latter features. Further experimental work
will certainly be required before the details of regrowth following
picosecond excitation are clearly understood.

ACKNOWLEDGMENTS

The authors wish to acknowledge enlightening discussions with G.L.
Olson and J.F. Lam. This work was supported in part by the Office of Naval
Research, The Robert A. Welch Foundation, and the North Texas State
University Faculty Research Fund.

REFERENCES

1. P.L. Liu, R. Yen, N. Bloembergen, and R.T. Hodgson, Appl. Phys. Lett.
   34, 864 (1979).

2. A.G. Cullis, H.C. Webber, and N.G. Chew, Appl Phys. Lett. 36, 547
   (1980).

3. J. Narayan and C.W. White, Appl. Phys. Lett. 44, 35 (1984).

4. M.O. Thompson, G.J. Galvin, J.W. Mayer, P.S. Peercy, J.M. Poate, D.C.
   Jacobson, A.G. Cullis, and N.G. Chew, Phys. Rev. Lett. 52, 2360
   (1984).

5. A.G. Cullis, N.G. Chew, H.C. Webber, and D.J. Smith, J. Crys. Growth
   68, 624 (1984).

218

6. R.F. Wood, D.H. Lowndes, and J. Narayan, Appl. Phys. Lett. 44, 770 (1984).

7. A.G. Cullis, in Energy Beam-Solid Interactions and Transient Thermal Processing, 1984, edited by D.K. Biegelsen, G.A. Rozgony, and C.V. Shank (Materials Research Society, Pittsburgh, 1985), p.15.

8. D.H. Lowndes, G.E. Jellison, Jr., R.F. Wood, S.J. Pennycook, and R.W. Carpenter, ibid., p.100.

9. P.S. Peercy and M.O. Thompson, ibid. p.53.

10. R.F. Wood, G.A. Geist, A.D. Solomon, D.H. Lowndes, and G.E. Jellison, Jr., ibid, p.150.

11. R.Tsu, R.T.Hodgson, T.Y. Tan, and J.E. Baglin, Phys. Rev. Lett. 42, 1356 (1979).

12. A.G. Cullis, H.C. Webber, N.G. Chew, J.M. Poate, and P. Baeri, Phys. Rev. Lett. 49, 219 (1982).

13. P.L. Liu, R. Yen, N. Bloembergen, and R.T. Hodgson, in Laser and Electron-Beam Processing of Materials, edited by C.W. White and P.S. Peercy (Academic Press, New, York, 1980), p.156.

14. C.V. Shank, R. Yen, and C. Hirlimann, Phys. Rev. Lett. 50, 454 (1983).

15. C.V. Shank, R. Yen, and C. Hirlimann, Phys. Rev. Lett. 51, 900 (1983).

16. M.C. Downer, R.L. Fork, and C.V. Shank, in Ultrafast Phenomena IV, edited by D. Auston and K.B. Eisenthal (Springer, New York, 1984), p. 106.

17. K. Gamo, K. Murakami, M. Kawabe, S. Namba, and Y. Aoyagi, in Laser and Electron-Beam Interactions and Materials Processing, edited by J.F. Gibbons, L.D. Hess, and T.W. Sigmon (North-Holland, Amsterdam, 1981), p. 97.

18. I.W. Boyd, S.C. Moss, T.F. Boggess, and A.L. Smirl, Appl. Phys. Lett. 46, 366 (1985).

19. M.F. Becker, R.M. Walser, J.G. Ambrose, and D.Y. Sheng, in Picosecond Phenomena II, edited by R.M. Hochsrasser, W. Kaiser, and C.V. Shank (Springer Verlag, Berlin, 1980), p.290.

20. I.W.Boyd, S.C. Moss, T.F. Boggess, and A.L. Smirl, Appl. Phys. Lett. 45, 80 (1984).

21. I.W. Boyd, S.C. Moss, T.F. Boggess, and A.L. Smirl, in Energy Beam Solid Interactions and Transient Thermal Processing, edited by J.C.C. Fan and N.M. Johnson (North-Holland, New York, 1984), p.203.

22. T.F. Boggess, A.L. Smirl, K. Bohnert, K. Mansour, S.C. Moss, and I.W. Boyd, IEEE J. Quant. Electron., to be published in 1985.

OBSERVATION OF SUPERHEATING DURING PICOSECOND LASER MELTING ?

N. FABRICIUS*, P. HERMES*, D. VON DER LINDE*, A. POSPIESZCZYK**, AND
B. STRITZKER***
* Fachbereich Physik, Universität-GHS-Essen, 4300 Essen, Fed. Rep. of
Germany
** Assoziation EURATOM-Kernforschungsanlage, Institut für Plasmaphysik,
Kernforschungsanlage Jülich, Jülich, Fed. Rep. of Germany
*** Institut für Festkörperforschung, Kernforschungsanlage Jülich, Jülich,
Fed. Rep. of Germany

ABSTRACT

The surface temperature of GaAs during pulsed laser heating is obtained
from the measured velocity distributions of evaporated atoms. For nanosecond
pulses the observed solid-liquid transition temperature agrees with the
equilibrium melting point T_M. With picosecond pulses the temperature rises
above T_M by several hundred degrees before melting can be detected. This
result is interpreted as evidence of superheating.

Introduction

The principal features of the interaction of short laser pulses with
absorbing materials [1] can be described in classical thermodynamic terms as
a rapid heating process. Substantial evidence has been accumulated over the
past several years that such thermal models provide a satisfactory description even if heating with laser pulses as short as 10^{-11} s is considered
[2]. However, an interesting new aspect of ultrafast heating is that the
solid-liquid phase transition (melting) can occur under highly superheated
conditions [3]. The heating rates attainable with intense ultrashort laser
pulses can be so large that the evolution of the phase transformation is no
longer limited by the heat supply - which is the typical "normal" situation
- but rather by the kinetics of the atomic rearrangement at the solid-liquid
interface. In such a situation the solid may be driven into a metastable
superheated state with a temperature greatly exceeding the equilibrium melt
temperature.

In this report we present measurements of the surface temperature of
GaAs during illumination with intense nanosecond and picosecond laser
pulses. Our results show that with nanosecond laser heating the surface
temperature at the onset of melting agrees with the equilibrium melt
temperature. On the other hand, we believe that our results with picosecond
heating pulses can be interpreted in terms of superheating of the solid by
several hundred degrees.

Experimental

Figure 1 gives a schematic illustration of our experimental method. The
surface of a (111) oriented GaAs crystal mounted in an ultrahigh vacuum
chamber (pressure less than 10^{-8} mbar) is irradiated with a laser heating
beam (λ = 532 nm) at an angle of approximately 45 degrees with respect to
the surface normal. For nanosecond heating the frequency doubled output of a

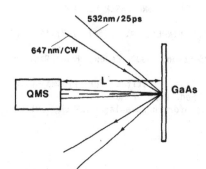

Fig. 1: Schematic of experimental arrangement.

single transverse mode, single frequency, passively Q-switched Nd-YAG laser with 10 ns pulse duration is used. Laser pulses of 25 ps duration for pico-second heating are obtained from the second harmonic of an actively and passively mode-locked Nd-YAG laser. The onset of melting is measured by monitoring the optical reflectivity of the GaAs surface with the help of a cw krypton ion laser beam at $\lambda = 647$ nm. The time resolution of the detection system for the reflectivity changes is approximately 500 ps.

The surface temperature of the GaAs crystal is determined from the velocity distribution of the atoms which are evaporated during laser heating. We use a quadrupole mass spectrometer (QMS) to detect the Ga and the As atoms and to measure the time of flight from the sample surface to the QMS detector. The actual time of flight is obtained from the measured time between the laser pulse and the signal output of the QMS. The drift times of the particles in the QMS - which are small in comparison with the time of flight proper - are well known and carefully taken into account in the evaluation of the data. Since the number of atoms which is detected for a single laser pulse is small, data accumulation over a large number of laser pulses is required, and the sample must be raster-scanned to avoid multiple exposure of the same surface area.

Results and Discussion

Figure 2 depicts examples of measured time of flight distributions of Ga atoms for nanosecond heating. Four different laser energies are shown: a) 156 mJ/cm^2, b) 189 mJ/cm^2, c) 257 mJ/cm^2, and d) 330 mJ/cm^2. Note that the distributions are normalized to unity. The actual number of detected particles is a strongly increasing function of the laser energy.

The important point to be noticed from Fig. 2 is that for the entire range of laser energies the measured time of flight distributions can be fitted by curves corresponding to Maxwell-Boltzmann velocity distributions (dashed lines). From these fits a temperature value is obtained, e.g. 900 K, 1400 K, 2300 K, and 3100 K, respectively for the four different laser energies of Fig. 2. The fact that the measured distributions can be represented by Maxwellians is indeed noteworthy. There is no temporal or spatial resolution in the present particle detection scheme. The measured time of flight distributions represent a temporal and spatial average of the surface temperature, which in general would not look like a Maxwell distribution.

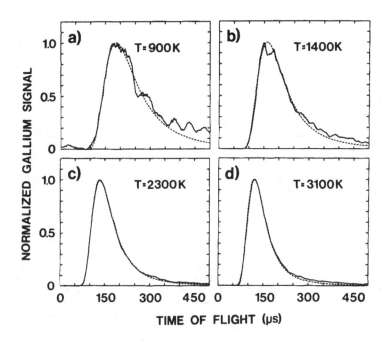

Fig. 2: Time of flight distributions of Ga atoms for different laser energies. a) 156 mJ/cm^2; b) 189 mJ/cm^2; c) 257 mJ/cm^2; d) 330 mJ/cm^2.

However, the particle current is a very strongly increasing function of the
temperature, and the temporal and spatial temperature average will be
weighted by this function. For example, if the particle current (evaporation
rate) is taken to be proportional to $\exp(-V/kT)$, where $V \simeq 4$ eV is related
to the heat of evaporation, it turns out from a computer simulation of our
experiment that the average temperature obtained from the temporally and
spatially integrated particle distribution is indeed very close to the maximum temperature. Thus the fact that the experimentally observed distributions are Maxwellian is taken as strong evidence that the temperature values
extracted from the distribution represent to a good approximation, the
maximum surface temperature.

In Fig. 3 we have plotted the experimental temperature versus peak
laser fluence (in the center of the Gaussian beam profile) for nanosecond
laser heating of GaAs. The data are obtained from measurements of Ga atoms.
The horizontal dashed lines represent room temperature (T_R) and the equilibrium melt temperature (T_M), respectively. The vertical dashed line marks
the threshold of melting, $E_{TH} = 200$ mJ/cm^2, as determined from the independent measurements of the optical reflectivity. What is important to note
from Fig. 3 is that the measured surface temperature at the onset of melting, i.e. at E_{TH}, is in excellent agreement with the known equilibrium melt
temperature of GaAs, $T_M = 1511$ K. In fact, for nanosecond laser heating of
GaAs superheating is expected to be less than our experimental accuracy of
the temperature measurement of ± 50 K. Thus, the agreement of the measured

222

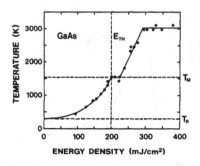

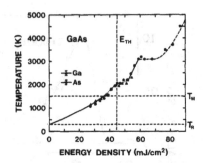

Fig. 3: Measured surface temperature vs laser energy for ns laser pulses.

Fig. 4: Measured surface temperature vs laser energy for ps laser pulses.

melting temperature with T_M is regarded as evidence of the adequacy and accuracy of our measuring technique.

The observed temperature plateau at $T = T_M$ just above E_{TH} reflects a condition in which the excess laser energy is used to supply the latent heat of melting. The temperature rise following the plateau at higher energies, on the other hand, signifies the overheating of the liquid surface layer.

Let us now turn to our picosecond results depicted in Fig. 4. For the entire range of pulse energies used in the picosecond experiments the velocity distributions of both Ga and As were observed to be Maxwellian. The ratio of Ga to As was measured to be approximately one, and temperature values obtained from Ga and As were the same. The temperatures extracted from Ga data (triangles) and As data (circles) are plotted as a function of peak laser fluence. The threshold energy of melting is $E_{TH} = 45$ mJ/cm^2, determined from the optical reflectivity measurements as mentioned above.

The striking feature of the measured temperature curve for picosecond laser heating is that at $T = T_M$ there is no change of the optical reflectivity and no temperature plateau. Instead, the temperature keeps going up, reaching a value in excess of 2000 K, before the onset of melting is established from the observed change of the optical reflectivity. For energies slightly above E_{TH} the slope of the temperature curve is significantly decreased, somewhat reminescent of the temperature plateau which is seen at $T = T_M$ in nanosecond laser heating (see Fig. 3). The temperature rise is resumed for laser energies greater than 51 mJ/cm^2. In this energy range the surface is covered with an overheated liquid layer of about 20 nm in thickness. The temperature plateau near 3200 K is probably due to the formation of a plasma and is not considered here. Comparing the measured temperatures corresponding to the respective melting threshold in the nanosecond and the picosecond experiment, we believe that the observed temperature difference of 540 K represents direct evidence of superheated conditions in picosecond heating of GaAs.

A few additional remarks may be appropriate. From the experimental limit for the detection of changes of the optical reflectivity we estimate that a liquid surface layer must have a thickness greater than a few nm to be detectable. Therefore it is possible that some partial melting in a surface layer less than the threshold thickness has already developed before the apperent "optical" melting threshold. Detailed numerical calculations based on recently published thermal data of GaAs [4] indicate that the temperature drop across this possible molten layer is expected to be small. Thus we argue that even if such partial melting has occured the measured surface temperature is close to the temperature of the solid-liquid interface.

Conclusions

Time of flight measurements of the atoms evaporated from a laser heated GaAs surface have been used to determine the surface temperature. For nanosecond laser heating the measured temperature at the onset of melting is in agreement with the equilibrium melt temperature of 1511 K. The higher temperature observed with picosecond pulses is interpreted in terms of superheating of the solid by several hundred degrees.

References

1. N. Bloembergen, in Laser Solid Interactions and Laser Processing, edited by S.D. Ferris, M.J. Leamy, and J.M. Poate, AIP Conference Proceedings, Number 50, (American Institute of Physics, New York 1978), p.1

2. R. Yen, J.M. Liu, H. Kurz and N. Bloembergen, Appl. Phys. A27, 153 (1982)

3. F. Spaepen and D. Turnbull, in Laser Annealing of Semiconductors, edited by J.M. Poate and J.W. Mayer, (Academic Press, New York 1982) p. 15

4. A.S. Jordan, J. Cryst. Growth 71, 551 (1985)

TIME-RESOLVED OPTICAL STUDIES OF PICOSECOND LASER INTERACTIONS WITH GaAs SURFACES

J.M. LIU*, A.M. MALVEZZI**, AND N. BLOEMBERGEN**
*GTE Laboratories, Incorporated, 40 Sylvan Road, Waltham, MA 02254
**Division of Applied Sciences, Harvard University, Cambridge, MA 02138

ABSTRACT

The interactions of picosecond laser pulses at 532 nm wavelength with GaAs surfaces have been studied with time-resolved reflectivity and transmission measurements at three probe wavelengths. At fluences below the melting threshold, the laser-generated electron-hole plasma is limited to a density below $\approx 10^{20} cm^{-3}$. The high reflectivities of molten GaAs observed at fluences above the threshold have a wavelength dependence inconsistent with a simple Drude model for a metallic GaAs molten layer. At high fluences, the evolution of a laser-induced vapor cloud is observed.

INTRODUCTION

A few experimental techniques have recently been applied to the investigation of picosecond laser interactions with GaAs surfaces at ultrahigh excitation levels. Akhmanov et al. [1,2] used the second-harmonic and sum-frequency generation in reflection from a Q-switched mode-locked Nd:YAG laser pulse train to probe the laser-induced phase transitions of GaAs surfaces with nanosecond resolution [1] and to study the structure of GaAs surfaces after pulsed laser irradiation [2]. With single picosecond pulses we have extended the second harmonic generation in reflection from crystalline GaAs surfaces to picosecond time resolution and have detected transient surface phase transition on the picosecond time scale [3,4]. Recently, Fabricius et al. performed time-of-flight measurements of As and Ga atoms evaporated from GaAs surfaces under nanosecond or picosecond pulse irradiation [5]. From the measured velocity distributions of evaporated atoms, they determined that with nanosecond pulse irradiation the observed solid-liquid transition temperature agrees with the equilibrium melting point whereas the solid GaAs surface could be superheated by as much as 540°K with picosecond pulses.

In this paper we summarize our recent time-resolved reflectivity and transmission studies on the interactions of picosecond laser pulses at 532nm with GaAs surfaces, probed at 532nm, 1.064μm, and 1.9μm wavelengths. In all the fluence regimes, it is interesting to observe many phenomena different from those observed in similar experiments on Si. GaAs is a much more complicated material than Si. Further analysis of our data will address more details on the processes involved during picosecond pulse laser interactions with GaAs.

EXPERIMENTAL

Semi-insulating GaAs wafers of both (100) and (110) one-side polished surfaces were used. Some samples were chemically etched on both sides for the transmission measurements to reduce scattering of the transmitted light by the unpolished back surface. In contrast to the second harmonic experiments [3,4], the reflectivity and transmission data do not depend on the surface orientation of the sample. Single pulses of 20ps duration at 532nm derived from the second harmonic of an active-passive mode-locked Nd:YAG laser were used as the pump pulses in all of the experiments described in this paper.

The surface melting threshold was previously determined to be at 30mJ/cm^2 [3,4]. For the probe, we used picosecond pulses at 532nm, 1.064µm, and 1.9µm wavelengths, derived form the same laser by harmonic generation or stimulated Raman scattering from an H_2 cell [6]. The ratio of the pump and probe spot diameters was always kept at larger than 5 to maintain good spatial resolution. The detailed experimental arrangement and procedures are similar to those described in References 7 and 8.

RESULTS

Crystalline GaAs has a 1.423eV bandgap at room temperature, which shrinks to 0.798eV at the melting point T_m = 1513°K [9]. GaAs is highly absorbing at 532 nm at all temperatures, but is almost transparent at 1.064µm up to 817°K, at which temperature the bandgap shrinks to the photon energy and GaAs starts to absorb 1.064µm light. At 1.9µm GaAs is transparent up to the melting temperature before it melts.

At pump fluences below the melting threshold of 30mJ/cm^2, the reflectivity at 532nm increases with pump fluences because of rapid heating of a thin GaAs surface layer within the pump pulse duration, as is shown in Figure 1. Model calculations show a very steep temperature gradient on the heated surface because GaAs has relatively strong optical absorption at 532nm and a relatively small thermal diffusion coefficient [10]. The increased reflectivity gradually decays as the heated surface cools down. Because the optical constants of GaAs at 532nm at elevated temperatures are not well known, quantitative information about the surface temperatures cannot be derived from these data. For wavelengths below the bandgap, it is known that the real part of the refractive index increases with temperature as $(1/n)(dn/dT) = 4.5 \times 10^{-5}/°\text{K}$, and the imaginary part k is negligibly small.

Figure 2 shows the reflectivity at 1.064µm as a function of pump pulse fluence at various probe time delays. Due to the large values of radiative and Auger recombination coefficients in GaAs [11] and the large surface recombination speed, changes in the reflectivity caused by an electron-hole plasma are not detectable at probe wavelengths at 532nm and 1.064µm. This result, which is in strong contrast to similar experiments in Si, sets an upper

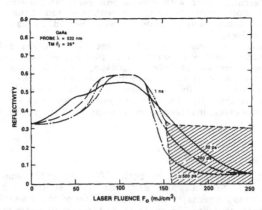

Figure 1: Reflectivity of GaAs probed at 532 nm as a function of pump pulse fluence at various probe time delays. The shaded area indicates distribution of scattered data at 1ns delay.

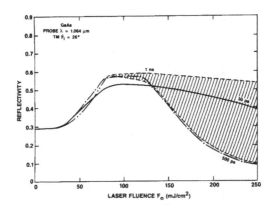

Figure 2: Reflectivity of GaAs probed at 1.064μm as a function of pump pulse fluence at various probe time delays. The shaded area indicates distribution of scattered data at 1ns delay.

limit of the carrier density at $\approx 10^{20} cm^{-3}$. However, the plasma-induced changes are observed in the reflectivity probed at 1.9μm for pump fluences below 30mJ/cm^2, as are shown in Figure 3. The absorption of the probe pulse by the electron-hole plasma is also observed at both 1.064μm and 1.9μm wavelengths [10]. At 20mJ/cm^2 pump fluence, the normalized absorption $A^* = 1 - (T/1-R)$ at zero time delay is $\approx 35\%$ and $\approx 10\%$ at 1.9μm and 1.064μm, respectively. Preliminary analysis [10] of the reflectivity and absorption data leads us to conclude the following results at 20mJ/cm^2: (1) the surface temperature is above 1000°K; (2) the averaged carrier density within the pump pulse duration is between $5 \times 10^{19} cm^{-3}$ and $10^{20} cm^{-3}$; (3) a plasma of this density is distributed in a layer about 2000Å thick, which is larger than the absorption depth of 532nm pump pulse at room temperature. It is very interesting to notice in Figure 3 that the carrier-induced reflectivity drop decays very fast at this carrier density, indicating a large Auger recombination coefficient on the order of $10^{-29} cm^6/sec$. Notice also that at 25mJ/cm^2 and 30mJ/cm^2, a minimum reflectivity occurs at a negaitve time delay and a complicated evolution of both plasma- and temperature-induced changes is observable. These data may indicate overheating of the surface near the end of the pump pulse, as is suggested by Fabricius et al. [5], and gradual melting of an extremely thin surface layer afterwards.

At fluences above the melting threshold, the reflectivity increases gradually, indicative of a very thin molten layer. The reflectivity reaches a relatively constant maximum value of molten GaAs when the pump fluence is between 80mJ/cm^2 and 120mJ/cm^2. This high reflectivity phase lasts for nanoseconds. Molten GaAs is metallic as has been determined by conductivity measurements [12]. The electrical conductivity jumps from $\approx 300 \ \Omega^{-1} \ cm^{-1}$ to 7900 $\Omega^{-1} \ cm^{-1}$ upon melting. However, the reflectivity values of the molten layer probed at 532nm, 1.064μm, and 1.9μm in our experiments are not consistent with a Drude model for a metallic system, which worked well for molten Si at various probe wavelengths. First, the high reflectivity values at all wavelengths are lower than expected for the molten GaAs calculated with a Drude model. More importantly, the maximum reflectivity of the molten GaAs

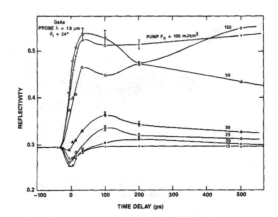

Figure 3: Reflectivity of GaAs probed at 1.9μm as a function of probe time delay at various pump pulse fluences. The lines are guides to the eyes.

layer decreases with increasing probe wavelength, which is in direct contradiction to the Drude model for a metallic system. Attempt to explain these anomalies with a molten layer thinner than or comparable to an optical absorption depth fails [10]. One possible explanation is that because of the compound structure of GaAs, the valence electrons are not fully ionized in molten GaAs and some covalent-bond nature is left. In fact, one may start to contemplate the real nature of molten GaAs.

When the pump fluence is increased above $120mJ/cm^2$, the reflectivity probed at 532nm and 1.064μm drops to very low levels due to rapid vaporization of GaAs surface, as is shown in Figures 1 and 2. However, this severe reduction of reflectivity is not observed at 1.9μm probe wavelength, although the reflectivity at 1.9μm becomes undulated as a function of time delay at these high pump fluences. Apparently the vapor cloud is relatively transparent to longer wavelengths. One interesting phenomenon is the fact that at a time delay of approximately 1ns after the pump pulse, the probe reflectivity data at 532nm and 1.064μm become very scattered, which are distributed in the shaded areas of Figures 1 and 2. Again, this phenomenon is not observed at 1.9μm. At longer time delays, the reflectivity gradually recovers to the high values, characteristic of the molten GaAs surface at each wavelength, with very small error bars in the data. This indicates that the vapor cloud atomizes and starts to cool down during expansion in about 1ns. Thereafter it gradually becomes transparent to the probe pulses.

CONCLUSION

We have performed a series of time-resolved reflectivity and transmission measurements at three different probe wavelengths on picosecond laser-excited GaAs surfaces. These new data consistently indicate that a very thin molten layer starts to develop at the surface at a threshold pump fluence of $30 \ mJ/cm^2$, in agreement with our previous second harmonic experiments [3,4]. At fluences below the melting threshold of $30mJ/cm^2$, an electron-hole plasma is detected at 1.9μm probe wavelength, but is not observable in the reflectivity at 1.064μm wavelength. These results set an upper limit of carrier density at about $10^{20} cm^{-3}$. At higher fluences, high reflectivities of molten GaAs are observed. However, the values of the high reflectivity as a function of probe wavelengths are not consistent with a simple Drude model for a metallic GaAs molten layer. This suggests further investigation of the real nature of molten GaAs. At even higher fluences, the evolution of a laser-induced vapor cloud is observed.

ACKNOWLEDGEMENTS

This research was supported by the U.S. Office of Naval Research under contract N00014-83K-0030, and by the Joint Services Electronics Program of the U.S. Department of Defense under contract N00014-75-C-0648.

REFERENCES

1. S.A. Akhmanov, N.I. Koroteev, G.A. Paitian, I.L. Shumay, M.F.Galjantdinov, I.B. Khaibullin, and E.I. Shtyrkov, Optics Commun. 47, 202 (1983).

2. S.A. Aknmanov, N.I. Koroteev, G.A. Paitian, I.L. Shumay, M.F. Galjautdinov, I.B. Khaibullin, and E.I. Shtyrkov, J. Opt. Soc. Am. B2, 283 (1985).

3. A.M. Malvezzi, J.M. Liu, and M. Bloembergen, Appl. Phys. Lett. 45, 1019 (1984).

4. J.M. Liu, A.M. Malvezzi, and N. Bloembergen, Mat. Res. Soc. Symp. Proc. 35, 137 (1985).

5. N. Fabricius, P. Hermes, D. von der Linde, A. Pospieszczyk and B. Stritzker, in this proceedings.

6. H.M. van Driel, L.A. Lompre, and N. Bloembergen, Appl. Phys. Lett. 44, 285 (1984).

7. J.M. Liu, H. Kurz, and N. Bloembergen, Appl. Phys. Lett. 41, 643 (1982).

8. L.A. Lompre, J.M. Liu, H. Kurz, and N. Bloembergen, Appl Phys. Lett. 43, 168 (1983).

9. J.S. Blakemore, J. Appl. Phys. 53, R123 (1982).

10. J.M. Liu and A.M. Malvezzi (unpublished).

11. C.B. Su and R. Olshansky, Appl Phys. Lett. 41, 833 (1982).

12. V.M. Glazov, S.N. Chizhevskaya, and N.N. Glagoleva, "Liquid Semiconductors" (Plenum Press, New York, 1969), Chapter 4.

PART IV

Laser-Induced Phase
Transformations in Graphite

PULSED LASER MELTING OF GRAPHITE[1]

G. BRAUNSTEIN[*], J. STEINBECK[*], M. S. DRESSELHAUS[*], G. DRESSELHAUS[*],
B. S. ELMAN,[**] T. VENKATESAN, [+] B. WILKENS, [+] and D. C. JACOBSON[+]

*Massachusetts Institute of Technology, Cambridge, MA
**GTE Laboratories, Waltham, MA
[+]Bell Communications Research, Murray Hill, NJ
[++]AT&T Bell Laboratories, Murray Hill, NJ

Abstract

Experimental evidence for laser melting of graphite, by irradiation with 30ns pulses from a ruby laser, is presented. RBS-channeling analysis, Raman scattering and TEM measurements reveal that the surface of graphite melts at a threshold energy density of about 0.6 J/cm². For laser pulse energy densities above 0.6 J/cm², the melt front penetration depth increases nearly linearly with increasing energy density. An intense emission of carbon particles during and after irradiation is observed. The thickness of the carbon layer removed in this process also increases nearly linearly with increasing pulse fluence. A dramatic redistribution of ion implanted impurities is also observed. Furthermore, the crystalline structure of the resolidified material is shown to depend on the energy density of the laser pulse. In order to explain these phenomena, a model for laser melting of graphite at high temperatures to form liquid carbon has been developed in which a free electron gas approximation is used to describe the properties of liquid carbon. The model is solved numerically to give the time and depth dependences of the temperature as a function of the laser pulse energy density. Very good agreement is found between the observed melt depth dependence on laser pulse energy density, as determined by RBS-channeling, and the model calculations. The redistribution of ion implanted impurities and the modification of the crystalline structure, caused by the pulsed laser irradiation, are also consistent with the model and permit the determination, for the first time, of interfacial segregation coefficients for impurities in liquid carbon. The model also predicts that liquid carbon at low pressure ($p < 1$ kbar) has metallic properties.

Introduction

The physical properties of liquid carbon have been the subject of intense debate. Scarce experimental information is available because of the experimental difficulties imposed by the high melting temperature of carbon. Consequently the present knowledge is based mainly on theoretical thermodynamical calculations for the phase diagram of carbon.

The proposed phase diagrams [1] are in good agreement with each other in the region of temperatures lower than about 3000°C, for which experimental data have been obtained, (see Fig. 1.[1]) but they disagree with each other in detail in the high temperature regime. The temperature range over which liquid carbon can exist, particularly at low pressures, is poorly established. The properties of liquid carbon are also strongly debated, some phase diagrams suggesting metallic behavior (an atomic liquid), other suggesting insulating behavior (a molecular liquid), and yet others having both metallic

[1]The MIT authors acknowledge NSF Grant #DMR 83-10482 for the support of their portion of the work.

234

Figure 1: Phase diagram for carbon according to Bundy.[1] Note
the prediction of a triple point
at low pressure at a temperature
~ 4200K (see inset).

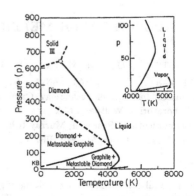

and insulating phases, depending on the temperature and the pressure of the liquid. Other properties of interest, such as the effect of the cooling rate on the structure of the resolidified material or the redistribution of impurities at a liquid–solid carbon interface, are completely unknown.

We have shown[2,3,4] that pulsed laser irradiation of graphite is an ideal technique to create and study liquid carbon. A large amount of energy can be coupled into the graphite lattice in a very reproducible and controllable fashion and in a very short time. A further advantage is that the graphite substrate will provide its own crucible, eliminating the problems of high temperature confinement and contamination from a noncarbon container.

In the present paper we review our experimental and computer simulation studies of the melting of graphite by high power laser irradiation [2,3] and give some examples of the important new information that has been obtained recently. Samples of highly oriented pyrolytic graphite (HOPG) have been irradiated in a direction perpendicular to the basal planes (i.e., along the c–axis) with pulses of a ruby laser (λ = 6943Å) of ~ 30 nsec FWHM duration and energy densities in the range 0.1 to 5.0 J/cm². Because of the large anisotropy in the physical properties of graphite,[5,6] due to its layered structure, and in particular the very low thermal conductivity along the c–axis direction, nearly all the energy of the incident laser pulse is deposited in the first ~ 1000Å below the surface of the sample. Using laser pulses with typical power densities on the order of 10^8W/cm², surface temperatures in excess of 4500 °K are obtained. At these elevated temperatures, the near surface region of the sample melts, making possible the study of the properties of liquid carbon.

Evidence for melting is obtained from the complementary information provided by RBS–channeling, Raman spectroscopy and TEM measurements. Rutherford backscat-tering(RBS)–channeling studies, on ion implanted and laser irradiated HOPG reveal a pronounced segregation of the impurity distribution towards the surface of the sample in much the same way as has previously been observed in the pulsed laser melting of silicon.[7] RBS–channeling studies on unimplanted HOPG show the formation of a disordered layer upon laser irradiation, above a reproducible threshold energy density. The width of the disordered region increases linearly with laser fluence (at least initially) and shows a sharp disorder/order interface with the underlying crystalline structure.

Transmission electron microscopy (TEM) and Raman spectroscopy, which were used to study the microstructure of the disordered layers, indicate that above but near threshold the near surface region is highly disordered, but at energy densities exceeding $\sim 3 J/cm^2$, randomly oriented graphite crystallites are seen to form. An intense emission of carbon atoms during and after irradiation has also been observed.[8]

The above experimental observations are unambiguously explained by the melting of the near surface region of the pulsed laser irradiated graphite. A computer model was developed to treat the melting of graphite and the evaporation of carbon atoms from the liquid as well as the redistribution of ion implanted impurities due to the melting process. By matching the model calculations to the experimental results, it has been possible to determine important properties of liquid carbon. Our experiments indicate that, at low pressures, graphite melts at about 4300 K and the liquid evaporates at 4700 K. Furthermore the calculation also predicts that liquid carbon has metallic properties. The analysis of the segregation of ion implanted impurities permits a determination of the interfacial segregation coefficients for impurities in liquid carbon. Finally the structure of the resolidified material is correlated to the velocity of the melt front during the resolidification process.

Experimental

Thin sections ($\sim 0.5 cm^2$ in area) of highly oriented pyrolytic graphite (HOPG) were implanted with [73]Ge and [75]As ions at energies of about 220 keV and 400 keV respectively with fluences of $1 \times 10^{15} cm^{-2}$. The ion implanted and similar unimplanted samples were then irradiated with pulses from a ruby laser ($\lambda = 6943$Å) of ~ 30 nsec duration and energy densities in the range 0.1 to 5.0 J/cm². RBS–channeling with 2 MeV He+ ions was used to determine the lattice disorder and the impurity redistribution induced by the laser irradiation. The microstructure of the irradiated regions was examined by TEM using a JEOL 200CX instrument operating at 200 keV and by Raman spectroscopy, in the backscattering geometry, using the 4880Å blue line of a CW–Argon laser. The samples for TEM were thinned to about 2000Å by repeated cleaving from the back surface, using scotch tape. The optical skin depth in the Raman experiments is about 1000Å for the 4880Å Ar+ laser line in graphite.

Results

The effect of pulsed laser irradiation on the crystalline structure of graphite was monitored by RBS–channeling measurements on laser irradiated, unimplanted, HOPG. For laser energy densities of ~0.6 J/cm² and above, a disordered region is seen to form. This region grows nearly linearly in depth with increasing laser energy density and eventually saturates at an energy density of 1.85 J/cm². As can be seen in Fig. 2, a sharp order/disorder interface exists between the disordered layers and the underlying crystalline structure. The extrapolation of the linear region to zero depth defines the threshold laser energy density of ~ 0.6 J/cm². The saturation in the thickness of the disordered region is attributed to the emission of carbon particles as described below.

One of the more distinctive features of laser annealing of semiconductors is the redistribution of impurities, previously introduced by ion implantation. We have observed a similar effect upon laser irradiation of ion implanted HOPG. Figure 3 shows the portion of the backscattering spectrum of Ge implanted HOPG, corresponding to the Ge peak, at various laser energy densities. At an energy density of 1.5 J/cm², the melt front partially penetrates the impurity distribution and those impurities which lie within the

Figure 2: (a) Channeling spectra of laser irradiated HOPG (unimplanted) showing the formation and growth of a disordered layer upon irradiation, and (b) thickness of the disordered region as a function of laser pulse energy density.

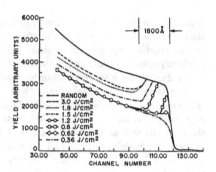

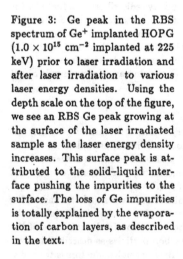

Figure 3: Ge peak in the RBS spectrum of Ge$^+$ implanted HOPG (1.0×10^{15} cm^{-2} implanted at 225 keV) prior to laser irradiation and after laser irradiation to various laser energy densities. Using the depth scale on the top of the figure, we see an RBS Ge peak growing at the surface of the laser irradiated sample as the laser energy density increases. This surface peak is attributed to the solid–liquid interface pushing the impurities to the surface. The loss of Ge impurities is totally explained by the evaporation of carbon layers, as described in the text.

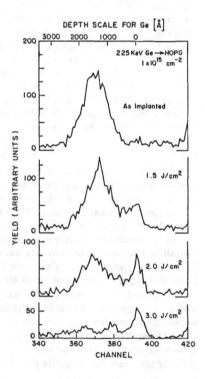

Figure 4: Shift of the As marker peak in the RBS spectrum of As$^+$ implanted HOPG (1.0×10^{15} cm^{-2} implanted at 400 keV), as a function of pulse energy density. The scale on the right indicates the corresponding thickness of the evaporated carbon layer.

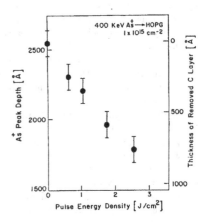

melted region are seen to segregate towards the surface of the sample. As the pulse fluence is increased, the melt front penetration increases until finally, for laser pulse energy densities of about 3 J/cm^2, nearly the entire impurity distribution falls within the molten region. At these high energy densities, the impurities are swept towards the surface, from which they eventually escape by evaporation.

An intense emission of carbon atoms during and after the irradiation is also observed. In order to determine the amount of carbon removed by the laser irradiation we have performed a marker experiment in which As$^+$ was implanted at 400 keV, resulting in a projected range of about 2550 Å, well beyond the probable penetration of the melt front. Indeed, we only observe a redistribution of As atoms located close to the sample surface. This redistribution, however, does not affect the position of the As peak, measured in the RBS spectrum. The shift of the As depth distribution in the RBS spectrum and the corresponding thickness of the removed carbon layer are presented in Fig. 4 as a function of pulse energy density. The RBS–channeling results again provide evidence for melting and resolidification of the near surface region of the HOPG sample. The intense emission of particles, which occurs with greater probability from the molten surface, is mainly responsible for the saturation in the thickness of the disordered region, observed in the RBS-channeling experiments. This intense emission of particles also provides an explanation for the reduction in the number of implanted impurities observed in the segregation experiments. (see Fig. 3)

Further evidence for the melting behavior was obtained from Raman spectroscopy and TEM studies which were used to study the microstructure of the disordered layers. The main features of the Raman spectrum of HOPG consist of a first–order zone center mode at ~ 1580 cm^{-1} and a second–order overtone at ~ 2720 cm^{-1} (see Fig. 5a). Upon irradiation with laser energy densities of about 1 J/cm^2, a disordered region is found as evidenced by the growth and line broadening of the disorder–induced ~ 1360 cm^{-1} line in the first–order Raman spectrum and the formation of a broad band in the second–order spectrum. As the pulse energy density is increased between 1 J/cm^2 and 3 J/cm^2, the line at ~ 1360 cm^{-1} grows at the expense of the ~ 1580 cm^{-1} line while the two lines further broaden and coalesce. This behavior is represented by Fig. 6b. A useful estimate of the degree of crystallinity of the lattice is provided by the relative intensity of the disorder induced ~ 1360 cm^{-1} line relative to the Raman–allowed ~ 1580 cm^{-1} line (I_{1360}/I_{1580}) which was shown to vary inversely with the crystallite size L_a of the graphite structure. In addition, the linewidths (FWHM) of the 1360 cm^{-1} and 1580 cm^{-1} lines, reflect the degree of disorder induced in the lattice[9] by the laser irradiation. A summary of the results for the first–order Raman spectra of the pulsed-laser irradiated HOPG is given in Figs. 6 where the intensity ratio I_{1360}/I_{1580} as well as the linewidths of the two lines are, respectively, plotted as a function of the energy density of the laser pulse. Figures 5 and 6 show that irradiation with pulse energy densities between 0.6 J/cm^2 and 3 J/cm^2 produce a highly disordered region with very small crystallite sizes. However, as the laser pulse energy density is further increased above 3 J/cm^2, the behavior changes significantly: in the first–order spectrum, the two lines at 1360 cm^{-1} and 1580 cm^{-1} narrow and become well defined, and in the second–order spectrum instead of the broad band, two peaks at 2720 cm^{-1} and 2945 cm^{-1} become clearly defined. Figure 5 and Figs. 6a and 6b show that after irradiation with pulse energy densities of about ~ 4J/cm^2 the Raman spectrum of the irradiated HOPG shows a resemblance to the Raman spectrum of glassy carbon. However, the analysis of the intensity ratio I_{1360}/I_{1580} in Fig. 6a indicates the formation of crystallites of about 100 Å in size. The structure of glassy carbon consists of randomly oriented, interleaved, short graphitic planar segments, less than 100 Å in size.

TEM studies support these observations.[2] Selected area TEM diffraction patterns show that above but near the melting threshold laser energy density the resolidified layer is highly disordered, but at energy densities exceeding 3 J/cm^2, randomly oriented graphite crystals are formed. This is evidenced by the appearance of the (002) reflections in the diffraction patterns, which indicate the formation of crystallites, some having their c–axis orthogonal to the original c–axes of the substrate. Further TEM studies are under way to determine the microstructure of the irradiated region, in particular for pulse energy densities larger than 3 J/cm^2.

We have developed a model[3] to explain the above experimental observations, based on a numerical solution of the heat equation, using the method of finite differences[10]. Since the model has been described in detail elsewhere[3], we summarize here briefly the main results and present some new results obtained after the first publication. This model uses experimental data for the physical properties of solid graphite([5,6] and references therein) and assumes the Ziman free electron gas model for liquid metals[11] to deduce the physical properties of liquid carbon. A comparison of the melt depth predicted by the calculation (including the evaporation of carbon atoms) and that experimentally determined by RBS–channeling measurements is shown in Fig. 7. The curve labeled "Solid Extension Model" corresponds to a calculation based on the extrapolation of the thermodynamical properties of solid graphite into the liquid region. Because of the low carrier concentration of graphite, the thermal conductivity in the "Solid Extension Model" is mainly by a phonon mechanism and therefore provides a maximum estimate of the melt depth vs. laser energy density if the liquid were insulating. The inaccuracy of this assumption is evident. In contrast, when the free electron gas model, including the surface evaporation, is used to account for the properties of liquid carbon, very good agreement is achieved, as can be seen in Fig. 7. The success of the Ziman free electron gas model for liquid metals in describing liquid carbon implies, in turn, that liquid carbon has metallic properties. In addition, to account for the intense emission of particles, observed experimentally, an evaporation temperature

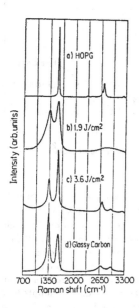

Figure 5: First and second order Raman spectra for pristine HOPG (a), laser irradiated HOPG (b) and (c), and pristine glassy carbon (d). Pulse energy densities for the laser irradiated samples are indicated.

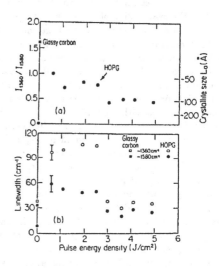

Figure 6: (a) Relative intensity of the disorder-induced ~ 1360 cm^{-1} line to the Raman-allowed ~ 1580 cm^{-1} line, as a function of laser pulse energy density. (b) Linewidths (FWHM) of the disorder-induced ~ 1360 cm^{-1} line and Raman-allowed ~ 1580 cm^{-1} line, as a function of laser pulse energy density. Note the change in behavior for pulse energy densities larger than 3 J/cm^2.

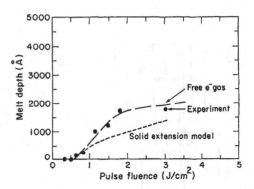

Figure 7: Comparison of the model calculations to the experimental data for the disordered region thickness as a function of incident laser pulse fluence. The experimental data points are from measurements using the RBS–channeling technique.

of $\sim 4700°$K is estimated. The total pressure exerted on the sample by the laser pulse and by the evaporation of carbon atoms is estimated to be less than 1 kbar.

The numerical solution of the heat equation provides time and depth dependencies of the temperature, as a function of the laser pulse energy density, from which melt front velocities and maximum melt depths can be determined.[3] Based on the melt front dynamics provided by the melting calculation, a second simulation was developed, to study the redistribution of ion implanted Ge impurities upon irradiation. A procedure similar to that previously used for laser annealing of ion implanted silicon was adopted.[12] Good agreement with the experimental results is obtained when a diffusion coefficient of 10^{-4}cm^2/sec is assumed for Ge in molten carbon (see Fig. 3). This rapid redistribution of impurities is typical of a liquid phase, and is inconsistent with the very low c–axis diffusion coefficient expected for crystalline HOPG. In this calculation the interfacial distribution coefficient $k' = C_S/C_L$, where C_S and C_L are the impurity concentration in the solid and liquid phases at the interface, is treated as a fitting parameter. In addition, the present model takes into account the experimentally determined evaporation of carbon particles, which results in a shift towards the surface of the impurity profiles, as seen in Fig. 8. Also the loss of Ge impurities shown in Figs. 3 and 8 is totally accounted for, in the calculations, by the evaporation of carbon particles. A detailed description of the segregation simulation will be presented elsewhere.[4] In a typical calculation, like the one presented in Fig. 8, two values of k' were used; one for the initial stage of the resolidification where the melt front recedes to the surface at relatively large velocities ($k'_f = 0.15$) and a smaller value of k' was used for the final stage of the resolidification when the melt front approaches the surface at very low velocities ($k'_s = 0.01$). A more detailed calculation would include a distribution of k' values between these two limits.[4] To our knowledge this is the first determination of interfacial distribution coefficients in liquid carbon.

The results of the melting modeling[3] indicate that the velocity of the melt front, during resolidification, decreases from about ~ 2 m/sec for pulse fluences of 1 J/cm^2 to about ~ 1 m/sec for pulse fluences of 3 J/cm^2. The melt front velocity results suggest an annealing or nucleation mechanism for the regrowth of larger graphite crystallites as the incident laser pulse fluence is increased. Since the liquid remains hotter for longer periods of time for larger laser pulse fluences, this means that the molten regions near

Figure 8: Comparison of the experimentally determined redistribution of Ge implanted impurities due to laser irradiation (a), to the results of the segregation simulation (b), for a laser pulse energy density of 1.5 J/cm².

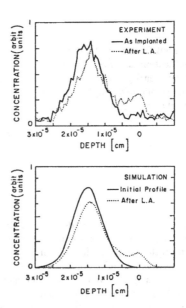

the liquid–solid interface remain at temperatures high enough for graphite annealing to take place. These annealing temperatures correspond respectively to the onset of two–dimensional (1500 K) and three–dimensional (2500 K) ordering in graphite.[9] Material remaining at these elevated temperatures for 150 ns residence time will thus produce larger graphite crystallites.

Similar results have been reported for silicon.[7] The final state of the resolidified melt has been shown to be strongly dependent on the velocity of the solid–liquid interface as the melt cools. If the velocity is too great, the molten region will form amorphous silicon. However, as the incident laser pulse fluence is increased, the velocity of the solid–liquid interface begins to slow down, allowing nucleation within the molten layer to occur, resulting in the formation of polycrystalline material.

This picture of the regrowth is consistent with the Raman scattering results which show that there is an increase in the restoration of two–dimensional ordering as the incident laser pulse fluence is increased above ∼ 3.0 J/cm². On the basis of this result, a residence time of 150 ns is estimated for annealing of graphite, based on the time necessary to initiate regrowth behavior, and corresponding to the total existence time of the liquid carbon created by the incident laser pulse. This behavior is shown schematically in Fig. 9.

Conclusions

Conclusive evidence for laser melting of graphite to form liquid carbon has been presented. When graphite is irradiated with 30 nsec ruby laser pulses, a threshold energy density of about 0.6 J/cm² is required to melt the graphite surface. For laser pulse energy densities above 0.6 J/cm², the melt front penetration increases nearly

242

Figure 9: Melt front velocity, during resolidification, as a function of pulse energy density. The structure of the resolidified graphite is indicated schematically.

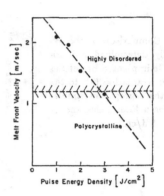

linearly with increasing energy density. During the molten phase an intense emission of particles occurs which results in the erosion of the sample surface. The removal of material from the sample surface also increases nearly linearly when the pulse fluence is increased above 0.6 J/cm², resulting in the saturation of the disordered region thickness (above 1.85 J/cm²) observed in RBS–channeling experiments. For laser pulse energy densities ≤ 1 J/cm², the melt front does not reach the region where the ion implanted impurities lie and the impurity distribution remains nearly unchanged. For energy densities between 1 J/cm² and 2 J/cm², the molten region partially penetrates the impurity distribution. Those impurities which lie within the molten region are seen to segregate towards the surface of the sample with a diffusivity of about 10^{-4} cm²/sec, once at the surface the impurities evaporate simultaneously with the surface carbon atoms. At these energy densities, the resolidification process is very fast and a highly disordered graphite layer is produced. Finally, for energy densities of 3 J/cm² and higher, for which the molten region extends throughout most of the impurity distribution, the impurities are swept towards the surface of the sample, from which they again escape. At these high pulse energy densities, the receding velocity of the melt front is relatively slow, allowing for the nucleation and growth of larger graphite crystallites. A computer model for laser melting of graphite has been developed which shows very good agreement with the experimental results.

Acknowledgments

The authors would like to thank Dr. K. Sugihara from MIT for many useful discussions. We also would like to acknowledge Dr. J.M. Gibson and Dr. J. Poate of AT&T Bell Labs. for their help and encouragement.

References

[1] F.P. Bundy, J. Chem. Phys. 38, 618 (1963).

[2] T. Venkatesan, D.C. Jacobson, J.M. Gibson, B.S. Elman, G. Braunstein, M.S. Dresselhaus and G. Dresselhaus, Phys. Rev. Lett. 53, 360(1984).

[3] J. Steinbeck, G. Braunstein, M.S. Dresselhaus, T. Venkatesan and D.C. Jacobson, J. Appl. Phys. (in press).

[4] G. Braunstein, J. Steinbeck, M.S. Dresselhaus, T. Venkatesan, B. Wilkens and D.C. Jacobson, (to be published).

[5] W.N. Reynolds, *Physical Properties of Graphite*, (Elsevier, New York 1968).

[6] B.T. Kelly, *Physics of Graphite*, (Applied Science Publishers, London 1981).

[7] J.M. Poate, G. Foti, D.C. Jacobson (eds.), *Surface Modification and Alloying by Laser, Ion, and Electron Beams*, Plenum Press, New York, 1983.

[8] T. Venkatesan, D.C. Jacobson, J. Steinbeck, G. Braunstein, B. Elman and M.S. Dresselhaus (submitted to this Symposium).

[9] B.S. Elman, G. Braunstein, M.S. Dresselhaus, T. Venkatesan and J.M. Gibson, *Phys. Rev.* B29, 4703 (1984).

[10] R.F. Wood and G.E. Giles, *Phys. Rev.* B23, 6 (1981).

[11] J.M. Ziman, *Phil. Mag.* 6, 1013 (1961).

[12] C.W. White in "Pulsed Laser Processing of Semiconductors", ed. by R.F. Wood, C.W. White and R.T. Young, vol. 23 of *Semiconductors and Semimetals* (Academic Press 1984), p.44.

TIME-RESOLVED PICOSECOND OPTICAL STUDY OF LASER-EXCITED GRAPHITE

C. Y. Huang,* A. M. Malvezzi,† J. M. Liu** and N. Bloembergen†
* Los Alamos National Laboratory, Los Alamos, NM 87545
† Division of Applied Sciences, Harvard University, Cambridge, MA 02138
** GTE Laboratories, 40 Sylvan Road, Waltham, MA 02254

ABSTRACT

The pump-and-probe technique is employed to perform picosecond time-resolved measurements of the reflectivity changes in highly oriented pyrolitic graphite excited by 0.532-μm pump pulses. At low pump fluences, the presence of a short-lived plasma and a high-temperature gradient gives rise to an increase in the reflectivity probed at 1.9 μm but causes a decrease at 1.064 μm. At the threshold fluence, 140 mJ/cm^2, the reflectivity drops abruptly, marking a phase transformation. Above the threshold, the reflectivity drops to ~0.2 from its original value of 0.42 at 1.064 μm and to ~0.4 from its ambient value of 0.50 at 1.9 μm. This new phase persists only for a few nanoseconds.

INTRODUCTION

It has been known for a long time that carbon at ambient pressure does not melt, but at high temperature it loses its mass by sublimination. As a result, the melting of carbon has been a subject of interest for more than a century, resulting in many investigations on this issue at various pressure and temperature ranges. The melting of carbon at approximately atmospheric pressures and at temperatures from 3600 to 4000 K was first reported more than a half century ago by Lummer [1] and Ryschkewitsch [2]. In 1981, Whittaker et al. [3], with the help of the pictures taken by a high-speed camera, reported the first observation of droplets of liquid carbon thrown from a spinning graphite sample heated by a CO$_2$ laser when the temperature reached 3800 K or higher. These authors also found that these droplets are transparent in the visible region. Very recently, by employing Rutherford-backscattering channeling spectroscopy, Venkatesan et al. [4] investigated the surface of pristine and As-implanted graphite modified by 30-ns, 694.3-nm laser irradiation. Based on the observation of the change of the channeling spectra, they concluded that graphite surface melts at the threshold fluence of 0.6 J/cm^2. Their data also provided an estimate of the melting point of ~4300 K. The experiments discussed above determine the melting on the basis of post-experimental examination of samples. In this paper, we report our time-resolved reflectivity measurements on highly oriented pyrolitic graphite (HOPG) excited by 20-ps, 532-nm laser pulses. The results indicate the onset of an ultrafast phase transition at a laser fluence of 140 mJ/cm^2 occurring during the exciting laser pulse duration.

EXPERIMENTAL RESULTS

We have performed a series of picosecond time-resolved reflectivity measurements by using an active-passive mode-locked Nd: YAG oscillator generating transverse single-mode light pulses of 30 ps at 1.064 μm. In order to pump our semimetallic HOPG samples obtained from Union Carbide, a nonlinear crystal has been utilized to produce 20-ps green (0.532 μm) pulses from the 1.064-μm pulses. In addition to the 1.064-μm probe, we have also employed 1.9-μm pulses, obtained from a hydrogen Raman cell excited by 1.064-μm pulses, to probe the sample surface excited or

modified by green pump pulses. The same experimental technique used in
Refs. 5 and 6 is employed here. The probe beam was incident at a small
angle of incidence (< 15°) on the sample surface which is normal to the
c-axis.

The optical properties of graphite have been investigated extensively [7].
It absorbs strongly at 0.532 µm. In our experiment, at a high green pump
fluence, a damage spot appears on the sample surface. Our optical micrographs show the appearance of a black ring marking the damaged area. The
formation of the black ring provides prima facie evidence for a structural
transformation. The damaged area strictly follows the variations of the
Gaussian fluence distribution of our laser beam on the sample, indicating
the thermal origin of the effect. We find a threshold value for the
surface damage, $F_{th} \sim$ 140 mJ/cm^2, in perfect agreement with the
visual and reflectivity observations.

Figure 1 shows our reflectivity data probed at 1.064 µm as a function
of the energy fluence of the green pump beam. At a delay of 30 ps after
the pump pulse, as displayed in Fig. 1(a), the reflectivity decreases with
increasing F at small F. In analogy to the case for Si [6], this decrease
may be attributable to the presence of a carrier plasma within the absorption depth at the surface which decreases the real part of the refractive
index. The reflectivity then remains constant from ~50 to ~140 mJ/cm^2.
This independence of F might result from the increase in the reflectivity
owing to the lattice heating which cancels out the further decrease in the
reflectivity by the increase in the carrier density at higher F. Above
~140 mJ/cm^2, the reflectivity decreases with increasing F and reaches
~0.26 at $F \sim$ 500 mJ/cm^2. The data for a delay of 50 ps are
presented in Fig. 1(b). The reflectivity for F < 140 mJ/cm^2 is roughly
the same as shown in Fig. 1(a). However, in this case, the reflectivity drops
rather sharply at $F \sim$ 140 mJ/cm^2. Like Si [6], we may identify a phase
transition taking place at the threshold $F_{th} \sim$ 140 mJ/cm^2, which is in
agreement with the threshold obtained from the post-experimental examination
of the damaged area. At the 200-ps delay, the probe pulse is temporally
completely separated from the pump pulse. Our data at this delay time are
depicted in Fig. 1(c). The reflectivity at low F is similar to that shown in
Figs. 1(a) and (b), but increases slightly at high F. At $F = F_{th}$, an abrupt
drop in the reflectivity is observed, providing prima facie evidence of the
occurrence of a first-order phase transition at F_{th}. For $F > F_{th}$, the
reflectivity remains constant around 0.26 up to F ~ 500 mJ/cm^2, and
decreases to ~0.20 at higher F. As shown in Fig. 1(d), this abrupt drop in
the reflectivity at F_{th} persists even up to the 700-ps delay. Nonetheless,
at this delay time, the reflectivity for $F > F_{th}$ has recovered to
~0.30. This recovery continues as the delay time is increased, as clearly
demonstrated in our data for 2.5 ns shown in Fig. 1(e). We have also made
some reflectivity measurements at F = 0 a few minutes after excitation
pulses. For F > 200 mJ/cm^2, these reflectivity values are almost the
same as those measured at 2.5-ns delay, indicating that the surface has
"recovered" in about 2.5 ns.

In Fig. 2 is displayed the reflectivity changes as a function of the
delay time at several pump fluences below the threshold. At F = 25 mJ/cm^2,
the reflectivity drops readily but recovers in less than 50 ps. This fast
transient suggests the presence of a short-lived electron-hole plasma. At
F = 100 mJ/cm^2, however, this plasma signature merges with a much slower
recovery of the reflectivity due to temperature contributions to the index
of refraction. Above the threshold, the decrease in the reflectivity is
much greater. As depicted, this decrease increases with increasing F, but
it seems to saturate around ~ 500 mJ/cm^2. The maximum decrease occurs
around 200 ps after the green pump pulse. The surface modification also
takes several nanoseconds to reach its final state. Our post-experimental
reflectivity measurements, as indicated by an arrow in the figure, shows
that the reflectivity never recovers completely to its original value,

suggesting the occurrence of a distorion of the lattice on the sample surface.

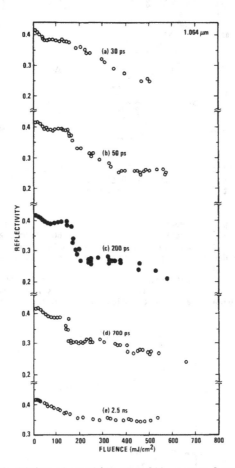

Figure 1.   Reflectivity of graphite at 1.064 μm as a function of 0.532-μm pump fluences for the delay times shown.

According to Ref. 7, the intrinsic carrier density in graphite is equivalent to a plasma frequency $\hbar\omega$ = 0.95 eV.   Based on the Drude model, if the probe frequency is lower than the plasma frequency, then the reflectivity increases with the carrier density. We have conducted the pump-and-probe measurement at 1.9 μm (0.65 eV). As shown in Fig. 3, below the threshold, the reflectivity does increase with F as the green pump generates more carriers. Around F_{th}, the reflectivity drops somewhat sharply. However, the drops are not as pronounced as those probed at 1.064 μm. The reflectivity-delay time curves for $F > F_{th}$ are similar to those measured at 1.064 μm shown in Fig. 2. Again, the reflectivity decreases rapidly; the minima occur at ~100 ps after the

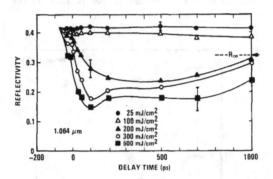

Figure 2.   Reflectivity of graphite at 1.064 μm as a function of probe
delay time at various 0.532-μm excitation fluences. R_∞
is the post-experimental reflectivity value.

green excitation pulse. However, the recovery to the final reflectivity
value takes several nanoseconds. For $F_{th} > F > 60$ mJ/cm^2, a sharp
peak at a short delay time appears, indicating a fast build-up and
subsequent rapid decay (~ 40 ps) of the carrier density as already
observed at 1.064 μm.

Probing at 0.532 μm also exhibits a sudden, more pronounced, drop in
the reflectivity for $F > F_{th}$. This again indicates a decrease in both
the real and imaginary parts of the dielectric constant at this wavelength.

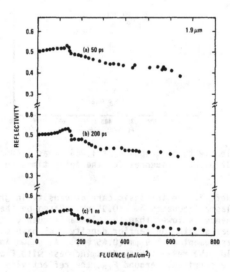

Figure 3.   Reflectivity of graphite at 1.9 μm as a function of
0.532-nm pump fluences for the probe delay times shown.

CONCLUSIONS

Below the threshold value F_{th} = 140 mJ/cm^2 for 20-ps pump pulses at
0.532 μm, the small decrease in the reflecticity can be explained by the
increase in the carrier density. This effect is counteracted by the rise in
the lattice temperature, which tends to increase the dielectric constant.
For F > F_{th}, the reflectivity decreases sharply. This decrease
requires a decrease in both the real and imaginary parts of the dielectric
constant, which is more marked as the probing wavelength is reduced from
1.9 μm via 1.064 μm to 0.532 μm.

The new phase, which may or may not be liquid carbon, becomes more
transparent than graphite at 0.532 μm. Consequently, the pump pulse can
cause the phase transition to proceed to increasing depths, comparable to
or larger than the probing wavelength. However, if we identify F_{th} as
the melting threshold, by using the temperature dependence of the heat
capacity [8,9] and the heat of fusion [9], we obtain a melting point of
~3900 K, which is consistent with that reported in Ref. 3. The high
temperature phase for T ≥ 3900 K definitely has a different electronic
structure than graphite. Our measurements indicate that the electronic
structure is more similar to that of glassy carbon or carbyne (an irregular
polymeric type structure with broken-chain elements C_n), than that of a
metal.

Clearly, better data on the reflectivity and transmission in thin
samples at several wavelengths are needed, to determine the electronic
structure of the high temperature carbon phase more precisely.

Acknowledgements

We would like to thank A. W. Moore for sending us the samples used in
this work and A. G. Whittaker for mailing us some of his unpublished
results. It is also a pleasure to acknowledge useful conversations with
G. Braunstein, G. Dresselhaus, M. S. Dresslelhaus, H. Ehrenreich, B.
Segall, J. Steinbeck, and D. Turnbull. This work is supported by the U.S.
Office of Naval Research under contract No. N00014-83K-0030 and the U.S.
Department of Energy.

References

1. O. Lummer, Verflussigung der Kohle und Hertellung der Sonnentemperatur,
   Druck and Verlag von Friedr. Vieweg and Sohn, Brunschweg, 1914.
2. E. Ryschkewitsch, Z. Electrochem. 27, 445 (1921).
3. A. G. Whittaker, P. L. Kintner, L. S. Nelson and N. Richardson,
   Aerospace Corp. Rept., SD-TR-81-60, 1981. See also A. G. Whittaker and
   P.L. Kintner, Carbon 23, 255 (1985).
4. T. Venkatesan, D. C. Jacobson, J. M. Gibson, B. S. Elman, G.
   Braunstein, M. S. Dresselhaus, and G. Dresselhaus, Phys. REv. Lett. 53,
   360 (1984).
5. J. M. Liu, R. Yen, H. Kurz, and N. Bloembergen, Appl. Phys. Lett. 39,
   755 (1981).
6. J. M. Liu, H. Kurtz, and N. Bloembergen, Appl. Phys. Lett. 41, 643
   (1982).
7. L. G. Johnson and G. Dresselhaus, Phys. Rev. B 7, 2275 (1973) and
   references therein.
8. J. E. Hove, in Industrial Carbon and Graphite (Soc. Chem. Ind., London
   (1958).
9. F. P. Bundy, J. Chem. Phys. 38, 618 (1963), and references therein.
10. J Steinbeck, G. Braustein, M. S. Dresselhaus, B. S. Elman and T.
    Venkatesan, Mat. Res. Soc. Symp. Proc. 35, 219 (1985).

NANOSECOND TIME RESOLVED REFLECTIVITY MEASUREMENTS AT THE SURFACE OF PULSED LASER IRRADIATED GRAPHITE

T. VENKATESAN,[†] J. STEINBECK,[‡] G. BRAUNSTEIN,[‡] M. S. DRESSELHAUS,[‡]
G. DRESSELHAUS,[‡] D. C. JACOBSON AND B. S. ELMAN
[†] Bell Communications Research, Murray Hill, New Jersey 07974
[‡] MIT, Cambridge, MA 02139
[*] AT&T Bell Laboratories, Murray Hill, New Jersey 07974
[**] GTE Laboratories, Waltham, MA 02254

ABSTRACT

Time resolved reflectivity measurements at the surface of pulsed laser irradiated graphite are found to be complicated by the evolution of carbon atoms from the surface which attenuates the probe beam. The evolution of the species occurs concomitant with the observation of the melt on a time scale of <2 nsec. Further, the emitted species have velocities of $\leq 2 \times 10^5$ cm/s. The experimental results suggest the formation of large clusters at the molten surface though more experiments are required to clearly distinguish the effect of clusters from those due to the formation of a randomly reflecting plasma.

Pulsed laser melting of diamond and graphite have been demonstrated recently.[1,2] There has thus been an increasing interest in understanding the properties of the molten carbon phase. There is considerable controversy about the nature of the molten phase of carbon with suggestions ranging from a carbene phase, to a metallic phase, to an insulating phase.[3,4] Solution of the heat equation for the conditions of the pulsed ruby laser experiments on graphite[5] is consistent with a liquid metallic phase. A time resolved reflectivity measurement with[6] tens of psec resolution indicates a decrease in the surface reflectivity which has been interpreted as due to the formation of an insulating phase on the surface. In the experiment reported here, time resolved measurements in the nsec regime indicate significant evolution of an optically attenuating species from the surface which interferes with the sampling of the phase-transformed surface by the probe beam. Our conclusion is that time resolved reflectivity measurements at the surface are complicated due to significant evolution of carbon species at the surface due to the laser pulse.

Strong evidence for evaporative loss comes from marker experiments performed on Ge implanted highly oriented pyrolytic graphite (HOPG), laser annealed at various fluences. The Rutherford backscattering spectra (Fig. 1) clearly reveal the effect of the melt, with the melt depth indicated by the arrow (as measured by channeling experiments[2]). The curves clearly show the segregation of the Ge atoms in the melt, as the melt bulldozes the Ge atoms towards the surface, leaving untouched the Ge atoms beyond the melt depth. Hence by overlapping the rear edge of the spectrum, an offset of the ordinate is observed which can only be explained by loss of material at the surface. Figure 2 summarizes our measurements of the evaporative loss in which we have used as a marker As[+] implants at various energies. Although the blow off of the surface occurs at pulse energy densities lower than 0.6 J/cm^2 (which we attribute to sublimation), the loss of the surface layers is more dramatic at energy densities larger than the melt threshold at 0.6 J/cm^2. Within the error limits of the experiments we must conclude that the evaporation is concomitant with the melt. The boiling point of the liquid carbon must be quite close to the melt temperature though to resolve this issue better more accurate measurements are needed.

252

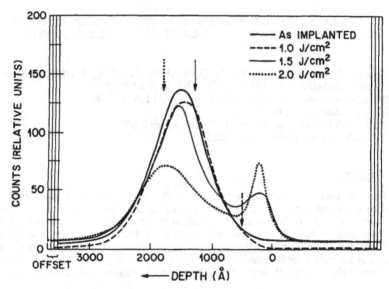

Fig. 1   Rutherford backscattering spectra of laser irradiated ion implanted HOPG. The Ge implant distribution as implanted and subsequent to irradiation at various laser intensities of 1.0 J/cm², 1.5 J/cm² and 2.0 J/cm². The ordinate is shifted to make the rear edge of the Ge spectra (depth of 3000-2500 Å) coincide. This gives a measure of the thickness of the layer that has evaporated from the surface. The depth of penetration of the melt measured by ion channeling is indicated by the position of the arrow, for each laser fluence.

The technique of time resolved reflectivity at the surface of semiconductors has been very useful in monitoring phase transitions induced by pulsed laser irradiation.[7] In the case of silicon, the molten phase is metallic and as a result the reflectivity of the surface to a cw probe beam increases rapidly when the surface melts and the reflectivity remains at the higher value until the melt refreezes (Fig. 3b). The same technique was applied to monitor the melt transition at the surface of HOPG. A cw He-Ne laser focussed to a spot size of 50 μm incident at an angle of ~75° with respect to the surface normal was detected by a fast photo detector. The combined response of the detector and the oscilloscope was ≲2 nsec. Since the ruby laser used to generate the high energy pulses produced a beam of nonuniform transverse intensity, a quartz homogenizer was used with a maximum exit diameter of 6 mm. The spacing between the homogenizer face and the sample surface was ~ 1mm and could be varied by a factor of two without significant change of intensity at the surface.

In Fig. 3 are shown the time resolved reflectivity traces of the reflected HeNe laser probe. Unlike the case of silicon (Fig. 3b), the reflectivity from a graphite surface goes to zero in a time scale of <2 nsec and recovers over a long time on the order of several 100 nsec (Fig. 3c). To ensure that this reflectivity change was due to the surface, the probe beam was launched parallel to the surface but close (~0.1 mm − 1.5 mm) to the surface. In this case also the probe beam was attenuated, but in a shorter time. The attenuation occurs with a time delay proportional to the height of the probe beam over the sample surface. Clearly, this experiment reveals that the reflectivity change is not due to the surface but due to the evolution of certain species from the surface.

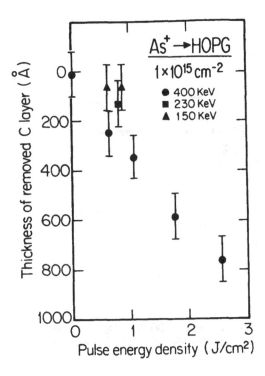

Fig. 2   Evaporative loss at the surface as a function of laser fluence measured by the
RBS study of   deep implant of   As⁺ into HOPG.

Going back to the reflectivity set up described earlier, the experiments were repeated for the homogenizer face held at different heights, h, above the sample surface. The reflectivity data of Fig. 4 contain information on the velocity of these species emitted from the surface. Looking at the reflectivity data, one sees a second attenuation dip in the reflectivity trace with a possible weak dip at a later time for small h values. The secondary dip is caused by the reflection of the emitted species by the face of the homogenizer rod with the weak dips attributed to more than one bounce between the homogenizer face and the sample surface. This is clearly seen in the figure as the dip occurs at a later time when the homogenizer is moved above the surface to different heights. An estimation of the velocity of the species from these data (Fig. 5) yields an average velocity for the emitted species of ~2 × 10⁵ cm/s which corresponds to ~0.15 eV per carbon atom (independent of the size of the emitted species). The reflectivity spectrum and the velocity of the species are essentially unaffected by experiments done in air or vacuum. This result would tend to support a cluster model with large cluster sizes in order to enable the cluster to retain its energy over such long distances. Though the density of the evolved species increases with laser fluence, the velocity of these species do not depend upon the laser pulse energy density, implying that increasing the laser fluence increases the duration of the melt at the surface without significantly raising the temperature of the melt.

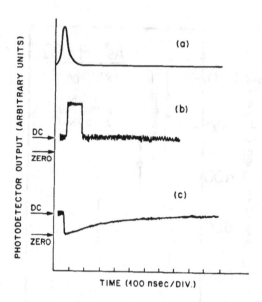

Fig. 3   The time resolved traces of (a) the ruby laser pulse. The reflected He-Ne probe at a (b) silicon surface and (c) graphite surface.

At a laser energy density of 1 J/cm^2, which is above the melt threshold, if one assumes that 100Å of the surface is removed in 10 nsec, this would account for a number density of $\approx 5 \times 10^{19}$ cm^{-3}, assuming an average velocity of 2×10^5 cm/s for the carbon atoms. A plasma density adequate to randomly reflect the probe beam[8] would require 6×10^{19} ions/cm^3 which implies that every evolved atom is ionized, which seems unlikely at these temperatures. Since the probe beam traverses a distance of ~6 mm over the sample, an absorption coefficient of ≥ 0.1 cm^{-1} is needed to explain the observed results. Non-resonant scattering is clearly ruled out since the Thompson cross section is much too small to produce an adequate attenuation. However, if one has large clusters formed, with a number density n and having a broad absorption band with an absorption cross section of 10^{-16} cm^2, then the minimum value of n required to produce an absorption coefficient of 0.1 cm^{-1} is 10^{15} cm^{-3}. For example to produce a ten atom cluster of such density, the fraction of evolved atoms that has to be in the form of such clusters is ~2×10^{-4}. This does not seem unlikely since experimental observation of significant amount of cluster formation has been reported in laser irradiated carbon by Kasuya[9] et al.

At the present stage the experimental results of time resolved reflectivity in the nsec range on graphite surface do not reveal information on the nature of the melt. However, there is evidence for the emission of species from the surface at energies of 0.12 eV per carbon atom. Whether these species are in the form of an ionized plasma or in the form of an agglomerated cluster has not been conclusively determined, and further work is underway to understand these effects in detail.

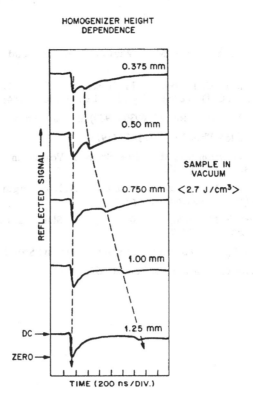

HOMOGENIZER HEIGHT
DEPENDENCE

0.375 mm

0.50 mm

SAMPLE IN
VACUUM
<2.7 J/cm³>

0.750 mm

1.00 mm

REFLECTED SIGNAL →

DC →

ZERO →

1.25 mm

TIME (200 ns/DIV.)

Fig. 4   The time resolved reflectivity traces for various homogenizer heights.

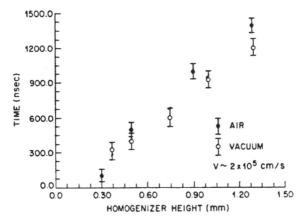

Fig. 5   The time delay between the primary and the secondary attenuation dips as a function of homogenizer height above the sample surface.

REFERENCES

1. J. S. Gold, W. A. Bassett, M. S. Weathers, and J. M. Bird, Science, *225*, 921 (1984).

2. T. Venkatesan, D. C. Jacobson, J. M. Gibson, B. S. Elman, G. Braunstein, M. S. Dresselhaus, and G. Dresselhaus, Phys. Rev. Lett. *53*, 360 (1984).

3. F. P. Bundy, J. Geophys. Res. *85*, 6930 (1980) and ref. therein.

4. J. A. Van Vechten, Phys. Rev. *B7*, 1479 (1973).

5. J. Steinbeck, G. Braunstein, M. S. Dresselhaus, T. Venkatesan and D. C. Jacobson, J. Appl. Phys. *58*, 4374 (1985).

6. C. Y. Huang, A. M. Malvezzi, J. M. Liu and N. Bloembergen, Mat. Res. Symp. proceedings (1986). Accompanying paper.

7. D. H. Auston, C. M. Surko, T. Venkatesan, R. E. Slusher and J. A. Golovchenko, Appl. Phys. Lett. *33*, 437 (1978).

8. J. D. Jackson, Classical Electrodynamics, John Wiley and Sons. Inc. New York.

9. A. Kasuya and Y. Nishina, Phys. Rev. *28*, 6571 (1983).

DYNAMICAL LASER DESORPTION PROCESS OF IONIC CLUSTERS
IN GROUP IV ELEMENTS

A. KASUYA and Y. NISHINA, The Research Institute for Iron, Steel and Other
Metals, Tohoku University, Sendai 980 Japan.

ABSTRACT

Emission of ionic clusters from surfaces of group IV elements has been
analyzed by means of time-of-flight measurements under nitrogen laser
irradiation in its peak intensity ranging from 50 to 200 MW/cm^2 with pulse
duration of 10 ns. The measurements on Si, Ge and β-Sn with a straight
flight tube show a continuous distribution of spectra in the flight time
corresponding to the mass-to-charge ratio between 1 and 2 in units of the
singly charged atom of respective elements. The line shape analysis
indicates that emitted ions immediately after laser excitation are coagulated in the form of large clusters having nearly an equal mass-to-charge
ratio between 5/4 and 3/2. They eventually decompose into monoatoms and
their ions in a time scale of 10 μs as confirmed by a quadrupole mass
analysis. These results are in remarkable contrast with the case of graphite
which exhibits a series of sharp spectral peaks. The clear distinction
between graphite and other group IV elements is attributed to the intrinsic
difference in their chemical bonds under highly excited conditions.

INTRODUCTION

Ion emission induced by laser excitation is presently interpreted
as sublimation of materials through thermal heating of the surface [1,2].
For high laser intensity over 1 GW/cm^2, materials evaporate explosively with
their kinetic energies in excess of 10 eV. Existing models account for the
energy transfer of laser beam to kinetic energies of emitted ions in order to
explain rapid temperature rise at the surface and high expansion velocity of
the blow-off materials. These models, however, give no details on dynamical
aspects of the transition from the condensed phase to the ionized vapor in
the process of laser desorption.
    This letter presents time-of-flight (TOF) analysis of ions emitted
from surfaces of Si, Ge and Sn under nitrogen laser irradiation with its peak
intensities up to 200 MW/cm^2 and pulse width of 10 ns. The results show a
band-like distribution of spectra on account of the fact that the emitted
ions are initially coagulated in the form of clusters which eventually
decompose through consecutive steps of fragmentation into individual ions and
atoms in the period up to 10 μs or so. Hence the emitted ions immediately
after laser excitation are not in a vapor phase of monoatoms nor of stable
molecular species. These characteristics of spectra are common in their
spectral shapes to all the above elements. This result suggests that the
initial value of mass-to-charge ratio as well as the decomposition process is
similar in these elements but quite different from carbon [3].

EXPERIMENTAL RESULTS AND DISCUSSIONS

Our TOF spectrometer consists of a straight drift tube of 62 cm at a
constant potential without any electrode arrangement for compensating the
initial kinetic energy spread of ions affecting the TOF spectrum. The
details of our experimental apparatus are given in Ref. 3. Figure. 1 shows
our typical spectra of ions emitted from Si, Ge and β-Sn for the acceleration
potential of 300 V applied across the distance of 2.5 cm between the sample

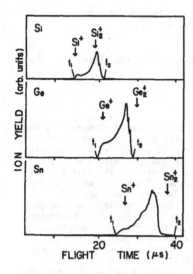

Figure 1. Time-of-flight spectra of ions emitted from Si, Ge and Sn by nitrogen laser excitation. The flight times corresponding to monomers and dimers of respective elements are denoted by arrows. The leading edge t_1 of the spectrum and the trailing edge t_2 are also indicated.

and the entrance aperture of the tube. The peak intensity of our nitrogen laser is varied from 50 to 200 MW/cm². These spectra are recorded for each laser pulse with a transient wave-form digitizer of 100 ns gate per channel. The ion yield for a given intensity is found to be higher in order of Si, Ge and Sn. These spectra exhibit continuous flight time distributions associated with a sharp rise (denoted by t_1) as well as a sharp cut-off (denoted by t_2). This distribution lies approximately in the flight time range corresponding to mass-to-charge ratio, M/q, between 1 and 2 as shown by arrows where M denotes the mass of ions in units of atomic mass of respective element and q the charge multiplicity. The spectral intensity in this figure represents the number of atomic ions in excess of 10^3. The characteristic feature of these spectra is that their profile changes from pulse to pulse of the laser shot. Figure 2 shows some of our measurements in Si. The following gives results for Si since Ge and Sn exhibit essentially similar spectral features. Though these spectra are recorded for a fixed intensity of laser beam , they exhibit a variety of distributions. It is emphasized, however, that the observed spectral shapes do not change at random but vary according to a definite rule as shown in Figs. 3 which is a plot of t_1 and t_2 vs. width $w = t_2 - t_1$. This figure shows a definite correlation between

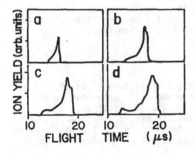

Figure 2. Time-of-flight spectra from Si sampled out of hundreds of measurements and displayed in order of increasing width from a through d.

the spectral positions (t_1 and t_2) with respect to w. It also shows a maximum distribution of data points for the value of w near 6 µs. The values of t_1 and t_2 for a given w are statistically reproducible within ±5%. Hence one can predict how often one finds spectrum having a particular width w out of a given number of laser desorption events. This set of spectra having a common value of w exhibit a unique value of both t_1 and t_2 within the above statistical probability.

One should note that the large spectral width comes neither from the initial kinetic energy of ions nor from space charge effect. For a TOF spectrometer employing a straight drift tube, the flight time t_f is given by $t_a + t_d$ where t_a is the time required for the emitted ions to be accelerated from the sample surface to the entrance aperture of the drift tube and t_d the time required to travel from the entrance to the end of the drift tube where the ions are detected. The kinetic energy gained at the time of desorption affects both t_a and t_d, while the space charge mostly influences t_a. The kinetic energy of ions can be measured from t_f with zero applied voltage (V_d=0) of the spectrometer and is found to be typically less than 30 eV (calculated for M/q = 1). In our TOF measurement with V_d = 300 V, this value of kinetic energy gives rise to broadening of the spectrum of less than 5 %. If the observed spectral width (Fig. 2-c for example) were to come entirely from the kinetic energy effect, the observed width corresponds to kinetic energy over 200 eV. The space charge effect also gives rise to negligible contribution, since its major effect is to modify the potential distribution of the acceleration field. In this case, the kinetic energy gained by the ions at the end of the field (entrance aperture of the drift tube) is unchanged and hence t_a is altered but not for t_d. In the extreme case where the motion of the ion current is space charge limited with potential variation in the 3/4 th power of the distance, a simple calculation shows that t_a is increased by 50 %. Since t_a is less than 1 µs for our spectrometer and $t_a \ll t_d$(=13.7 µs for M/q=1), error in the total flight time amounts to less than 4 %. Hence the wide distribution of our TOF spectrum is not due to the spread in initial kinetic energy of ions or to the space charge effect.

Since the wide spectral width is not caused by extrinsic effects as shown above, it has to come from decomposition of emitted ions within the period of acceleration, t_a in the spectrometer [4]. Though our TOF spectra appear to be complex at a glance, spectral shapes are essentially similar to that shown in Fig. 1-a having respective values of t_1, t_2, w and total area. Hence the TOF spectra can be represented by sharp rising edge and cut-off. These edges are connected by smooth tails on which weak peaks and shoulders are superposed. Chait and Field [4] have shown that TOF spectral peaks exhibit tails if emitted ions decompose within the period t_a after their emission. The tails appear between the peak of parent and that of its

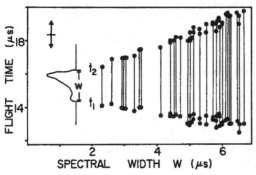

Figure 3. Plot of the leading edge t_1 and the trailing t_2 vs w=t_2-t_1. The vertical line connects t_1 and t_2 of the same spectrum. The double-headed arrow denotes the statisical fluctuation range (±5%) of t_1 and t_2.

fragment and extend from both of these peaks toward each other. The spread in spectrum depends on the decomposition time, t_c, of parent ion with respect to t_a ($\geq t_c$). Thus, the observed profile can be explained in terms of a decomposition process, $Si_k^{a+} \rightarrow Si_m^{b+} + Si_n^{c+}$. Here the inequality $n/c \geq k/a \geq m/a$ holds for an auxiliary condition $n/c \geq m/b$, because of the conservation of charge, $a=b+c$ and of mass, $k=m+n$. Suppose this process occurs with N of Si_k^{a+} ions at the rate given by $dN/dt = -N/t_c$. Those ions which decompose into Si_m^{b+} and Si_n^{c+} immediately after their emission by laser excitation contribute to the spectral leading edge at the flight time for Si_m^{b+} and to the trailing for Si_n^{c+}. Because of the inequality on mass-to-charge, these edges always appear on both sides of the spectral position of the parent ion, Si_k^{a+} which would exhibit a peak if it did not decompose in the acceleration field before reaching the entrance aperture of the drift tube. Those ions which decompose subsequently in the acceleration field give rise to a continuous distribution of tails from these edges toward Si_k^{a+}. The spectrum, therefore, exhibits a sharp leading edge at Si_m^{b+} as well as a trailing edge at Si_n^{c+}, both connected with smooth tails toward Si_k^{a+}. These two edges always appear as a pair and are closely correlated with each other because they are caused by the fragments from a parent ion. This particular property of the line profile predicted in this decomposition process is consistent with the observed features in our TOF spectrum.

On account of the above decomposition process, the spectral shape depends on three parametric conditions: A) M/q of the parent ion, k/a, B) M/q's of fragments, C) time evolution of decomposition. Figure 3 shows that the distribution of points for t_1 and that for t_2 can be extrapolated to a common point around $t_0 = 15$ µs at $w = 0$. The smooth distribution of data points shows that the pair of points t_1 and t_2 observed for each spectrum take positions close together or far apart by an out-of-phase displacement with respect to t_0. If the condition A), k/a of parent, were to vary for each event of laser desorption, t_1 and t_2 in the spectrum would shift more or less together in phase (not out-of-phase) in the same direction as the shift of t_0 (parent). If B), M/q's of fragments, were different from event to event for a given k/a, t_1 and t_2 would change their values independently of each other according to the way the parent ions break up. Neither of these features appears in the results shown in Fig. 3. The distribution of data points in Fig. 3 can take place only if the conditions A) and B) of the decomposition process are common to all of the data points in Fig. 3, leaving only C) varying from event to event of laser desorption.

The time evolution of decomposition process (condition C) can be identified clearly by considering two extreme cases, where thousands of atoms/ions are emitted upon laser excitation: Process I) Atoms/ions are emitted as a large number of small molecules like ionized gas consisting of dimers or trimers of Si, Process II) they are initially divided up into a relatively small number of large clusters, like a droplet. Since the number of emitted ions is large in the Process I, the TOF spectrum would exhibit a distribution corresponding to the statistical average of the decomposition processes so that one should observe an identical spectrum for each desorption event. In the Process II on the other hand, the TOF spectrum does not represent the statistical average of the fragments, because it reflects the decomposition processes of only a small number of ionic clusters at a time. Suppose a cluster consists of 10 Si atoms. Mass and charge conservations require that there are 22 possible fragmentation channels which can give rise to the spectral feature of Fig. 3. Hence for large clusters consisting of more than 50 Si atoms or so, hundreds of examples of our measured spectra are still insufficient to exhaust all the possible profiles. According to this interpretation, wide variation in our TOF spectrum in Fig. 1 have a definite physical significance. The spectral width w depends on how rapidly the fragmentation of clusters proceeds toward their ultimate form of monoatomic ions. The more rapidly the fragmentation proceeds, the more readily do M/q's of fragments differ from that of the parent in the acceleration field of TOF spectrometer, and thereby the wider does the spectral width w become. In the limiting case w=0, the condition $t_c \gg t_a$ would be required for decomposition so that the corresponding $t_1 = t_2 = t_0$ represents M/q of the parent. From the value of t_0 at w = 0, one can deduce M/q of the parent cluster in the plot of Fig. 2 to be in the range $5/4 \leq M/q \leq 3/2$. The parent cluster with this value of M/q is present within a period of the first 100 ns after laser excitation, because our measurement can detect decomposition process in that time scale.

Further evidence for the decomposition of the Process II can be found through the V_d dependence of the TOF spectrum. Since the increase in V_d is to reduce the flight time of ions passing through the acceleration field between the sample and the entrance aperture of the drift tube, the spectrum reflects only earlier stages of decomposition process in its distribution. In the Process I, one expects to observe the peak for parent between the spectral edges since there would be always a finite probability for some of the parent ions to pass through the acceleration field of the spectrometer without experiencing decomposition. This probability would increase with increasing V_d so that the parent peak would increase in its intensity relative to the contribution from fragments. Our spectrum, on the other hand, does not show any sign of such a peak at $V_d = 300$ V as shown in Fig. 1 nor at $V_d = 2000$ V, the highest voltage of measurement. This must be the case with Process II, because the parent cluster ceases to exist at the very first step of decomposition, leaving no trace on the spectrum.

Our independent measurements by a quadrupole mass spectrometer shows that Si^+ is the only dominant species and Si_2^+ and others are at least one order of magnitude smaller than Si^+ in the spectral intensity [5]. This decomposition continues over 10 μs as indicated in the time variation of the detector current. This result is also consistent with a similar measurement for flight time distribution of emitted clusters passing through an electrostatic energy analyzer.

The continuous distribution of spectra shown in Fig. 1 is in remarkable contrast with the series of sharp spectral peaks observed in the case of graphite under similar experimental condition as shown in Fig. 4 [3]. The chemical species are identified as $(C_{3i})^{2\pm}$ (i = 1,2,3,....). Figure 5 shows the result of our quadrupole mass analysis [3]. According to a theoretical calculation by Pitzer and Clementi [6], the carbon molecules are stable in the form of linear chains in which C_3 is most stable. The decomposition time of C_3^+ to C^+ is estimated to be 9.8 μs which is much longer than t_a (.63 μs for C^+) [5]. On the other hand, the decomposition time is comparable to t_a in the case of Si. This is the reason why the spectrum of graphite exhibits sharp lines while that of Si shows continuous distribution. The definite distinction in the TOF line profile may be attributed to the

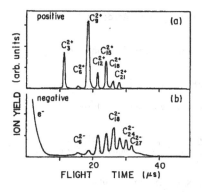

Figure 4. Time-of-flight spectra of (a) positive ions and (b) of negative ions emitted from graphite.

262

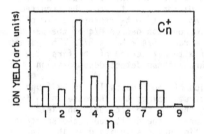

Figure 5. Quadrupole mass spectrum of ions emitted from graphite upon nitrogen laser excitation.

intrinsic difference in the chemical bond energy of C with respect to the rest of group IV elements.

CONCLUSION

Our time-resolving analysis of TOF spectra shows that ionic cluster with an initially common value of mass-to-charge ratio ($5/4 \leq M/q \leq 3/2$) are emitted from Si, Ge and Sn by pulsed nitrogen laser excitation. Our method of analysis has a novel feature of deducing their subsequent decomposition process in a submicrosecond scale from the line profile of our TOF spectrum measured in the total flight time range of 10 μs. One should note that the apparently random nature of TOF spectra in Fig. 2 are seen to have a strong systematic dependence if t_1 and t_2 are plotted as a function of w as in Fig. 3. The laser desorption process in our range of excitation intensity and energy density is distinctively different from any other desorption phenomenon described in terms of a single particle excitation or ordinary evaporation through thermal heating [1,2].

AKNOWLEDGMENTS

Authors would like to acknowledge Professor K. Ohno of Hokkaido University, and Professor C. Horie and Professor M. Tachiki of Tohoku University for valuable discussions.

REFERENCES

1. J. F. Ready, Effects of High Power Laser Radiation (Academic Press, New York, 1971).

2. N. Furstenau and H. Hillenkamp, Int. J. Mass. Spectrom. Ion Phys. 37, 135 (1981).

3. A. Kasuya and Y. Nishina, Phys. Rev. B28, 6571 (1983).

4. B. T. Chait and F. H. Field, Int. J. Mass Spcetrometry and Ion Phys. 41, 17 (1981).

5. A. Kasuya and Y. Nishina, Secondary Ion Mass Spectroscopy SIMS IV, Springer Series in Chemical Physics 36, ed. A. Benninghoven (Springer, Berlin, 1984), p. 164.

6. K. S. Pitzer and Clementi, J. Am. Chem. Soc. 81, 4477 (1959).

MICROSTRUCTURAL STUDIES OF PULSED–LASER IRRADIATED GRAPHITE SURFACES[1]

J. S. SPECK, * J. STEINBECK, * G. BRAUNSTEIN, * M. S. DRESSELHAUS, *
and T. VENKATESAN+
*Massachusetts Institute of Technology, Cambridge, MA 02139
†Bell Communications Research, Murray Hill, NJ 07974

Abstract

The surface structures of laser-irradiated samples of highly oriented pyrolytic graphite (HOPG) have been investigated using optical microscopy, scanning electron microscopy (SEM), transmission electron microscopy (TEM), and scanning transmission electron microscopy (STEM). The samples were irradiated with a 30 ns ruby laser ($\lambda = 6943$ Å) pulse with energy fluences ranging from 0.1 to 3.0 J cm^{-2}. Optically, the specimens show a damage region approximately 5 mm in diameter. The surface structure displays three characteristic regions: an outer boundary characterized by submicron carbon spheroids resting on the surface; an inner boundary characterized by both submicron spheroids and 1 to 5 μm 'torn' carbon layers which appear to have broken away from the graphite surface after irradiation; a central region characterized by a uniform density of spheroids and a pattern of surface upheavals which trace out a grain pattern similar to that of the pristine substrate. Electron diffraction patterns taken on the irradiated region indicate an ultra-fine grain 2–dimensionally ordered carbon. Qualitative trends in the areal density of different microstructural features are presented. In addition, a simple model explaining the observed features is given. All observations are consistent with the rapid solidification of liquid carbon.

Introduction

The high temperature phases and properties of the most refractory element, carbon, have both behooved and eluded the research community for a number of years. Recent advances in laser technology have led to the development of short time/high energy laser pulses which are capable of melting surface layers of metals and semiconductors. Surfaces of HOPG have been irradiated with high energy laser pulses to create and study the properties of liquid carbon. It is anticipated that after irradiation, the surface structure of the samples will be modified – most radically if the surface has melted.

Previous work on this problem using Rutherford backscattering spectrometry, Raman spectroscopy, and ion channeling has led to values of the disordered layer thickness as a function of laser pulse fluence [1], shown in Fig. 1. These investigators concluded that liquid carbon formed when samples were irradiated with laser fluences greater than 0.6 J cm^{-2}. Further evidence for melting can be obtained by direct microscopic examination of the irradiated region and then drawing conclusions from the observed structure. It will be shown below that the results of this study are entirely consistent with melting of graphite.

Experimental Procedure

Samples of pristine HOPG were irradiated in air or vacuum with 30 ns ruby laser

[1]The MIT authors acknowledge NSF Grant #DMR 83-10482 for the support of their portion of the work, and JSS acknowledges NSF for financial support in the form of a graduate fellowship.

Figure 1: The effect of laser pulse
fluence on melt depth as determined by RBS (after [1])

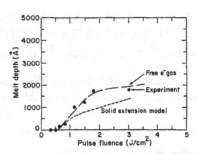

pulses (λ = 6943 Å) with energy fluences ranging from 0.1 to 3.0 J cm^{-2}. A quartz
homogenizer was used to provide uniform light intensity on the sample surface. The
samples were oriented such that the incident laser beam was normal to the graphite
basal planes which constituted the sample surface. The laser pulse was focused to form
a ~5 mm diameter spot on the specimen surface.

Scanning electron microscopy on the irradiated surfaces was performed using a JEOL
200 CX electron microscopy operating at 80 kV. TEM was also performed on a JEOL 200
CX TEM, in this case operating at 200 kV. Bright field TEM, selected area diffraction,
and high resolution TEM were performed. Chemical analysis, by x-ray fluorescence,
was performed on a VG HB5 STEM. Specimens for TEM and STEM were prepared by
peeling the HOPG from the unirradiated side down to an electron transparent thickness
(~1000Å) and then repeatedly bathing, ~20 times, the thin specimen in either methanol
or acetone solutions.

Results and Discussion

Any single technique for evaluating a complex structure will provide only limited
information. By using several different microstructural techniques, a deeper understanding of a structure can be developed. This will be shown to be the case for the laser
irradiated samples discussed below.

For samples irradiated with laser energy fluences less than 1.0 J cm^{-2}, SEM and
optical microscopy results indicate no alteration of the surface structure of the HOPG.
If melting does occur at these low laser energy fluences, then the melt solidifies in a
'quiet' and uneventful manner.

For energy fluences in excess of 1.0 J cm^{-2}, the structure becomes rather complicated.
The following three characteristic regions are observed in all samples irradiated with
energy fluences in the range 1.0 to 3.0 J cm^{-2}:

●A region exterior to direct irradiation which is littered with small, ~0.1 μm, carbon
spheroids. The number density of the spheroids decays with increasing distance from
the perimeter of the irradiated region.

●A boundary region showing both spheroids and 1 – 5 μm torn layers of carbon on
the surface.

●A central region with spheroids and surface upheavals which trace out a grain
pattern comparable to that of pristine HOPG.

The characteristic regions can easily be distinguished with optical microscopy. In
Fig. 2a, an optical micrograph of a sample irradiated with a 2.0 J cm^{-2} pulse, the

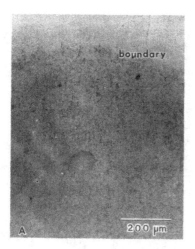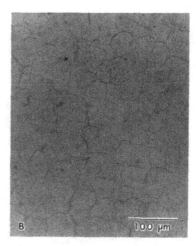

Figure 2: Optical micrographs of: (a) a sample irradiated with a 2.0 J cm^{-2} laser pulse, (b) the central region of a sample irradiated with a 3.0 J cm^{-2} pulse (see text).

boundary is clearly visible. The rough background contrast near the boundary is due to 1 – 5 μm torn layers of carbon resting on the surface, whereas the fine background contrast is due to ~0.1 μm carbon spheroids on the surface. More detail of the typical central region is shown in Fig. 2b. Here a grain defining pattern is observed on the surface. The linear features, as will be shown below, are actually surface upheavals. A more detailed examination provides further structural information on these features.

The torn layers of carbon have maximum density adjacent to the laser irradiated perimeter. The general structural appearance of the torn layers at the boundary is shown by SEM in Figs. 3a and 3b. In Fig. 3a, it is readily seen that the torn layers are pulled back and elevated from the surface. In addition, there is a fine dispersion of spheroids on the surface. The torn layer in the lower right hand corner of Fig. 3b shows two important features. The first is that the torn layers are quite thin, at most they are several hundred Å thick. Second, spheroids are observed on both sides of the torn layers of carbon. This indicates that the spheroids impacted the layer after it had peeled away from the surface.

Further examination of the spheroids leads to interesting surface morphologies as shown in Fig. 4. In this instance, in the center of a sample irradiated with 3.0 J cm^{-2}, there are three spheroids forming a stack in the center of the micrograph. The only plausible explanation for such a morphology is that the carbon in the spheroids makes an excursion into the atmosphere prior to reattachment on the surface. Small carbon black spheres have frequently been observed near carbon arcs [2]. Nevertheless, a full understanding of the mechanism of forming the spheroids is still lacking.

Surface upheavals are the other structural feature of significance in the center of the irradiated samples. These upheavals are clearly observed on the surface of a sample irradiated with a 2.5 J cm^{-2} pulse as shown in Fig. 5. The upheavals are a result of surface buckling of the substrate, presumably at grain boundaries during constrained rapid thermal expansion.

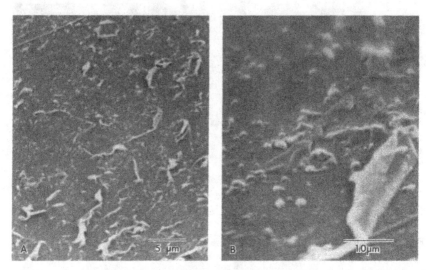

Figure 3: Scanning electron micrographs of the boundary region of a sample irradiated with a 2.5 J cm^{-2} laser pulse. (a) Low magnification view showing general structure. (b) High magnification view showing detail of torn layers and spheroids.

Figure 4: The surface structure in the central region of a sample irradiated with a 3.0 J cm^{-2} laser pulse. Notice the roughness of the surface.

Figure 5: Scanning electron micrograph of spheroids and surface upheavals in the central region of a sample irradiated with a 2.5 J cm^{-2} laser pulse. Note that the surface upheavals meet to form a three grain junction in the center of the figure.

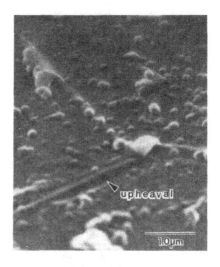

At this juncture, two additional observations must be made. The first is that samples which were irradiated in air had structures which were indistinguishable from samples irradiated in vacuum with similar energy fluences. Further, the spheroids are not an artifact of the homogenizer position. Samples which were pulse laser irradiated without a quartz homogenizer still displayed spheroids on the surface. High spatial resolution x-ray fluorescence measurements on the spheroids indicate that they are nearly pure carbon. TEM analysis of these structures provides further proof that the observed structures are not artifacts.

All of the general features of the structure are seen by TEM and can be directly attributed to solidification of laser melted carbon. Figure 6, a transmission electron micrograph of the damaged region of a sample irradiated with a 3.0 J cm^{-2} pulse, shows a torn layer, spheroids and one surface upheaval. Selected area diffraction patterns from areas as small as 1 μm^2 display a fully developed ring pattern as shown in Fig. 7. The periphery of the bright area corresponds to a (001) reflection for a 2–dimensionally ordered carbon. The outer two rings correspond to $(100)_{graphite}$ and $(110)_{graphite}$. This diffraction pattern is entirely consistent with previously published work [3]. This diffraction pattern corresponds to a randomly oriented 2–dimensional carbon structure. The thin resolidified layer corresponds to the chill zone which is observed in macroscopic castings. The only possible mechanism for forming such a structure, considering the precursor structure, is by solidification of molten carbon.

The dominant dependence of the surface microstructure on laser energy fluence is that the spheroid density in the sample center decreases with increasing energy. On the sample surface, for a given laser energy, the density of torn layers is maximum at the boundary, whereas the density of spheroids is constant across the irradiated region, then falls off with distance from the boundary. Due to constrained thermal contraction, it is anticipated that the surface stress is maximum at the perimeter of the irradiated

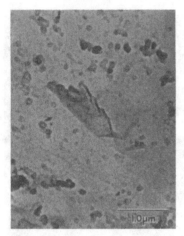 

Figure 6: General structure, as seen by TEM, of a sample irradiated with a 3.0 J cm^{-2} laser pulse. All general features of the microstructure are readily observed.

Figure 7: Selected area electron diffraction pattern from the central region of a sample irradiated with a 2.5 J cm^{-2} laser pulse.

region. Hence, tearing occurs near the boundary on cooling – leading to torn layers of carbon on the surface. Further work is in progress to establish quantitative relationships between surface features and laser energy fluence.

Conclusion

The observed structural features of the irradiated surfaces can be explained as follows:

(1) The spheroids result from carbon which has made an excursion into the atmosphere above the melt surface.

(2) The surface upheavals in the sample are due to constrained rapid thermal expansion of the substrate on heating.

(3) The torn layers on the surface are due to constrained thermal contraction of both the substrate and the resolidified layer.

The only possible explanation for the damaged layer is that it results from solidification of liquid carbon. No other model can account for the randomly oriented ultra-fine grain structure which is observed after irradiation using microstructural and other techniques [1].

References

[1] G. Braunstein, J. Steinbeck, M.S. Dresselhaus, G. Dresselhaus, B.S. Elman, T. Venkatesan, B. Wilkens, and D.C. Jacobsen, current symposium proceedings.

[2] J. Abrahamson, *Carbon* 12, 111 (1974).

[3] T. Venkatesan, D.C. Jacobson, J.M. Gibson, B.S. Elman, G. Braunstein, M.S. Dresselhaus, and G. Dresselhaus, *Phys. Rev. Lett.* 53, 360 (1984).

PART V

Laser-Induced Phase
Transformations in Metals

AMORPHOUS GALLIUM PRODUCED BY PULSED EXCIMER
LASER IRRADIATION

J. FRÜHLINGSDORF AND B. STRITZKER
Institut für Festkörperforschung, Kernforschungsanlage Jülich, D-5170 Jülich,
Fed. Rep. Germany

ABSTRACT

Pure crystalline Ga films (α-Ga, β-Ga) have been irradiated at low
temperatures (≤ 20 K) with an Excimer laser. By measuring the superconducting
transition temperature T_c and the residual resistivity ρ_0, the resulting
Ga phases (α-Ga, β-Ga, a-Ga) can be identified.
Both crystalline Ga phases can be transformed into the amorphous phase.
The threshold energy density for the $\beta \rightarrow$ a transition depends on the film
thickness, whereas the $\alpha \rightarrow$ a transition occurs always at about 225 mJ/cm^2.
This behavior is in agreement with earlier observations that a-Ga can
grow on top of the α-phase but not on the β-phase.
The results of laser quenching are compared with other non-equilibrium
techniques for the production of a-Ga, such as vapor quenching and low
temperature ion iradiation.

INTRODUCTION

During recent years investigation of metastable metallic systems
has attracted great interest because of their remarkable electrical,
mechanical or superconducting properties [1]. Several different non-equilibrium methods have been used to produce metastable phases. Experiments
have shown that vapor quenching and low temperature ion irradiation result
in similar systems because of comparably high quenching rates (10^{14}K/sec)
[2] whereas laser quenching with 40 nsec laser pulses is several orders
of magnitude slower [3]. Therefore a comparison of laser quenching experiments
to vapor quenching or ion beam irradiation is very interesting.
For such a test Ga appears to be an ideal system for several reasons.
First, there is no interfering influence of a second species since it
is a monocomponent system. In contrast to typical metals, a high degree
of homopolar bonding exists in the lattice of Ga, and therefore it differs
rather strongly from a close packed structure. Since vapor quenching results
in close packed arrangements 'typical' metals are built up in a crystalline
form even at He-temperatures, but Ga can be frozen in as an amorphous
layer [4]. In addition, it is known that amorphous Ga films (a-Ga) can
be produced by low temperature ion irradiation [5,6,7].
An important aspect is that three phases can be clearly distinguished
(Fig. 1). The first one is the amorphous phase formed by vapor condensation
onto a substrate at 4.2 K. This phase, with a resistivity ρ_0 of 29 $\mu\Omega$cm
and a superconducting transition temperature T_c = 8.5 K, is stable until
about 16 K. It then transforms into the metastable crystalline β-phase
(β-Ga, monoclinic [8]) with a resistivity of 3 $\mu\Omega$cm and a T_c value of
6.3 K. At about 60 K β-Ga transforms into the stable crystalline α-phase
(α-Ga, orthorhombic [8]) with a resistivity of 12 $\mu\Omega$cm and a T_c of 1.07 K.
From these data it is clear that the different phases can be rather well
identified by their resistivity and T_c values, but they also show that
any laser quenching experiment intended to produce a-Ga must be performed
at low temperatures because of the low crystallization temperature of
a-Ga (16 K).

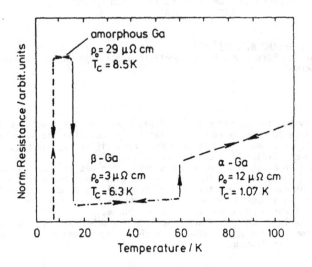

Fig. 1 Normalized resistance of vapor quenched gallium [4]. The different
phases are characterized by the residual resistivity ρ_0 and the
superconducting transition temperature T_c.

EXPERIMENTAL

Ga metal (Koch-light Lab. Ltd., grade 6 N) was evaporated onto quartz
substrates at liquid He temperature, resulting in a-Ga. Samples with thicknesses ranging from 15 nm to 70 nm were mounted onto a substrate holder
of a cryostat. The resistance was determined by a standard four-point
probe technique. The temperature was measured by a calibrated Ge- resistor
(1 K $\leq$ T $\leq$ 50 K) and a Pt-resistor (50 K $\leq$ T). Laser irradiation could
be performed in situ immediately after quench condensation and suitable
annealing to obtain β- and α-Ga (see Fig. 1). The laser used in the
experiments was a KrF-Excimer laser with 248 nm wavelength, 40 nsec pulse
duration and output power of about 0.5 J/cm² (without any focusing element).

RESULTS

β-Ga phase

Ga in the β-phase was obtained from a-Ga by heating to 20 - 40 K.
Laser irradiation at low temperatures resulted in transitions β   a-Ga
at certain threshold energy densities that depended on the film thickness.
The transition was detected by an increase of the residual resistivity
and the occurrence of a superconducting transition at 8.5 K, both typical
for a-phase. The resulting threshold energy densities for different film
thicknesses are plotted in Fig. 2. The energy necessary to amorphize the
β-Ga film increases with increasing thickness of the film.

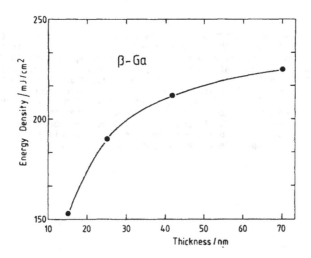

Fig. 2   Threshold energy density for the amorphization of β-Ga versus
film thickness

Another surprising result can be seen in Fig. 3. Transformation into the
amorphous state as detected by ρ-increase can be achieved even at substrate
temperatures as high as 20 K. However, in contrast to laser quenching
at 4 K, this a-Ga is not stable and has a certain lifetime that decreases
with increasing temperature.

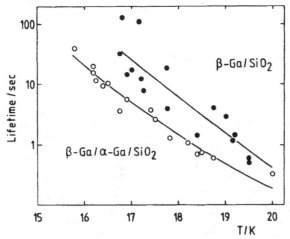

Fig. 3   Lifetime of a-Ga as obtained by laser quenching (300 mJ/cm^2) of
α+β-Ga and β-Ga (42 nm film thickness) at different substrate temperatures

α-Ga phase

α-Ga was produced by warming β-Ga-films up to a temperature of about
150 K. Laser irradiation of this stable Ga phase also results in a transition
into the amorphous state. The results as shown, in Fig. 4, indicate that
the α → a transition occurs at about 225 mJ/cm², independent of the film
thickness. Laser quenching with lower energy densities leads to transitions
α → α + β but no regular behavior has been observed. Laser quenching
of films containing both α- and β-Ga (i. e. after several pulses without
heating to 70 K) changes the onset of amorphization to that of very thin
pure β-Ga films (∼160 mJ/cm²). In addition, lifetime measurements of
the a-phase as obtained from α + β → a transformations exhibit a temperature dependence comparable to that of β-Ga (see Fig. 3).

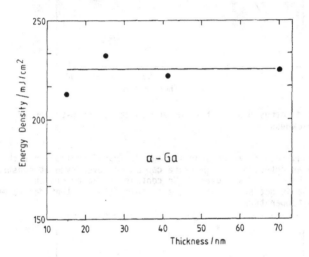

Fig. 4: Threshold energy for amorphization of α-Ga versus film thickness

DISCUSSION

Since these results will be compared to that of low temperature ion
irradiation, the main features of ion beam amorphization of Ga films are
briefly reported here [5,6,7]. Only Ar irradiation of α-Ga at temperatures
below 10 K results in a-Ga. Only ∼10¹⁴ ions/cm² are sufficient to produce
ρ and T_c values typical for the amorphous state. However, a-Ga prepared
in this way is unstable against further irradiation. It transforms into
β-Ga at larger fluences. The irradiation of β-Ga does not lead to an
amorphization at comparable fluences. The amorphous phase obtained after
fluences of 2·10¹⁶ ions/cm² is caused and stabilized by ∼5 at% O-impurities.
Although the results of laser quenching apparently disagree with those
of ion irradiation they can be interpreted in close analogy if the main
differences are considered. Low temperature ion irradiation produces collision
cascades (spikes) with energy densities that locally are high enough to
exceed the melting energy, so the material is quasi molten and quenched

because of rapid energy dissipation. X-ray measurements [8] verify that
the short-range order of molten Ga (similar to a-Ga) is closer to that
of β-Ga than to that of α-Ga. Thus α-Ga is supposed to act less efficiently
as crystallization nuclei than β-Ga and only α-Ga can be amorphized by
ion irradiation. The different behaviour of α- and β-Ga concerning irradiation
induced amorphization is in good agreement with vapor quenching experiments
[9]. The amorphous phase can be obtained by quench condensation onto α-Ga
as a substrate, but not onto β-Ga. Again this can be interpreted in terms
of the similar short range order of the 'liquidlike' amorphous phase and
the β-phase.
    The results of laser quenching confirm these results. The onset of
amorphization of β-Ga depends on the film thickness, i.e. the whole β-film
must be molten to avoid recrystallization on β - nuclei. This is not the
case for α-Ga where the amorphization does not depend on film thickness
for energy densities ≳ 225 mJ/cm². Laser quenching of α-Ga with lower
energy densities (≲ 215 mJ/cm²) results in transitions α → α + β, probably
leading to a thin β-film on top of an α-substrate. In a further laser
quenching experiment this β-layer on top of the α-phase can be amorphized
at substantially lower energy densities of ~160 mJ/cm², the value for
thin β-film (see Fig. 2)
    In addition the lifetime measurements (Fig. 3) compare the stability
of the a-phase on different substrates, i.e. quartz and α-phase. In the
first case (upper line) the whole Ga film (β-phase) is quenched into the
a-phase. In the latter case (lower line), only the β-film on top of the
α-layer on the quartz is transformed into the amorphous state. Fig. 3
shows that the amorphous phase on the quartz substrate is more stable
than on the α-phase. This results demonstrate clearly that the α-layer
acts more efficiently as crystallization nuclei than the quartz surface.
In further experiments details of the amorphization and recrystallization
shall be investigated using transient conductance and reflectivity measurements.

CONCLUSION

    The experiments give evidence for the first time of the possibility
to amorphize a pure metallic element only by laser irradiation, i.e. without
a second stabilizing component. Transitions of crystalline (α,β) to
amorphous Ga even at substrate temperatures as high as 20 K demonstrate
that low temperatures are necessary not only to provide high quenching
rates but also to stabilize the amorphous phase. These results of laser
quenching are compared to vapor quenching and low temperature ion irradiation
as indicated in Fig. 5. The internal energy without any temperature dependence is plotted in arbitrary units versus temperature. Equilibrium
phases (vapor , liquid, α-Ga) are indicated by solid lines, whereas metastable phases (amorphous and β-Ga) are shown by dashed lines. The similarity of the liquid and amorphous phases is demonstrated by identical
values of internal energy. Included in Fig. 5 are all applicable non-
equilibrium techniques. Vapor quenching (dashed arrow) transforms the
vapor directly into the amorphous phase; 4 K ion and laser irradiation
(thin arrow) transform α-Ga into the amorphous phase, whereas only 4 K
laser irradiation (thick arrow) results in the amorphization of β-Ga.

276

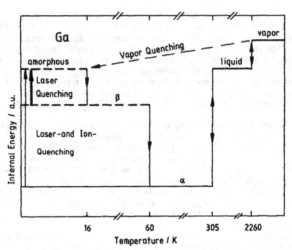

Fig. 5 Schematic plot of the internal energy of the different Gallium
phases as a function of temperature [10].

REFERENCES

[1] R.R. Appleton, B. Sartwell, P. Peercy, R. Schaefer, and R. Osgood,
Materials Science and Engineering 70 (1985), 23.

[2] B. Stritzker in: Surface Modification and Alloying by Laser, Ion
and Electron Beams, J. M. Poate, G. Foti, and D.C. Jacobsen (eds.),
Plenum Publ. Comp., New York (1983), 165.

[3] J. Narayan, W.L. Brown, and R.A. Lemons (eds.), Laser-Solid Interactions
and Transient Thermal Processing of Materials, Materials Research
Society Symp. Proc., Vol. 13, Elsevier, New York (1983).

[4] W. Buckel, and R. Hilsch, Z. Physik 138 (1954), 109.

[5] U. Goerlach, P. Ziemann, and W. Buckel, Nuclear Instr. and Methods
209/210 (1983), 235.

[6] M. Holz, P. Ziemann, and W. Buckel, Phys. Rev. Letters Vol. 51 No. 17
(1983) 1584.

[7] U. Goerlach, M. Hitzfeld, P. Ziemann, and W. Buckel, Z. Physik
B 47 (1982), 227.

[8] A. Defrain, J. Chim. Phys. - Chim. Biol. 74 (1977), 851.

[9] A. Bererhi, L. Bosio, and R. Cortes, J. Non-Cryst. Solids 30 (1979),
253.

[10] B. Stritzker in: Laser and Electron-Beam Interactions with Solids,
B.R. Appleton, and G.K. Keller (eds.), Elsevier, New York (1982),
363.

PICOSECOND TRANSIENT REFLECTANCE MEASUREMENTS
OF CRYSTALLIZATION IN PURE METALS

C.A. MacDONALD, A.M. MALVEZZI AND F. SPAEPEN
Division of Applied Sciences, Harvard University, Cambridge, MA 02138

ABSTRACT

The kinetics of rapid crystallization in metals is investigated by measuring transient optical property changes in laser irradiated gold and copper films in the picosecond and nanosecond regimes. Melt thresholds are determined by examining concentration profiles of multi-layered films before and after irradiation.

INTRODUCTION

Pulsed laser irradiation has been shown[1] to result in the formation and extremely rapid quenching of thin molten layers of both semiconductors and metals. The case of silicon has been extensively investigated. However, the solidification of silicon is accompanied by a change in the character of the atomic bonding, from metallic in the liquid to covalent in the crystal. In metals the transition on freezing from a disordered state to one of long range order occurs without a change in bonding. Thus it is to be expected that solidification is more rapid in metals than in silicon. In fact it has been proposed[2] that the limiting velocity for solidification in metals is the speed of sound. Monitoring of transient changes in optical and electrical properties has been employed extensively in the study of the ultrarapid crystallization of silicon under the conditions of pulsed laser irradiation. These experiments were facilitated by the dissimilarity of the crystal and its melt; the reflectance differs by a factor of two and the dc conductivity by several orders of magnitude. In metals the reflectance changes by only about 5% on melting and the dc conductance drops by at most a factor of two. The present work employs a standard pump/probe technique to monitor transient changes in the reflectance during pulsed laser irradiation.

THEORY

An estimate of crystal-liquid interface velocity can be obtained from the following kinetic analysis, based on the discussions by Turnbull[2,3,4]. For any material, simple rate theory yields an expression for the crystal growth/melt velocity, u, as a function of the driving free energy of crystallization, ΔG_C:

$$u = f k_i \lambda (1 - e^{\frac{\Delta G_C}{RT_i}}) \tag{1}$$

where k_i is the attempt frequency, f the fraction of sites in the interface at which incorporation into the crystal/melt is successful and T_i the temperature of the interface. In the approximation that the (negative) entropy of crystallization, ΔS_C, is constant, one obtains:

$$\Delta G_C = \Delta S_C (T_m - T_i). \tag{2}$$

Thus for small undercooling (or overheating) the velocity is proportional to the deviation from the melting temperature:

$$u = u_0 \frac{\Delta S_C}{R} \frac{(T_m - T_i)}{T_i} \tag{3}$$

where $u_0 = f k_i \lambda$ is the limiting velocity. In general in metals the site fraction, f, is nearly unity (the crystal/liquid interface can be considered atomically rough). For silicon, the site fraction is small, $f \sim .06$, corresponding to a large entropy of crystallization. Thus the interfacial velocity in silicon is small for low undercoolings.

If the attempt frequency is thermally activated the growth velocity exhibits a fairly sharp maximum at some finite undercooling, as shown in figure 1. Usually, the thermally activated process is identified with a diffusive jump similar to that for diffusion in the bulk liquid. In this case the growth velocity is limited by the rate of diffusion in the liquid. For metallic liquids, crystal growth may not be thermally activated, but rather collision-limited. The attempt frequency, k_i, is then equal to the frequency of thermal vibration

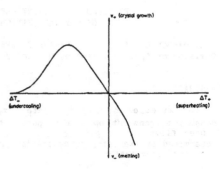

Figure 1. Schematic plot of the crystal-melt interface velocity as a function of deviation from the melting temperature for a thermally activated attempt frequency.

and the prefactor, u_0, can be approximated by the speed of sound, u_s, in the liquid. (This assumes that u_s is constant for all vibrational wavelengths.) In this case the growth velocity continues to increase with decreasing temperature and remains non-zero down to very low temperatures.

A variety of evidence exists to support the hypothesis that crystal growth in pure metals is collision-limited. Nickel dendrites have been observed[2] to grow into the melt at velocities up to 50 m/s. The measured ambient reduced undercooling of .1 is an upper limit to the value at the liquid-crystal interface. Since for metals $\Delta S_c \approx R$, equation (3) with u_0 equal to u_s (3000 m/s) predicts a velocity of ~300 m/s. Comparison between this estimate and the measured value of 50 m/s sets a lower limit on the velocity prefactor, $u_0 = k_i \lambda f$, of $u_s/6$. In addition, Turnbull[5] has noted that crystallization of dilute amorphous metal alloys at temperatures as low as 30K[6],[7] far below the diffusional kinetic freezing temperature, T_g, implies the existence of a faster process for crystallization than diffusion. There also exists indirect evidence for rapid regrowth velocities in metals after pulsed laser melting.[8] Large (~1 μm) grains were observed in a previously fine grained (~500Å) iron sample after irradiation by a 30ps pulse. If the regrowth is assumed to have taken place during 1-2 ns, as predicted by a simple heat flow analysis, the regrowth velocity must have been 500-1000 m/s.

EXPERIMENTAL TECHNIQUE

In this work the melt duration and therefore crystal-melt interface velocity is determined from the time-resolved reflectance measurements of a metal film following pulsed laser irradiation. In a similar experiment on silicon, Liu[9] observed a large (factor of 2) transient increase in the reflectivity corresponding to the metallic melt, followed by a return to the original value as the surface crystallized. The measurements of Thompson et al.[10] of transient conductance in silicon are consistent with these results and also with heat flow calculations. The largest measured crystallization velocity in silicon is ~15 m/s.[11]

Smaller changes in reflectivity are expected in the case of metals due to the similarity of the liquid and solid phases. The Drude theory for an ideal electron gas can be used to predict the trends of reflectivity with heating and melting. In both cases the reflectivity is expected to decrease slightly. The quantitative value of the change on melting is difficult to compute, and variations occur between samples with different

preparation.[12],[13],[14],[15] It is
thus important to establish
whether a reflectivity
decrease is due to surface
melting or solid state heating
alone. A technique to
accomplish this is discussed
in the section on sample
preparation.

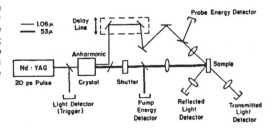

Figure 2. Experimental set-up for
transient reflectance measurements
employing 20ps Nd:YAG laser pulses.

Laser System

A 20ps Nd:YAG laser with
output at 1.06μm was employed
to provide the large thermal
gradients required for this
experiment. The set-up for
this laser is shown in figure
2. Part of each pulse is frequency doubled (to .53μm) and
used to melt the sample. The
remainder at the fundamental
frequency is reduced in intensity, delayed relative to the
pump pulse and used to
probe the sample. The probe
pulse is also used alone to
measure the sample reflectance
before and after each pump
pulse. The initial value is
used to normalize the subsequent reflectivity changes.
A typical plot of normalized
reflectivity versus pump
fluence for a delay of 2ns is
shown in figure 3.

In addition these
experiments were performed on
a nanosecond time scale
using a 40 ns pulse ruby laser
to verify the shape of the
reflectivity response.

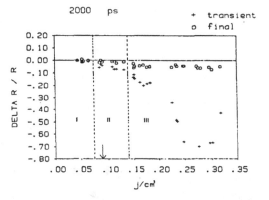

Figure 3. Plot of the relative change in
reflectivity at 2ns (+) and several
seconds (o) after irradiation by a 20ps
laser pulse for a variety of laser
fluences on the sample of figure 4.
Fluence regime I corresponds to solid
surface heating, regime II to melting,
and regime III to permanent surface
damage. The mark corresponds to the
fluence of the shot profiled in figure 4.

Sample Preparation

To establish that melting
has occurred, composionally
modulated thin films with
alternate 25-100Å layers
of gold and copper (chosen for
complete miscibility in the
solid and liquid states.) were
produced on polished glass
substrates by ion beam sputtering[16]. Because the solid state diffusion rates are slow, mixing of the layers can only occur in the liquid
state. Mixing of the layers is thus a signature of melting as a result of
pulsed laser irradiation. Composition versus depth profiles were obtained by
Auger spectroscopy on the surface with simultaneous ion beam sputtering to
mill away the layers. A typical profile is shown in figure 4. The repeat
distance of the as-deposited modulation was verified by X-ray diffraction.

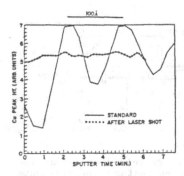

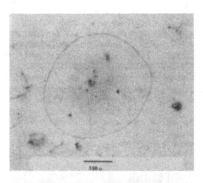

Figure 4. Auger spectroscopy profile of copper concentration versus depth in a copper/gold modulated film of 100Å repeat length.

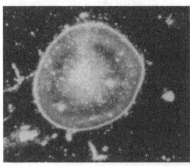

Figure 5. Optical micrograph (upper) and dark field optical micrograph (lower) of a ps laser irradiated damage spot on a Au/Cu film with 50Å repeat length and 1000Å overall thickness on a silicon substrate. The shot is in the third fluence regime.

In addition to the modulated films this experiment was performed on several combinations of pure metals and substrates. Homogeneous polycrystalline gold films of 200-2500Å thickness were produced by rf sputtering on a variety of highly polished substrates (silicon, sapphire, glass) in a low vacuum chamber designed for coating SEM samples. Thin copper and aluminum films were produced by vapor deposition. Mechanically polished bulk copper and aluminum samples were also studied. All of the metals exhibited similar reflectivity behavior.

EXPERIMENTAL RESULTS

Reflectivity results

Figure 3 displays a plot of reflectivity versus fluence obtained on the picosecond laser described above for a modulated Au/Cu film of 100Å repeat length. The data can be divided into three fluence regimes. In the first (below .04 J/cm^2) no reflectivity change is observed. In the second (between .04 and .15 J/cm^2) there is a transient reduction in the reflectivity, followed by complete recovery. Above .15 J/cm^2 there is permanent damage. These features were common to all the materials studied. Auger depth profiles before and after irradiation of the film of figure 3 are shown in figure 4. The irradiation fluence is indicated by the mark in figure 3, and is in the intermediate regime. The mixing of the layers confirms that melting has occurred in this regime.

Microscopy

Figure 5 is an optical micrograph of typical permanent damage spots (fluence regime III) for a modulated Au/Cu film. The major feature of figure 5 is a sharp circle surrounded by a light ring appearing at lower fluences. Lateral scanning Auger profiling indicates that the circle may be the edge of a pit from which one or more of the 50Å layers have been removed. If a surface damage feature can be associated with a particular fluence threshold, then the damage radius, r_d, can be expressed as a function of the applied fluence, f. For a Gaussian beam intensity profile this becomes:

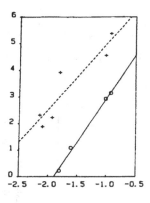

Figure 6. The square of the radius of the sharp circular feature seen in figure 5 versus the log of the applied fluence (o). The crosses (+) are for the light area outside the circle. The lines are least squares fits.

$$f_{th} = f e^{\dfrac{-r_d^2}{2\sigma^2}} , \qquad (8a)$$

or:

$$r_d^2 = 2\sigma^2 \ln(f) - 2\sigma^2 \ln(f_{th}) \quad (8b)$$

where f_{th} is the threshold fluence. If r_d^2 is plotted versus $\ln(f)$ for the sharp

LOG FLUENCE

circular feature displayed in figure 5, the result, shown in figure 6, is reasonably linear, and yields a threshold (.02 J/cm^2) equal to the start of the third, permanent damage, fluence regime (as described in figure 3). This implies that circle is the boundary of the damage region, consistent with the lateral scanning Auger profiling results. The damage may be associated with evaporation. Similarly the second line in figure 6 is a plot for the faint circular region seen in figure 5. In this case, the threshold corresponds to the beginning of the intermediate fluence regime (see figure 3) This feature, therefore, corresponds to melting.

Velocity results

To produce plots of reflectivity, R, and transmissivity, T, versus time the fluence regions such as those shown in figure 3 were divided into bins. A plot of the relative change in R(t) from the initial value at each spot of a 200Å thick gold film on a sapphire substrate is shown in figure 7 for two fluences in the intermediate regime. For the lower fluence bin, the reflectivity begins to recover in 400ps, and recovery is complete in 600ps. If this recovery is interpreted as due to the movement of the crystal-melt interface, then the regrowth velocity is:

$$u = \dfrac{200 \times 10^{-10} \text{ m}}{200 \times 10^{-12} \text{ s}} = 100 \text{ m/s.} \qquad (12)$$

Figure 8 displays similar results for a bulk copper specimen. Here recovery begins at 600ps and ends by 1ns. Assuming that the reflectivity samples a depth equal to the skin depth (~250Å) allows the interfacial velocity for the last stage of crystallization to be estimated at roughly 60m/s. This calculation is probably an underestimate.

CONCLUSIONS

Transient reflectivity measurements can be used to monitor the interfacial regrowth velocities in metals. Optical microscopy and Auger analysis of compositionally modulated films were used to establish that the transient decays in sample reflectivity were accompanied by surface melting. The fastest crystallization velocity measured in this experiment was 100 m/s, in

282

general agreement with a collision-limited growth computer model. This value is the highest known to the authors for any material and is substantially larger than the maximum measured for silicon. This observation supports the theory that crystallization in metals is collision-limited.

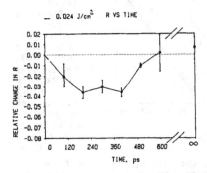

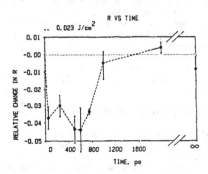

Figure 7. Relative change in the sample reflectivity, $(R(t)-R_0)/R_0$, for 20ps laser irradiation at two different fluences on a 200A gold film with a sapphire substrate.

Figure 8. Relative change in sample reflectivity versus time for a bulk copper specimen. The fluences (here and in figure 7) were in the melting regime (II) for the samples.

ACKNOWLEDGEMENTS

The authors acknowledge useful discussions with Prof. D. Turnbull and introduction to the Crank-Nicolson technique by J. Fröhlingsdorf. This work has been supported by the Office of Naval Research under contract number N00014-83-K-0030.

REFERENCES

1. See for example: J.C. Fan and N.M. Johnson, eds., Energy Beam-Solid Interactions and Transient Thermal Processing, North Holland, NY 1984.
2. D. Turnbull, in Physical Processes in Laser-Materials Interactions, NATO-ASI 1980 Pianore, Italy, Plenum Press 1983.
3. S.R. Coriell and D. Turnbull, Acta Metall., 30, p2135 (1982).
4. D. Turnbull, J. de Physique C4, p1.
5. J.L. Walker, pp114-5, in B. Chalmers, Principles of Solidification, Wiley, NY 1964.
6. M.R. Bennett and J.B.Wright, Phys. Stat. Sol. A., 13, p135 (1972).
7. W.Buckel, Z. Physik, 138, p.136 (1954).
8. C.J. Lin and F. Spaepen, in Chemistry and Physics of Rapidly Solidified Materials, B.J. Berkowitz and R.O. Scattergood, eds., TMS-AIME, 1983.
9. P.L. Liu, et al., Appl. Phy. Lett., 34, p.864 (1979).
10. M.O. Thompson et al., Phys. Rev. Lett., 50, p.896 (1983).
11. W.L. Brown in ref.1, p.9.
12. K. Ujihara, J. Appl. Phys. 43, no.5, p2376 (1972).
13. S.D. Pudkov, Zh. Tekh. Fiz 47, p649 (1977).
14. G.S. Arnold, Applied Optics, 23, no.9, p1434 (1984).
15. H.G. Dreehsen, J. Appl. Phys. 56, no.1, p238. (1984)
16. F. Spaepen, A.L. Greer, K.F. Kelton, and J.L. Bell, Rev. Sci Instrum., 56, no.7, p1340 (1985).

TRANSIENT CONDUCTANCE MEASUREMENTS AND HEAT-FLOW ANALYSIS OF
PULSED-LASER-INDUCED MELTING OF ALUMINUM THIN FILMS

J. Y. TSAO[*], S. T. PICRAUX[*], P. S. PEERCY[*] AND MICHAEL O. THOMPSON[**]
*Sandia National Laboratories, Albuquerque, New Mexico 87185
**Cornell University, Ithaca, NY 14853

ABSTRACT

We report time-resolved electrical-resistance measurements during
pulsed-laser melting of a metal, aluminum. The resistances are correlated
with the thresholds for partial and full melting. We describe a semi-
analytic solution, based on an impulse-response function, for the response,
to a heating pulse, of a thin film of good thermal conductor supported by an
infinite substrate. The results agree well with the resistance
measurements, and confirm our interpretation of the data. In addition,
time-resolved reflectance measurements establish that, in this geometry,
melting and solidification proceed via the motion of a well-defined, planar
liquid/solid interface, whose position can be deduced from the resistance
measurements. These measurements permit the first real-time determinations
of melt-depths and quenching histories during rapid-solidification
processing of metals.

INTRODUCTION

Rapid-solidification processing (RSP) is an increasingly important
technology for modifying both surface and bulk properties of metals and
metal alloys [1]. The kinetics of cooling and solidification during RSP,
however, have been difficult to measure directly. As a result, knowledge of
the kinetics is heavily dependent on numerical calculations, making it
difficult to extract detailed quantitative information.

Recently, it has been shown that detailed information on melting and
freezing kinetics during pulsed-laser melting of silicon can be deduced from
transient (electrical) conductance [2-3] and reflectance [4] measurements.
Here, we describe the application of these techniques to the pulsed-laser
melting of a metal, aluminum [5]. Interpretation of these measurements for
metals is more complicated than for semiconductors, because of the much
smaller conductivity difference between the liquid and solid phases, and
because of the temperature-dependent conductivity of the pure phases.
Nevertheless, we are able to precisely correlate the measured conductances
both with temperature and with melt depth, and hence to monitor, for the
first time, the quench rate and freezing velocity during RSP of a metal.

EXPERIMENTAL

The samples were prepared by e-beam evaporation of metal, through a
meander-pattern stencil mask, onto silicon wafers which had been thermally
oxidized to form an oxide ~ 0.75-μm thick. The films consisted of a 2-nm Cr
adhesion layer followed by an Al layer of thickness d_{Al} = 297 nm. The
measured room-temperature resistances were consistent with that expected for
high-purity Al [6]. Transient conductances (resistances) were measured
using a charge-line configuration [3]. For transient reflectance
measurements, a TM-polarized, single-longitudinal-mode 488-nm Argon-ion
probe laser beam was focussed at grazing (85°) incidence onto a 1.5 x 1.5-mm
Al square formed adjacent to the meander pattern.

The samples were irradiated by spatially homogenized [7] 31 ns (τ_{FWHM}), 694-nm-wavelength laser pulses derived from a Q-switched ruby laser. On the time scale of these pulses, the thermal-diffusion length in the Al film ($2\sqrt{D_{Al}}\tau_{FWHM}$ ~ 3.5 μm) is much greater than the film thickness, so that to first order the film temperature can be considered uniform. Because of the temperature dependence of the resistivity of Al, a uniform film temperature greatly simplifies the interpretation of the measured film resistances. In addition, on the same time scale, the thermal-diffusion length in the SiO_2 film ($2\sqrt{D_{SiO_2}}\tau_{FWHM}$ ~ 230 nm) is less than the SiO_2 thickness, so to first order the SiO_2 layer can be considered semi-infinite.

TCM MEASUREMENTS

Time dependences of the resistances during irradiation with various laser fluences F are shown in Fig. 1(a). To suppress resistance variations due to slight sample-to-sample differences in thickness and in L/W ratio, the resistances were normalized by the measured room-temperature resistances. The major features are a rapid increase in normalized resistance r during the laser pulse, consistent with an increased resistivity as the sample is heated, followed by a slower decrease in r as the Al film cools due to thermal conduction into the substrate.

The dependence of the maximum transient normalized resistance r_{max} on F is shown in Fig. 1(b). Three distinct regimes, which can be identified as solid, partial melt, and full melt, are noted on the figure. In the solid and liquid regimes, the resistance increases linearly with fluence, due to the linear dependence of the solid and liquid resistivities on temperature. In the partial-melt regime, the resistance increases nonlinearly from that of the solid to that of the liquid at the Al melting temperature, due to a nonlinear dependence of total resistance on melt fraction.

Based on these identifications, we interpret the highest-fluence time-resolved curve in Fig. 1(a) to represent a fully melted film. The initial, slow decrease in resistance in time is caused by cooling of the liquid. The sharper decrease is caused by solidification, and the final slow decrease is caused by cooling of the solid. The interpretation of the fully liquid regime has been confirmed by measurements of the diffusion of an implanted solute, Zn, during pulsed laser irradiation [5].

HEAT-FLOW ANALYSIS

The interpretation described above is supported by calculations of the response, to heating, of a thin skin of perfect thermal conductivity on a semi-infinite substrate of much poorer thermal conductivity. Because this geometry is useful, and can be solved partly analytically, we briefly sketch the derivation here. In essence, our method replaces the usual numerical simulation of temperature in both space and time, with an analytic solution in space followed by a numerical simulation in time. In this simplified calculation, which agrees well with experiments, we also assume temperature-independent thermal properties.

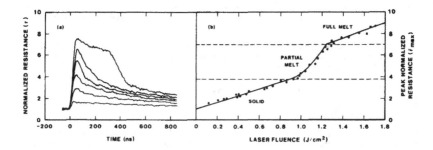

Fig. 1. (a) Time-resolved normalized resistances ($\tilde{r}$) during pulsed laser heating. Fluences, from bottom to top, are 0.20, 0.65, 0.99, 1.14, 1.24 and 1.39 J/cm^2. The peak of the laser pulse occurs at t=0. (b) Peak normalized resistances ($\tilde{r}_{max}$), taken from time-resolved measurements similar to those in Fig. 1(a), versus fluence. The melting threshold occurs at 0.85 J/cm^2, for $\tilde{r}_{max}$ = 3.58; full melting of the film occurs at 1.24 J/cm^2, for $\tilde{r}_{max}$ = 7.08. The solid line is the result of heat flow calculations, using the parameters listed in Table 1.

We start by writing the rate of change of the thin film temperature as

$$c_1 d_1 \dot{T}(t) = J_{in}(t) - J_{out}(t), \qquad (1)$$

where J_{in} is the heat entering, and J_{out} is the heat leaving, the thin film. The heat input J_{in} is due both to laser irradiation and to latent-heat liberation during the phase transformation:

$$J_{in}(t) = J_{laser}(t) + J_{\Delta H}(t) \qquad (2)$$

The latent-heat contribution is calculated via $J_{\Delta H}(t') = \beta \cdot \Delta H \cdot [T_m - T(t')]$, where β is the (assumed large) slope of the interface velocity vs superheating/supercooling response function. The heat leaving the film is calculated by numerically convolving the heat that entered the film at all previous times with an impulse response function:

$$J_{out}(t) = \int J_{in}(t') \cdot G(t-t') \cdot dt'. \qquad (3)$$

This impulse response function represents the heat flow out of the thin film at time t due to a heat impulse at time t=0, and can be solved for analytically. The advantage of this approach over a full numerical simulation, of course, is that this "kernal" needs to be calculated only once. The impulse response function is the solution to a pair of coupled heat diffusion equations for the thin film and the substrate:

$$\frac{\partial^2 T_1}{\partial x^2} - \frac{1}{D_1}\frac{\partial T_1}{\partial t} = \frac{\partial^2 T_2}{\partial x^2} - \frac{1}{D_2}\frac{\partial T_2}{\partial t} = 0, \qquad (4)$$

where T_1 and T_2 are the temperatures relative to room temperature in the thin film and in the substrate, respectively, D_1 and D_2 the thermal diffusivities, and κ_1 and κ_2 the thermal conductivities. The initial conditions are $T_1 = 1/(c_1 d_1)$ (the temperature rise corresponding to unit heat content) and $T_2 = 0$, where d_1 is the film thickness. The boundary conditions are $\kappa_1 [\partial T_1/\partial x]_{x=0} = \kappa_2 [\partial T_2/\partial x]_{x=0}$, $[T_1]_{x=0} = [T_2]_{x=0}$, and $[\partial T_1/\partial x]_{x=-d_1} = 0$, where $x = 0$ is the plane separating film from substrate.

Using Laplace transform techniques the time dependence of the temperature distribution, and hence the heat flow into the substrate (i.e., the impulse response function), can be calculated to be:

$$G(t) = \frac{2\kappa_2}{c_1 d_1 \sqrt{\pi D_2 t}} \cdot \frac{\alpha_1}{\alpha_1 + \alpha_2} \cdot \left[1 - \frac{2\alpha_2}{\alpha_1 - \alpha_2} \cdot \sum_{n=1}^{\infty} \left[\frac{\alpha_1 - \alpha_2}{\alpha_1 + \alpha_2} \right]^n \cdot e^{-n^2 d_1^2/D_2 t} \right] \qquad (5)$$

where $\alpha_1 = \sqrt{\kappa_1 c_1}$ and $\alpha_2 = \sqrt{\kappa_2 c_2}$. In practice, we have found that for the calculations described here, the infinite sum in Eq. (5) converges rapidly enough to allow truncation after 50 terms.

The solid line passing through the data points shown in Fig. 1(b) is the result of such simulations. The parameters followed by asterisks in Table 1 were determined in the following way: $R_{A\ell}^{sol}$ and $R_{A\ell}^{liq}$ were determined by the measured partial and full melt fluence thresholds, $\Delta r_{A\ell}^{-s\ell}$ by the measured resistance jump from partial to full melt, and $\alpha_{A\ell}^{liq}$ by the measured dr_{max}/dF slope in the full-melt regime.

Table 1: **Values of Parameters Used in Heat-Flow Calculations**

Parameter	Definition	Value	[Ref]
$c_{A\ell}$	Heat capacities of sol-Al, liq-Al (assumed equal)	2.92 J/cm^{3}K	[6]
$\kappa_{A\ell}$	Thermal conductivity of sol-Al, liq-Al (assumed equal)	1.5 W/cm K	[6]
$D_{A\ell}$	Thermal diffusivities of sol-Al, liq-Al (assumed equal)	0.51 cm^2/s	[6]
c_{SiO_2}	Heat capacity of sol-SiO$_2$	2.7 J/cm^{3}K	[8]
κ_{SiO_2}	Thermal conductivity of sol-SiO$_2$	0.014 W/cm K	[8,9]
D_{SiO_2}	Thermal diffusivity of sol-SiO$_2$	0.0044 cm^2/s	[8,9]
$\alpha_{A\ell}^{sol}$	Temperature coefficients of sol-Al resistivity, normalized to room-temperature solid resistivity	0.00434 K^{-1}	[6]
$\alpha_{A\ell}^{liq}$	Temperature coefficients of liq-Al resistivity, normalized to room-temperature solid resistivity	0.00447 K^{-1}	[*]
$R_{A\ell}^{sol}$	Reflectance of sol-Al at 694 nm	0.915	[*]
$R_{A\ell}^{liq}$	Reflectance of liq-Al at 694 nm	0.904	[*]
$\Delta r_{A\ell}^{-s\ell}$	Al resistivity increase upon melting, normalized to room-temperature solid resistivity	3.32	[*]
$\Delta H_{A\ell}$	Latent heat of Al melting	1020.48 J/cm^3	[6]
T_M	Melting temperature of Al	933 K	[6]
T_R	Room temperature	296 K	

In Fig. 2(a) is shown the time-resolved results of a simulation (solid line) superimposed over measured data (dashed line) for a partial melt. For the time scale of major interest here, the fit is quite good, although at longer times, the simulation predicts a somewhat higher temperature than is observed. Also shown in Fig. 2(b) is the corresponding temperature profile, from which can be deduced the quench rate (≈ 3.3 K/ns for this data).

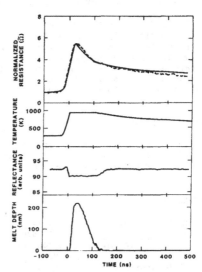

Fig. 2. Time-resolved (a) measured (dashed line) and calculated (solid line) normalized resistance, (b) calculated film temperature, (c) glancing-angle reflectance and (d) melt depth inferred from the resistance measurements. The laser fluence was 1.14 J/cm².

REFLECTANCE MEASUREMENTS

Thus far, we have established that the resistance maxima can be correlated with the thresholds for partial and full melts, and in the case of partial melts, with the molten liquid fraction. In order to be able to correlate the dynamically varying resistances with the motion of a planar liquid/solid phase boundary, transient-reflectance measurements were performed. To illustrate, the time dependence of the glancing-angle reflectance for the partial melt in Fig. 2(a) is shown in Fig. 2(c). Two important points are established by this figure. First, the onset and termination of melting, as deduced by comparison with Fig. 2(b), correspond, within the accuracy of the experiment, to the plateau of reduced reflectance. This indicates that melt initiates and terminates at the surface. Second, the flatness of the reflectance plateau during solidification indicates that freezing occurs by the motion of a liquid-solid interface planar to within 60 nm [5].

With this interpretation, and with precise knowledge, derived from Fig. 1b, of the resistances, r_s and r_ℓ, associated with solid and liquid Aℓ at the melting temperature, we can now associate measured resistances with melt depths via

$$d(t) = d_{A\ell} \cdot \frac{1 - \bar{r}_s / \bar{r}(t)}{1 - \bar{r}_s / \bar{r}_\ell} . \tag{6}$$

This equation is valid if thermal gradients across the film are negligible, as described above. The melt depth d(t) inferred from the experimental data in Fig. 2(a) using Eq. (6) is shown in Fig. 2(d). From this melt depth, velocities can be readily deduced. The melt-in velocity is approximately 16 m/s; the peak solidification velocity is approximately 3 m/s.

CONCLUSIONS

In conclusion, three features of this experiment deserve special emphasis. First, due to the spatial homogeneity of the incident laser intensity, the heat-flow isotherms are, to first order, planar. The resulting planar solidification front can simplify greatly the measurement of liquid/solid interfacial phenomena, such as solute segregation, which might otherwise be obscured by a poorly defined growth geometry. Second, the liquid/solid interface position is measured, rather than calculated. This is particularly important for pulsed laser annealing of metals, since small sample-to-sample variations in optical reflectivity can lead to large changes in absorbed energy, and hence in melt-depth history. Third, the solidification velocity measured here is in a range of direct interest to fundamental studies of RSP, e.g., studies of the morphological stability of liquid/solid interfaces at high solute concentrations [10].

ACKNOWLEDGEMENTS

We thank D. Buller for assistance with sample preparation. This work performed at Sandia National Laboratories was supported by the U.S. Department of Energy under contract DE-AC04-76DP00789.

REFERENCES

1. R. Mehrabian, ed., Rapid Solidification Processing: Principles and Technologies, III (National Bureau of Standards, Gaithersburg, 1982).
2. G. J. Galvin, M. O. Thompson, J. W. Mayer, R. B. Hammond, N. Paulter and P. S. Peercy, Phys. Rev. Lett. 48, 33 (1982).
3. M. O. Thompson, Liquid-Solid Interface Dynamics During Pulsed Laser Melting of Silicon-on-Sapphire, Ph.D Thesis (Cornell, 1984).
4. D. H. Auston, C. M. Surko, T.N.C. Venkatesan, R. E. Slusher and J .A. Golovchenko, Appl. Phys. Lett. 33, 437 (1978).
5. J. Y. Tsao, S. T. Picraux, P. S. Peercy and M. O. Thompson, to be published in Appl. Phys. Lett.
6. Y. S. Touloukian, ed., Thermophysical Properties of High Temperature Solid Materials, Vol. 1: Elements (Macmillan, New York, 1967), pp. 7-15.
7. A. G. Cullis, H. C. Webber and P. Bailey, J. Phys. E 12, 688 (1979).
8. W. E. Beadle, J. C. C. Tsai and R. D. Plummer, Quick Reference Manual for Integrated Circuit Technology (Wiley, New York, 1985), pp. 1-9.
9. B. H. Armstrong, in The Physics of SiO_2 and its Interfaces, edited by S. T. Pantelides (Pergamon, New York, 1978).
10. M. Cohen and R. Mehrabian, in Ref. 1, pp. 1-27.

TIME-RESOLVED LASER INDUCED TRANSFORMATIONS IN CRYSTALLINE Te THIN FILMS

E. E. Marinero, W. Pamler and M. Chen

IBM Almaden Research Center, 650 Harry Road, San Jose, CA 95120-6099

ABSTRACT

We report on time-resolved measurements of laser-induced transformations in Te thin films. Rapid thermal transformations are effected
utilizing a KrF excimer laser (12 ns duration, 249 nm). The kinetics of
the transformation is investigated utilizing transient conductance and
infrared absorption measurements as well as reflectivity and transmissivity changes at 633 nm. The threshold for melting is accurately
determined as well as melt velocities pertinent to the recrystallization

kinetics of the molten films. Criteria for glass formation of Te are also
discussed.

INTRODUCTION

Tellurium belongs to the trigonal system (1) and the structure of the
crystalline phase, can be described as a network of spiral chains linked by
van-der-Waal forces (2). On melting at 452°C, the electrical conductivity
of single crystals increases by approximately 800 times (3). The physical
changes accompanying melting of Te are summarized by Glazov (3), unlike Si,
at the melt, Te remains semiconducting and the covalent bonding in the
spiral chains is retained. Analysis of pair-distribution functions points
to a structural evolution from the melt to elevated temperatures. The
coordination number is about two near the melting point; it increases
rapidly to reach three at 500°C, then rises more slowly and tends to six
at very high temperatures (T° 1700°C) (Refs. 4,5). The optical and
electrical properties in the crystalline and amorphous phase are reviewed
by Tutihasi et al (6) and Stuke (7), similarly, Hodgson has measured the
optical properties of liquid Te and found that the reflectivity decreases
as a function of temperature (8).

The experiments reported in this work permit time-resolved monitoring
of the physical changes induced in Te as a result of laser annealing.

EXPERIMENTAL

Crystalline Te thin films are prepared by evaporating Te in a vacuum
of 10^{-7} mbar onto 1.2 mm thick glass substrates held at ambient temperature.
For transient conductance experiments, complimentary evaporation masks are

utilized to generate 1 mm x 9 mm Te stripes. Electrical contacts are provided in a second evaporation stage which deposits Al pads at both ends of the Te stripe.

The annealing laser is a commercial KrF excimer laser whose energy fluence at the sample can be varied between 0 to 100 mJ/cm^2 utilizing a variable attenuator. Three independent time-resolved techniques are employed to monitor the laser-induced transformations: a) reflectivity and transmissivity (R,T) measurements at 633 nm b) pump-and-probe infra-red absorption experiments. In this case, laser pulses (4 ns duration) at 3.6 micron are sent copropagating with the excimer laser onto the sample and the degree of transient absorption monitored either as a function of delay time between the uv and the IR pulses or for a fixed delay, as a function of annealing laser fluence. c) transient conductance measure-ments. The electrical circuit and apparatus for time-resolved conductance experiments are described in Ref. (9). As discussed by Thompson (10), direct measurements of melt-depth, peak melt-in and regrowth velocities during laser annealing can be obtained by monitoring resistance changes of the sample under study.

Measurements above the melt-threshold are single pulse experiments and a fresh sample is utilized for each fluence value.

RESULTS

Typical conductance signals as a function of laser fluence are dis-played in Fig. (1). Below 10 mJ/cm^2, transient profiles reproducing exactly the temporal characteristics of the excimer laser are obtained. These originate from photoconduction in Te and based on the speed of the signal response it is inferred that the carrier lifetime in the thin films must be a fraction of a nanosecond. Above 10 mJ/cm^2, a dominant feature pro-gressively develops with increasing fluence, this is indicative of a growing melt-depth and its maximum is attained at the same point in time, implying that the melt-in velocity is dramatically increasing. A series of parameters relevant to the kinetics of melting and recrystallization of Te thin films have been measured directly from these profiles and by numerical differentiation of their time-resolved evolution. These include melt-depths, peak melt-in and regrowth velocities as well as melt duration. In Ref. (9), detailed information on these measurements is presented, some of the key results are outlined in the discussion subsection.

Transient Conductance

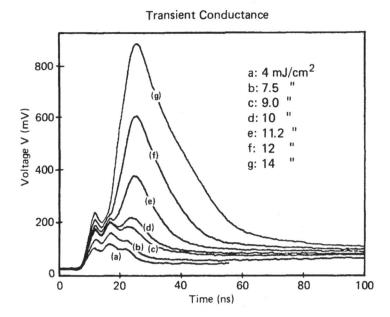

a: 4 mJ/cm^2
b: 7.5 "
c: 9.0 "
d: 10 "
e: 11.2 "
f: 12 "
g: 14 "

Figure 1 - Transient conductance profiles in Te polycrystalline thin films as a function of annealing laser fluence.

Figure 2 presents time resolved reflectivity profiles for three laser fluences. The reflectivity decreases as a consequence of laser heating and a sharp feature followed by a slower and longer tail is obtained. The amplitude and width of the fast component are strong functions of the absorbed laser fluence. Below 10 mJ/cm^2 the transient does not exhibit this sharp spike, its appearance is correlated to the onset of melting of the irradiated area. Increasing the laser fluence beyond a critical threshold in the case of the 75 nm film leads to the formation of the amorphous phase of Te. This is evident in the profile obtained when the absorbed fluence equals or exceeds 18.5 mJ/cm^2. Increasing the fluence even further leads to ablation. This is strongly dependent on the nature of the substrate and film thickness. The 250 nm Te films on glass cannot be amorphized with the laser pulse characteristics utilized in this work, ablation takes place before the glassy state can be formed. The amorphous state of Te at room temperature is not stable and the irradiated area recrystallizes within tens of seconds to minutes leading to the formation of crystallites with grain sizes exceeding a few microns.

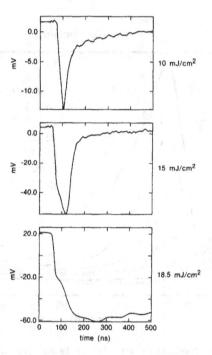

Figure 2 - Time-resolved reflectivity profiles (633 nm) in Te (75 nm thick) at three excimer laser fluences. The amorphous state is obtained when the absorbed fluence exceeds 18.5 mJ/cm^2.

The effect of laser fluence on transient conductance and optical properties of Te are summarized in Fig. 3. The 249 nm reflectivity of Te depends on the incident power, it is measured to be 20% at room temperature and close to 46% at the film temperature attained for an incident fluence of 40 mJ/cm^2. The fluence values given in the figure have been corrected for this non-linear dependence. The transient infrared absorption was measured with the probe laser delayed 2 ns from the rising edge of the excimer laser. Longer delays probe the resolidification phase of the melt.

Conductivity profiles have been converted to peak melt-depth using the procedure described in Ref. 9. A comparison of melt-depth is given in the figure for 75 and 250 nm thick Te films.

Examination of the three set of curves, indicates that beyond a critical absorbed fluence (10.5 ± 0.5 mJ/cm^2) the visible and infrared properties vary quite rapidly.

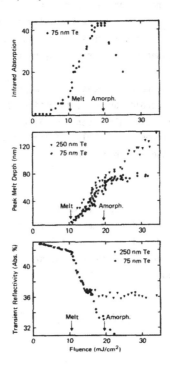

Figure 3 - Summary of IR absorption, 633 nm reflectivity and melt-depth amplitude as function of absorbed fluence. Threshold for melting and amorphization are indicated in the figure.

Similarly, the conductivity analysis indicates that at this fluence value the film surface begins to melt. A finite element temperature model calculation of the thin film structure yields a melt-threshold value of 11.0 mJ/cm^2, in excellent agreement with the experimental findings. The infrared absorption saturates at higher fluences and further increase leads to a decrement in its magnitude. A similar effect cqn be seen in the conductivity curve where a sharp slope change is evident around 18.5 ± 0.7 mJ/cm^2. This is indicative that the 75 nm film has melted completely through and very little change in conductance can be observed beyond this value. No saturation is observed for the thicker film and ablation results before it can be completely melted.

The 633 nm reflectivity (maximum of fast component) decreases as the sample is heated but the changes above the melt do not follow a simple linear relationship with fluence. The reflectivity for the 250 nm film saturates beyond 14 mJ/cm^2. This is possibly due to the fact that a final temperature is attained by the surface layers that constitute the skin-depth.

The 75 nm film for absorbed fluences 18.5 mJ/cm^2 quenches into the amorphous phase. This threshold for glass formation is also marked in Fig. 3. Comparison of the peak melt-depth profile to the reflectivity curve indicates that the threshold for glass formation coincides with complete melt-through of the absorbing structure.

DISCUSSION

Laser annealing of Te films leads to melting which can result on recrystallization of the melt or in the formation of the amorphous state. This is clearly illustrated in Fig. 4 where time-resolved conductance and reflectivity profiles are displayed for two sample thickness irradiated by the same fluence.

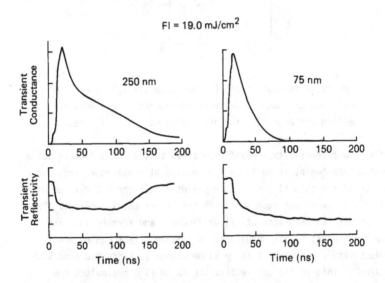

$$FI = 19.0 \text{ mJ/cm}^2$$

Figure 4 - Comparison of transient conductance and reflectivity profiles for 75 and 250 nm films. The amorphous state of Te is formed directly from the melt in the case of the 75 nm film (see text for details).

Following the melt duration, the thicker sample returns to the crystalline phase as evidenced by the reflectivity and conductivity values obtained after 150 ns. The thinner film, on the other hand, is characterized by a much shorter melt-duration and the reflectivity value attained within 60 ns of the melt-onset is drastically different from the value measured prior to melting. As mentioned in the previous section, the amorphous state is not stable at room temperature but it necessitates tens of seconds to minutes for recrystallization. Two important conclusions can be drawn from these observations, first the glassy state of Te is formed without the need of a first-order transformation, this is inferred from the speed of the transformation and second, the nucleation time is signi-ficant.

Comparing experimental measurements of melt duration to those pre-dicted by the temperature model, large discrepancies are encountered. We suggest that these large differences (factors of 8-10 times) are indicative of large undercooling of the Te melt. The maximum melt-in velocity follows a linear dependence on absorbed laser fluence, it is close to 8 m/s at the amorphization threshold for 75 nm films. The approximate slope of this parameter is 0.8 m/s per mJ/cm^2 and is indepen-dent of sample thickness. The regrowth velocity is also independent of thickness and it reaches a maximum of 2 to 3 m/s at 16 mJ/cm^2, it remains flat to 20 mJ/cm^2 thereafter decreasing to a value of 1 m/s at 30 mJ/cm^2. Full details on these measurements are given in Ref. 9.

One of the most important results of this work is the observation that a necessary criterium for glass formation of Te is complete melting of the thin film. This implies that all critical nucleation sites must be destroyed. Hence the recrystallization process prior to amorphization must be epitaxial in nature. This will be investigated in future studies.

ACKNOWLEDGEMENTS

It is a pleasure to acknowledge many fruitful discussions with M. O. Thompson and R. Barton. Thanks to Bill McChesney for preparing the films and to IBM Corporation for granting W. P. a one-year-research fellowship.

REFERENCES

[1]   R. W. E. Wyckoff, "CRYSTAL STRUCTURES", Vol. I, Chap II, (Inter-science, New York, 1960) pp. 17.
[2]   A. J. Bradley, Phil. Mag (Ser 6), 48, 477, (1924)

296

[3]   V. M. Glazov, S. N. Chizhevskaya and N. N. Glagoleva "Liquid
      Semiconductors", Chap III, (Plenum Press, New York 1969), pp 84.
[4]   G. Tourand and M. Brevil, J. Phys, Paris, 32 813, (1971)
[5]   R. Barrue and J. C. Perron, Phil Mag B, 51, 317, (1985)
[6]   S. Tutihasi, G. S. Roberts, R. C. Keezer and R. E. Drews, Phys.
      Rev, 177, 1143, (1969)
[7]   H. Keller and J. Stuke, Phys. Stat. Sol. 8, 831, (1965)
[8]   J. N. Hodgson, Phil. Mag (Ser 8), 8, 735, (1963)
[9]   W. Pamler, E. E. Marinero and M. Chen, to be published.
[10] M. O. Thompson, PhD Thesis, Cornell (1984); M. O. Thompson, G. J.
      Galvin, J. W. Mayer and R. B. Hammond, Appl. Phys. Lett, 42, 445,
      (1983); M. O. Thompson, G. J. Galvin, J. W. Mayer, P. S. Peercy,
      J. M. Poate, D. C. Jacobson, A. E. Cullis and N. E. Chew, Phys.
      Rev. Lett 52, 2360, (1984).

NUCLEATION OF ALLOTROPIC PHASES DURING PULSED LASER ANNEALING OF MANGANESE

J. H. PEREPEZKO,* D. M. FOLLSTAEDT** AND P. S. PEERCY**
*University of Wisconsin, Madison, Wisconsin 53706
**Sandia National Laboratories, Albuquerque, New Mexico 87185

ABSTRACT

   Manganese has four allotropes with an equilibrium melting point of the
high temperature δ-phase at 1517 K and calculated metastable melting points
for the Υ, β and α phases at 1501 K, 1481 K and 1395 K, respectively. Our
observations for Mn irradiated with a pulsed laser and supporting estimates
of maximum allotropic transition rates indicate that transformations between
allotropes are suppressed during heating with ~ 25 ns laser pulses, as well
as during subsequent cooling. Upon pulsed heating of α-Mn to the melt
threshold, the melt is undercooled 122 K below the δ-Mn melting point. For
incident laser pulse energy densities near the melting threshold,
resolidification involves regrowth of α-Mn from the substrate. At energy
densities well above threshold, the Υ-Mn phase forms by separate nucleation
and growth from the undercooled melt, and is retained upon rapid
solidification. From these results and analyses, we conclude that
significant melt undercooling, which may exceed 100 K, can occur during
pulsed laser melting of metallic crystals and that the resulting crystalline
structure is determined by both thermodynamics and nucleation kinetics.

INTRODUCTION

   A variety of rapid solidification treatments have been used in numerous
studies to produce microstructural modifications, including novel metastable
phases and amorphous structures [1]. A few techniques such as pulsed laser
annealing allow quantitative assessment of some of the kinetic conditions
operating during structural modification [2-5]. However with pulsed surface
melting, the resulting structural modifications are under the potential
influence of the underlying crystalline substrate and the outermost liquid
surface, which may be coated with an oxide film. Indeed, most metastable
phases formed with pulsed laser annealing are supersaturated solid solutions
of the substrate structure or thin amorphous layers. These products are
more the result of a competitive crystal growth condition [2] than the
nucleation of new crystalline structures. Only limited reports on the
nucleation of metastable crystalline phases following pulsed melting appear
to be available [6].
   Against this background it is apparent that a clearer evaluation of
nucleation kinetics may be possible in a material which has several phases
with well-defined thermodynamic properties. We report the development of a
strategy based on this approach which involves pulsed melting of a pure
metal with allotropic phases [7]. This approach can be extended to
intermetallic structures and ordered alloy phases. With this method, the
phase quenched-in following pulsed melting reflects the nucleation and
growth kinetics of the competing phases and the degree of melt undercooling.
In this regard Mn offers four distinct phases, and we use it to
experimentally demonstrate the method.
   The temperature dependence of the free energy differences between the
four allotropes of Mn and the liquid, $\Delta G^i = G_i - G_L$ (where i = α, β, Υ or
δ), calculated as discussed in detail elsewhere [7,10], is presented in
Fig. 1. The calculated melting points for the allotropes are: α - 1395 K,
β - 1481 K and Υ - 1501 K (all metastable), and δ - 1517 K (equilibrium).
The range of undercooling possible with this experimental strategy is

298

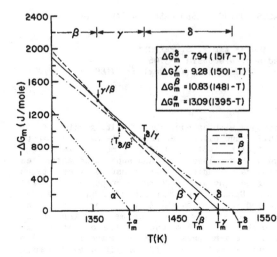

Fig. 1. Free energy differences relative to the liquid for α, β, γ and δ Mn, as functions of temperature [7]. Melting points (T_m), solid-phase transformation temperatures ($T_{i/j}$), and equilibrium solid phases (top) are indicated.

The equations shown in the figure:

$$\Delta G_m^\delta = 7.94\,(1517 - T)$$
$$\Delta G_m^\gamma = 9.28\,(1501 - T)$$
$$\Delta G_m^\beta = 10.83\,(1481 - T)$$
$$\Delta G_m^\alpha = 13.09\,(1395 - T)$$

indicated by the differences in melting points of these phases. If a substrate of α-Mn is heated to 1395 K so rapidly that solid state phase transformations are suppressed, then upon melting the liquid in contact with α-Mn will be undercooled by 122 K with respect to the high temperature δ-Mn phase. A similar behavior would be expected with other metals, such as Ti, Zr and Be, exhibiting allotropic transformations. In these three metals the low temperature phase is hcp and high temperature phase is bcc; calculated undercoolings which could be produced by rapidly melting the low-temperature phase are 152 K, 302 K and 5 K, respectively. More generally, relationships like those in Fig. 1 could also apply to an intermetallic structure or ordered alloy phase that exhibits a structural transformation or first-order disordering transition with increasing temperature. Pulsed melting offers a unique probe for such metastable structures that can develop during ultra rapid quenching.

EXPERIMENTS WITH MANGANESE

The pulsed melting experiments reported here involve Mn in two sample configurations with different initial crystal structures. In the first case, the room temperature equilibrium α phase was used as the substrate. The second configuration was produced by annealing a sample encapsulated under vacuum to 900°C where the β phase is stable, and quenching it into iced water. This produced predominantly β-Mn, but with some α-Mn, as a substrate material for pulsed laser annealing.

Samples of α-phase Mn (99.99% purity) were metallographically and electrochemically polished to obtain a flat surface. Irradiation was provided by Q-switched ruby laser (25-ns full width at half-maximum) through a quartz homogenizer [12] to assure beam uniformity ($\leq \pm 5\%$). An argon probe laser (488 nm) was also incident on the sample surface at glancing angle for real-time measurement of the reflectance during irradiation. The transient reflectance measurements and subsequent changes in the surface appearance were used to determine when melting occurred. Samples for transmission electron microscopy (TEM) were prepared by back thinning using jet electropolishing [7] to expose the irradiated near-surface for observation to a depth ~ 0.1 μm.

Fig. 2. Electron diffraction patterns taken with TEM at 120 kV from laser-irradiated α-Mn. a) <111> zone pattern of α-Mn obtained after 0.77 J/cm^2, 25 ns laser pulse. b) <110> zone pattern of Y-Mn obtained after 1.16 J/cm^2, 25 ns laser pulse, with superimposed (111) Y-Mn ring at d = 0.22 nm.

After irradiation of an α-phase sample at an incident density of 0.77 J/cm^2, which just exceeded the melt threshold energy density, the α-phase was retained in the near surface region of the substrate, as shown by the electron diffraction pattern in Fig. 2a. At an incident energy density of 0.88 J/cm^2, which was clearly above the melt threshold, the electron diffraction pattern from the melt layers was again predominantly α-phase, although some weak diffraction rings which may be due to Y phase were observed in isolated regions of the sample. At 1.16 J/cm^2 there was clear evidence for the formation of Y phase in the melted layer as indicated by the <110> fcc pattern shown in Fig. 2b. Additional Y-phase diffraction patterns similar to Fig. 2b, some with <112> orientations, were also obtained at 1.48 J/cm^2. The Y phase had grain sizes of order of one micron; no other Mn allotropes were detected with the α-Mn substrates. Examination of the second type of sample with a predominantly β-Mn substrate which contained some α, showed the α, β and Y phases after pulsed melting; δ was not detected. Additional details of the TEM observations are presented elsewhere [7]. To our knowledge, this is the first time the Y-phase has been quenched to room temperature in pure Mn.

DISCUSSION

During pulsed laser treatment of a metal with allotropic changes, several different reaction paths are possible during heating to the melt temperature and during rapid solidification and cooling, as illustrated in Fig. 3 for a metal with two allotropes, α and β. For example, along path (I) the low temperature α phase can transform to the high temperature β phase which then melts, resolidifies, and then transforms in the solid state back to α upon cooling. In this case all possible phase transformations occur during heating and cooling. Along path (II), however, the α → β reaction is suppressed during rapid heating of α and melting occurs at $T_m^α < T_m^β$, yielding a liquid which is undercooled with respect to β. During cooling along (II) several solidification paths are possible. For example, resolidification of α may occur as in reaction #1. Alternately, β may nucleate in the undercooled liquid and resolidify as noted in reaction #2 or β may nucleate on the surface as in reaction #3. For both reactions #2 and #3 subsequent solid-state transformation of the β phase surface layer to α can be suppressed by sufficiently rapid quenching, and β is retained. It is

300

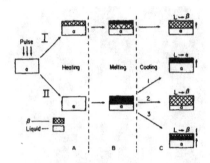

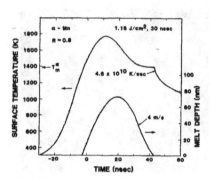

Fig. 3. Diagram of the pathways and reaction sequences possible for solid-state transformation, melting and resolidification in a metal with two allotropes, α and β.

Fig. 4. Calculated surface temperature and melt depth versus time for α-Mn irradiated with a 1.16 J/cm², 30 ns pulse.

worthwhile to note that only along path (II) during reaction #1 is epitaxial resolidification likely. Consequently, the observation of a non-epitaxial surface layer need not imply a solid state allotropic transformation [14] as in path (I), but can result from nucleation of the high-temperature allotrope in the undercooled liquid and resolidification as in path (II), reaction #2 or #3. The evaluation of the different reaction paths for the case of Mn can be considered according to the example of Fig. 3, but the existence of four allotropes increases the number of possible reactions.

For α-Mn, if the pulsed heating is sufficiently rapid to suppress the solid state transformations to any of the other allotropes, melting occurs when the near-surface reaches T_m^α = 1395 K. The melt front then propagates into the solid until the interface temperature cools to the melting point of the substrate. If the β, γ and δ phases are absent upon reaching the α-L metastable equilibrium, then the melt zone is undercooled by 122 K with respect to the stable δ-L equilibrium.

Several experimental observations indicate that allotropic transformations do not occur with laser treatment of the α phase. First, when β is formed during furnace annealing, it can be retained to room temperature by quenching at ~ 10³ K/s. Thus, if β had formed during any of the irradiations, it would have been retained during the subsequent rapid quench, calculated to be ~ 10¹⁰ K/s. Second, the retention of γ following its nucleation from the liquid with 1.16 J/cm² establishes that if an α → γ reaction had occurred in the 0.77 J/cm² sample, γ would have been observed. Finally, if α had transformed to δ, then the γ phase observed with 1.16 J/cm² would have to have been formed by solid-phase transformation. Arguments presented immediately below and the observed 1 μm grains of γ-Mn render an α → δ transformation prior to melting, as well as the possibility that δ nucleated from the liquid and transformed to γ, very unlikely.

For solid state reactions between pure metal phases, the rate of incoherent interface propagation can be described by [15]

$$v = v_0 [1 - \exp(-\Delta G_{i/j}/RT)] , \tag{1}$$

where v_0 is a product of terms involving the interfacial jump frequency, jump distance and growth site fraction, and can be estimated by D_b/a, where D_b is a boundary diffusivity and a is the atomic spacing. For Mn, limited diffusion measurements [16] suggest v_0 ~ 10-100 cm/s. Estimates of the maximum transformation velocities based on an evaluation of Eq. (1) give values of $\sim 10^{-1}$ - 1 cm/s, which are typical for thermally activated solid state reactions [17]. Finite-element heat flow calculations for the surface temperature and melt depth versus time for a 1.16 J/cm² laser pulse are shown in Fig. 4. The near-surface layer spends only $\sim 10^{-7}$ sec at elevated temperatures, and thus a negligible amount of solid-state reaction between any two allotropes (either growth or decomposition) is expected during heating and subsequent cooling. The development of a 1 μm grain size during the brief period at elevated temperature requires a γ phase growth rate ~ 500 cm/s, which is consistent with growth into a liquid, but inconsistent with the solid-phase transformation velocities. Thus both observations and calculated kinetics indicate that transformation to other possible allotropes does not occur before the onset of melting at $T^{\alpha}_m = 1395$ K.

Several features of the γ phase formation are notable. Since the melt time increases with incident energy density, the obvious presence of the γ phase at 1.16 and 1.48 J/cm², and its absence at 0.77 J/cm², show that its formation is related to the molten state. Under this assumption, these results can be interpreted to estimate nucleation kinetics of γ in the undercooled melt. To obtain an upper limit on the time needed for nucleation, numerical heat flow calculations were performed for melting and resolidification of α-Mn. The calculation used the reflectivity (0.8) obtained by fitting the observed melt threshold (~ 0.8 J/cm²) for the samples used to obtain the data in Fig. 2. The major uncertainty in applying these calculations to the experiment results from the difficulty in preparing reproducible, high reflectivity surfaces. For 1.16 J/cm², we obtain a melt duration of 45 ns, which places an upper limit on the nucleation time of fcc γ-Mn. This time is comparable to the measured nucleation time of 15 ± 10 ns for cubic AlSb in molten Al [5]. Uncertainties in the reflectivity and thermal properties of Mn make the 45 ns upper limit tentative, and prohibit our obtaining a lower limit. With a grain size of one micron, the limiting nucleation time implies a nucleation rate for the γ phase of $\sim 10^{18}$ cm^{-3}s^{-1}.

Nucleation of γ requires the liquid to be undercooled below $T^{\gamma}_m = 1501$ K ($\Delta T > 16$ K); however nucleation may have occurred at temperatures down to or below $T^{\alpha}_m = 1395$ K ($\Delta T \geq 122$ K). The dominant phase to nucleate is determined to a large extent by the lowest values for the activation barrier for nucleation [10]:

$$\Delta G^* = K \, \sigma^3/[\Delta G_m/V_m]^2 \, , \qquad (2)$$

where K is related to the nucleus geometry and possible heterogeneous nucleant potency, σ is the liquid-solid interface energy and V_m is the molar volume. In terms of free energy, the phase with the largest $(\Delta G_m/V_m)$ value is favored. The use of temperature-corrected V_m values does not significantly change the temperature range favoring γ from that shown in Fig. 1 (1410 K - 1360 K). If cooling into this range is needed for γ nucleation then the undercooling will exceed 117 K, which can be achieved with pulse melted α-Mn. However, the preliminary observation of γ and the absence of δ following pulsed melting of β-Mn, where expected liquid undercooling to $T^{\beta}_m = 1481$ K ($\Delta T = 36$ K) favors δ formation from free energy considerations (Fig. 1), suggests that the specific nucleation site details (i.e., the $K\sigma^3$ prefactor) probably control the phase selection. Indeed, since the temperature of the melt zone is fairly uniformly at T^{α}_m near the end of the melt time, it is possible that γ forms independently of the α-L interface, perhaps at the outermost melt surface in contact with an oxide film. In this case, two solidification fronts could be active in the laser irradiated layer (i.e., α-L and γ-L), as noted in reaction #3 of path (II) in Fig. 3. This possibility and others are under further study.

In summary, a procedure has been developed for generating liquid undercooling during pulsed laser surface melting. This procedure is based upon rapid melting of a metal system with allotropic transitions. Well defined thermodynamic relationships between the allotropes allow numerical evaluation of the undercooling. This technique can be used for novel studies of phase selection, nucleation kinetics and growth competition during the controlled rapid solidification of surface layers.

ACKNOWLEDGMENTS

We thank A. Van Donsel and M. Moran for skillful assistance with sample preparation. This work performed at Sandia National Laboratories supported by the U.S. Department of Energy under contract number DE-AC04-76DP00789.

REFERENCES

1. See, e.g., Proc. of the 4th International Conference on Rapidly Quenched Metals, eds. T. Matsumato and K. Suzuki (Japan Inst. Metals, Sendai, 1982).
2. Michael O. Thompson, J. W. Mayer, A. G. Cullis, H. C. Weber, N. G. Chew, J. M. Poate and D. C. Jacobson, Phys. Rev. Lett. 50, 896 (1983).
3. Michael O. Thompson, G. J. Galvin, J. W. Mayer, P. S. Peercy, J. M. Poate, D. C. Jacobson, A. G. Cullis and N. G. Chew, Phys. Rev. Lett. 52, 2360 (1984).
4. P. S. Peercy, Michael O. Thompson, J. Y. Tsao and M. J. Aziz, Appl. Phys. Lett. 47, 244 (1985).
5. D. M. Follstaedt, S. T. Picraux, P. S. Peercy, J. A. Knapp and W. R. Wampler, Mat. Res. Soc. Symp. Proc. 28, 273 (1984).
6. M. Laridjani, P. Ramachandrarao and R. W. Cahn, J. Matl. Sci. 7, 627 (1972).
7. D. M. Follstaedt, P. S. Peercy and J. H. Perepezko, submitted to Appl. Phys. Lett.
8. R. Hultgren, P. D. Desai, D. T. Hawkins, M. Gleiser, K. K. Kelly and D. D. Wagman, Selected Values of the Thermodynamic Properties of the Elements, (ASM, Metals Park, OH, 1973), p. 301.
9. W. B. Pearson, Handbook of Lattice Spacings and Structures of Metals and Alloys (Pergamon, Oxford, 1958), Vol. I, p. 734.
10. J. H. Perepezko and W. J. Boettinger, Mat. Res. Soc. Symp. Proc. 19, 223 (1983).
11. J. H. Perepezko and J. S. Paik, J. Non-Cryst. Solids 61, 113 (1984).
12. A. G. Cullis, N. C. Webber and P. Bailey, J. Phys. E12, 688 (1979).
13. Z. S. Basinski and J. W. Christian, Proc. Roy. Soc. London, A223, 554 (1948).
14. Examinations of laser pulse-melted Ti and Be have been done by L. Buene, E. N. Kaufmann, C. M. Preece and C. W. Draper, Mat. Res. Soc. Symp. Proc. 1, 591 (1981).
15. J. W. Christian, The Theory of Transformations in Metals and Alloys, (Pergamon Press, Oxford, 1965).
16. J. Askill, Phys. Stat. Sol. 33, K105 (1969).
17. J. H. Perepezko, Met. Trans. A, 15A, 437 (1984).

LASER-MATERIAL INTERACTIONS IN PHASE CHANGE OPTICAL RECORDING

ROGER BARTON AND KURT A. RUBIN
IBM San Jose Research Laboratory, 650 Harry Rd., San Jose, CA 95120-6099

ABSTRACT

We have studied laser induced crystallization and amorphization reactions in tellurium and tellurium alloy thin films, which are of interest for use in optical data storage. The writing characteristics of these films are determined by plotting changes in film reflectivity against the parameters of laser power and pulse width. The laser pulse conditions necessary for film melting, crystallization, and amorphization are identified and then checked against the results of a numerical heat diffusion model. In this way we determine the critical quenching rate for formation of the amorphous phase in pure tellurium (2×10^9 C/sec). Minimum crystallization times are determined and found to increase by orders of magnitude with addition of alloying elements such as germanium or selenium. TEM analysis reveals significant differences in the crystallization mechanisms of different alloy systems.

INTRODUCTION

The phenomena of laser quenching and laser annealing are usually considered for their applications in metal and semiconductor processing, respectively. Each phenomenon, however, finds an additional and critically important role in the technology of phase change reversible optical data storage. In the phase change scheme for optical recording (1) a focused laser beam writes on a disk by melting small spots in a thin film and allowing them to quench rapidly into an amorphous state. Erasure is achieved by applying a lower power laser pulse which raises the local temperature to near the melting point and holds it there over times long enough for crystallization to occur. A sequence of amorphous spots in a polycrystalline background is easily distinguished by large (sometimes 40%) differences in reflectivity.

The disk drives employed in optical recording incorporate GaAs lasers along with diffraction limited optics. With written spots that are less than one micron in diameter, these devices achieve recording densities of hundreds of megabits per square inch. The techniques for maintaining accurate focus and tracking on a rotating disk have already been proven in the marketplace for compact audio disks.

Tellurium alloys have been considered for phase change erasable recording media since these materials can be melted with modest laser power densities and remain relatively stable in the amorphous state. Current research centers on the factors which control crystallization rate, both at room temperature (where it should not occur) and under laser irradiation (where it should occur very quickly).

We have undertaken a systematic study of the laser conditions necessary for both writing and erasing on typical optical recording thin films (3). These results, when summarized in Phase Transformation Kinetics (PTK) diagrams, illustrate physical mechanisms that control the recording process. PTK diagrams are also useful for comparing the recording properties of different thin film materials. Transmission electron microscopy (TEM) has recently been employed to explore more fundamental issues involving crystallization mechanism.

EXPERIMENTAL

Most of our measurements are performed on an automated coupon tester which is designed to write large arrays of one micron spots onto a thin film at a variety of laser powers and pulse widths. The tester is described in detail in reference 2. Briefly, the tester employs a Kr ion laser (647 nm) which is acousto-optically modulated and focused to a 0.8 micron FWHM spot via a 0.5 numerical aperture objective. Focus is maintained on the sample via automatic servo control. The pulse modulator, the sample translation stage, and reflectivity detectors are all interfaced to an IBM Series/1 minicomputer.

The thin film samples for our study are prepared by vacuum evaporation onto glass substrates. Alloy films are obtained by simultaneous evaporation from separate Knudson cell heaters. Dielectric overcoats, such as SiO_2, are then applied via electron beam evaporation in order to increase the resistance of the film to laser-induced ablation. Samples for TEM analysis are deposited onto carbon-coated mica substrates.

By systematically varying the parameters of laser power and pulse width, we have been able to determine the laser conditions that are necessary for either amorphization or crystallization. Phase changes are identified by characteristic changes in reflectivity and summarized in a plots such as Figure 1. Here "crystallized" marks a region in pulse width and power where reflectivities were observed to increase by at least 5%. (The experiment in Figure 1 was done on a pure tellurium film which was initially crystalline. A 15.9 mW, 50 ns pulse was applied first in order to produce an amorphous spot, then followed by a laser pulse of varying width and power.) The region marked amorphous is characterized by little or no change in sample reflectivity; ablation is characterized by sharp decreases in reflectivity.

Not all of the boundaries in Figure 1 are sharply defined. Instead the contrast can change continuously over regions that are nearly 2 mW in power or 100 nsec in pulse width. Optical microscopy reveals that these graduations correspond to changes in the diameter of a written spot. The choice of a 5% reflectivity contour in Figure 1 corresponds to a crystallization reaction that is 50% complete.

In other experiments we have used a more powerful cw Argon laser for writing spots with diameters up to 30 microns. A separate HeNe laser is employed along with the Ar laser for monitoring reflectivity differences. Samples prepared in this fashion are particularly useful for TEM analysis.

THERMAL MODELING

As an aid to interpretation of various laser-induced phase transformations, we have performed numerical simulations of the laser heating process. We use an explicit finite difference routine for calculating heat flow in two dimensions: vertically into the substrate and horizontally along the radius of a written spot (4). The program assumes a gaussian distribution of incident power, a certain fraction of which is absorbed in the active (tellurium) layer. Bulk values are assumed for the material constants: density, heat capacity, thermal conductivity, and heat of fusion. The constants are assumed to be temperature independent. The program has been tested carefully to insure that results are independent of the cell size, sample size, or calculation precision.

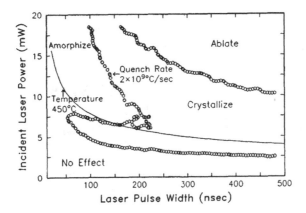

Figure 1: A PTK diagram for the sample: 750 A tellurium on glass with 2000 A of SiO$_2$ overcoat. "Crystallize" marks conditions in laser power and pulse width where reflectivities were observed to increase. "Ablate" marks reflectivity decrease. "Amorphize" and "No Effect" indicate no significant reflectivity change. The sample, initially crystalline, was preconditioned with a 15.9 mW, 50 nsec amorphizing laser pulse. The solid line shows minimum conditions necessary for reaching the Te melting point (450 C). These temperatures as well as quench rates were determined via numerical simulation.

RESULTS AND DISCUSSION

Basic mechanisms of the recording process are revealed in PTK diagrams such as Figure 1. These can be illustrated by following a line in that figure of constant pulse width and gradually increasing power (for instance at 150 nsec). An irradiated spot is initially amorphous, and at lower pulse powers the material is not heated sufficiently for any significant reflectivity change to be observed. At a particular threshold power (3 mW), however, temperatures are reached which allow crystallization to occur within the 150 nsec pulse period. The reflectivity increases. At higher powers (7 mW) the spots are left again in the amorphous state, presumably because temperatures are reached which are sufficient for melting.

In order to confirm a "melting threshold" at this boundary in the PTK diagram we have used our thermal model to determine the conditions under which heated spots reach at least the melting temperature (450 C for Te). The model generally agrees with the location of the melting boundary to within 30% in pulse power. Often the simulation predicts melting at slightly lower powers. The difference can be attributed to measured spots which are slightly larger than their assumed diffraction limit, or to thin film material constants that differ from their bulk values. The simulated laser spot size has been increased by 21% in order to give the fit shown in Figure 1.

Continuing up in power at 150 nsec we come again to a region of reflectivity increase or crystallization. In this case the spots must have been heated to well above the melting point, but did not cool at rates fast enough to retain the amorphous state. At higher pulse widths and higher powers the cooling rate is reduced by increased quantities of absorbed energy. In this region (along the upper amorphization boundary) our thermal simulation indicates that maximum temperatures vary widely (from 600 to 1200 C). Cooling rates along this boundary, however, when simulated in the vicinity of the melting temperature, are identical to within 5 percent.

306

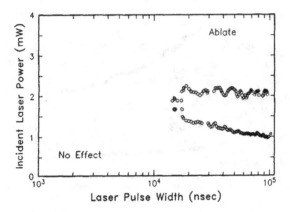

Figure 2: A PTK diagram for Te60Se40 on glass covered with 2000 A of SiO2. A minimum crystallization time of 10 microseconds is indicated.

Finally, laser pulses at maximum power in Fig. 1 impart sufficient energy to the material to cause ablation and a sharp reflectivity decrease. Ablation is the mechanism utilized most frequently in "write-once" optical recording, but must be avoided in the design of an erasable system.

We have used the thermal calculations to determine a semi-quantitative "critical quench rate" for tellurium which must be exceeded in order for amorphization to occur. We have assumed that a material must acheive a certain critical undercooling without crystallizing if it is to eventually reach the glass transition temperature and remain stable in the amorphous state. This critical temperature, corresponding to the "nose" in a crystallization TTT diagram (time-temperature-transformation), is assumed to lie at a temperature ten to twenty percent below the melting temperature T_m. Our thermal model determines how much time is required in cooling the center of a molten spot from T_m to the critical temperature after a specified writing pulse. The critical cooling rate determined along the amorphization threshold in Figure 1 is $(1.8 +/- 0.8) \times 10^9$ C/sec. The quoted error represents our uncertainty in the magnitude of the required undercooling. Errors in the thermal modeling, or in the experimental determination of the quenching threshold, fall well within this result. Quench rates on the order of 10^9 to 10^{10} C/sec have been discussed previously in the literature (5), but rarely have they been determined, as in this case, to within an order of magnitude.

The PTK diagram has been particularly useful for comparing the recording properties of various thin film active layers and substrates. Some of these effects can be anticipated from inspection of Figure 1. If plastic, which has a very low thermal conductivity, is used instead of glass for the substrate material, then one encounters difficulty in obtaining quench rates sufficient for amorphization. In this case, the amorphous region in Figure 1 is reduced to a small area on the left hand side.

The other important parameter derived from the PTK diagram is the minimum crystallization time (65 nsec in Fig. 1). This parameter depends on nucleation and growth kinetics that are particular to the active layer material. We find that addition of alloying elements can dramatically increase the crystallization time. Figure 2, for instance, shows a PTK diagram for an alloy of $Te_{60}Se_{40}$. In this case the minimum crystallization time is 10 microseconds. Crystallization is only observed at temperatures below the melting threshold. After melting the material either cools into the amorphous state or is ablated. It is

Table I: Important optical recording parameters for a variety of tellurium alloys and compounds.

Composition	Minimum Crystallization Pulse Width	Critical Quench Rate (°C/sec)
Te	65 nsec	$1.8 \pm 0.8 \times 10^9$
$Te_{90}Se_{10}$	300 nsec	3×10^8
$Te_{60}Se_{40}$	10 μsec	1×10^7
$Te_{98}Ge_2$	1.10 μ sec	9×10^7
$Te_{90}Ge_{10}$	80 μ sec	1×10^6
GeTe	<30 nsec	>3 $\times 10^9$

possible to use the measured crystallization time in Figure 2 to estimate a critical quench rate for this material. This is accomplished by dividing our assumed critical supercooling $(0.15 \times T_m = 110 \text{ C})$ by the minimum crystallization time, yielding 1×10^7 C/sec. A similar estimate, when applied to pure Te, yields excellent agreement with the thermal simulation result.

In Table I we summarize the parameters for crystallization time and quench rate determined for a selection of Te alloy materials. Large differences in crystallization times are evident. In order to acheive reasonable data rates in an optical recording device it is useful to have a material that can be erased in less than one microsecond. Pure Te fits this criterion; unfortunately Te also crystallizes spontaneously in a matter of seconds at room temperature. Recently it has been demonstrated that the compound material GeTe (rhombehedral crystal structure) can combine the attributes of fast erase time with room temperature data stability (6).

More detailed understanding about crystallization mechanisms may be gained via examination of crystalline microstructures. Figure 3 presents some of our initial TEM photographs of laser-crystallized grains in the alloys $Te_{95}Ge_5$ and $Te_{80}Se_{20}$. These spots were heated with an Ar laser beam 20 microns in diameter using pulse widths of 4 and 30 microseconds respectively. The pulse widths were chosen so that the resulting crystal-

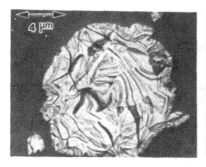
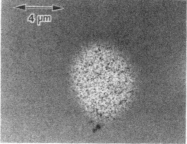

Figure 3: TEM bright field micrographs for Te80Se20 (left) and Te95Ge5 (right). Each spot was annealed for 30 and 4 microseconds respectively with a defocused 5 Watt Argon ion laser. Crystallization conditions were near their minimum in pulse width. Diffraction patterns indicate the presence of crystalline Te in each photo.

308

lization reactions would be 90% complete. Diffraction patterns taken from each sample were indicative of the tellurium crystal structure. The grain structures, however, are markedly different.

Te/Se alloy microstructures appear similar to those observed in pure tellurium films. They are characterized by very large grains which are not uniform in size. In some cases only one or two grains appear, each larger than 5 microns. These microstructures suggest that crystallization is a nucleation controlled event. The similarities between microstructures in Te and Te/Se alloys are understandable since complete miscibility exists in this alloy system (7).

The Te/Ge system, by contrast, is characterized by a eutectic reaction and immiscibility between Te and GeTe phases (7). This may account for the extremely fine grain structure seen in Figure 3 (right) since phase separation generally retards crystal growth rate. Smaller sized grains are indicative of an increased ratio of nucleation rate to crystal growth rate.

SUMMARY

We have presented a PTK diagram for pure tellurium and $Te_{60}Se_{40}$ and have described how these diagrams can be used to characterize the optical recording properties of thin film materials. The tellurium diagram, when combined with thermal modeling, allows a determination of the critical quench rate for formation of the amorphous phase. Minimum crystallization times were also measured and found to increase rapidly with the concentration of alloying elements such as germanium or selenium. TEM analysis reveals significant differences in the crystallization mechanisms of these two alloys.

ACKNOWLEDGEMENTS

The authors would like to thank V. Jipson for use of the thermal simulation program. They also thank M. Chen, D. Rugar, and E. Marinero for help with the optical testing systems as well as for useful discussions.

REFERENCES

1. S.R. Ovshinsky, U.S. Patent 3,530,441 (1970).

2. M. Chen, K. A. Rubin, V. Marrello, U.G. Gerber, V.B. Jipson, Appl. Phys. Lett. 46 , 734 (1985).

3. K.A. Rubin, R. Barton, M. Chen, V. Jipson, and D. Rugar, presented at the MRS symposium on Mass Memory Technologies, San Francisco, CA, April, 1985.

4. V. Jipson and C.R. Jones, J. Vac. Sci. Technol. 18 , 105 (1981).

5. M. von Allman, "Laser Quenching", in Glassy Metals II , H. Beck and H-J Guentherodt eds., (Springer, Berlin, 1983).

6. M. Chen, K.A. Rubin, and R.W. Barton, presented at the Optical Society of America, Washington D.C. (October 1985).

7. M. Hansen, ed., Constitution of Binary Alloys , (McGraw-Hill, N.Y., 1958).

AU-CLUSTER REDISTRIBUTION DURING NANOSECOND LASER-ANNEALING
OF METAL/INSULATOR MATRICES

W. PAMLER, E. E. MARINERO, M. CHEN AND V. B. JIPSON
IBM Almaden Research Center, 650 Harry Rd., San Jose, CA   95120-6099

ABSTRACT

We report on the growth and redistribution of Au clusters caused by nanosecond laser interaction of $Au_x(TeO_2)_{1-x}$ thin films with intense excimer laser radiation. This laser-induced phenomenon is studied in a time-resolved manner using transient reflectivity and transmissivity techniques. Structural and compositional changes are investigated using Rutherford Backscattering, XPS depth profiling, x-ray diffraction and conductivity measurements. Our studies indicate that melting of the binary structure initializes segregation, growth and coalescence of Au crystallites in the amorphous TeO_2 matrix.

INTRODUCTION

Laser-thin film interactions are currently a subject of great scientific interest and technological importance. Extensive studies on Si and a few other semiconductors has led to an understanding of phenomena such as melting[1], explosive recrystallization[2], amorphization[3], excess dopant incorporation[4] and enhanced diffusion of ion-implanted impurities.[5]

In this work we report on laser-induced growth and redistribution of metal clusters incorporated in an oxide matrix.

Laser heating of the absorbing film leads to melting, which in turn induces irreversible changes of physical properties, composition and structure. For example, conductivity changes in excess of four-orders-of-magnitude are obtained. A variety of time-resolved studies and micro-structural techniques are utilized to characterize and understand the nature of the laser-induced transformation.

EXPERIMENTAL

Au was incorporated into TeO_2 matrices by coevaporation of high purity targets onto a variety of substrates held at ambient temperature. The Au content x (mole fraction) varied from 0.4 to 0.8 and typical film thickness were 75nm and 250nm.

During deposition the composition was controlled by monitoring the individual evaporation rates of Au and TeO_2 using independent quartz-

310

microbalances and calibrated using RBS.[6]

Films were exposed to single pulses of a KrF excimer laser (12ns FWHM, λ= 249nm) whose fluence at the sample (2.5mm diameter circle) could be varied between 0 to 120 mJ/cm^2.

A description of the time-resolved experiments is given in Ref. 6. It suffices to mention that the technique using transient-reflectivity, permits one to measure the kinetics of the transformation in the ns time regime.

Following laser irradiation a variety of measurements were conducted to probe the resulting physical changes. These include sheet resistance, optical transmission spectra and XPS depth-profiling. Structural modifications were studied by x-ray diffraction and in some cases utilizing cross-sectional TEM analysis.

RESULTS

The most striking effects induced by laser exposure are large changes in the optical and electrical properties of the irradiated areas. The nature of the change depends on the Au concentration as illustrated in Figure 1.

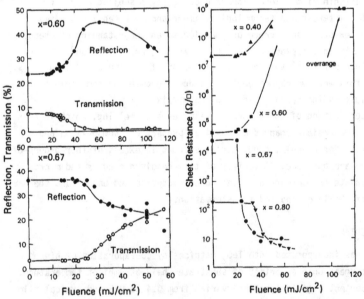

Fig. 1 - Laser-induced optical and electrical changes in Au$_x$(TeO$_2$)$_{1-x}$ matrices. Irradiated areas are exposed to single laser pulses. Threshold for irreversible change clearly visible (see text for details).

In the figure, the functional dependence of reflectivity and transmissivity (633nm) as well as sheet resistance is plotted vs laser fluence. It is evident that a minimum laser energy needs to be exceeded in order to drastically modify the optical and electrical properties of the matrices. The threshold depends on sample composition and both changes in resistance and reflectance tend to saturate at higher fluences. Beyond the maximum change, further increase in fluence results in ablation of the structure.

There is a clear compositional dependence in the behavior of the resistance and reflectance plots. The sign of the change reverses for a critical composition around x=0.65. The samples containing an Au fraction x=0.67 change upon irradiation by 10^4 times to become more conducting. The exact opposite phenomenon is observed for samples containing an Au fraction x=0.6. Clearly, irradiation of Au-rich samples (x $\geq$ 0.67) decreases the film's sheet resistance, whereas for x $\leq$ 0.6 the reverse is true. The fluence threshold for both resistance and reflectivity changes is measured to be the same within experimental errors.

X-ray diffraction studies reveal that the as-deposited matrix consists of small Au crystallites (5-10nm) homogeneously distributed in an amorphous TeO_2 matrix. As the Au concentration increases, the nearest-neighbor distance between clusters decreases leading to enhanced conductivity.

Further influence of the laser irradiation on the Au-TeO_2 films was observed in the composition depth profiles measured by XPS. Figure 2 shows depth profiles of a $Au_{0.67}(TeO_2)_{0.3}$ sample, before (a) and after (b) exposure. The depth distribution of the atoms in the as-deposited film is essentially homogeneous, but after laser exposure a wide region near the surface is depleted in Au and consists of almost pure TeO_2. Since the sputtering rates of the Au-TeO_2 mixtures and pure TeO_2 are different, the sputtering time is not proportional to the sputtered thickness. However, calibration of the sputtering rate by means of an evaporated TeO_2 film yields a thickness of the top TeO_2 layer of $\simeq$ 12nm for a 40 mJ/cm^2 laser pulse.

The segregation effect depends on laser fluence and sample composition. It vanishes for fluences at or near the threshold and increases with higher pulse energies. Among the compositions studied, the films with x=0.67 showed the most prominent segregation. For x=0.80 films the segregated TeO_2 layer thickness was observed to be approximately one third of that observed in the x=0.67 sample. For samples with x<0.61 almost no difference in the depth profiles could be observed, even for the highest laser fluences utilized.

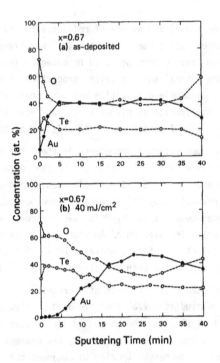

Fig. 2 - XPS depth profile of a $Au_{0.67}$ $(TeO_2)_{0.33}$ film prior (a) and after (b) excimer laser exposure. The single pulse fluence utilized was $\simeq$ 40 mJ/cm^2.

The TeO$_2$ surface layer is not due to preferential ablation of Au from the binary mixture since RBS analysis revealed that the number of Au, Te, and O atoms is not altered in the fluence regime below ablation. Further thickness profile measurements across the exposed area showed no thickness variations in excess of the instrument's resolution of $\simeq$ 5nm. Therefore, we conclude that the modification of the depth profile is caused by re-distribution of Au and TeO$_2$ within the film itself: the Au atoms move away from the surface and accumulate possibly near the substrate.

DISCUSSION

The irreversible processes described in this work occur only above a critical fluence threshold. This threshold depends on sample composition and the nature of the substrate. Two experimental findings establish that this threshold corresponds to the fluence needed to melt the binary mixture. First, the transformation takes place within $\leq$30ns (as measured by time-

resolved measurements), which can only be consistent with liquid phase
diffusion phenomena. Second, Au and TeO_2 are not miscible in the molten
state and phase-separate. This accounts for the behavior evident in
Fig. 2. A full account of these two findings and a more extensive report
of this work is given in Ref. 7.

X-ray diffraction analysis confirms that laser melting of the binary
mixtures induces coalescence and growth of the Au clusters in the
amorphous TeO_2 matrix. Thus, the altered size and distribution (segregation)
of the Au clusters dramatically alters the electrical and optical characteristics of the matrices. Cross sectional TEM analysis of samples prior
and after laser exposure further confirms the x-ray diffraction and the
XPS analysis.[7]

The compositional dependence of the electrical changes can be understood utilizing a percolation model which imposes a critical volume
fraction of the metallic clusters to become interconnected. Utilizing a
similar analysis discussed in Refs. 8 and 9 we determine that in the
present case the percolation threshold lies between x=0.65 and 0.77. Below
the percolation value, after laser exposure the coalescence of Au clusters
leads to wider separation between crystallites which consequently causes
an increase in sheet-resistance. Above the percolation threshold, on
resolidification, the Au crystallites grow and interconnect, hence dramatically decreasing the sheet resistance. The influence of vertical phase-
segregation further aids the coalescence of the clusters[7].

The size and distribution of metal particles in binary matrice has a
large effect on the optical properties of the structure. Fig. 3 compares
the transmission spectra of a pure 50nm Au film (a) to matrices containing
x=0.6 and 0.67 prior (b) and after (c) laser exposure.

Work by Perrin et al[10] and Cohen et al[11] on similar matrices indicates
that the spectral features are controlled by the composition of the sample.
Although the spectra in 3b look very similar, laser exposure dramatically
modifies the transmission spectra. As mentioned before, x=0.67 is above
the percolation threshold whereas x=0.6 lies below the critical value. The
extensive study of Ref. 7 clearly demonstrates that for the various compositions, different cluster size and distributions are obtained as a consequence of laser annealing. This radically alters the electrical and
optical characteristics of the binary matrices here described.

314

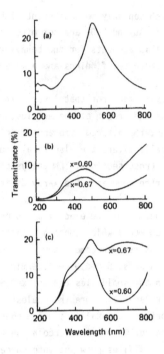

Fig. 3 - Optical transmission spectra for a 50 nm pure Au film (a) and for $Au_x(TeO_2)_{1-x}$ as-deposited (b) and laser-irradiated (c) matrices.

ACKNOWLEDGEMENTS

Many thanks to R. Davis, S. Gutierrez, J. Leavitt, G. Lim for helping with the structural analysis, to R. Barton and E. Kay for fruitful discussions and W. McChesney for fabricating the structure. W. P. thanks the IBM Corporation for financial support.

REFERENCES

1) R. F. Wood and G. E. Giles, Phys. Rev. B23, 2923, (1981)

2) H. J. Leamy, W. L. Brown, G. K. Celler, G. Foti, G. H. Gilmer and
    J. C. C. Fan, "Laser and Electron-Beam Solid Interactions and Materials
    Processing", ed by J. F. Gibbons, L. D. Hess and T. W. Sigmon (North-
    Holland, New York, 1981) pp 89.

3) M. O. Thompson, J. W. Mayer, A. G. Cullis, H. C. Weber, N. G. Chew,
    J. M. Poate and D. C. Jacobson, Phys. Rev. Lett. 50, 896, (1983)

4) P. Baeri, G. Foti, J. M. Poate, S. V. Campisano, E. Rimini and A. G.
    Cullis in Ref. 2, pp 67.

5) R. B. Fair, "Energy Beam-Solid Interactions and Transient Thermal Processing", ed. by D. K. Biegelsen, G. A. Rozgonyi and C. V. Shank (MRS, Pittsburgh 1985) pp 381.

6) E. E. Marinero, W. Pamler, M. Chen, V. Jipson and W. Y. Lee, J. Vac. Sci. Technol $\underline{B3(5)}$, 1560, (1985).

7) W. Pamler, E. E. Marinero and M. Chen, Phys. Rev $\underline{B}$, in press.

8) J. Perrin, B. Despax, V. Hanchett and E. Kay, J. Vac. Sci. Technol. $\underline{A3}$, 823, (1985).

9) S. P. McAlister, A. D. Inglis and P. M. Kayll, Phys. Rev. $\underline{B31}$, 5113, (1985).

10) J. Perrin, B. Despax and E. Kay, Phy. Rev. $\underline{B32}$, 719, (1985)

11) R. W. Cohen, G. D. Cody, M. D. Coutts, B. Abeles, Phys. Rev. $\underline{88}$, 3689, (1973).

PART VI

Ion-Semiconductor Interactions and Thin Film Growth

KINETICS, MICROSTRUCTURE AND MECHANISMS OF ION BEAM INDUCED
EPITAXIAL CRYSTALLIZATION OF SEMICONDUCTORS.

R.G. ELLIMAN[1], J.S. WILLIAMS[2], D.M. MAHER[3], W.L. BROWN[3]

1) CSIRO Chemical Physics, P.O. Box 160, Clayton, 3168, Australia.
2) RMIT Microelectronics Technology Centre, Melbourne, Australia.
3) AT&T Bell Laboratories, Murray Hill, N.J., U.S.A.

ABSTRACT

     Ion-beam induced epitaxy is shown to be essentially athermal over
the temperature range 200-400°C, and to exhibit no dependence on
substrate orientation and little dependence on doping in this regime. On
the other hand, the formation and propagation of defects during growth
and the interaction of the advancing crystal-amorphous interface with
implanted impurities is essentially identical for both thermally induced
and ion-beam induced epitaxy. These observations lead to a simple model
for ion-beam induced epitaxial crystallization in which epitaxial growth
is nucleated by defects generated at, or near, the crystal-amorphous
interface by the ion beam. Comparisons of ion-beam induced epitaxy and
thermally induced epitaxy suggest that the 2.7 eV activation energy
associated with the latter process is dominated by a   2.0 eV nucleation
step.

1: INTRODUCTION

     Solid phase epitaxial crystallization of amorphous semiconductor
layers is an important low temperature process for the removal of
radiation damage in integrated circuit fabrication. Thin amorphous
silicon layers produced by ion implantation can be recrystallized
epitaxially from an underlying crystalline substrate at temperatures
$\gtrsim$500°C [1,2], as shown schematically in figure 1. Crystallization
proceeds in the solid phase and is characterized by a thermal activation
energy of 2.7 eV [3]. The crystal growth rate is, therefore, very
sensitive to substrate temperature, varying from 10nm/min at 550°C to $\sim$
10^{-7} nm/min at 318°C, for growth on (100) oriented substrates. The
crystal growth rate is also strongly influenced by the crystallographic
orientation of the underlying crystal [4], being 25 times faster on (100)
substrates than on (111) substrates; and is significantly enhanced by the
presence of small concentrations of electrically active dopants [5,6]. A
widely accepted model of thermally induced solid phase epitaxial
crystallization (SPEC) is based on a Si-Si bond breaking nucleation event
and its propagation along the crystal-amorphous interface [7,8]. The
interface is envisaged to resolve itself into a minimum free energy
configuration consisting of (111) terraces bounded by <110> ledges. The
average orientation of the interface, determined by the orientation of
the initial substrate, is maintained by the size and number of such
terraces. Epitaxial growth is believed to proceed by atomic
rearrangement at kink sites along the <110> ledges, with the
experimentally determined 2.7 eV activation energy resulting, primarily,
from the bond breaking (2.0 eV) activation events. Epitaxy can, thus,
be envisaged as a nucleation and growth process which is nucleation
limited. In this model, the epitaxial growth rate is determined by the
concentration of nucleation sites, or kinks, at the interface, and is
influenced by those factors which change the kink concentration such as
the substrate orientation [7] and doping [8,9].

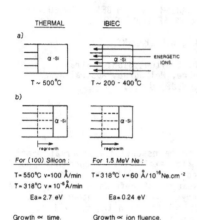

Figure 1: Schematic diagram comparing the annealing conditions and dependencies of a) thermally induced epitaxy and b) IBIEC.

Solid-phase epitaxial crystallization can also be stimulated by energetic ion irradiation at temperatures as low as 200°C [10-22], (see figure 1). Investigation of this process dates back to the early 1970's when Holmen et al [10,11] reported epitaxial crystallization of amorphous germanium layers at temperatures as low as 228°C for samples irradiated with 40 keV Ge$^+$ ions. Subsequent studies [12-22] have demonstrated similar effects in silicon. An understanding of the processes by which ion irradiation promotes epitaxial crystallization has only emerged more recently with the systematic studies by Linnros et al [14,16,17,22] and by the present authors [15,18-21]. In this paper we review the current understanding of ion-beam induced epitaxial crystallization (IBIEC).

2: KINETICS

Recent studies [19-21] have concentrated on measuring the dependencies of IBIEC, most notably, the dependence of growth on ion fluence, substrate temperature, ion energy, substrate orientation and doping. In this section these IBIEC dependencies are illustrated by appropriate examples taken from studies of 0.6-3.0 MeV Ne$^+$ ion irradiation of silicon samples with preamorphized surface layers. Unless otherwise stated, amorphous silicon layers were first prepared by implantation at 77 K with 50 keV Si$^+$ ions to a fluence of 2×10^{14} cm^{-2}. These samples were then preannealed at typically 450°C for 15 minutes to provide a well defined crystal-amorphous interface for subsequent IBIEC measurements. Figure 1 summarises some of the key observations and figures 2 and 3 illustrate typical effects of ion fluence, ion energy and irradiation temperature on IBIEC.

The aligned RBS spectra of figure 2 [20] depict IBIEC of samples irradiated at a) 318°C with 1.5 MeV Ne$^+$ ions and b) 450°C with 600 keV Ne$^+$ ions. At 318°C, epitaxial crystallization is observed to proceed linearly with ion fluence, advancing 6nm for each 10^{16} Ne$^+$ cm^{-2}. This corresponds to approximately 3 silicon atoms crystallized for each incident Ne$^+$ ion. At this temperature thermally induced epitaxy is negligible [3], with a predicted growth rate of only 1×10^{-7}nm/min. At 450°C the initial growth 'rate' is 40nm/10^{16} Ne$^+$ cm^{-2}, significantly higher than that for the sample irradiated at 318°C. In this case,

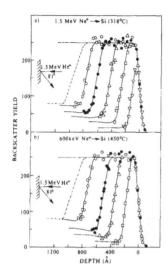

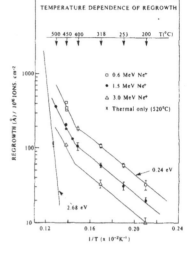

Figure 2: RBS/channeling spectra depicting the thickness of an amorphous silicon layer, as implanted (– – –), following a preanneal at 450°C for 15 minutes (0), and after successive equal fluence Ne$^+$ ion irradiations. Irradiation conditions are shown on the figure, with fluence increments of a) 3×10^{16} cm^{-2} and b) 5×10^{15} cm^{-2}.

Figure 3: Temperature dependence of beam induced regrowth as a function of Ne$^+$ irradiation energy. The temperature dependence of thermally induced epitaxy, labelled 2.68 eV, has been included for comparison.

––––––––––– // –––––––––––

however, the extent of crystal growth decreases with increasing ion fluence [20,21]. This non-linearity of growth with ion fluence results from the higher irradiation temperature and is not a consequence of the reduced Ne$^+$ ion energy. [20,21]. The dependence of IBIEC on substrate temperature and ion energy is more clearly delineated in figure 3.

Figure 3 plots the epitaxial growth induced by a fluence of 1×10^{16} Ne$^+$ cm^{-2} as a function of 1/T for three different ion energies [20]. The data clearly indicate two temperature regimes. Between 200-400°C epitaxy is observed to proceed linearly with ion fluence, being well characterized by a thermal activation energy of 0.24 eV. Above 400°C, non linearities in growth are observed and the thermal activation energy appears to be somewhat higher (0.5 eV). These measured activation energies for IBIEC are particularly small when compared to the 2.7 eV activation energy of thermally induced epitaxy. This indicates, particularly for the temperature range 200-400°C, that IBIEC is an essentially athermal process. The temperature dependence of thermally induced epitaxy has also been included in figure 3 for comparison. (The constant ion flux employed for IBIEC was used to convert between ion fluence and time.)

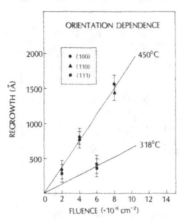

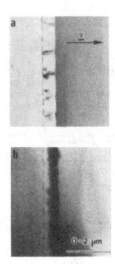

| xtal | | amorphous |

Figure 4: Substrate orientation dependence of IBIEC induced by 1.5 MeV Ne$^+$ ions, at two temperatures. Results obtained from RBS/channeling analysis.

Figure 5: TEM x-section micrographs depicting IBIEC induced under identical irradiation conditions on a) (100), and b) (111) oriented silicon substrates. IBIEC was induced at 318°C by 1.5 MeV Ne$^+$ ion irradiation. (I) - initial crystal-amorphous interface, (S) - surface.

Figure 3 also indicates that increasing the Ne$^+$ irradiation energy decreases the extent of growth [19,21]. In fact, growth has previously been shown to scale linearly with the energy deposited into nuclear collisions at, or near, the crystal-amorphous interface [16,19,21]. The significance of this observation will be discussed later with regard to IBIEC mechanisms.

The orientation dependence of IBIEC is illustrated in figures 4 and 5. Figure 4 depicts the growth induced on different orientation substrates by 1.5 MeV Ne$^+$ irradiation. The growth data are derived from RBS/channeling analysis and, as such, are limited to relatively thin layers in the case of (111) oriented substrates due to the formation of twins (see figure 5). Figure 4 quite clearly indicates that the extent of growth is independent of substrate orientation for IBIEC of these layers. However, as shown in figure 5, the microstructure of the recrystallized layers varies markedly between (100) and (111) orientations. In particular, IBIEC of layers on (100) substrates results in single crystal growth, whereas growth on (111) substrates gives rise to a high density of twins evident, as a dark band in figure 5b. Such microstructural aspects of IBIEC are discussed in more detail in section 3. The significant feature to note in figure 5 is that the total extent of growth (60 nm) is the same for both (100) and (111) substrates, even when twins form in the latter case.

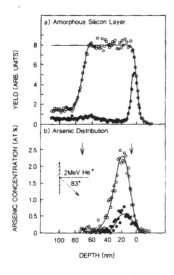

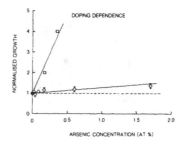

Figure 7: The effect of doping on thermally induced epitaxy (□) and IBIEC (◇).

Figure 6: RBS/channeling spectra depicting complete IBIEC of a sample previously implanted with 50 kV As to a fluence of $3x10^{15}$ cm^{-2}. a) Amorphous silicon layer before (0) and after (●) IBIEC. b) Arsenic distribution after IBIEC, random (0) and aligned (●). IBIEC was induced at 350°C with 1.5 MeV Ne$^+$ ion irradiation.

———————————————//———————————————

Figures 6 and 7 show the influence of electrically active impurities on IBIEC. In particular, figure 6 shows IBIEC of an amorphous layer previously implanted with 50 keV As$^+$ to a fluence of $3x10^{15}$ cm^{-2}. Figure 6a demonstrates the complete epitaxial crystallization of the amorphous silicon layer following irradiation with 1.5 MeV Ne$^+$ at 350°C to a fluence of $2x10^{17}$ cm^{-2}. The implanted arsenic, figure 6b, is observed to be approximately 80% substitutional after IBIEC. The substitutional arsenic fraction is similar to that obtained following thermal annealing, however, the sheet resistivity of the layer is 350 ohm/□, significantly higher than the 70 ohm/□ measured for thermally annealed samples. It is not clear at this stage whether the higher resistivity of ion-beam annealed samples is a consequence of a lower active fraction of arsenic or a lower carrier mobility. Clearly, further studies are required to clarify this point.

Figure 7 shows the effect of arsenic on the extent of epitaxial growth for thermally induced epitaxy [5] and IBIEC. Even allowing for the higher resistivity of the ion-beam annealed layers, it is apparent that IBIEC is at best weakly dependent on doping when compared to the 400% effect observed for thermally induced epitaxy. the implications of these observations are discussed further in section 4.

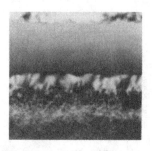 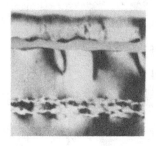

a   b

Figure 8: TEM x-section micrographs of a buried amorphous silicon layer, a) pre-annealed at 680°C for 5 secs, and b) following subsequent IBIEC at 318°C with 1.5 MeV Ne⁺ ions. (S) - surface, (A) - amorphous silicon - (The amorphous silicon layer in a) is 200nm thick).

―――――――――――――――――――――― // ――――――――――――

3: GROWTH PHENOMENA

In the temperature range 200-400°C, IBIEC of thin (<100 nm) amorphous silicon layers can result in crystalline material essentially free of extended defects [21,23]. However, a band of defects is observed at the projected range of the annealing ions [21,23]. For example, 1.5 MeV Ne⁺ irradiation at 318°C results in a narrow band (~ 80 nm) of extended defects centred at ~ 1.6μm, beneath an essentially extended-defect-free surface region which extends to ~ 1.4μm [23]. Nevertheless, surface amorphous layers recrystallized by IBIEC can be less than perfect as illustrated in figure 5. Spanning dislocations are observed to propagate during the crystallization of thick (>300nm) amorphous layers on (100) substrates (figure 5a). For these amorphous layers which were produced by Si self implantation, the amorphous-crystalline interface is less abrupt as a consequence of increased ion straggling. This leads to the nucleation of spanning dislocations. IBIEC of amorphous layers on (111) substrates invariably results in the formation of twins after approximately 20 nm of single crystal growth [24], as illustrated in figure 5b. The microstructure outlined above for IBIEC is consistent with the observed growth behaviour during thermally induced epitaxy. Furthermore, the growth-related microstructure for both IBIEC and thermally induced epitaxy is controlled by the initial state of the crystal-amorphous interface. We further illustrate the similarities between the growth related behaviour of the two crystallization processes in figures 8,9 and 10.

Figure 8a depicts a buried amorphous layer which has been partially annealed thermally at 680°C for 5 seconds. The initial amorphous layer was formed by Si⁺ implantation at room temperature. Annealing produces epitaxy from both interfaces but results in the formation and propagation of extended defects in the form of spanning dislocations and twins [24]. This initially defective structure was subsequently beam annealed at 318°C with 1.5 MeV Ne⁺ ions to produce the structure shown in figure 8b. The defects (spanning dislocations and twins) have continued to propagate with the advancing crystal-amorphous interface precisely as if the sample were annealed thermally.

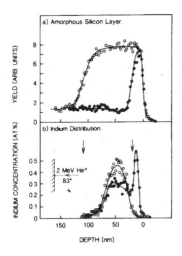

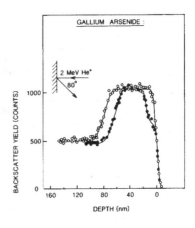

Figure 9: RBS/channeling spectra showing the thickness of the amorphous silicon layer and implanted indium distribution before (0) and after (●) IBIEC, for a sample previously implanted with 80 kV In to a fluence of 1×10^{15} cm^{-2}. IBIEC was induced at 350°C by 1.5 MeV Ne$^+$ ion irradiation.

Figure 10: RBS/channeling spectra of a GaAs sample before (0) and after (●) IBIEC induced at 100°C with 1.5 MeV Ne ions. The total Ne$^+$ ion fluence was 2×10^{17} cm^{-2}.

Figure 9 illustrates IBIEC of a sample previously implanted with 80 keV In$^+$ ions to a fluence of 1×10^{15} cm^{-2}. Implanted indium is clearly redistributed by the advancing crystal-amorphous interface which has recrystallized 80 nm following IBIEC. The excess indium is essentially 'zone refined' by the advancing crystal-amorphous interface. Following IBIEC, ~ 15% of the implanted indium has been segregated towards the surface. We speculate that this phenomenon is a result of rapid diffusion in amorphous silicon and indium solubility differences in amorphous and crystalline silicon [25,26]. Further details of the segregation process and its influence on solid phase epitaxial crystallization are given elsewhere [26]. It is intriguing to note that indium segregation during epitaxial growth appears to be identical for both IBIEC and thermal annealing [2].

We have undertaken preliminary studies of IBIEC of GaAs, which again indicate epitaxial growth at temperature well below that at which epitaxy is induced thermally. Figure 10 depicts aligned RBS spectra from a GaAs sample before and after irradiation with 1.5 MeV Ne$^+$ ions at 100°C. In this case partial epitaxial growth has occurred from both the crystal-amorphous interface and the surface. Epitaxial growth from the surface suggests that the initial amorphous layer was buried below a thin crystalline layer at the surface. RBS/channeling measurements show that epitaxy has ceased after 16 nm of growth. Additional data indicate that the layer is completely crystalline following IBIEC suggesting that a highly defective crystalline layer, indistinguishable from amorphous GaAs

326

TABLE 1: Comparison of the dependencies of thermally induced epitaxy
and IBIEC.

	DEPENDENCE	THERMAL	IBIEC
N U C L E A T I O N	Substrate temperature	Strong dependence characterized by an activation energy of 2.7 eV.	Essentially athermal with an activation energy of only 0.24 eV.
	Substrate orientation	Strong dependence, being 25 times faster on (100) than (111).	Extent of growth shows no dependence on substrate orientation.
	Substrate doping	Growth rate dramatically enhanced by small concentrations of dopant impurities.	Extent of growth shows no dependence on substrate doping.
	Microstructure	SIMILAR	
G R O W T H	Defect propagation	SIMILAR	
	Impurity segregation	SIMILAR	

as observed by channeling analysis, results from IBIEC. This situation
is consistent with thermally annealed GaAs layers and further indicates
the similarities between the epitaxial growth behaviour induced by ion
irradiation and thermal annealing.

4: MECHANISM

Table 1 compares the dependencies and annealing behaviour of
thermally induced epitaxy and IBIEC, summarizing the observations
outlined above. The table is divided into two sections: processes which
are believed to be controlled by nucleation; and processes which are
related to crystal growth. The crucial point to note from this table is
the dramatic difference between nucleation related processes for IBIEC
and thermally induced epitaxy: the similarity of growth related behaviour
is more readily understood given the epitaxial nature of the
crystallization process in each case.

As pointed out in the introduction, thermally induced epitaxy is
believed to be limited by a high energy (2.0 eV) nucleation step. The
rate of epitaxy is controlled by the concentration of nucleation sites
and consequently exhibits a strong dependence on factors which influence
this concentration, namely: substrate temperature, orientation and
doping. On the other hand, IBIEC exhibits essentially no dependence on
these parameters, suggesting that crystallization is dominated by the
athermal activation of growth by the ion beam.

· The available IBIEC data provides further insight into the nature of the beam induced activation process. In particular, it has previously been demonstrated [16,19-21] that the growth induced by IBIEC is proportional to the energy deposited into nuclear collisions at, or near, the crystal-amorphous interface. This suggests that defects generated at, or near, the interface are responsible for nucleating growth. Additional measurements support this view. Specifically, growth induced by IBIEC has been shown to be independent of the amorphous layer thickness for layers as thin as ～5nm, see for example figure 2a. Furthermore, IBIEC of buried amorphous silicon layers, in which the incident annealing ions were channeled in the crystalline surface layer, show a reduction in the extent of growth[21]. This suggests that defects generated at, or near, the crystalline side of the interface play a role in IBIEC. More detailed studies are required, however, to determine the precise nature of the nucleation process.

In conclusion, IBIEC and its dependencies have been reviewed. The data are consistent with a model in which epitaxial crystallization is activated by atomic collisions at, or near, the crystal-amorphous interface. It is speculated that subsequent defect migration and/or crystallization are responsible for the measured 0.24 eV activation energy of IBIEC. Furthermore, if the nucleation event and/or the subsequent defect migration/crystallization processes are the same for both IBIEC and thermally induced epitaxy, the nucleation energy for thermally induced epitaxy can be estimated to be 2.1-2.5 eV, consistent with a Si-Si bond breaking model.

ACKNOWLEDGEMENTS

The Australian Special Research Centres Scheme is acknowledged for partial support of this project. RGE also acknowledges the CSIRO Postdoctoral Award Scheme and CSIRO division of Chemical Physics for financial support.

REFERENCES

1. L. Csepregi, E.F. Kennedy, S.S. Lau, J.W. Mayer and T.W. Sigmon. Appl. Phys. Lett., 29, 645 (1976).

2. J.S. Williams. Chapter 5, Surface Modification and Alloying by Laser, Ion and Electron Beams, edited by J.M. Poate and G. Foti. (Plenum Press, N.Y., 1983).

3. G.L. Olson, S.A. Kokorowski, J.A. Roth and L.D. Hess. Mat. Res. Soc. Symp. Proc. 13, 141 (1983).

4. L. Csepregi, E.F. Kennedy, J.W. Mayer and T.W. Sigmon. J. Appl. Phys. 49, 3906 (1978).

5. L. Csepregi, E.F. Kennedy, T.J. Gallagher, J.W. Mayer and T.W. Sigmon. J. Appl. Phys. 48, 4234 (1977).

6. I. Suni, G. Goltz, M.G. Grimaldi, M.A. Nicolet and S.S. Lau. Appl. Phys. Lett. 40, 269 (1982).

7. F. Spaepen and D. Turnbull, in Laser Annealing of Semiconductors, edited by J.M. Poate and J.W. Mayer. (Academic Press, New York 1981). p.15.

8. J.S. Williams and R.G. Elliman.
   Phys. Rev. Lett. 51, 1069 (1983).

9. I. Suni, G. Goltz and M.A. Nicolet.
   Thin Solid Films, 93, 171 (1982).

10. G. Holmen, A. Buren and P. Hogberg.
    Rad. Effects 24, 51 (1975).

11. G. Holmen, S. Peterstrom, A. Buren and E. Bogh.
    Rad. Effects. 24, 45 (1975).

12. I. Golecki, G.E. Chapman, S.S. Lau, B.Y. Tsaur and J.W. Mayer.
    Phys. Lett. 71A, 267 (1979).

13. J. Nakata and K. Kajiyama.
    App. Phys. Lett. 40, 686 (1982).

14. B. Svensson, J. Linnros and G. Holmen.
    Nucl. Instr. Meth. 209/210, 755 (1983).

15. R.G. Elliman, S.T. Johnson, K.T. Short and J.S. Williams.
    Mat. Res. Soc. Symp. Proc. 27, 229 (1984).

16. G. Holmen, J. Linnros and B. Svensson.
    Appl. Phys. Lett. 45, 1116 (1984).

17. J. Linnros, B. Svensson and G. Holmen.
    Phys. Rev. B30, 3629 (1984).

18. R.G. Elliman, S.T. Johnson, A.P. Pogany and J.S. Williams.
    Nucl. Instr. Meth. 7/8, 310 (1985).

19. J.S. Williams, R.G. Elliman, W.L. Brown and T.E. Seidel.
    Mat. Res. Soc. Symp. Proc. 37, 127 (1985).

20. J.S. Williams, R.G. Elliman, W.L. Brown and T.E. Seidel.
    Phys. Rev. Lett., 55, 1482 (1985).

21. J.S. Williams, W.L. Brown, R.G. Elliman, R.V. Knoell, D.M. Maher and
    T.E. Seidel. Presented at the 1985 MRS Spring Meeting, San
    Francisco, U.S.A. To be published.

22. J. Linnros and G. Holmen.
    Presented at the 1985 Ion Beam Analysis Conference, Berlin.
    Nucl. Instr. Meth. B. To be published.

23. R.G. Elliman, J.S. Williams, S.T. Johnson and A.P. Pogany.
    Presented at the 1985 Ion Beam Analysis Conference, Berlin.
    Nucl. Instr. Meth. B. To be published.

24. D.M. Maher, T.E. Seidel, J.S. Williams and R.G. Elliman.
    To be presented at the 1986 Meeting of the Electrochemical Society.

25. R.G. Elliman, J.M. Poate, J.S. Williams, J.M. Gibson, D.C. Jacobson
    and D.K. Sood, Mat. Res. Soc. symp. Proc. This proceedings.

26. R.G. Elliman, J.M. Poate, K.T. Short and J.S. Williams.
    To be published.

SELF AND ION BEAM ANNEALING OF P,Ar, AND Kr IN SILICON

S. CANNAVO',* A.LA FERLA*, S.U.CAMPISANO*, E.RIMINI*, G.FERLA+, L.GANDOLFI+,
J.LIU⁰ and M.SERVIDORIᐃ
* Dipartimento di Fisica, Università di Catania - 57 Corso Italia -
I95129 Catania - Italy
+ S.G.S. Microelettronica - Stradale Primosole 50 - I95100 Catania - Italy
o Materials Science and Engineering - Cornell University - Ithaca N.Y.14853
ᐃ Lamel-CNR, Via dei Castagnoli 2 I 40126 Bologna Italy

ABSTRACT

The damage produced by high current density ∿10μA/cm² implants of
120 keV P+ into <111> and <100> silicon wafers, 500 μm thick, has been investigated in the fluence range 1x10¹⁵/cm²-1x10¹⁶/cm² by ion channeling
and by transmission electron microscopy. For both orientations the
thickness of the damage layers increases with the fluence up to 2x10¹⁵/cm²
and then decreases. The rate of regrowth is a factor two faster for the
<100> with respect to the <111> oriented Si crystals. Similar ratios have
been found in pre-amorphized samples and irradiated with Kr+ ions in the
temperature range 350°C-430°C. The TEM analysis reveals the presence of
hexagonal silicon and of twins in small amounts for both orientations. The
beam induced epitaxial growth depends also on the species present in the
amorphous layer. A comparison between self-annealing and beam annealing in
Si <100> preamorphized with Ar+ or P+ shows a noticeable retardation of the
growth rate in the presence of Ar+.

INTRODUCTION

Ion implantation is routinely performed with high current machines to
minimize the time and to increase the output. A large amount of power is
delivered to the wafers, and their temperature can increase considerably
according to the thermal contact with the holder. The damage caused by
the implant may therefore be partly annealed by the temperature rise.
The interplay between disordering and ordering during high current
implants is of relevance not only for the technological implantations
but also for understanding the basic phenomena. The fact that the
self-annealing which takes place during high current implant cannot be
attributed only to a pure thermal regrowth of the amorphous layer has
been shown several years ago [1-3]. Recently [4] it has been found
that just a slight amount of an amorphous layer, under a masked region,
regrows by solid phase epitaxy during the temperature rise caused by
the implant.
The ion beam annealing has been investigated quantitatively [5-6] by
irradiation of self-ion implanted silicon samples with Ne+ in the energy
range 0.6-3 MeV and in the temperature range 200°C-500°C. The regrowth is
governed by an activation energy of 0.24 eV, i.e. a factor 10 lower than that
of pure thermal growth [7] for the temperature range 200°C-400°C.
At a given temperature the thickness of the regrown layer increases
linearly with the energy deposited into nuclear collisions. These results
were used [8] to explain quantitatively the self-annealing during high
current density implants.
The beam induced regrowth is explained [9] in terms of defects(kink-like)
generated by the beam at the amorphous-to-single crystal interface and by
their migration with a low activation energy. To investigate in more
detail the beam annealing mechanism and its relation with the thermal one,

we report in the present work experiments performed on both <100> and <111> oriented Si wafers and on amorphous layers with high concentrations of P and of Ar atoms.

EXPERIMENTAL PROCEDURE

Silicon <100> and <111> oriented crystals, 500μm thick, were implanted with 120 keV P^+ ions using a Varian-Extrion implanter. The 1.1mA beam was scanned electrostatically with an average current density of 9.0 μA/cm^2 over the wafer fixed to the holder by springs. The fluences ranged between 0.1 and 1.0x10^{16}/cm^2. Some samples were pre-amorphized at a lower current density with P^+ or Ar^+ ions and subsequently irradiated at temperatures in the 350°C-430°C range with Kr$^+$ ions of energy in the range 350-700 keV.
The samples were analyzed by 2.0 MeV He$^+$ Rutherford backscattering in combination with the channeling effect technique. The grazing detection geometry was used to enhance the depth resolution. Some samples were thinned with ion milling and examined with transmission electron microscopy.

RESULTS

The channeling analysis of damage created during 120 keV P^+ implants into 500 μm thick silicon 4" wafers of both <100> and <111> orientation is shown in Fig.1 for different fluences. The aligned spectra indicate that at the 4x10^{15}/cm^2 dose the same damage is found in both orientations. It consists of a buried amorphous layer with a thin surface crystalline layer. The peak at a depth of ∿2000Å, equal to the sum of the projected range, R_p, and of the standard deviation, ΔR_p, is a common feature of high current implants. It is associated to the clustering of mobile point defects, probably Si interstials.

At higher fluences the buried amorphous layer shrinks on both sides with a higher rate for the <100> orientation with respect to the <111> orientation. At 7.5x10^{15}/cm^2 no amorphous layer remains in the <100> sample while a layer 1000Å thick is still present in the <111> sample. A further increase in dose causes an increase in the damage at $R_p+\Delta R_p$ and in the surface peak for the <100> samples.

The beam annealing in the <111> samples takes place at a reduced rate. A complete sequence of channeling spectra is shown in Fig.2 for the <111> Si samples implanted up to a fluence of 1.5x10^{16}/cm^2. The buried amorphous layer disappears at high doses and is replaced by a deep disordered region.

The processes which occur during high dose implantation are reported schematically in Fig.3. As the ion slows it loses energy in both nuclear and electron collisions. The energy deposited into nuclear collisions has a maximum at a depth of about 0.8 times the projected range. Where the deposited energy density overcomes the threshold value for the amorphization a buried amorphous layer is formed. With increasing dose the energy deposited into nuclear collisions increases too and the amorphous layer increases in thickness. At high dose-rate implants and in the absence of a good thermal contact between the wafer and the holder the temperature rises and the epitaxial regrowth of the α-layer can take place.

The residual disorder at the finish of the implant results then from the occurrence of damage generation and amorphous regrowth. In the present experiments the temperature rise occurs mainly in the fluence range 4x10^{14}-4x10^{15}/cm^2 and then saturates to a value which depends only on the emissivity of the wafer, of the surrounding, on the surrounding temperature and on the power [10]. A steady state is reached when the impinging power equals the irradiated one. In our condition the temperature saturates at a calculated value of about 480°C.

The measured thickness of the regrown amorphous layer starting from

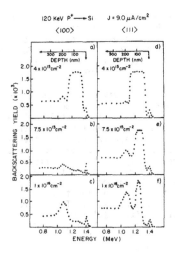

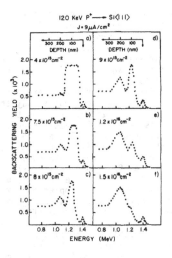

Fig.1.Aligned spectra depicting ion beam self-annealing for 500μm thick <100> and <111> Si wafers implanted with 120 keV P⁺ at an average current density of 9.0μA/cm² and for different fluences.

Fig.2.Channeling analysis of 500μm thick Si <111> wafers implanted with 120 keV P⁺ at an average current density of 9.0μA/cm² and for fluences ranging between $4\times10^{15}/cm^2$ and $1.5\times10^{16}/cm^2$.

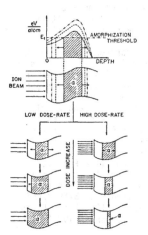

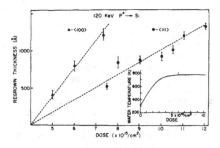

Fig.3.Schematic representation of the disordering-ordering mechanisms induced by ion beam bombardament.

Fig.4.Regrown thickness of α-Si as a function of 120 keV P⁺ dose for Si <100> and <111>.The inset in the lower right-hand part shows the calculated wafer temperature vs dose.At about $4\times10^{15}/cm^2$ the saturation value of 480°C is reached.

$4x10^{15}/cm^2$ fluence allows a comparison of experimental data taken at the same calculated [10] temperature value (as shown by the inset of Fig.4). The data are shown in Fig.4 for the <111> and for the <100> oriented sample. The growth rate is linear with dose for both orientations but for the <111> orientation is approximately half of the <100> orientation. This result is in agreement with previous measurements [2] performed with 2.5 MeV As^+.

So far the experiments were performed using the same beam to heat the sample. In a more controlled way the ion beam annealing has been studied using low dose-rate of Kr^+ ions to irradiate pre-amorphized samples at temperatures in the 350°C-430°C range. A comparison between the two orientations is shown in Fig.5 where the aligned spectra are reported for the Si as-implanted (60 keV P^+-$7.5x10^{15}/cm^2$) and after 600 keV Kr^+ $2x10^{15}/cm^2$ irradiation at 400°C. The thickness of the regrown amorphous layer is also in this case a 50% higher for the <100> orientation with respect to the <111>.

The structure of the beam stimulated regrowth has been analyzed for both <100> and <111> samples by TEM. In both samples dislocation loops, small amounts of twins and hexagonal silicon [11] are seen. The diffraction pattern of the <111> sample after implantation with $9x10^{15}/cm^2$ is shown in Fig.6. The halo indicates the presence of an amorphous layer, the diffraction spots labelled as (1) are associated to the hexagonal silicon while the satellites (2) are due to twins. The amount of twins has been estimated to be about 10%-20% of volume. In a comparison with thermal annealing it must be noted that the regrowth of the amorphous layer on Si <111> occurs [12] at a slower rate than that of Si <100> and is not linear with time with a non planar amorphous-to-single crystal interface. The break in the thickness-annealing time is obtained after a regrown thick layer of about 1000 Å, and then twins are formed. The thickness of the regrown layers in our experiment due to the presence of two amorphous-single crystal interfaces is lower than 1000 Å so that the formation of twins is reduced. A thermal annealing at 600°C for 1 hr of the $4x10^{15}/cm^2$ P^+ implanted sample causes a nearly perfect regrown layer with a small density of dislocation loops and a negligible amount of twins.

The hexagonal silicon phase is associated with hot implants [11]. These defects are present also in silicon single crystal implanted with Kr^+ ions at temperature of 400°C. No hexagonal phase has been detected at a target temperature of 320°C.

The influence of the species present in the amorphous layer on the regrowth has been measured during either beam heating or irradiation at a heated target. In the first experiment <100> Si samples were preamorphized at a low dose-rate with 60 keV P^+ or with 70 keV Ar^+. During implantation the temperature rise was negligible. The implant energy and dose was chosen in such a way to produce an amorphous layer ∿1300 Å thick and nearly the same impurity concentration profile for both ions. These samples were then implanted with 120 keV P^+ at a current density of 9.0 $\mu A/cm^2$ for several fluences. The amorphous layer regrows with a reduced rate in the silicon samples preamorphized with Ar. The regrowth stops when the interface reaches the location of the maximum Ar concentration. (see Fig.7).

Similar results are also obtained with 600 keV Kr^+ irradiation at a target temperature of 400°C. The channeling analysis performed in the samples preamorphized with P^+ or Ar^+ ions and then irradiated with $3x10^{15}/cm^2$ Kr^+ ions indicates (see Fig.8) that the amorphous layer, ∿1200 Å thick, is nearly all completely epitaxially regrown in the Si(P) while only ∿400 Å has regrown in the Si (Ar) sample.

DISCUSSION AND CONCLUSION

Ion beam annealing follows the same trend found in the thermal annealing of ion implanted Si samples. The epitaxial regrowth of the amor-

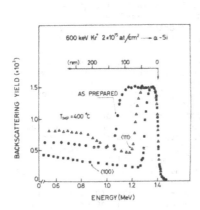

Fig.5.Channeling analysis of <111> and <100> Si wafers pre-amorphized with 60 keV 7.5×10^{15}/cm^2 P$^+$ and irradiated at 400°C with 600 keV 2×10^{15}/cm^2 Kr$^+$ ions.

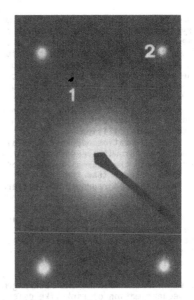

Fig.6.Electron diffraction pattern of <111> Si implanted with 120 keV P$^+$ at a dose of 9×10^{15}/cm^2. The halo indicates the presence of an amorphous region,the spots are associated with hexagonal silicon. The streaks near the other diffraction spots are due to twins

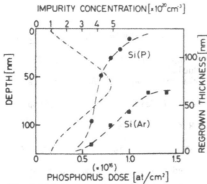

Fig.7.Thickness of regrown amorphous layer (right) in <100> Si pre-amorphized with 60 keV P$^+$ or with 70 keV Ar$^+$ at low dose-rate and subsequently implanted with 120 keV P$^+$ at a current density of 9.0 μA/cm^2.The concentration profile of Ar or P in the as prepared samples is shown in the left hand side.

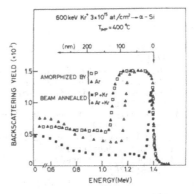

Fig.8.Channeling analysis of 600 keV Kr$^+$ 3×10^{15}/cm^2 beam induced regrowth at 400°C in <100> Si pre-amorphized with Ar$^+$ or with P$^+$ irradiation at low dose-rate.

phous layer occurs [9] by a planar movement of the amorphous to single crystal interface. The activation energy of the process has so far only been measured [5] with accuracy in the Ne^+ irradiation case and has been found to be 0.24 eV in the temperature range 200°C-400°C. At higher target temperatures it increases. The amount of regrowth increases linearly with the energy deposited into nuclear collision.

The same trend has been found using Kr^+ ions in the energy range 350-700 keV. A quantitative analysis is not straightforward due to the overlap of the preexisting damage with that created by the irradiation, and to the uncertainty in the depth distribution of the energy deposited into nuclear collision.

The orientation dependence is also found during self-ion and beam annealing, but with respect to thermal annealing the effect is much weaker. For instance the thermal regrowth of amorphous layers on <111> Si is a factor 20 slower than that on <100> Si [12]. In the present experiments on ion beam induced regrowth the difference is less than a factor 2.

The influence of impurity on the thermal regrowth amorphous layer is well established, it is species and concentration dependent. The enhancement of the regrowth rate with P^+ and the retardation with Ar^+ respect to the self-ion amorphous case has also been found in the beam annealing case but the change is considerably reduced. During thermal annealing the difference in rates is about two orders of magnitude while in the ion beam annealing it is less than an order of magnitude.

Several models are reported in literature to explain the epitaxial growth: bond breaking and rearrangement mechanism [14], generation and subsequent motion of kink-like defects on interfacial ledges and terraces [15], and point defect migration [16]. At this stage a detailed discussion of the mechanism responsible of ion beam stimulated regrowth is premature. The role played by the energy density deposited into nuclear collisions and the absence of long range migration are in agreement with a kink-like regrowth mechanisms. The generation of these defects under irradiation becomes athermal and the low activation energy of the process is associated to their migration. The orientation and impurity dependence may imply also an influence on the migration of the defects.

REFERENCES

[1]   I. Golecki,G.E.Chapman,SS.Lau,B.Y.Tsaur, and J.W.Mayer Phys.Lett. 71 A 267(1979)
[2]   J.Nakata,M.Takahashi and K.Kajiyama - Jap.J.Appl.Phys. 20,2211(1981)
[3]   S.Svensson,J.Linnros and G.Holmen - Nucl.Instr.Meth. 209/210,755(1983)
[4]   S.Cannavò,M.G.Grimaldi,E.Rimini,G.Ferla and L.Gandolfi Appl. Phys.Lett/ 47,138(1985)
[5]   J.S.Williams,R.G.Elliman,W.L.Brown, and T.E.Seidel, Mat.Res.Soc.Symp. Proc. 35,127(1985)
[6]   J.S.Williams,W.L.Brown,R.G.Elliman,V.R.Knoell,D.M.Maher and T.E.Seidel MRS-Spring Meeting 1985
[7]   L.Csepregi,E.F.Kennedy,S.S.Lau,J.W.Mayer and T.W.Sigman, Appl. Phys.Lett. 29,645(1976)
[8]   S.Cannavò,A.La Ferla,E.Rimini,G.Ferla and L.Gandolfi, subm. to J.Appl. Phys.
[9]   J.S.Williams,R.G.Elliman,W.L.Brown and T.E.Seidel - Phys.Rev.Lett. 55, 1482(195)
[10]   P.D.Parry,J.Vac.Sci.Techn. 13,622(1976)
[11]   T.Y.Tan,H.Foll and S.M.Hu - Phil.Mag.B 10 127(1981)
[12]   L.Csepregi,E.F.Kennedy,J.W.Mayer, and T.W.Sigmon, J.Appl.Phys. 49,3906 (1978)
[13]   J.S.Williams, in Surface Modification and Alloying - ed. by J.M.Poate,

G.Foti and D.C.Jacobson (Plenum Press,N.Y. 1983) p.133
[14] F.Spaepen and D.Turnbull, in Laser and Electron Beam Processing of Semiconductor Structure, edited by J.M.Poate and J.W.Mayer (Academic Press,
N.Y.1981) p.15
[15] J.S.Williams and R.G.Elliman - Phys.Rev.Lett. 51,1069 (1983).
[16] I.Narayan J.Appl.Phis. 53,8607 (1982).

ENHANCEMENT OF GRAIN GROWTH IN ULTRA-THIN
GERMANIUM FILMS BY ION BOMBARDMENT

HARRY A. ATWATER, HENRY I. SMITH, and CARL V. THOMPSON
Massachusetts Institute of Technology
Cambridge, Mass. 02139

ABSTRACT

We report the enhancement of grain growth in 50 nm-thick Ge films during Ge ion bombardment, or self-implantation, at 500-600 °C. Conventional and cross-sectional transmission electron microscopy (TEM) indicate that normal grains grow to a columnar structure at considerably lower temperatures than during ordinary thermal annealing. Furthermore, self-implantation-enhanced normal grain growth is found to be very weakly dependent on temperature. The time dependence and temperature dependence of grain growth during self-implantation were compared with data for thermal annealing experiments and suggest that the enhancement results from elastic collisions between ions and atoms located at the grain boundaries.

INTRODUCTION

Ion beams are being increasingly recognized as effective tools for enhancing kinetic processes in solids. Beginning in the 1970's, intensive research in particle beam-altered phase stability accompanied the development of nuclear reactor structural materials[1]. In addition, research in the electronics field has led to the observation and development of ion-beam-enhanced impurity diffusion [2-4], ion- beam-induced solid phase epitaxial regrowth [5-8], ion-beam-induced epitaxy [9], and, of course, ion implantation. In the work on ion beam-enhanced diffusion and beam-induced solid phase regrowth there is general consensus that the kinetic enhancement is non-thermal at sufficiently low ion fluxes. Ion-beam-enhanced diffusion is thought to result from a beam-induced vacancy-interstitial population in excess of thermal equilibrium values which enhances substitutional diffusion[2]. In ion-beam induced epitaxial regrowth several growth mechanisms have been proposed. One mechanism is beam-induced vacancies and interstitials which migrate to, and are incorporated at, the amorphous-crystalline interface[6]. Another is the generation by the ion beam of nucleation sites for growth[7].

The objective of the present research is to develop non-thermal methods for enhancing the atomic and grain boundary mobilities in ultra-thin polycrystalline films on amorphous substrates. Because the driving force for grain growth can be very large in ultra-thin films, such enhancement can lead to grain growth at temperatures well below those required for conventional thermal annealing. Moreover, destructive microstructural changes such as thermal grooving or beading [10] are competitive with grain growth during thermal anneals. Because it is possible to employ lower substrate temperatures in self-implantation enhanced grain growth, these destructive processes can be avoided. We report here the first investigation of self-implantation-enhanced normal grain growth at elevated temperatures. Germanium was chosen as the film material because of extensive recent work on grain growth in ultra-thin Ge films [11], and because radiation-enhanced annealing of Ge has already been studied [8,12]. Ge $^+$ ions were implanted, rather than dopant species in order to avoid confusion between

338

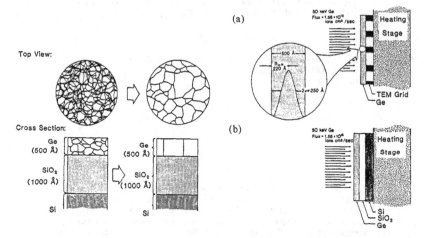

Figure 2. The two experimental configurations for self-implantation enhanced grain growth. A 50 keV Ge beam has a calculated range equal to one half of the film thickness. A flux of 1.56 x 10 [12] ions/cm [2] -sec was used. In (a), a freestanding 50 nm-thick Ge film is supported at the edges by a TEM grid which is attached to a heating stage. In (b), the Ge film is on a thermally oxidized Si wafer.

physical and chemical kinetic enhancement. Dopant effects on grain growth have been described extensively elsewhere[13,14]. The growth regime of interest was the development of a columnar normal grain structure in the ultra-thin Ge film, as shown in Figure 1. In this regime, the principal driving force for growth is the minimization of grain boundary energy. The time and temperature dependences of the grain growth during self-implantation were determined and compared with grain growth data obtained in thermal annealing.

EXPERIMENTAL

Samples were prepared in two forms. Unsupported 50 nm-thick Ge films were prepared by room-temperature electron beam evaporation of Ge onto NaCl substrates. The Ge films were floated off the NaCl, onto TEM grids and cleaned in deionized water. Other samples were prepared by room-temperature electron beam evaporation of 50 nm-thick Ge films onto thermally oxidized Si substrates.

Samples for thermal annealing experiments were all taken from the same wafer, sealed in clean quartz ampoules which had been pumped to 5 x 10 [-7] Torr, and then annealed in a constant temperature furnace. For the self-implantation experiments, samples were implanted at 10° tilt with [74]Ge [+] at 50 KeV and a current density of 0.23-0.28 uA/cm [2] on a resistively heated stage whose temperature was varied from 500 - 600 °C, as depicted in Fig. 2. The ion dose was varied between 5 x 10 [13] and 5 x 10 [15] ions/cm [2]. Freestanding Ge films and films on thermally oxidized Si wafers exhibited similar grain growth in both thermal anneals and self-implantation.

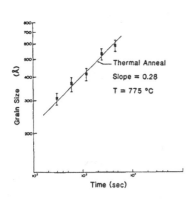

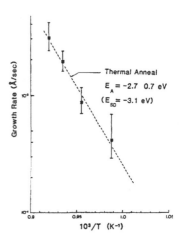

Figure 3. Time dependence of grain growth for a 50 nm-thick Ge film on SiO $_2$ for thermal annealing at 775 °C. The data show that the grain size r∝t$^{0.28}$.

Figure 4. Temperature dependence of grain growth rate for a 50 nm-thick Ge film on SiO $_2$ between 750 °C and 815 °C. An activation energy for boundary migration of 2.7 +/- 0.7 eV is seen. By comparison, the activation energy for self-diffusion in Ge is 3.1 eV.

The grain size and morphology were examined by bright field and dark field transmission electron microscopy (TEM), using both conventional and cross-sectional techniques. The grains imaged in bright field and dark field micrographs were digitized to facilitate computer generation of grain size distributions. Grain size data were fitted to lognormal distributions. Because individual grains in a non-columnar film are difficult to resolve in bright field, grain size measurements in the non-columnar films were taken from dark field micrographs. The grain size measurements in the columnar films were derived from both bright field and dark field micrographs, which gave identical results.

RESULTS AND DISCUSSION

The time dependence of grain growth to a columnar structure during ordinary thermal annealing at T=775 °C is shown in Fig. 3. The data indicate that the grain radius, r, is proportional to t $^{0.28}$, where t is time. Figure 4 is an Arrhenius plot of the growth rate of a 50 nm-thick Ge film. The growth rate dr/dt is estimated by dividing the final columnar grain size, by the time required to achieve a columnar structure. Data were taken between T=750 and 815 °C. The activation energy for grain growth is estimated to be E $_A$ =2.7 +/- 0.7eV. The large uncertainty in this measurement is due to thermal grooving of the Ge film, which complicated grain size measurement. However, the measurement does indicate that the activation energy is between two-thirds of and approximately equal to the activation energy of self-diffusion in Ge (E $_{SD}$ =3.1eV)[15].

During self-implantation, the time dependence of grain growth, for a constant ion beam current, is similar to that seen during thermal annealing. However, an equivalent growth rate is achieved at a much lower temperature. Furthermore, the growth rate is only very weakly temperature dependent, as illustrated in Fig. 5. The variation of grain size, r, with time shows that r∝t $^{0.31}$ for self-implantation

340

Figure 5. Time dependence of self-implantation enhanced grain growth for a constant ion flux of 1.56×10^{12} ions/cm^2-sec. The growth rate is identical at 600°C and 500°C, and the grain size varies as $r \propto t^{0.31}$. The data of Fig. 3 are plotted for comparison.

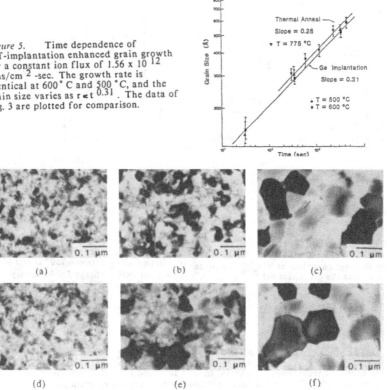

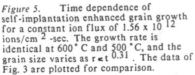

(a) (b) (c)

(d) (e) (f)

Figure 6. Figure 6(a) is a transmission electron micrograph of a freestanding 50 nm-thick Ge film implanted with 5×10^{13} ions/cm^2 at 600°C. In 6(b), a similar film has been implanted with 5×10^{14} ions/cm^2 at 600°C; in 6(c) is a similar film with a dose of 5×10^{15} ions/cm^2. In 6(d)-6(f) similar films were implanted at 500°C with 5×10^{13}, 5×10^{14}, and 5×10^{15} ions/cm^2, respectively.

thermally annealed at T=775°C is plotted in Fig. 5. The grain growth rate is, within experimental error, identical at T=600°C and T=500°C. This indicates that the activation energy for the rate-limiting process in self-implantation-enhanced grain boundary migration is very low. We estimate it to be less than 0.15 eV. Figures 6(a) - 6(c) illustrates the changes in grain morphology as a function of ion dose at 600°C. The increase in grain size is apparent. A columnar grain structure is achieved at a dose of 5×10^{15} ions/cm^2. Close inspection reveals that the number of defects within grains is also reduced as the ion dose is increased. Transmission electron diffraction indicates that the crystallographic texture is random, as expect for normal grain growth. Figures 6(d) - 6(f) illustrate the film morphology during self-implantation at T = 500°C. The morphology variation with ion dose is identical to that seen in 6(a) - 6(c).

Figure 7(a) is a cross-section of a 50 nm-thick Ge film on 100 nm SiO$_2$ thermally annealed at T=600°C for 60 min. Small non-columnar grains are seen. In Fig. 7(b), a 50 nm-thick Ge film on 100 nm SiO$_2$ has been self-implanted at 600°C for 53 min. The grain size is considerably larger, the development of a columnar structure is evident, and no grain boundary grooving is observed.

(a)   (b)

Figure 7.   In (a), a cross-section of a 50 nm-thick Ge film on 100 nm of thermal SiO_2 is shown. The sample has been thermally annealed at 600 °C for 60 min. In 7(b), a similar sample has been implanted at 600 °C for 53 min, with a dose of 5 x 10 [15] ions/cm [2] at a flux of 1.56 x 10 [12] ions/cm [2] -sec.

Several possible mechanisms can be proposed for self-implantation-enhanced grain growth. One possible mechanism is an enhancement of the grain boundary mobility resulting from elastic collisions between primary and knock-on ions and atoms located at or near grain boundaries. The observation of an activation energy of less than 0.15 eV is consistent with this hypothesis, since the elastic collisions between ions and atoms remove an activated jump across the grain boundary from the migration process. The time dependence of normal grain growth, according to simple theory, is $r \propto t^{1/2}$. During self-implantation, the same dependence is expected, since the grain boundary mobility, although enhanced, is not dependent on grain size or time. The driving force is proportional to $1/r$, so $r \propto t^{1/2}$ is expected. The observed time dependence during self-implantation ($r \propto t^{0.31}$) is similar to that seen in ordinary thermal annealing ($r \propto t^{0.28}$) and both are consistent with other experimental results [16,17]. This suggests that the grain boundary mobility is not a function of grain size or time.

Enhancement of grain growth might also arise from vacancy-interstitial pairs which are generated throughout the volume of grains and which diffuse to grain boundaries where they enhance grain boundary dislocation motion. We would expect, however, that the activation energy of grain boundary migration would be at least equal to the energy for vacancy migration. Hence, the observed temperature independence of grain growth is not consistent with this model. Furthermore a different time dependence would be expected in which $r \propto t$, since in this case the grain boundary mobility would be linearly dependent on r, the grain size .

CONCLUSION

Self-implantation-enhanced grain growth at 500-600 °C has been observed in 50 nm-thick Ge films. Grain growth is apparently independent of temperature in this range. It is noteworthy that self-implantation promotes grain growth at approximately 0.6 T_m rather than the 0.9 T_m required for ordinary thermal annealing. The time dependence and temperature dependence of enhanced grain growth suggest that elastic collisions between ions and atoms at grain boundaries are responsible for an enhancement of the grain boundary mobility.

342

ACKNOWLEDGMENTS

The authors wish to thank J. Woodhouse and M. McAleese of the MIT Lincoln Laboratory for expert, careful assistance in the ion implantation work. Helpful discussions with and computer work by J.E. Palmer are also gratefully acknowledged.   One of us (H.A.A.) wishes to acknowledge fellowship support from Exxon Corp. and Sohio Corp. during different portions of this work. This work was supported by the National Science Foundation under Grant no. ECS-8506565, and the Air Force Office of Scientific Research under Grant no. AFOSR-85-0154.

REFERENCES

1.   K.C. Russell, to be published in Progress in Materials Science , J. Christian, P. Haasen, and T.B. Massalski eds., Oxford Univ. Press.

2.   H. Strack, J. Appl. Phys. *34* , 2405 (1963).

3.   R.L. Minear, D.G. Nelson and J.F. Gibbons, J. Appl. Phys. *43* , 3468 (1972).

4.   E.W. Maby, J. Appl. Phys. *47* , 830 (1976).

5.   I. Golecki, G.E. Chapman, S.S. Lau, B-Y. Tsaur and J.W. Mayer, Physics Lett. *71A*, 267 (1979).

6.   S. Cannavo, M. G. Grimaldi, E. Rimini, G. Ferla and L. Gandolfi, Appl. Phys. Lett. *41* , 138 (1985).

7.   J.S. Williams, R.G. Elliman, W.L. Brown and T.E. Seidel, Phys. Rev Lett. *55* , 1482 (1985).

8.   G. Holmen, S. Peterstrom, A. Buren and E. Bogh, Rad. Effects, *24* , 45 (1975).

9.   P.C. Zalm and L.J. Beckers, Appl. Phys. Lett. *41* , 167 (1982).

10.   D.J. Srolovitz and C.V. Thompson to be published in Thin Solid Films.

11.   J.E. Palmer S.M. Thesis, Dept. of Electrical Eng. and Computer Sci., MIT, August 1985.

12.   G. Holmen, P. Hogberg, and A. Buren, Rad. Effects. *24* , 39 (1975).

13.   Y. Wada and S. Nishimatsu, J. Electrochem Soc. *125* , 1499 (1978).

14.   D.A. Smith and T.Y. Tan, in Grain Boundaries in Semiconductors, edited by H.J. Leamy, G.E. Pike and C.H. Seager (Elsevier Science Publishers, New York, 1982) p. 63.

15.   A. Seeger and K.P. Chik, Phys. Stat. Sol. *29* , 455 (1968).

16.   P. Gordon and T.A. El-Bassyouni, Trans. Am. Inst. Min. Engrs. *233* , 391 (1965).

17.   J.P. Drolet and A. Galibois, Acta Met. *16* , 1387 (1968).

MICROSTRUCTURE OF NIOBIUM FILMS ORIENTED BY NON-NORMAL INCIDENCE ION BOMBARDMENT DURING GROWTH

J.M.E. HARPER, D.A. SMITH, L.S. YU* and J.J. CUOMO
IBM Thomas J. Watson Research Center
P.O. Box 218, Yorktown Heights, NY 10598

ABSTRACT

We demonstrate that non-normal incidence ion bombardment applied during thin film growth has a pronounced alignment effect on crystallographic orientation. Restricted fiber texture is achieved in Nb films deposited at room temperature onto amorphous silica substrates with simultaneous 200 eV Ar+ ion bombardment at 20 degrees from glancing angle. Xray pole figure measurements and transmission electron diffraction show that the alignment direction is a channeling direction for the incident ions between (110) planes. The degree of alignment increases linearly with the fraction resputtered by the ion beam. Recommendations are given for optimizing this ion beam orientation effect.

INTRODUCTION

Thin film properties depend greatly on the energy and flux of energetic particle bombardment occurring during film growth. Many of these effects are reviewed in a recent article [1]. Here, we examine the effect of non-normal incidence ion bombardment on microstructure, and demonstrate a pronounced alignment effect causing highly restricted fiber texture in Nb films on amorphous substrates at room temperature. The experimental method has been summarized [2], and is described in more detail in a forthcoming publication [3], which also gives references to other related work. In this paper, we describe the microstructure of these films, and discuss methods to optimize this ion beam orienting effect in conjunction with the strong influence of the substrate direction.

Niobium thin films were deposited by argon ion beam sputtering from a Nb target onto fused silica substrates at room temperature, and onto thin silicon nitride membranes for transmission electron microscope (TEM) examination. Simultaneously, a 200 eV Ar+ ion beam was directed at the growing film at 20 degrees from glancing angle. This ion flux was adjustable from zero up to 1.3 times the arriving Nb atom flux. The degree of orientation of the resulting films was measured by Xray pole figure analysis, and verified by TEM.

---

* Present address: Hypres Corp., Elmsford, NY

XRAY ANALYSIS OF DEGREE OF ORIENTATION

The degree of orientation was measured by Schulz's Xray pole figure technique [3], using the (110) Bragg diffraction peak. The pole figure of a Nb film deposited on fused silica without ion bombardment is shown in Fig. 1. The central spot and concentric ring correspond to diffraction from grains oriented with (110) planes parallel to the substrate surface. The 60 degree separation between the ring and central spot represents the angular distance between (110) planes. This pattern indicates a strong (110) fiber texture perpendicular to the substrate plane, with the uniform intensity of the ring indicating a random azimuthal orientation, typical of bcc films on amorphous substrates.

A representative pole figure of a film prepared under simultaneous ion bombardment at room temperature is shown in Fig. 2, obtained with an ion/atom arrival rate ratio of 1.0. Non-uniform intensity in the 60 degree ring has developed, indicating a non-uniform azimuthal distribution of grains. Thus, a restricted fiber texture has been induced by ion bombardment during film growth.

The direction of the incident ion beam is indicated by an arrow in Fig. 2. Analysis of the (110) intensity peaks shows that the ion beam induces a film orientation in which the incident ions see a planar channeling direction between (110) planes. This direction is shown in the (110) stereographic projection (Fig. 3), in which it is evident that a mirror reflection about the ion beam direction is equally likely as the orientation shown. Thus, the broad (110) intensity peaks at +90 and -90 degrees from the ion beam direction (Fig. 2) are a merging of two sets of (110) peaks obtained from Fig. 3 and its reflection. The narrowness of the resulting orientation distribution should therefore be judged from the width of the narrow peaks in the ion beam direction in Fig. 2. Also evident in Fig. 2 is a tilting of the fiber axis towards the ion beam direction, reaching a value of 5-7 degrees at an ion/atom ratio of 1.0.

A quantitative estimate of the degree of orientation is obtained by averaging the highest peak intensities along the diffraction ring and normalizing this value by the central peak intensity. This degree of orientation is plotted in Fig. 4 as a function of ion/atom arrival rate ratio. The degree of orientation increases substantially with ion flux, up to values for which about half of the film grains are aligned to within 5 degrees of the ion beam direction, for an ion/atom ratio of 1.3. At this level, approximately 75% of the depositing film is resputtered during deposition, with an average sputtering yield of 0.6 Nb atoms per Ar+ ion. The degree of orientation increases linearly with the fraction resputtered [3].

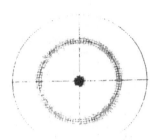

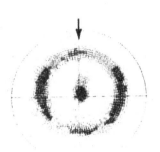

Fig. 1. Pole figure of Nb film deposited at room temperature without ion bombardment, showing (110) fiber texture with no azimuthal orientation.

Fig. 2. Pole figure of Nb film deposited in the presence of 200 eV Ar+ ion bombardment, at an ion/atom ratio of 1.0, showing restricted azimuthal orientation. The arrow shows the ion beam direction.

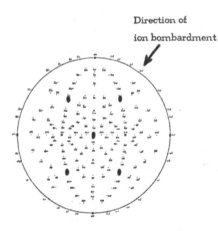

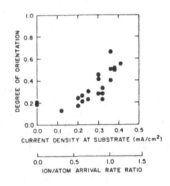

Fig. 3. (110) stereographic projection, showing grain orientation selected by ion beam. The mirror reflection about the ion beam direction is also selected.

Fig. 4. Degree of orientation of Nb films vs. arrival rate ratio of ions/atoms during deposition.

TEM ANALYSIS OF MICROSTRUCTURE

Transmission electron microscopy was used to examine the grain size and preferred orientation of these ion bombarded films. The bright field image of a film prepared under an ion/atom ratio of 1.0 is shown in Fig. 5. The polycrystalline Nb film has a grain size of about 100-300 angstrom, with no clear grain shape anisotropy. This morphology remained relatively unaffected by the degree of ion bombardment. Dark field images from the (110) diffraction ring also showed no obvious grain shape anisotropy.

The transmission electron diffraction image of the same film, obtained with the electron beam perpendicular to the substrate, is shown in Fig. 6. The observed diffraction rings, in order of increasing diameter, are (110), (200), (211), (310) and (222). The relative weakness of the (310) ring indicates the presence of (110) fiber texture, and the slight anisotropic intensity variation in the rings shows the presence of azimuthal anisotropy induced by the ion bombardment. This is most evident in the (110) ring. The azimuthal intensity variation is not as distinct as in the Xray pole figure (Fig. 2), indicating that the restricted fiber texture in the TEM films (500 angstrom thick) may not be as fully developed as in the thicker films used for pole figures (3000 angstrom).

The presence of strong (110) fiber texture is further confirmed by tilting the TEM sample relative to the electron beam. The diffraction image from the same sample as in Fig. 6, tilted 40 degrees, is shown in Fig. 7. Here, the diffraction rings break up into distinct arcs, confirming the highly anisotropic fiber texture. Further details on the degree of orientation as a function of ion bombardment are given in Ref. [3].

DISCUSSION

The microstructure of Nb films oriented by non-normal ion bombardment during deposition is characterized by fine-grained polycrystalline morphology, with a high degree of orientation induced by the ion beam. The ion beam direction selects a channeling direction for growth. In the case described here, the ion beam is incident at 20 degrees from glancing angle, and causes the selection of two equivalent grain orientations for which the ion beam sees a planar channeling direction between (110) planes. We also observe that the ion beam causes a tilt in the fiber axis as the film structure accomodates this level of ion flux. We expect that the dominant orienting mechanism is the variation of sputtering yield of film grains with azimuthal angle of incidence, and that channeling directions are favored by their lower sputtering yields.

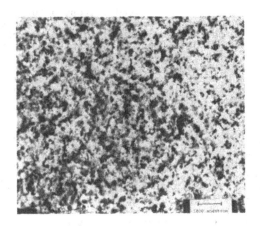

Fig. 5. Bright field TEM
micrograph of Nb film
prepared with ion/atom
ratio of 1.0.

Fig. 6. Electron
diffraction image from
film of Fig. 5, taken with
electron beam perpen-
dicular to plane of
sample.

Fig. 7. Electron
diffraction image from
film of Fig. 5,6, taken
with electron beam at 40
degrees from sample
normal.

In contrast with the tilted microstructure often observed with non-normal incidence vapor flux, the orientation of the films described here should be relatively insensitive to the direction of the arriving vapor flux, as evidenced by the well-defined fiber texture obtained without ion bombardment. This level of preferred orientation is normally only achieved at high substrate temperature in evaporation or sputtering, and indicates the high kinetic energy of arrival of ion beam sputtered atoms.

To optimize this ion beam orienting effect, we suggest choosing an angle of incidence such that the ion beam selected growth direction is compatible with the strong orienting influence of the substrate plane. For example, with a bcc film, a fiber texture with (110) perpendicular to the substrate plane is strongly favored even in the absence of ion bombardment. By choosing the angle of ion bombardment of 45 degrees from normal incidence, the ion beam should select the (100) axial channeling direction while not disturbing the (110) perpendicular orientation. Together, the orienting effect of both substrate and ion beam should produce a highly aligned growth direction. By such a careful choice of beam direction, the tilting of the fiber axis should be eliminated, since the angle between (100) and (110) is 45 degrees, allowing even better definition of the growth orientation. The technique of ion beam oriented film growth described here should be applicable to many materials, and may provide a useful process in controlling thin film orientation at low temperatures.

ACKNOWLEDGMENTS

We thank S. Herd for help in identifying this effect, R. Reif for supporting the M.S. thesis project (L.S.Y.) under which much of this study was performed, A.P. Segmuller, J.M. Karasinski and J. Angilello for help with the Xray pole figure technique, and B. Bumble, C.V. Jahnes and G.N. Hyer for help with the ion beam deposition system.

REFERENCES

1. J.M.E. Harper, J.J. Cuomo, R.J. Gambino and H.R. Kaufman, in Ion Bombardment Modification of Surfaces: Fundamentals and Applications, p. 127, ed. by O. Auciello and R. Kelly (Elsevier, Amsterdam, 1984).

2. L.S. Yu, J.M.E. Harper, J.J. Cuomo and D.A. Smith, Appl. Phys. Lett. 47, 932 (1985).

3. L.S. Yu, J.M.E. Harper, J.J. Cuomo and D.A. Smith, J. Vac. Sci. Technol. (to be published 1985).

THE CRYSTALLINE-TO-AMORPHOUS PHASE TRANSITION IN IRRADIATED SILICON

D.N. SEIDMAN*,**,R.S. AVERBACK**,P.R. OKAMOTO**AND A.C. BAILY*,**
*Northwestern University, Materials Science Dept., Evanston,IL 60201
**Argonne National Laboratory, Materials Science and Technology
Division, Argonne, IL 60439

ABSTRACT

The amorphous(a)-to-crystalline (c) phase transition has been studied in electron(e⁻) and/or ion irradiated silicon (Si). The irradiations were performed in situ in the Argonne High Voltage Microscope-Tandem Facility The irradiation of Si, at <10 K, with 1-MeV e⁻ to a fluence of 14 dpa failed to induce the c-to-a transition. Whereas an irradiation, at <10 K, with 1.0 or 1.5-MeV Kr^+ ions induced the c-to-a transition by a fluence of ≈0.37 dpa. Alternatively a dual irradiation, at 10 K, with 1.0-MeV e⁻ and 1.0 or 1.5-MeV Kr^+ to a Kr^+ fluence of 1.5 dpa -- where the ratio of the displacement rates for e⁻ to ions was ≈0.5--resulted in the Si specimen retaining a degree of crystallinity. These results are discussed in terms of the degree of dispersion of point defects in the primary state of radiation damage and the mobilities of point defects.

INTRODUCTION

The amorphization of silicon (Si) as a result of irradiation by energetic particles has been studied for over two decades, but the exact mechanism(s) by which the transition from the crystalline(c)-to-amorphous(a) phase takes place has remained unresolved[1-12]. In this paper we present new experimental results on the c-to-a phase transition for electron (1- MeV e⁻) and/or ion-irradiated Si (1.0 or 1.5-MeV Kr^+ions). The experiments were performed employing the unique capabilities of the Argonne National Laboratory High Voltage Electron Microscope-Tandem Facility. The experimental results show directly that the c-to-a phase transition depends on the degree of dispersion of the point defect distribution in the primary state of radiation damage. That is, a 1 MeV e⁻ irradiation--which produces a random distribution of Frenkel pairs[13]--to a fluence of 14 dpa failed to induce the c-to-a transition. Whereas, an irradiation with 1.0 or 1.5-MeV Kr^+ ions--which produces dense displacement cascades[14]-- to a fluence of 0.37 dpa induces the c-to-a transition. Moreover, it has been shown for the first time that surprisingly the dual irradiation of a region with 1-MeV e⁻ and 1.0 or 1.5-MeV Kr^+ ions can strongly retard the c-to-a transition, if the ratio (R) of the displacement rates (dpa s⁻¹) for electrons-to-ions exceeds ≈0.5. Atomistic models are presented for the observations.

EXPERIMENTAL PROCEDURE

The experiments consisted of irradiating <100>-p-type Si specimens in the 1-MeV transmission e⁻ microscope (TEM) with 1-MeV e⁻ and/or 1.0 or 1.5-MeV Kr^+ions at a specimen temperature of <10 K. The first experiment consisted of irradiating a small region-- ≈2 μm diam.-- of the Si specimen with 1.0-MeV e⁻ to a high fluence, at <10 K, to see if it could be amorphized under these extreme conditions. The thermal conductivity of c-Si is similiar to that of pure metals and therefore c-Si presents no special beam heating problems. The second experiment involved irradiating Si with 1.0 or 1.5-MeV Kr$^+$ions, at <10 K, to determine the fluence at which the c-to-a phase transition occurs. And the third experiment consisted of a simultaneous irradiation, of a small region, with 1.0-MeV e$^-$ and 1.0 or 1.5-MeV Kr$^+$ions at a fixed value of R. The last experiment was used to study directly the effect of the spatial distribution of point defects on the c-to-a transition.

Electron Current Density Profile and Electron Irradiations

The electron current density as a function of position in the specimen has been shown to be given by a Gaussian expression [15]. The expression has the form:

$$I_e(r)=I_0\exp[-(r/r_0)^2];$$

where I_0 is the value of the electron flux at r=0 and r_0 is given by

$$r_0=(I_T/\pi I_0)^{1/2};$$

where I_T is the total electron current. The values of I_0 and I_T employed were 3.63×10^{19} e$^-$cm^{-2}s^{-1}and 168.5 nA, respectively. The value of r_0 was first determined by measuring I_T and I_0 with the aid of a Faraday cup. The quantity $2r_0$ is the effective beam diameter (D_0) and this value was 1.92 μm for all the e$^-$ irradiations.

The e$^-$ irradiations were performed by continuously irradiating the specimen and intermittently monitoring the state of the irradiated volume by observing bend extinction contours in bright-field transmission electron micrographs and selected area diffraction patterns (SADPs). The SADPs were taken employing an aperture whose diameter was smaller than the diameter of the e$^-$ irradiated area.

Kr$^+$ Ion Irradiations and the Dual Irradiations

All the 1.0 or 1.5-MeV Kr$^+$ irradiations were performed in situ employing the tandem accelerator. The dual irradiations were performed by irradiating a small region--D_0= 1.92 μm--of the specimen with e$^-$ and a much larger region which included this small region with the 1.0 or 1.5-MeV Kr$^+$ion beam. At the latter energies no Kr$^+$ ions were deposited in the specimen. The e$^-$ flux was maintained at a constant value and the Kr$^+$ ion flux was systematically increased in steps. The state of the dual irradiated region was also monitored intermittently as described above.

EXPERIMENTAL RESULTS

1.0-MeV Electron Irradiation

Figure 1 shows the effect of the1-MeV e$^-$ irradiation of a specimen maintained at <10 K. The bright field transmission electron micrographs are on the left-hand side and their corresponding SADPs are on the right-hand side. Figures 1(a) and 1(b) are for fluences of zero and 14 dpa, respectively. The displacement of the bend extinction contour in the e⁻-irradiated area demonstrates that there are stresses associated with this region. Note the absence of diffuse scattering rings in the SADP even after 14 dpa had been accumulated. Small secondary point-defect clusters--≈50 Å diameter-- were found in the e⁻-irradiated area.

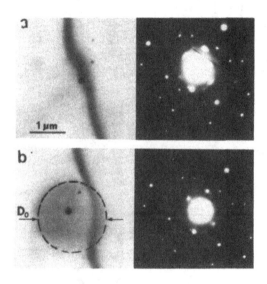

Figure 1: The effects of 1 MeV e⁻ irradiation of a Si speciman at <10 K. In (a) the specimen had received zero dpa and in (b) 14 dpa. The irradiated area is denoted by D_0. The corresponding SADPs are on the right-hand side.

1.0 or 1.5-MeV Kr⁺Ion Irradiation

Silicon specimens irradiated at <10 K with either 1.0 or 1.5-MeV Kr⁺ became amorphous at <0.4 dpa. The lack of crystallinity was determined from the absence of bend extinction contours, and the absence of sharp diffraction spots and the presence of diffuse scattering rings in the SADPs.

Dual Electron and Ion Irradiations

Figure 2 demonstrates the effects of the dual irradiation, at <10 K, on the degree of crystallinity of the irradiated volume of material. The accumulated fluences due to the 1.0-MeV Kr⁺ion irradiation in Figs. 1(a), 1(b) and 1(c) are 8.57×10^{12} cm⁻², 4.26×10^{13} and 1.45×10^{14} cm⁻² (0.02, 0.11 and 0.37 dpa), respectively. The value of the e⁻ flux was

$3.63\times10^{19}\text{cm}^{-2}\text{s}^{-1}(2.64\times10^{-3}\text{ dpa s}^{-1})$.

    The displacement of the bend extinction contour in going from Fig.2(a) to 2(c) is quite clear. This displacement is due to the stresses associated with the dual irradiated region. The presence of the bend extinction contours after 0.37 dpa of 1.0-MeV Kr^+ had been accumulated demonstrates that the Si retains a large degree of crystallinity. The corresponding SADP [Fig. 2(c)] exhibits sharp diffraction spots as well as some diffuse scattering. The latter is indicative of a-Si in this region. It is important to note that by 0.37 dpa of 1.0-MeV Kr^+ the surrounding Si is essentially completely amorphous. The dashed circle in Fig. 2(c) is denoted the critical diameter (D_c) and is defined by the displaced bend extinction contour(s). The value of R at D_c is ≈ 0.5. Physically this is the minimum value of R necessary to maintain a degree of crystallinity. The same silicon specimen was irradiated to an accumulated Kr^+ fluence of 1.5 dpa by increasing the Kr^+ ion flux in discrete steps to a value of 4.5×10^{-3}dpa s⁻¹. The value of D_c decreased but the value of R at D_c was constant at ≈ 0.5. A 1-MeV e^- irradiation of a partially a-Si region, at <10 K, failed to induce crystallization of the a-Si.

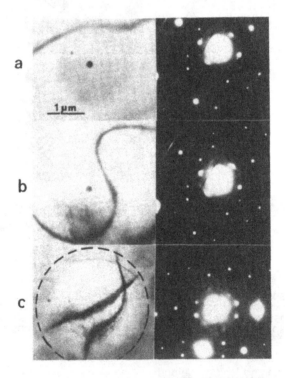

Figure 2: This figure shows the effect of a dual irradiation--1.0-MeV e^- and 1.0-MeV Kr^+ --on a small (D_0=1.92 μm) area. Figures 1(a),1(b) and1(c) were irradiated to fluences of 0.02, 0.11 and 0.37 dpa, respectively. The e^- flux was 2.64×10^{-3} dpa s⁻¹. The SADPs (right hand side) were taken with an aperture whose diameter was <1.92 μm. Note

the bend extinction contours in the irradiated area.

DISCUSSION

1 MeV Electron Irradiation

The irradiation, at <10 K, to a fluence of 14 dpa with 1 MeV e⁻ demonstrates that even under these extreme conditions Si can not be amorphized. The only other low-temperature in situ e⁻ irradiation experiment is due to Foell[16]. Foell e⁻-irradiated Si specimens between 15 and 60 K, employing a beam voltage between 400 and 650 keV; the maximum value of the fluence was ≈10.8 dpa (5×10^{23} cm^{-2}). Foell also found no evidence of a-Si. However, he observed small dislocation loops. These loops are presumably interstitial in character, as some form of the self-interstitial atom (SIA) is known to be mobile in Si at these low temperatures[12]. The existence of a highly-mobile SIA at <10 K implies that this temperature is in the so-called Stage II regime. The neutral vacancy (v) becomes mobile at ≈70 K and the negatively doubly-charged v at ≈160 K[12]; hence all charge states of the v are immobile at <10 K. To date no experimental evidence has been obtained for the stimulated athermal migration of v's in Si by the e⁻ beam via, for example, the Bourgoin-Corbett mechanism[17,18]. Hence, the only possible origin of the dislocation loops observed by Foell and ourselves is due to the clustering of highly mobile SIAs--as a result of random-walk encounters--which convert into dislocation loops when the clusters exceed a critical size. The formation of dislocation loops from the clustered SIAs suppresses the c-to-a transition in the case of the e⁻ irradiation. Thus, the conditions for the suppression of the c-to-a transition appear to be: (a) a random distribution of Frenkel pairs; (b) a highly mobile SIA; and (c) an immobile v.

It should be noted that measurements of the production rate of damage in e⁻-irradiated metals, in Stages I and II, are consistent with the preceeding ideas. Electron irradiation, below Stage I, produces a random distribution of immobile Frenkel pairs. In Stage II--where SIAs are highly mobile--clustering of SIAs and trapping of SIAs at impurities are very important reactions. The net result is that the production rate of radiation damage--caused by MeV e⁻ irradiation-- in Stage II is very low compared to that in Stage I and therefore the supersaturation of Frenkel pairs in Stage II is small compared to the value in Stage I--this reduces the the tendency towards amorphization. In Stage II the concentration of isolated v's is greater than that of SIAs, since the clustering of SIAs as a result of long-range thermally activated migration reduces the concentration of isolated SIAs.

Dual Electron and Kr⁺ Ion Irradiations

The retardation of the c-to-a transition, at <10 K, as a result of a dual irradiation can be understood qualitatively employing the following argument. The 1.0-MeV Kr⁺ produce displacement cascades with a v-rich core[19,20], surrounded by a mantle of SIAs whose local concentration is several at.%[21]. Embryos of a-Si form in this region as a result of essentially dynamic SIA-SIA reactions--that is, little or no thermally-activated long-range migration. This is analogous to small Stage I recovery in metals which had been irradiated with heavy ions.

354

The 1.0-MeV e⁻-irradiation in Stage II produces a state of damage, as described above, which has a strong surplus of isolated v's over SIAs. Hence, the v's dynamically shrink and/or destroy the a-Si embryos around the displacement cascades--this retards the c-to-a transition. With increasing fluence the volume fraction of the a-Si phase increases as a result of the conversion of the a-Si embryos to the a-Si phase. Once the latter conversion has occurred it can not be converted back to the c-Si phase at <10 K, as we have observed that a 1-MeV e⁻ irradiation of a partially a-Si region, at <10 K, failed to induce crystallization of the a-Si. The growth and shrinkage of the a-embryos is essentially the random walk problem with absorbing barriers. The effect of the e⁻-irradiation is a strong one, as in the presence of only the Kr⁺ flux the c-to-a transition takes place at <0.4 dpa, while for the dual irradiation Si retains crystallinity at a Kr⁺ fluence of 1.5 dpa. The ratio R is the control variable and for larger values of R the value of the ion fluence to which Si retains crystallinity is increased. [The details of this model will be published elsewhere.]

ACKNOWLDGEMENTS

This research was supported by the U.S. Department of Energy. Dr. Hartmut Wiedersich is thanked for kind support and encouragement. DNS wishes to thank Dr. Georges Martin for stimulating his interest in the c-to-a phase transition problem and for useful discussions during a most pleasant stay at Saclay.

References

1. F.L. Vook and H.J. Stein, Radiat. Effects 2, 23 (1969).
2. F. F. Morehead, Jr. and B.L. Crowder, Radiat. Effects 6, 27 (1970).
3. L.T. Chadderton, Radiat. Effects 8, 77 (1971).
4. M.L. Swanson, J.R. Parsons and C.W. Hoelke, Radiat. Effects 9, 249 (1971).
5. J.F. Gibbons, Proc. IEEE, 60, 1062 (1972).
6. J.R. Dennis and E.B. Hale, Radiat. Effects 30, 219 (1976); J. Appl. Phys 49, 1119 (1978).
7. R.S. Nelson, Radiat. Effects 32, 19 (1977).
8. F. Cembali, L. Dori, R. Galloni, M. Servidori and F. Zignani, Radiat. Effects 36, 111 (1978).
9. N.A. Sobolev, G. Götz, W. Karthe and B. Schnabel, Radiat. Effects 42, 23 (1979).
10. R. Kalish, T. Bernstein, B. Shapiro and A. Talmi, Radiat. Effects 52, 153 (1980).
11. R.P. Webb and G. Carter, Radiat. Effects 59, 69 (1981).
12. J.W. Corbett, J.P. Karins and T.Y. Tan, Nucl. Instum. & Meth. 182/183, 457 (1981).
13. J.W. Corbett, Solid State Physics:Suppl.7 (Academic Press, New York,1966),Chapt. 1, pp.1-5; M.W. Thompson,Defects and Radiation

Damage in Metals (Cambridge Press, England, 1969),Chapt. 4, pp. 89-142.
14. J.A. Brinkman, Am. J. Phys. 24, 246 (1956).
15. P.R. Okamoto and N.Q. Lam, Mat. Res. Soc. Sympos. Proc. 41,241 (1985).
16. H. Föll, Inst. Phys. Conf. Ser. 23, 233 (1975).
17. J.W. Corbett and J.C. Bourgoin, Trans. IEEE NS-18, 11 (1971).
18. J.C. Bourgoin and J.W. Corbett, Phys. Lett. 38A, 135 (1972).
19. C.-Y. Wei and D.N. Seidman, Appl. Phys. Lett. 34, 622 (1979); C.-Y. Wei,M.I. Current and D.N. Seidman, Philos. Mag. A, 44, 459 (1981); M.I. Current, C.-Y. Wei and D.N. Seidman, Philos. Mag. A, 47, 407 (1983).
20. R.S. Averback, R. Bendek, K.L. Merkle, J. Sprinkle and L.J. Thompson, J. Nucl. Mater. 113, 211 (1983).
21. L.A. Beaven, R.M. Scanlan and D.N. Seidman, Acta Metall. 19, 1339 (1971); C.-Y. Wei and D.N. Seidman, Philos. Mag. A, 43, 1419 (1981).

TRAPPING OF INTERSTITIALS DURING ION IMPLANTATION IN SILICON*

R.J. CULBERTSON AND S.J. PENNYCOOK
Solid State Division, Oak Ridge National Laboratory, Oak Ridge, TN 37831

ABSTRACT

The solid phase epitaxial regrowth of silicon implanted with a group V dopant, such as antimony, results in excellent incorporation of the dopant atoms into silicon lattice sites. However, annealing at higher temperatures or longer times results in transient dopant precipitation with a diffusion coefficient up to five orders of magnitude above that of tracer diffusion and with a reduced activation energy.
This precipitation is accompanied by the nucleation of dislocation loops that are interstitial in nature, and the transient ceases as the dislocation loops develop. It is believed that Si interstitials are trapped in a stable defect complex during the implantation process. Although they survive SPE these complexes dissolve at higher temperatures and release a large supply of interstitials which serve to promote dopant migration via an interstitialcy mechanism until they condense to form the observed dislocation loops. By following the Sb implantation with an implantation of B to an equivalent concentration profile the loop formation is efficiently suppressed. For higher B concentrations the Sb precipitation is no longer observed. Results for As implantation are similar to Sb except that As precipitates can not be directly observed. Calculations of the dopant and interstitial concentration depth distributions were also performed.

INTRODUCTION

Ion implantation of Si has been an important technique for achieving dopant profiles and concentrations unobtainable by other means. Especially important has been the ability to greatly exceed equilibrium solubility limits while still maintaining the dopants in the electrically active substitutional sites. Damage due to the implantation process has been well-studied. For medium to high dopant concentrations the Si surface layer is amorphized, and only damage at the end of the ion range was thought to be important. However, it has recently been demonstrated that defect complexes for group V dopants may be formed within the amorphous layer which are sufficiently stable to survive the conditions of SPE but dissolve at higher temperatures or longer heating times.[1,2] The resulting release of a large supply of Si interstitials gives a large transient enhancement of the dopant diffusion via an interstitialcy mechanism characterized by a reduced activation energy. The dopant atoms leave the electrically active sites and form precipitates. The Si interstitials condense into dislocation loops at the mean projected ion range, and the transient dopant diffusion reduces as the dislocation loops develop. The present results show that the formation of the loops can be suppressed by following the group V dopant implantation with a group III dopant (B) implanted to a similar concentration. Furthermore, for higher concentrations of the group III dopant the enhanced precipitation of Sb can be completely suppressed.

*Research sponsored by the Division of Materials Sciences, U.S. Department of Energy under contract DE-AC05-84OR21400 with Marietta Energy Systems, Inc.

EXPERIMENT

Both n-type and p-type Si(100) samples with resistivities of 2.5 to 8 ohm-cm were used. Group V dopant ions (Sb and As) were then implanted to the desired concentrations. In some cases, the samples were also implanted with B at an energy selected to yield a similar concentration distribution. The samples were annealed in a furnace at typically 600°C for 60 minutes in order to regrow the implanted layer via solid phase epitaxy. The substitutional fractions[3] of the group V-Si alloys were measured by 2 MeV He+ ion scattering/channeling techniques. Additional annealing for fixed durations at higher temperatures were also performed.

The annealing temperatures and times were chosen so that dopants formed a band of small precipitates. Profile broadening, or the lack of it, was determined by Rutherford backscattering analysis of Sb and As implanted Si with a depth resolution of 100-200 Å. The boron depth distribution was calculated by MARLOWE[4] since boron is not easily detected by ion scattering. The results of MARLOWE calculations for the Sb and As distributions were in quantitative agreement with the ion scattering measurements. The existence of the precipitates was determined by transmission electron microscopy (TEM), and the dopant diffusion coefficients were determined from the mean square radii of the precipitates.[2] In addition the presence of dislocation loops within the band was established, and their size, number density, and their nature were determined.

RESULTS

A depth distribution determined by Rutherford backscattering analysis of Si(100) implanted with 5×10^{15} Sb-cm^{-2} at 200 keV and furnace annealed at 600°C for 60 minutes, is shown in Fig. 1. Calculations using MARLOWE for 200 keV Sb in amorphous Si are also shown. The calculated dopant profile is in excellent agreement with the experimental results. With a density of 5×10^{22} lattice sites cm^{-3} the calculated interstitial profile shows that many interstitials are generated for each lattice site from the surface to well beyond the peak of the dopant profile.[5] Less than one displacement per atom is sufficient to amorphize Si at room temperature.[6]

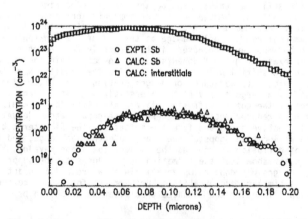

Fig. 1. Concentration profile of implanted Sb and interstitial Si for 200 keV Sb implanted to a dose of 5×10^{15} cm^{-2} into Si(100). The experimental data is from Rutherford backscattering analysis using 2 MeV He+, and the calculations are from MARLOWE.

After SPE the Sb substitutional fraction was typically greater than 0.90 indicating that most of the Sb was incorporated into electronically active substitutional sites. No broadening of the Sb profile was observed for the annealing conditions used in this study. However, a band of precipitates started to form over the depth of the highest Sb concentration after a 20 min annealing at 650°C. Considerably more precipitation is seen at slightly higher temperatures, and dislocation loops were observed to nucleate at the same depth as the precipitates. The Sb diffusion coefficients were found to be approximately five orders of magnitude above tracer values, and the activation energy was 1.8 eV, less than half of the tracer value of 4.08 eV.[7] This activation energy is low since point defects are not formed by the usual process, and it corresponds to a migration energy. The transient precipitation ends with only 15–20% of the dopant in the precipitates, although the remainder in excess of the solubility limit can be precipitated out by higher temperature annealing.

Samples implanted with both Sb and a comparable amount of B showed similar Sb precipitation. Well developed precipitates are shown in Fig. 2a. An Arhennius plot of the Sb diffusion coefficients obtained from the annealing of Sb/B implanted Si samples is shown in Fig. 3. The large enhancement of the diffusion coefficients and the reduction of the activation energy compared to tracer diffusion are clearly evident. However, unlike the pure Sb implantation no loop formation occurred in this case. Even after a 20 min annealing at 875°C loop formation was not observed, as shown in Fig. 2a. This indicates that the presence of boron has prevented the interstitials from condensing into loops, perhaps by aiding the interstitial diffusion to other sinks, such as the surface.

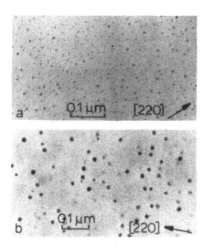

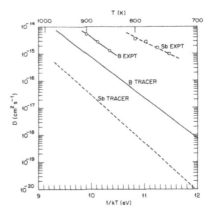

Fig. 2. TEM results for 5×10^{15} Sb-cm^{-2} implanted at 200 keV into Si(100). In (a) the sample was also implanted with 5×10^{15} B-cm^{-2} at 35 keV and annealed at 875°C for 20 min. In (b) the sample was implanted with 2×10^{16} B-cm^{-2} at 35 keV and annealed at 900°C for 20 min.

Fig. 3. Arhennius plot of Sb and B diffusion coefficients for Si(100) implanted with 5×10^{15} Sb-cm^{-2} at 200 keV. The Sb data is for a sample that was also implanted with 5×10^{15} B-cm^{-2}, as in Fig. 2, and the B data is for a sample that was also implanted with 2×10^{16} B-cm^{-2}, as in Fig. 2b.

For a significantly higher dose of implanted B the precipitation of Sb is suppressed. In Fig. 2b only B precipitates, readily distinguished from Sb, are seen. This is consistent with tracer diffusion; tracer diffusion predicts that even for annealing at 1000°C for one hour precipitates with only 10 Å radius will form, which is close to the visibility limit of diffraction contrast TEM. As in the Sb case, no loops were observed. It is interesting to note that the B diffusion showed a slight enhancement over tracer values, as shown in Fig. 3.

Implantation of As and B into Si samples was also performed. Without the B implantation substantial loop formation was seen at the projected range of As after annealing at 800°C for 20 min, as shown in Fig. 4a. No As precipitates were seen although As precipitation cannot be ruled out since if As precipitates as small clusters[8] they would be invisible to the TEM techniques employed. By implanting B after the As implantation and following with an identical annealing the loops are not observed, as shown in Fig. 4b. As in the Sb case the B implantation inhibits the formation of interstitial loops. A series of As depth distributions is shown in Fig. 5 for samples regrown by SPE by annealing at 600°C for 60 min. In Fig. 5a no B had been implanted. The relatively poor substitutionality correlates with TEM observations that a number of small projected range defects had already formed after SPE growth. By implanting a moderate amount of B prior to SPE a marked improvement in substitutionality is seen (Fig. 5b). TEM observations showed that no projected range defects had formed at this stage, as expected from Fig. 4b. Maximum substitutionality is seen for a high B dose implantation where we expect no trapped interstitials to survive SPE growth (Fig. 5c).

Fig. 4. TEM results for 1×10^{16} As-cm^{-2} implanted at 220 keV into Si(100). In (a) the sample was annealed at 800°C for 20 min. In (b) the sample was implanted with 5×10^{15} B-cm^{-2} at 35 keV and annealed at 800°C for 20 min.

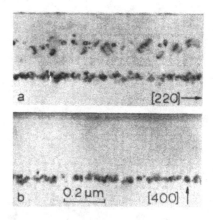

DISCUSSION

The dopant-Si trapping complex responsible for the transient diffusion and projected range loops is thought to be a simple dopant interstitial in the form proposed by Tan et al.[9] and Frank.[10] By extending their model to group V dopants, the relative stability of defect complexes for group V dopants and instability of complexes involving group III dopants are explained.[2] The basic feature of the model is that the dopant-Si-interstitial pair are bonded together and occupy a single tetrahedral site. The complex is formed during ion implantation and although the environment is amorphous the local structure still involves tetrahedral bonding similar to the crystalline case. The model predicts that the most stable configuration is for a neutral group V interstitial which is expected to be formed only in an n-type sample. The dopant-Si atoms are bound together by both a double covalent bond and an electrostatic bond due to polarization.

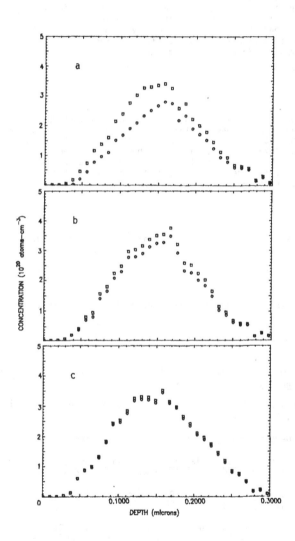

Fig. 5. As concentrations as a function of depth as determined by ion scattering/channeling analysis. The As was implanted at 220 keV for a dose of 1×10^{16} As-cm^{-2}. In (a) no B was implanted; in (b) and (c) B was implanted at 35 keV for doses of 5×10^{15} cm^{-2} and 2×10^{16} cm^{-2}, respectively. The boxes and circles represent the total concentration and substitutional concentration, respectively.

By following an Sb implantation with a B implantation at a higher dose the sample is expected to become p-type, and the neutral group V interstitial type of bonding would not form. The only possible group V interstitial would be a positively charged one, with no electrostatic bonding contribution. This would not be significantly more stable than self or group III interstitials and therefore would not survive SPE growth. The results for As are consistent with these ideas. However, the role of B in suppressing the loop formation in both the Sb and As case is not yet completely understood.

CONCLUSION

The present results for double dopant implantation support the trapping model for dopant-interstitial complexes formed during ion implantation. The double implantation technique, involving both group V and group III dopants, prevents the formation of the projected range defects, although this effect is not yet fully understood. Higher concentrations of the B component for the Sb-B double dopant implantation case prevents transient Sb precipitation, which is also consistent with the trapping model.

ACKNOWLEDGMENTS

We gratefully acknowledge the technical assistance of J. L. Moore, C. W. Boggs, and J. T. Luck.

REFERENCES

1. S.J. Pennycook, J. Narayan, and O.W. Holland, J. Crystal Growth 70, 597 (1984).

2. S.J. Pennycook, R.J. Culbertson, and J. Narayan, J. Mat. Res. (in press).

3. L.C. Feldman, J.W. Mayer, and S.T. Picraux, in Materials Analysis by Ion Channeling (Academic Press, NY, 1982).

4. M.T. Robinson and I.M. Torrens, Phys. Rev. B9, 5008 (1974); M.T. Robinson, Phys. Rev. B27, 5347 (1983).

5. R.J. Culbertson and S.J. Pennycook, Nucl. Instrum. Methods (in press).

6. B.R. Appleton, in Ion Implantation and Beam Processing, ed. by J.S. Williams and J.M. Poate (Academic Press, Sydney, 1984) p. 197.

7. R.B. Fair, in Impurity Doping Processes in Silicon, ed. by F.F.Y. Wang (North-Holland, NY, 1981) p. 315.

8. W.K. Chu, in Laser and Electron Beam Processing of Electronic Materials, ed. by C.L. Anderson, G.K. Celler, and G.A. Rozgonyi, Electrochem. Soc. Proc. 80-1 (1980) p. 361.

9. S.I. Tan, B.S. Berry, and W. Frank, in Ion Implantation in Semiconductors and Other Materials, ed. by B.L. Crowder (Plenum Press, NY, 1973) p. 19.

10. W. Frank, Rad. Effects 21, 119 (1974).

MEV ION INDUCED MODIFICATION OF THE
NATIVE OXIDE OF SILICON

R.L. HEADRICK* and L.E. SEIBERLING**
*Department of Materials Science and Engineering, University
of Pennsylvania, Philadelphia, PA 19104
**Department of Physics, University of Pennsylvania, Philadelphia,
PA 19104

ABSTRACT

We have studied the native oxide of silicon (110) and the changes
produced by MeV ion bombardment using transmission ion channeling of 5.9 MeV
^{9}Be, and Elastic Recoil Detection Analysis (ERDA). Transmission channeling
was used to measure interfacial nonregistered Si and adsorbed C and O.
ERDA was used to measure the surface concentration of H. MeV ions were
found to cause an increase in the interfacial nonregistered silicon
which saturates at approximately one monolayer. Rapid desorption of
hydrogen was observed. The effect of 2 keV electrons on the silicon
native oxide was also studied by Auger Electron Spectroscopy.

INTRODUCTION

Several discoveries of surface and interface phenomena induced by
high energy ions have recently led to a renewed interest in the modification
of solids by energetic ions in the MeV/amu energy range. Earlier studies
focused on beam damage caused by point defects produced by high momentum
transfer collisions between ions and stationary target atoms [1].
More recently, sputtering of solid surfaces has been studied extensively
[2,3], and desorption from solid surfaces has been reported [4]. In many
cases these effects appear to be initiated by electronic ionization
effects at or near the surface. Enhanced adhesion of thin metal films
to insulator, semiconductor, and metal substrates by MeV ions indicate
probable physical and chemical modification of interfaces induced by
the passage of these ions [5]. Direct information about the basic interfacial processes that accounts for these results is still lacking, although
either ionization or knock-on processes appear to be sufficient to produce
enhanced adhesion [6,7].
    In this paper we report on an investigation of the native oxide of
silicon (110) and its modification by 5.9 MeV ^{9}Be ion bombardment. The
basic structure of the native oxide has previously been established by
Elipsometry and ESCA as less than 1 monolayer of oxygen on the surface
in a chemical state distinct from SiO_2, covered by a physisorbed hydrocarbon layer [8]. Our results indicate that ion irradiation of the
surface induces a reaction of interfacial silicon atoms with the adsorbed
H, C and O. Hydrogen is found to desorb quickly. Bombardment of the
surface with 2 keV electrons significantly alters the Si (LVV) Auger
electron lineshape.

EXPERIMENTAL

    A device grade 1000 ohm-cm p-type (110) Si wafer was doped with
boron at 1000°C in N_2 for 90 min. The wafer was then cut into rectangular
pieces and a selective etch [9] was used to produce a 5000Å thick,
0.6 cm diameter single crystal window in the center of one piece. The
sample was cleaned using a method similar to the one described in

364

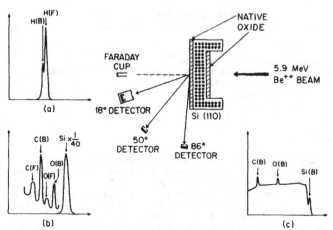

FIGURE 1   Schematic diagram of the scattering geometry. Representative
spectra (counts vs. energy) from the three surface barrier
detectors are shown. The spectra are from: a) the 18° detector,
b) the 50° detector, and c) the 86° detector.

reference 10, involving successive chemical oxidation and etching steps
with a final etch in dilute HF. The sample was then mounted on a two
axis goniometer and loaded into the UHV scattering chamber within 30
min. The chamber was baked at 120°C for 24 hours. After cooling, the
pressure in the scattering chamber was 3 x 10^{-9} Torr. A 5.9 MeV ^{9}Be
ion beam was produced by the Penn tandem Van de Graaff accelerator and
collimated to less than 0.15 degrees angular divergence before entering
the scattering chamber. No increase in the pressure was observed when
the beam was put on target, however it should be noted that in a separate
run in which a base pressure of 2 x 10^{-10} Torr was obtained (by a
longer bakeout) the pressure in the scattering chamber increased to
approximately 1 x 10^{-9} Torr when the beam was put on target. ^{9}Be ions
passing through the sample were collected in a Faraday cup and used for
current integration. The charge state of the ^{9}Be ions increased from
+2 to an average of +3.35 after passing through the sample. The beam
spot size was 2.3 mm^2 and the beam current was less than 10 particle
nA. The sample was aligned in the ⟨110⟩ crystallographic direction to
a minimum yield of less than 5%. After alignment, the beam was moved
to a fresh spot and the sample was bombarded with consecutive doses of
7 x 10^{15} ions/cm^2. The spectra collected for each dose were used to
analyze the surface modifications produced by the ^{9}Be ion bombardment.

Three surface barrier detectors were used simultaneously to measure
the amounts of nonregistered Si and adsorbed H, C and O on the back
(beam exit) surface of the thin window. Figure 1 shows a schematic
drawing of the scattering geometry, along with representative spectra
obtained from each of the three detectors. The detector at 86° was used
to measure the number of Si atoms that have moved moved laterally from
registered to nonregistered positions. In the transmission channeling
geometry, ions scattered from nonregistered Si atoms on the back surface
have a higher energy than those scattered from atoms in the bulk because
of the difference in energy loss of channeled versus nonchanneled ions
[11]. Thus, the silicon surface peak is at a higher energy than the
bulk signal. The detector at 50° was used to measure the amount of
adsorbed C and O on both the front and back surface of the thin window.
Finally, the detector at 18° was used to measure the hydrogen elastic

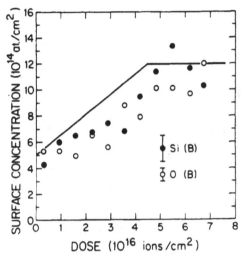

FIGURE 2   The concentration of nonregistered Si and adsorbed O on the
back surface as a function of beam dose. The solid line shows
the behavior of the Si peak during ion bombardment at a higher
pressure [13].

recoils. A thin Havar [12] foil was placed in front of the detector to
stop the (far more abundant) scattered ^{9}Be ions. The H on the front
and back surfaces of the thin window are barely resolved.

Auger measurements were performed in a separate surface analytical
chamber with a base pressure of 5×10^{-11} Torr which is equipped with a PHI
model 10-155 cylindrical mirror analyzer (CMA) and an integral electron
gun. An electron energy of 2 keV and beam current of 1 uA were used.
The same chemical cleaning procedures used in the ion scattering experiments were used to prepare the sample before loading into the vacuum
system. Auger Electron Spectroscopy was used to investigate the chemical
state of the silicon surface both before and after electron irradiation.

RESULTS

In figure 2, we present the concentration of Si and O on the back
surface as a function of beam dose. The closed circles represent the
number of Si atoms that have moved laterally from registered positions
to nonregistered locations, and the open circles represent the number of O
atoms (assumed to be in random locations) on the surface. The solid
line shows the behavior of the Si peak during ion bombardment at a pressure
of 2×10^{-7} Torr (from reference 13). At the higher pressure, a clear
saturation was observed above a dose of 5×10^{16} ions/cm^2. It is not
obvious from the present data whether or not a saturation has occured.
The O and Si concentrations increase at approximately the same rate, the
ratio between them being approximately one throughout the experiment.
The behavior of the C and H concentration during ion bombardment is
displayed in Figure 3. The solid circles represent the number of C
atoms on the back surface. It is clear that the C concentration remains
constant at roughly 2.9×10^{15} atoms/cm^2. The open circles represent
the sum of the number of H atoms on the front and the back surfaces
divided by two. This procedure was used because the front and back H
peaks were not completely resolved (see Figure 1). An analysis of the H

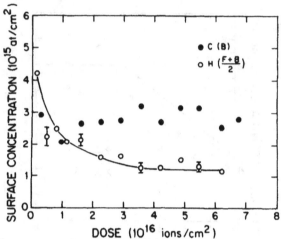

FIGURE 3   The concentration of adsorbed C on the back surface, and the sum of the concentration of H on the front and back surfaces divided by two, as a function of beam dose.

peaks separately revealed roughly twice as many counts in the front as in the back peak. It is not known whether the amounts of H on the two surfaces are actually different, or whether the difference in the peak areas is caused by channeling effects. The scattering cross section for Be ions with energies near 5.9 MeV on H has been measured to be 1.82 times the the Rutherford cross section [14], and is 11% larger on the back surface.

The Auger results are presented in figure 4. Figure 4a is the Si LVV spectrum of a UHV clean Si surface. Figure 4b is the Si LVV spectrum of a chemically cleaned and etched Si surface, and 4c is the spectrum of the same surface after exposure to 2 keV electrons for 60 min. Several changes have occured during the electron irradiation, including the appearance of a new peak at about 71 eV. No change was observed in the carbon or oxygen KLL peak heights even after a 120 minute electron exposure. This differs from the ion channeling results, since a significant amount of excess oxygen was observed to accumulate on the surface during ion bombardment. However, the Auger measurements were carried out at a much lower pressure, and took less time than the ion scattering measurement. Finally, 4d is the spectrum of a chemically oxidized silicon surface. The spectrum is characteristic of SiO_2.

DISCUSSION

Ultra-high vacuum studies of oxygen adsorption onto silicon surfaces using Low Energy Electron Diffraction (LEED) and Auger Electron Spectroscopy (AES) have shown that the oxygen coverage saturates at slightly more than one monolayer, and that the LEED pattern is slowly extinguished with increasing oxygen coverage. The ion channeling results presented above, when extrapolated to zero dose, show that for the case of the native oxide, 5×10^{14} at/cm^2 of silicon is disordered at the surface with one oxygen for every nonregistered silicon. Thus, the coverage of oxygen in the native oxide is only half of the saturated coverage of O_2 onto clean silicon and the amount of displaced silicon is significantly less than expected for simple O_2 adsorption.

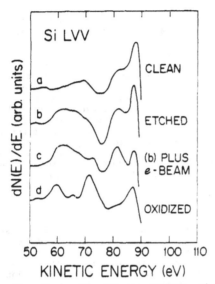

FIGURE 4 Si LVV spectra of several silicon surfaces. The preparation of the individual surfaces is described in the text.

The Auger results presented above can be compared directly with those of Garner et al. [15] for oxygen adsorption onto UHV cleaved silicon. Curve (b) in figure 4 corresponds closely to a very low exposure (10^6 L), and does not resemble the spectrum for the highest coverages (10^{12} L). It was proposed in references 15 and 16 that the low exposure data of Garner et al. corresponds to oxygen adsorbed on the surface in a molecular state (either O_2 or OH from water vapor contamination). Since the surface of our sample was exposed to a wet etch, it is reasonable to expect that the surface is saturated with dissociated H_2O. Ibach et al. [16] proposed that H_2O dissociates into OH and H on the silicon surface. Si in SiH is not expected to be displaced significantly from its lattice site [17]. If the Si in SiOH is displaced significantly, then 1/2 monolayer and 1 monolayer of oxygen and hydrogen, respectively, are expected to be present on the surface, with 1/2 monolayer of silicon displaced from lattice sites, consistent with our ion channeling data. This model of the native oxide, however, is likely to be an oversimplified one in light of the very large hydrocarbon concentration present on the surface.

The native oxide is sensitive to both electron and MeV ion irradiation. The Si LVV Auger spectrum of the electron irradiated silicon native oxide surface (fig. 4c) shows significant differences from the unirradiated surface (fig. 4b). The new peak observed at about 71 eV has been observed in connection with atomic rather than molecular adsorption of oxygen on Si [15]. Since no change in the carbon or oxygen peak heights was observed, this modification must have been caused by oxygen already present on the surface changing from a molecular to an atomic state.

Ion bombardment of a thin thermal oxide on Si does not produce the rapid initial increase of nonregistered Si observed for the case of a native oxide [13]. Also, ion bombardment of a native oxide of Si at 2×10^{-7} Torr shows a clear saturation of nonregistered Si above a dose of approximately 5×10^{16} ions/cm^2 [13]. These facts argue against a knock-on process being responsible for the production of nonregistered Si. Furthermore, beam heating is not thought to be important because beam currents differing by a factor of two were previously shown to produce the same rate of disorder production [13]. We suggest that ionization induced release of hydrogen allows subsequent reaction of silicon with adsorbed molecules present on the surface.

In the present study, H loss is associated with an increase of the surface Si and O peaks. Hydrogen bonded directly to the Si surface may be released and free up Si bonds that subsequently re-form with neighboring atoms or oxygen adsorbed from the vacuum. Hydrogen release from OH groups on the surface may also explain the apparent change of adsorbed oxygen from a molecular to an atomic state. It is interesting to note that the threshold dose for ion beam enhanced adhesion for most metals on Si is less than 1×10^{16} ions/cm^2 [5]; at this dose only the H concentration has changed significantly. H released from the hydrocarbon layer may allow bonding to occur across the metal-silicon interface and improve adhesion. Thus, ionization induced hydrogen release may account for the observations reported in this paper as well as the ion beam enhanced adhesion effect.

CONCLUSIONS AND ACKNOWLEDGEMENTS

In this paper, we have shown that the native oxide of silicon (110) contains 5×10^{14} oxygen atoms/cm^2 on the silicon surface, with one nonregistered silicon atom for each oxygen atom. The data are consistent with each surface silicon atom being bonded to either OH or H from dissociated H_2O. Both MeV ion and keV electron irradiation cause silicon to react with adsorbed molecules. Ionization induced hydrogen desorption may be responsible for the surface reactions.

This work was supported by the NSF [PHY-8213598] and the IBM corporation.

REFERENCES

1. G.H. Kinchin and R.S. Pease, Rep. Prog. Phys. 18, 1 (1955).
2. W.L. Brown, L.J. Lanzerotti, J.M. Poate and W.M. Augustyniak, Phys. Rev. Lett. B 40, 1027 (1978).
3. J.E. Griffith, R.A. Weller, L.E. Seiberling and T.A. Tombrello, Rad. Eff. B 51, 223 (1980).
4. J.P. Thomas, M. Fallavier, J. Tousset, Nucl. Instr. Meth. B 187, 537 (1981).
5. M.H. Mendenhall, Ph.D. Thesis, Caltech (1983).
6. I.V. Mitchell, J.S. Williams, D.K. Sood, K.T. Short and S. Johnson, Mat. Res. Soc. Symp. Proc. B 25, 189 (1984).
7. J.E.E. Baglin, G.J. Clark and J. Bottiger, Mat. Res. Soc. Symp. Proc. B 25, 179 (1984).
8. S.I. Raider, R. Flitsch and M.J. Palmer, J. Electrochem. Soc.: Solid State Science and Technology, p. 413, March, 1975.
9. N.W. Cheung, Rev. Sci. Instrum. 51, 1212 (1980).
10. A. Ishizaka, K. Nakagawa and Y. Shiraki, 2nd International Symposium on Molecular Beam Epitaxy, Tokyo 1982, p. 183.
11. L.C. Feldman, P.J. Silverman, J.S. Williams, T.E. Jackman and I. Stensgaard, Phys. Rev. Lett. B 41, 1396 (1978).
12. Havar is a product of Hamilton Technology Inc., Lancaster, PA.
13. L.E. Seiberling and R.L. Headrick, to be published in the Proceedings of the American Chemical Society, Pottsdam, NY, (June 1985).
14. F.S. Moser, Phys. Rev. B 104, 1386 (1956).
15. C.M. Garner, I. Lindau, C.Y. Su, P. Pianetta and W.E. Spicer, Phys. Rev. B 19, 3944 (1979).
16. H. Ibach, H.D. Bruchmann and H. Wagner, Appl. Phys. B 29, 113 (1982).
17. R.J. Culbertson, L.C. Feldman, P.J. Silverman and R. Haight, J. Vac. Sci. Technol., 20 (1982) 868.

ION BEAM DEPOSITION OF MATERIALS AT 40–200 eV: EFFECT OF ION ENERGY AND
SUBSTRATE TEMPERATURE ON INTERFACE, THIN FILM AND DAMAGE FORMATION*

N. HERBOTS, B.R. APPLETON, S.J. PENNYCOOK, T.S. NOGGLE, AND R.A. ZUHR
Solid State Division, Oak Ridge National Laboratory, Oak Ridge TN 37831

ABSTRACT

Ion beam deposition (IBD), the process whereby magnetically analyzed
ions are directly deposited on single crystal substrates, has been studied
for ^{74}Ge and ^{30}Si ions on Si(100) and Ge(100). The effects of sputter-
cleaning prior to deposition and substrate temperature during deposition
were investigated. Three analytical techniques were systematically used to
obtain information on the deposited films: (1) Rutherford backscattering
combined with ion channeling, (2) cross-section TEM, and (3) Seeman-Bohlin
X-ray diffraction. In the energy range explored (40–200 eV), the width of
the interface between the IBD film and the substrate was found to be always
less than 1 nm. Each IBD layer was highly uniform in thickness and composition for deposition temperatures from 300 K to 900 K. Without prior
sputter-cleaning and annealing of the Si(100) and Ge(100) substrates, no
epitaxy was observed. UHV conditions were found to be a requirement in
order to grow crystalline Si films presenting bulk-like density. This was
not the case for Ge films which showed bulk-like density for IBD at higher
pressures. Results on the first Si/Ge superstructure grown by IBD are also
shown.

INTRODUCTION

The direct deposition of low energy ions (IBD) on single crystal
substrate has come into attention in recent years in several laboratories.
[1-7] However, the central issue has often been a reliable, efficient,
and uniform source of low energy ions, and the properties of the deposits
varied often according to the type of apparatus used.
In this paper we present results recently obtained with the first
IBD apparatus using both: (1) an intense ion source provided by an ion implanter, combined with an electrostatic lens that decelerates the ions to
the selected energy and, (2) a UHV deposition chamber that allows in-situ
surface cleaning of the substrates, and maintains a base pressure of 10^{-10}
Torr. More details about this system can be found elsewhere.[7,8]

INTERESTS IN THE IBD CONCEPT

The ability to deposit materials in UHV directly from an ion beam
that is mass and energy analyzed presents several attractive features. The
well-defined low energies (<200 eV) ensure minimal damage, and favor a
high uniformity of the deposit. The precise selection of the deceleration
voltage allows optimization of the ion energy to avoid detrimental effects
such as ion self-sputtering[7] or to favor desirable ones like epitaxial
growth. For the latter case, the use of the ion energy rather than thermal energy to promote layer reordering may permit epitaxial growth to occur
at lower temperatures than other techniques, such as molecular beam epitaxy
(MBE). The use of an ion implanter also provides complete versatility in
selecting the deposited species, and accurate control of the deposited dose.

*Research sponsored by the Division of Materials Sciences, U.S. Department
of Energy under DE-AC05-84R21400 with Martin Marietta Energy Systems, Inc.

An obvious application of IBD is the preparation of semiconductor devices, since the high selectivity of the ion source combined with UHV conditions ensures very high purity deposits. The possibility of switching from one ion to another during deposition, the accuracy of the dose, and the occurrence of epitaxial growth at low temperatures make IBD especially suited for the preparation of quantum wells for electronic and optical devices.[9] On a more fundamental point of view, the fact that IBD films are monoisotopic provides new opportunities to explore major issues in material research. For instance, IBD of the Si isotope of mass 30 on a Si single crystal (mass = 28.086 a.m.u.) allows mass separation by ion scattering between the deposit and the substrate, as can be seen in Fig. la. This unique feature provides the opportunity for the study of mass transport and atomic redistribution at the ^{30}Si/Si(100) interface during the epitaxial growth occurring while deposition is taking place, or the solid phase regrowth induced under subsequent annealing.

The purpose of this paper is to present experimental evidence of some of the many possibilities of the IBD technique described above, and to clarify the experimental conditions necessary to obtain specific properties in IBD films.

EXPERIMENTAL RESULTS

^{74}Ge and ^{30}Si ions were deposited at energies ranging between 40 eV and 200 eV on Si(100) and/or Ge(100). In initial studies the surface oxide naturally present on the substrates was not removed. Two base pressures were studied to examine the effect of UHV conditions on IBD thin film growth: 10^{-10} Torr and 10^{-9} Torr (mainly residual hydrogen). The pressure in each case rose by one order of magnitude during deposition because the ion beam, with a cross-section of 1.5 cm^2, created an important gas load. Thin film growth was observed as a function of ion energy and substrate temperature.

Fig. 1b (^{30}Si on Si) and Fig. 2 (^{74}Ge on Si) show the typical characteristics of IBD films examined by cross-section TEM: highly uniform, perfectly continuous, with a sharp interface. Black streaks known as extinction lines, are often observed below the deposits. These are typically observed by TEM in good quality single crystals because the extensive thinning required for transparency induces macroscopic strains in the section.

Ge films deposited at 10^{-9} Torr and 10^{-10} Torr and Si films grown at 10^{-10} Torr grew amorphous at 300 K and presented a polycrystalline, columnar growth at 700 K and 900 K. Preferential orientation along the <110> direction in the polycrystalline films were similar to what is observed on layers grown by chemical vapor deposition on silicon dioxide. Fig. 3 shows the effect of substrate temperature on the IBD film microstructure for the case of ^{30}Si IBD on Si(100) at a base pressure of 10^{-10} Torr, measured by X-ray diffraction under grazing incidence (Seeman-Bohlin geometry, also called thin film camera).

The IBD film density was calculated by dividing the deposited dose, measured by ion scattering, by the geometric width of the layer, measured by cross-section TEM. The density of the Ge films was found in all cases to be equal to the bulk density of single crystal Ge, even for layers as thin as 3 nm.[7] No difference was observed between ^{74}Ge films deposited at the two base pressures studied. In the case of Si, however, the density and the microstructure of the films were markedly dependent on the deposition pressure, as shown in Fig. 4. For a base pressure of 10^{-9} Torr rising during the deposition to 10^{-8} Torr, the atomic density was found to be 3.75 × 10^{22} at/cm^3 in deposits made at 300 K (curve a) as well as at 900 K (curve b). This value represents only 75% of Si bulk density and is typically found in amorphous, hydrogenated Si films. X-Ray diffraction revealed that the IBD films were indeed amorphous. Since these Si films

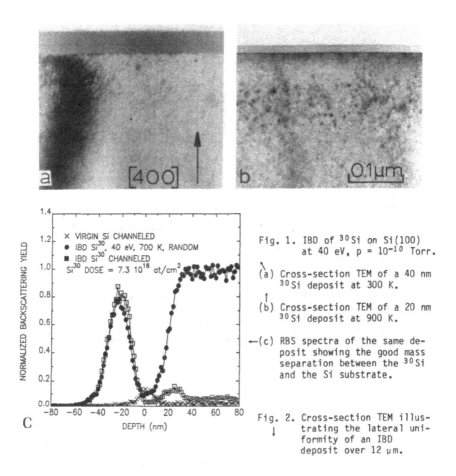

Fig. 1. IBD of ^{30}Si on Si(100)
at 40 eV, p = 10^{-10} Torr.

(a) Cross-section TEM of a 40 nm
^{30}Si deposit at 300 K.

(b) Cross-section TEM of a 20 nm
^{30}Si deposit at 900 K.

←(c) RBS spectra of the same deposit showing the good mass
separation between the ^{30}Si
and the Si substrate.

Fig. 2. Cross-section TEM illustrating the lateral uni-
formity of an IBD
deposit over 12 µm.

ION BEAM DEPOSITION AT 40 eV OF ^{74}Ge ON Si (100) AT ROOM TEMPERATURE

had been deposited at a higher base pressure, Auger spectroscopy combined
with Ar$^+$ 500 eV Ar$^+$ sputter-profiling performed at a base pressure of 5 ×
10^{-11} Torr was used to study possible contaminants such as carbon, oxygen
and stainless steel. No trace of contaminants was found in the IBD Si
films, except for oxygen at the ^{30}Si/Si interface, due to the native oxide
at the original Si surface. When the IBD depositions were performed at a
base pressure of 10^{-10} Torr, the Si films exhibited an atomic density of
5 × 10^{22} at/cm^3 equal to the bulk density of single crystal Si, and their
microstructure was then dependent on the deposition temperature, as shown
in Fig. 3. These observations are consistent with the hypothesis that Si
deposited at a base pressure of 10^{-9} Torr reacts with the hydrogen found to
be the main residual gas at this vacuum, so that Si:H films are formed

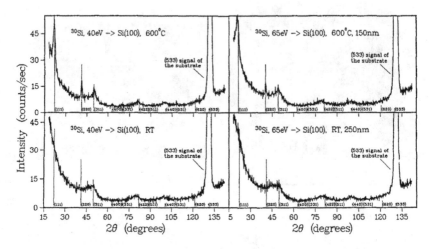

Fig. 3. X-Ray diffraction on ^{30}Si films on Si(100) deposited at a base pressure of 10^{-10} Torr. In (a), the ion energy was 40 eV, and two deposits are shown: one at 300 K and 700 K. In (b), the energy used was 65 eV. For comparable deposition rates, the grain size along the <111> direction is 50% larger (9 nm) at 40 eV (a) than at 65 eV (b) where it is only 6 nm. In the <110> direction the average grain size is 12 nm for both (a) and (b).

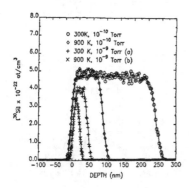

Fig. 4.

RBS spectra of IBD ^{30}Si on Si, illustrating the effect of the deposition pressure on the ^{30}Si film density. For each pressure, 300 K and 900 K deposits have been performed, showing that the substrate temperature does not affect the density.

instead of pure Si deposits. This indicates that compound films can be formed by leaking one of the components into the chamber during ion beam deposition. None of these effects were observed with ^{74}Ge.

The interface width between the deposits and the substrate, measured by cross-section TEM, was found to be less than 1 nm for all the IBD conditions used, as can be seen in Fig. 1a, 2 and 5. These characteristics were found to be the same for films grown at 40 eV and 200 eV. This is in agreement with the range straggling values obtained with the Monte-Carlo simulation code TRIM. The number of defects created by Si and Ge ions was found to decrease rapidly with energy, but anomalous damage behavior was observed at 700 K and 900 K. (A detailed analysis of the damage creation in single crystal substrates for the present IBD experiments can be found in [7]).

Fig. 5 shows the first Ge/Si multilayer structure achieved by IBD. Cross-section TEM and ion channeling measurements established that the properties found for the deposits mentioned above were also obtained in this more complex structure: sharp interface definition, good lateral uniformity, accuracy of the dose, and high compositional purity.

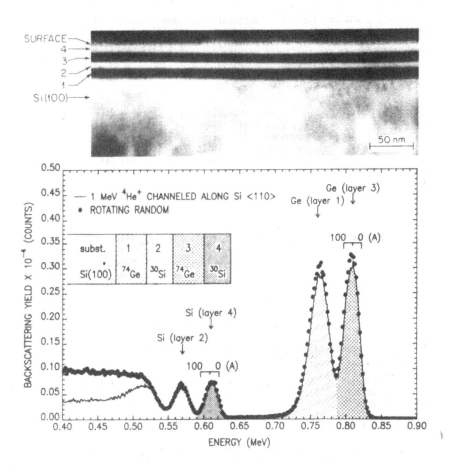

Fig. 5. Ge/Si multilayer obtained by IBD with 65 eV ions at 300 K. The structure of the sample is shown in the lower left. The cross-section TEM (upper part) illustrates the sharpness of the several interfaces. RBS spectra below show the modulation in composition obtained by switching from ^{74}Ge to ^{30}Si during IBD.

CONCLUSIONS

The present experimental results illustrate some of the attractive properties of IBD for material preparation. For base pressures equal to 10^{-9} Torr and below, highly uniform, continuous IBD layers can be grown, and the coalescence seen for films deposited at 10^{-7} Torr[2] is not observed. The data also establish that the presence of a native oxide impedes epitaxial growth of IBD of Ge and Si on Ge(100) and Si(100) respectively, even at temperatures as high as 900 K under UHV conditions.
Further IBD experiments performed on sputter-cleaned and annealed substrates provided epitaxial growth of ^{30}Si and ^{74}Ge on Si(100) and Ge(100). They will be presented elsewhere.[8,10,11] All these experiments confirm the data of P. C. Zalm et al.[3] on Si, which contradicted the results claimed in [5] and [6].
The present results also demonstrate the possibility of forming multilayered structures by IBD. Experiments combining the present IBD results on the uniformity of very thin films, the feasibility of multi-layer deposition, and the later results on epitaxy are presently underway.

ACKNOWLEDGMENTS

It is a pleasure to acknowledge G. Carter, I.H. Wilson, and P.C. Zalm for helpful discussions. The authors wish to thank D.M. Zehner and G.W. Ownby for giving them access to their Auger facility. They also wish to thank J.Z. Tischler and J.D. Lewis the use of their X-ray camera. They also gratefully acknowledge J.T. Luck and C.W. Boggs for their patience and skill during the preparation of the cross-sections for the TEM and J.L. Moore and D.K. Thomas for their active help in the construction and maintenance of the IBD apparatus.

REFERENCES

1. T. Tsukizoe, T. Nakai, N. Ohmae, J. Appl. Phys. 48, 4770-4776 (1976).

2. J. Amano, et al., J. Vac. Sci. Technol. 13 (2), 591-595 (1976).

3. P.C. Zalm and L.J. Beckers, Appl. Phys. Lett. 41 (2), 167-169 (1982).

4. D.G. Armour, P. Bailey, P.A. Judge, G. Sharples, P. Byers, to be publ.

5. K. Yagi, S. Tamura, T. Tokuyama, Jap. J. Appl. Phys. 16, 245-251 (1977).

6. T. Tokuyama, K. Yagi, K. Miyake, M. Tamura, N. Natsuaki, and S. Tachi, NIM 182/183, 241-250 (1981).

7. N. Herbots, B. R. Appleton, T.S. Noggle, R. A. Zuhr, and S. J. Pennycook, Proc. 11th Int. Conf. At. Coll. in Sol., Georgetown Univ., Washington DC, Aug. 4–9, to be published in NIMB (1986).

8. T.S. Noggle, B.R. Appleton, R.A. Zuhr, and N. Herbots, to be published.

9. L. Esaki, Semiconductor Superlattices, in MBE and Heterostructures, ed. by L.L. Chang and K. Ploog (Martinus Nijhof Publ., 1985) p.p. 1–36.

10. N. Herbots, T.S. Noggle, B.R. Appleton, R.A. Zuhr, and S.J. Pennycook, 5th Int. Conf. on Ion Beam Modif. of Mater., 9-13 June 1986, Catania.

11. B.R. Appleton, R.A. Zuhr, S.J. Pennycook, T.S. Noggle, and N. Herbots, ibid. [10].

TEMPERATURE AND DOSE DEPENDENCE OF AN AMORPHOUS
LAYER FORMED BY ION IMPLANTATION

Eliezer Dovid Richmond and Alvin R. Knudson
Naval Research Laboratory, Code 6816, Washington, D.C.

ABSTRACT

A model is formulated to predict the width of an amorphous layer in Si produced by ion implantation. The dependency of the amorphous Si layer width on the ion implantation energy, dose, and temperature is computed.

INTRODUCTION

Over the past 15-20 years, an extensive amount of research has been performed on the production and annealing of ion implantation damage in Si.[1,3b] This research focused on understanding the dependency of the ion implantation damage on various experimental parameters such as ion mass, substrate temperature, ion energy, and dose rate, and on the transition to the amorphous state. The prediction of the position and width of an amorphous layer formed by the ion implantation has not been formally addressed.

In today's microelectronics, an amorphous layer is used in many technological processes. For silicon on sapphire, an amorphous Si (a-Si) layer formed by self-implantation is used to significantly improve the crystalline quality of the silicon film.[2] Forming a pre-amorphous layer, before performing the doping implantation for electronic devices, reduces the secondary defects and increases the dopant activation efficiency.[3] To optimize these techniques requires predicting the position and width of the amorphous layer as a function of substrate temperature and of ion implantation fluence.

In this paper, we will present a model for predicting the width of the a-Si layer formed by ion implantation. The a-Si layer width will be parameterized as a function of the substrate temperature and dose for two ion energies. All the results reported here apply to self-implantation in Si, but easily can be extended to other implantation species. The calculations are in excellent agreement with experimental results. The position of the peak of the deposited damage and its profile shape will be briefly discussed.

EXPERIMENT

The substrates consisted of 2" Czochralski grown intrinsic (100)Si wafers. The wafers were implanted with [28]Si[+] ions at an energy of 55 keV or 120 keV using a Research Model Veeco 300kV ion implanter. The current density in all cases was less than $0.22\mu a/cm^{-2}$. The fluence was either $1x10^{15}cm^{-2}$ or $3x10^{15}cm^{-2}$, depending on the substrate temperature. The substrates were thermally affixed to the substrate holder to ensure good thermal contact. The substrate temperature was directly sensed by attaching a 0.001" diameter chromel-alumel thermocouple to the front surface of the substrate. The thin diameter thermocouple wire was used to avoid localized temperature variations. The temperature of the substrate varied from 94°K to 342°K. The samples were analyzed by ion channeling of 1 MeV ^{3}He[+] ions performed with the Naval Research Laboratory Van der Graaff generator. The ^{3}He[+] beam was incident normal to the sample surface so that channeling occurred along the (100) axis. A silicon surface barrier detector counted the backscattered ions at an angle of 165°. The offset of the detector was near zero. The amorphous layer widths were measured from the dechanneling spectra by extrapolating to the random level as discussed in ref. [2].

THEORETICAL MODEL:

To predict the width of the amorphous layer as a function of temperature and dose requires three components: 1) a criterion for determining the occurrence of the crystalline to amorphous (c -> a) phase transformation; 2) a means of computing the deposited damage energy profile at absolute zero; and 3) a method to account for the temperature dependent vacancy outdiffusion shrinkage of the amorphous regions formed along each ion track. Each of these components will be discussed below.

The criterion at which a c -> a phase transformation occurs is predicted by the critical energy density model[4]. This model states that a c -> a transformation occurs from the increase in lattice energy caused by the displacement of the lattice atoms. At a critical deposited energy, Ec, a phase transformation of the highly damaged region becomes energetically favorable. This critical deposited energy[4b] equals $6 \times 1023 eV/cm-3$ for all bombarding ions ranging from B to Sb in Si.

The energy deposited into structural damage, $D_n(E,z)$, is referred to as deposited damage energy, damage energy, or just sometimes damage. E is the energy of the incident ion, and z is the penetration into the target in a direction normal to the surface. For semiconductors, it is assumed that the energy transferred to the electronic processes does not cause subsequent structural damage. Nonetheless, the electronic energy loss may result in enhanced annealing, enhanced diffusion, or an irradiation temperature dependence of the damage production. The tables[5] used to calculate the damage profiles presented here are based on an extension of the LSS[7] theory by Brice[8],[9]. These results apply only to amorphous targets or for ions incident along non-channeling directions in crystalline targets, and for cases in which almost all of the ions penetrate well into the target.

The temperature dependence of the a-Si layer width is accounted for by the model of Morehead and Crowder[6]. The model assumes that the implanted ions form amorphous cylindrical regions which overlap to form a continuous amorphous layer. The radius of the cylinder is temperature dependent. The temperature dependence results from the enhanced diffusion of vacancies during a time of the order of 10^{-9} sec. Using this model a temperature factor $F(T)$ is defined by

$$F(T) = \{1 - K' (dE/dx)^{-1/2} e^{-U/kT}\}^{-2} \tag{1}$$

where $K' = 115 \ (keV/\mu m)^{1/2}$ and $U = 0.06$ eV for Si[6]. dE/dx is the energy independent nuclear energy loss per unit path length estimated from the theory of Nielson[6]. For ^{28}Si, $dE/dx = 474$ kev/μm. $F(T)$ equals 1 at absolute zero in the abscence of vacancy outdiffusion, and goes to infinity at some critical temperature. For self-implantation into Si, this critical temperature equals 418°K.

Combining these three theories, the width W of the a-Si layer can be computed from

$$W = z_2 - z_1 \qquad z_2 > z_1 \tag{2}$$

where z_1 and z_2 are determined from

$$D_n(E, z_{1,2}) = F(T) E_c/\phi \tag{3}$$

and ϕ is the fluence of the ion implantation. When D_n has its maximum value, eq. (3) determines the fluence at absolute zero ($F(T)=1$) at which no amorphous layer is formed ($z_1 = z_2$).

RESULTS AND DISCUSSION:

The deposited damage energy is shown in Figure 1 for the $^{28}Si^+$ incident ion energies of 55 keV and 120 keV. The peak of the damage occurs at .59 Rp = 454Å for E = 55 keV and at 0.62 Rp = 1148Å for 120 keV. R_p is the projected range. The peak height of the damage increases with decreasing ion energy. Therefore, at lower ion energies, a finite amorphous layer results both for higher substrate temperatures and for lower doses. In addition, the damage profile becomes narrower with decreasing ion energy. Thus, a thinner amorphous layer width occurs at lower ion energies.

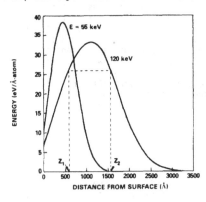

FIGURE 1
Deposited damage energy distributions for ^{28}Si ion implantations into Si at incident ion energies of 55 and 120 keV.

The temperature factor F(T) as a function of temperature is shown in Figure 2 for ^{11}B, ^{28}Si, and ^{31}P. The change in nuclear energy loss between ^{28}Si and ^{31}P is sufficiently large that significant errors occur in the predicted a-Si layer width, especially above room temperature.

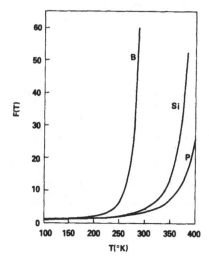

FIGURE 2
Temperature factor F(T) as a function of T for ^{11}B, ^{28}Si and ^{31}P.

Figure 3 shows the amorphous layer width as a function of temperature for the incident ion energies of 55 keV and 120 keV and fluences of 1×10^{15} cm^{-2} and $3 \times 10^{15} cm^{-2}$. The dashed curve in Figure 1 illustrates the solution of eqs. (2) and (3) to determine the a-Si layer width. For the case illustrated in Figure 1, T_S = 300°K and ϕ = 1x10¹⁵cm⁻². The a-Si layer width increases sharply as the temperature decreases, and then asymtotically approaches its value at absolute zero. As discussed above, a nonzero amorphous layer width exists at higher temperatures as the implantation ion energy is lowered.

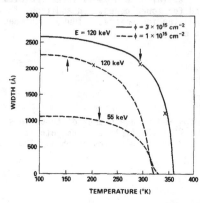

FIGURE 3

Amorphous layer width as a function of temperature for the incident ²⁸Si ion energies of 55 and 120 keV and for fluences of 1x10¹⁵cm⁻² and 3x10¹⁵cm⁻². The X's represent experimental data. The temperature below which the amorphous layer reaches the surface is indicated by the vertical arrow for each fluence and ion energy.

The damage depends more strongly on the incident ion energy than on the fluence. For example, at 200°K, the amorphous layer width increases from 1022Å to 2079Å, when the incident ion energy is doubled. On the other hand, when the dose at 120 keV is tripled, the layer width increases only by a factor of 0.2. These results are valid below the temperature where the a-Si layer width sharply increases.

The layer width expands as the temperature is lowered for a given ion energy and fluence. At a certain point, the layer meets the surface and expands only toward the interior of the target. The arrow in Figure 3 indicates this temperature in each case. For the 120 keV implantation, this temperature is 148°K for ϕ = 1x10¹⁵cm⁻² and 284°K for ϕ = 3x10¹⁵cm⁻². At 55 keV and a fluence of 1x10¹⁵cm⁻² the amorphous layer reaches the surface for T_S = 211°K.

The layer width as a function of fluence is plotted in Figure 4. The substrate temperature is 300°K. The a-Si layer width increases with fluence. For high fluences, the a-Si layer width increases slowly owing to the long tail in the damage profile. The lower energy implantations have finite layer widths for lower fluences, owing to the higher peak of the damage profile. The amorphous layer reaches the surface (indicated by the arrow) for a fluence of 2.8x10¹⁵cm⁻² and 3.9x10¹⁵cm⁻² for an ion energy of 55 keV and 120 keV, respectively.

To verify the model, intrinsic (100)Si wafers were self-implanted at various temperatures. Figure 5 illustrates the results. The width of the amorphous layer was measured by extrapolating the edge of the dechanneled spectra to the random level, as indicated in Figure 5b. For comparison with the model, these experimental results are indicated in Figure 3 by the "X's". Excellent agreement is found except at the highest temperature where the accuracy of the temperature factor F(T) is most critical.

Directly sensing the substrate temperature was invaluable. For example, the substrate holder temperature was 77°K, but the sample temperature was 94°K. Additionally, beam heating effects could be directly monitored. For all cases, the sample temperature never rose more than 2°K during the implantation.

The model contains certain assumptions. One assumption presumes that

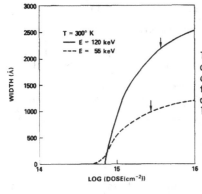

FIGURE 4

The amorphous layer width as a function of fluence for incident ^{28}Si ion energies of 55 and 120 keV at a substrate temperature of 300°K. The vertical arrows indicate the fluences above which the a-Si layer reaches the surface.

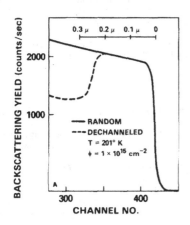

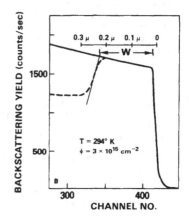

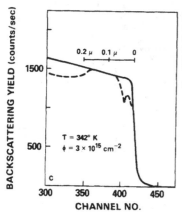

FIGURE 5

Dechanneling spectra of damage layers formed by self-implantation at substrate temperatures of a) 201°K, b) 294°K, c) 342°K. The incident ion energy was 120 KeV. Determination of the amorphous layer width W by extrapolation to the random level is illustrated in b).

the deposited damage energy profile is unchanged as function of temperature, and that its position and shape are identical to that shown in Figure 1. The electronic losses may contribute to enhanced annealing or diffusion, and, therefore, may alter the position or shape of the damage profile. Secondly, the model for the formation of amorphous regions by ion implantation by Morehead and Crowder disagrees with more recent results[10]. The model proposed here is insensitive to these assumptions in light of the excellent agreement with experiment shown in Figure 3.

The critical dose, which defines the fluence at the onset of the formation of an a-Si layer, has been examined as a function of energy and temperature. Our model predicts[11] that the critical dose is lower and is a function of energy in contrast to an earlier model.[6] Further results and development of the model will appear elsewhere.[11]

CONCLUSION:

A model has been formulated which successfully predicts the width of the amorphous layer formed by ion implantation in Si as a function of temperature and dose. This is useful in semiconductor material preparation to improve the crystalline quality of the material, or in device fabrication techniques to enhance dopant activation efficiency and reduce secondary structural defects. Excellent agreement is found between a-Si layer widths predicted by the model and amorphous layer widths formed by self-implantation.

REFERENCES

1. James F. Gibbons, Proc. IEEE 60, 1062 (1972); S.T. Picraux and F.L. Vook, Rad. Eff. 11, 179 (1971); John R. Dennis and Edward B. Hale, J. Appl. Phys. 49, 1119 (1978); Ion implantation and Beam Processing, eds. J.S. Williams and J.M. Poate (Academic Press, New York, 1984), chapts 1 and 2.
2. Eliezer Dovid Richmond, Alvin R. Knudson, T.J. Magee, H. Kawayoshi, and C. Leung, J. Vac. Sci. Technol. A2, 569 (1984).
3. W. Maszara, C. Carter, D.K. Sadana, J. Liu, V. Ozguz, J. Wortman, and G.A. Rozgonyi, Energy Beam-Solid Interactions and Transient Processing, eds. J.C.C. Fan and N.H. Johnson (North Holland, New York, 1984), p. 285.
4. a) H. Muller, K. Schmid, H. Ryssel, and I. Ruge, Ion Implantation in Semiconductor and Other Materials, ed B.L. Crowder (Plenum Press, New York, 1973), p.203; b) F.L. Vook, Radiation Damage and Defects in Semiconductors, ed. J.E. Whitehouse (Inst. of Phys., London, 1972), p. 60.
5. D.K. Brice, Ion Implantation Range and Energy Deposition Distributions, Vol. 1: High Energies (Plenum Press, New York, 1975).
6. F.F. Morehead Jr. and B.L. Crowder, Rad. Effects 6, 27 (1970).
7. J. Lindhard, M. Scharff, and H.E. Schiott, Mat. Fys. Medd. Dan. Vid. Selsk. 33, No. 14 (1963).
8. D.K. Brice, Rad. Effects 11, 227 (1972).
9. D.K. Brice, Rad Effects 6, 77 (1970).
10. J. Narayan, 32nd National AVS Symposium, 19-22 November 1985, PTC-WeM7; Jack Washburn, Cheruvu S. Murty, Devendra Sadana, Peter Byrne, Ronald Gronsky, Nathan Cheung, Roar Kilaas, J. Nucl Instr. and Meth. 209/210, 345 (1983).
11. Eliezer Dovid Richmond, unpublished.

TEMPERATURE DEPENDENT AMORPHIZATION OF SILICON DURING SELF-IMPLANTATION[1]

W.P. MASZARA[*2], G.A. ROZGONYI[*], L. SIMPSON[**] AND J.J. WORTMAN[**]
[*] Materials Engineering Dept., North Carolina State Univ., Raleigh NC 27695
[**] Dept. of Electrical and Computer Engineering, NCSU, Raleigh NC 27695

ABSTRACT

We have investigated the damage which results from silicon self-implantation for the range of doses from 2E14 to 1E16 cm^{-2} for temperatures from 82 to 296 K for 150 and 300 keV implants. Cross-sectional TEM was used to evaluate the nature of the amorphous layer. The experimental results were correlated with computer calculated damage distributions using a Monte Carlo simulation program. The depth of amorphous-crystalline interface(s) was evaluated as a function of dose and temperature. An experimental damage energy density curve was constructed. Using the curve, a critical energy density for amorphization, E_c, was calculated for the samples implanted at different temperatures. The energy was found to depend on depth and implant energy, and it increased with temperature. A study of a-c interface morphology shows no dependance on temperature within the range considered. Kinetics of dynamic annealing are discussed in conjunction with the above findings.

INTRODUCTION

The damage produced during implantation of ions into semiconductors is a subject of considerable importance in practical applications of the process. It is especially useful to know the proper fluence of the ions being implanted and the temperature of the substrate to render its surface region amorphous, since lower annealing temperature and more complete activation of dopant are achieved for amorphized semiconductor. If the amorphous state is obtained in a separate implantation process (pre-amorphization) preceding dopant introduction, its extent into the substrate can be chosen to exceed the depth of the dopant distribution profile, thus also providing elimination of a channeling effect. Self-implantation seems to be a natural choice for preamorphization of silicon substrates when precise control of shallow junctions in VLSI-compatible technology is required.

In this work we have investigated the damage which results from silicon self-implantation for the range of fluences from 2E14 to 1E16cm^{-2}. The temperature of the substrate was carefully controlled from LN to RT. Kinetics of the amorphization process and morphology of the damage are discussed.

In what seems to be the most universal approach to explaining amorphization, the model suggested by Stein et al. [1] proposes that a critical energy density (E_c) deposited into atomic processes sets the threshold for amorphization which was evaluated to be about 12 eV/atom regardless of the projectile species. This value applies to low temperature implants, about LN$_2$ [1] and below (Narayan et al.[2] for 4 K).

Morehead and Crowder [3] developed a model (subsequently, referred to as M-C) taking explicitly into consideration temperature effects on the

---

[1] Partially supported by Semiconductor Research Corporation
[2] Present address: Allied Bendix Aerospace, Columbia MD 21045

overall amount of damage sustained in the substrate. The model assumes that a cylindrical damage cascade radially collapses due to thermally assisted defect interaction (vacancy outdiffusion from cascade center and their annihilation by interstitials) into an amorphous one of smaller radius. For a given critical dose a collective amount of amorphous zones produces a continuous amorphous layer. Despite apparent disagreement of the model's assumptions with observed lack of direct amorphization for light ions [2,4], predictions of Morehead and Crowder have been found to coincide surprisingly well with experiment for light and heavy ions implanted in silicon [3,5,6] as well as germanium [7].

EXPERIMENTAL

A series of 10-20 ohmcm, 3" p-type (100) wafers were implanted at two different energies 150 and 300 keV with doses of $^{28}Si^+$ ranging from 2E14 to 1E16 cm^{-2} at a dose rate equal to .25 μA/cm^2. Quarters of wafers were mounted on a steel sample holder with the help of highly heat conductive silicon-based silver paste and spring clips. A sheathed thermocouple was soldered into a small groove on the surface of the holder with its tip located directly under the center of the sample. The temperature of some of the samples was additionally monitored by an infrared thermometer (spectral response 7-18 μm) placed inside the implantation chamber. The holder was cooled by liquid nitrogen (LN$_2$), dry ice + acetone, ice + water, and water at room temperature (RT) to maintain the temperature of the sample at 82, 197, 274 and 296 K, respectively. These temperatures are slightly higher than the temperatures of respective coolants due to the temperature drop across the wall of the holder-dewar. During implantation the temperature increase as recorded by both thermocouple and pyrometer was never greater than 1-1.5 °C above its initial value.

The damage structure of implanted samples was determined using the cross-sectional transmission electron microscopy (XTEM). The XTEM technique is ideally suited for differentiation between amorphous and crystalline regions while maintaining resolution within a nanometer order. The amorphous layer thickness was measured directly from TEM micrographs assuming that only the width of a layer containing no visible crystallites be considered.

DISCUSSION OF RESULTS

The depth of the amorphous-crystalline (a/c) interface(s) as measured from XTEM micrographs is shown in Fig.1 as a function of $^{28}Si^+$ dose at 150 and 300 keV and at four different temperatures (82, 197, 274, and 296 K). Using data from Fig.1, damage energy density curves:

$$dE_d/dx = E_c/D_c(x) \tag{1}$$

(where D_c -critical dose for crystalline-to-amorphous transition, E_c is in eV/cm^3; to express E_c in eV/atom divide it by 5E22 cm^{-3} - the atomic density of Si) were constructed for both implantation energies at LN$_2$ (Fig.2). The assumption was made that a temperature of 82 K is low enough to prevent any appreciable annihilation of individual unstable defects - mostly single vacancies outdiffusing from the cascade center capturing interstitials - before stable damage is created. Lack of dose rate effects at LN$_2$ temperatures [8], which, we believe, occur only when a second cascade can overlap the first one before that collapses, supports the above assumption. It was also assumed that because there is virtually no migration of point defects at LN$_2$, the transition from damaged to amorphous state within the implanted layer will require deposition of the same amount of E_c regardless of its location (i.e. presence of potential point defect sinks, such as wafer surface, is ignored). The ordinate of Fig.2 was scaled to match the peak of a theoretical curve computed with the help of TRIM84 [9], a computer program using Monte Carlo simulation. The scaling factor, which is E_c, was found to be equal to 12 eV/atom for the 300keV curve - in excellent agreement with the literature [1,2]. The same value was obtained for the 150 keV implant where higher dE_d/dx peak was proportionally matched by lower doses observed for the amorphization threshold. The depth of the

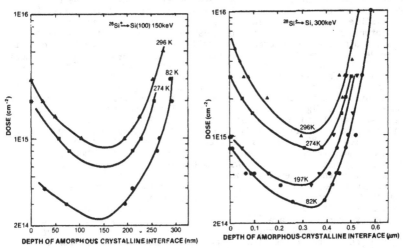

Fig.1. a   b

Relation between dose and depth of amorphous–crystalline interface(s) as measured from cross-sectional TEM micrographs for samples implanted with $^{28}Si^+$ at temperatures from 82–296 K. Implantation energies: a) 150 keV, b) 300 keV.

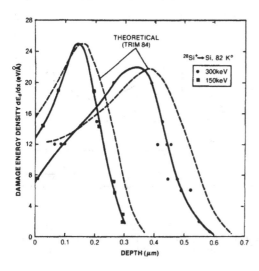

Fig.2.   Damage energy density (eV/Å) for 150 and 300 keV $^{28}Si^+$ ions implanted into Si; calculated using Monte Carlo simulation (dashed lines) and experimental results from LN curves of Fig.1 (solid lines).

experimental dE_d/dx curve peak was found to be shallower than the
theoretical one for both implantation energies indicating a need for some
revision of the theoretical model. The nonsymmetrical shape of the curve
closely matched that of the theoretical with some deviation toward lower
values near the surface of the sample. Detailed discussion of the
descrepancies between the two curves is beyond the scope of this paper.

As data for LN_2 sample in Fig.1 shows, higher critical doses are
required to render silicon amorphous at the depths away from the peak of
the damage curve - an obvious implication of the shape of the curve. It is
also visible that amorphization during implantation at elevated
temperatures requires increased fluences - effect of dynamic annealing.
Given the distribution of damage energy deposited per average ion, which is
independent of temperature, an increase of E_c is necessary with temperature
to satisfy relation (1). Such a trend was observed earlier by Vook [10] who
analyzed data from several labs for B, P, Sb, Ne and O implanted in Si.

An attempt has been made to quantify E_c for the implant energies
investigated as a function of depth and temperature. Using the experimental
damage curves of Fig.2, critical doses from Fig.1 and eqn.(1) critical
energy was calculated for samples implanted at 197, 274 and 296 K and
plotted versus depth in Fig.3 for 300 and 150 keV. Shown also is a straight
line relation for 82 K samples (earlier assumed constant with depth). As
the plots indicate the energy required to turn silicon amorphous increases
monotonically with temperature. At room temperature E_c corresponding to the
peak of damage energy density reaches a value about four times higher than
that of low temperature (82 K), for the 300 keV implant. Energies
calculated for the 150 keV at corresponding temperatures are somewhat
lower. Lower critical doses for less energetic Li, N and Ne ions were
observed by Dennis and Hale [6]. The results of Fig.3a and 3b are redrawn
in Fig.4 to show directly relation between values of E_c (for meaningful
illustration chosen at the damage peak depths) and temperature for both
implantation energies. As it can be seen, somewhat lesser critical energies
are required deposited per atom (i.e. lower doses) at lower implant
energies to render silicon amorphous at elevated temperatures.

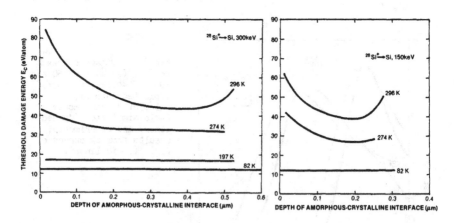

Fig.3.   a   b
Threshold damage energy per atom for amorphization of silicon substrate
with $^{28}Si^+$ vs. depth at which amorphous-crystalline transition is observed
at temperatures from 82-296 K. Implantation energies: a) 150keV, b) 300keV.

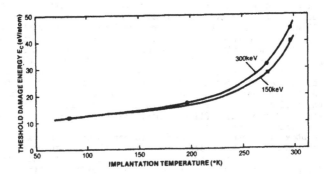

Fig.4. Threshold damage energy per atom vs. implant temperature for 150 and
300 keV silicon implants. The data corresponding to the depth of damage
peak (0.14 μm for 150 keV, 0.34 μm for 300 keV) is obtained from Fig.3.

Depth (experiment and TRIM84, see Fig.2) and lateral (TRIM84)
distribution of damage energy density for average single cascade shows that
the defects are created more densely by 150 keV beam than the 300 keV
implant at LN. This in turn leads to higher survivability of the damage
against dynamic annealing at elevated temperatures. Such a mechanism would
account for lower critical energies (or doses) observed for the 150 keV
process throughout the implanted volume.

The increase of implant temperature also causes a marked increase of
threshold energy above the value of that observed at damage peak location,
at the depths away from the peak. The higher the temperature the stronger
the difference in energy. As the data indicates, the required amount of
damage energy deposited per atom at the surface of 300 keV sample at room
temperature is about twice as high as that necessary to amorphize regions 0.3μm below it. This points at a higher rate of defect annihilation there. Similarly to the behavior of E_c, as measured at damage peak,
this additional increase in value at peripheries of the peak is less
pronounced for 150 than for 300 keV implant.

A difficulty in amorphizing the surface regions of the sample at
elevated temperatures was also observed by others and ascribed to thermally
assisted outdiffusion of point defects (vacancies) toward the surface which
supposedly plays the role of a sink for these defects [11]. This is rather
unlikely. First, because the increase of E_c, as our results indicate,
happens in fact on both sides of damage peak implying that the phenomenon
is not uniquely associated with surface. Secondly, individual vacancies can
only migrate until they produce stabile complexes or become annihilated by
interstitials and their diffusion length is then, according to M-C model,
only a few angstroms. The enhanced annihilation of near and far end of a
cascade, as suggested by Poate and Williams [4] and observed in Fig.3,
reflects an actual distribution of damage within cascade. Less densely
distributed vacancies there will be more prone to dynamic annealing
(interaction with fast moving interstitials) before they are able to form a
stable damage by clustering with neighboring vacancies (the simplest
cluster, the divacancy, once formed remains stable up to about 550 K [12]).

One may notice that the increase of E_c is more pronounced at the near side
than at the far one of the damage peak location, for the same values of the

deposited damage energy density dE_d/dx. Coincidence of this effect with the fact that most of ion energy lost to electronic interactions happens closer to the surface than the damage, indicates that ionization events may be responsible for enhancement of dynamic annealing near the surface. In particular, free electrons produced during collisions, when captured by point defects (e.g. neutral vacancy), may increase their mobility thus leading to an increase of dynamic annealing. Another possible explanation involves distribution of the damage: although the deposited damage energy (as specified per incremental depth) may be the same for two different depths along the cascade, the lateral spread of damage at the shallower location is likely to be greater due to more energetic recoils produced at early stages of projectile flight. This, again, would make the defects more likely to become annihilated.

The above indicates that in order to obtain exact description for the temperature dependent amorphization the M-C method has to be applied to individual incremental cylindrical slices of the cascade since its near and far end regions contain damage characterized by different kinetics: the faster collapsing ends of cascade resemble lighter ion average damage, whereas its central section is more heavier-ion-like. These observations lead to further conclusion: the ions, for which a larger fraction of initial defects produced in the cascade survives the temperature assisted quenching, i.e. the heavy ions, will not only exhibit lesser dependance of amorphization on temperature, as observed before [3], but also the "peripherial effect" i.e. inhibited amorphization (or enhanced dynamic annealing) at the surface and deep side of damage range will be smaller.

A morphology of amorphous-crystalline interface, or two interfaces in the case of buried amorphous layer, has been observed to vary with dose. Due to straggling of damage distribution and dynamic annealing a transition layer containing amorphous and crystalline material rather than a planar interface is observed. The width of such a layer has been measured and plotted for all samples implanted at 300 keV versus distance of the center of the layer from surface. As Fig.5 shows regardless of the temperature of the sample the data generally follows one relationship: a peak centered about where the corresponding damage peak has been found. Values lying close to the peak reflect a broad transition region found in low dose

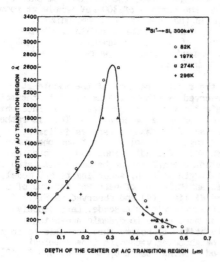

Fig.5. Width of amorphous-crystalline transition region as a function of its distance from wafer surface. The widest transition is observed when buried amorphous layer begins to develop. No dependance on temperature is observed.

implanted samples just below or at doses corresponding to continuous buried layer formation. Transition becomes much sharper for the samples with wide or no buried layer at all. The behavior of the layer width is believed to be related to the shape of the damage peak. The steeper the dE_d/dx curve, the less depthwise straggling is observed and, in consequence, the sharper the interface is formed. A striking lack of thermal dependence of the interface morphology seems to imply that although the buildup of the damage proceeds with smaller steps at higher temperatures, for the same amount of stable damage created the structure of the damage is very similar. This suggests that the dynamic annealing is confined to a single cascade quench and does not affects already existing damage (within the temperature range considered). Beam induced annealing of existing damage was observed at higher ($>200°C$) temperatures [13]. Practical importance of the morphology of a-c interface was pointed out by Maszara et al.[14] who found that its increased roughness causes, in subsequent solid phase epitaxial regrowth of amorphous layer, increasing density of "hair-pin" defects spanning the regrown layer. The defects can be potentially detrimental to device performance.

REFERENCE

1. H.J. Stein F.L. Vook, D.K. Brice, J.A. Borders and S.T. Picraux, Proc. 1st Int. Conf. on Ion Implantation, 1971, edited by L.T. Chadderton and F.H. Eisen (Gordon and Breach, London), p.17.
2. J. Narayan, D. Fathy, O.S. Oen and O.W. Holland, Mat. Lett. 2 (3), 211 (1984).
3. F.F.Morehead, Jr. and B.L.Crowder, Rad. Effects. 6, 27 (1970).
4. J.M. Poate and J.S. Williams, in Ion Implantation and Beam Processing edited by J.S. Williams and J.M. Poate (Academic Press, 1984),p.13.
5. J.R. Dennis, G.K. Woodward and E.B. Hale, in Lattice Defects in Semiconductors, 1974 (Inst.Phys.Conf.Ser.No.23 ,1975), p.467.
6. J.R. Dennis and E.B. Hale, J. Appl. Phys. 49 (3), 1119 (1978).
7. V.M. Gusev, M.I. Guseva and C.V. Starinin, Rad. Effects 15, 251 (1972).
8. O.W. Holland, D. Fathy and J. Narayan, in Advanced Photon and Particle Techniques for the Characterization of Defects in Solids (Mat.Res.Soc. Symp.Proc. vol.41, 1985), edited by J.B. Roberto, R.W. Carpenter and M.C. Wittels.
9. J.P. Biersack and L.G. Haggmark, Nucl. Instr. and Meth. 174, 257 (1980).
10. F.L. Vook, in Radiation Damage and Defects in Semiconductors, 1972 (Inst.Phys. Conf.Ser.No.16, 1973), edited by J.E. Whitehouse, p.60.
11. O.W. Holland, J. Narayan and D. Fathy, presented at 1984 Conf. on Ion Beam Modification of Materials, Cornell University, July 16-20, 1984.
12. J.W. Corbett, J.P. Karins and T.Y. Tan, Nucl. Instr. and Meth. 182/183, 457 (1981).
13. J.S. Williams, W.L. Brown, R.G. Elliman, R.V. Knoell, D.M. Maher and T.E. Seidel, in Ion Beam Processes in Advanced Electronic Materials and Device Technology (Mat.Res.Soc.Symp.Proc., vol.45, 1985), edited by B.R. Appleton, F.H. Eisen and T.W. Sigmon, p.79.
14. W. Maszara, D.K. Sadana, G.A. Rozgonyi, T. Sands, J. Washburn and J.J. Wortman, in Energy Beam-Solid Interactions and Transient Thermal Processing,1984 (Mat.Res.Soc.Proc.,vol 35), edited by D.K. Biegelsen, G.A. Rozgonyi and C.V. Shank, p.277.

IMPURITY DIFFUSION, CRYSTALLIZATION AND PHASE SEPARATION
IN AMORPHOUS SILICON.

R.G. Elliman[1], J.M. Poate[2], J.S. Williams[3], J.M. Gibson[2], D.C. Jacobson[2]
and D.K. Sood[3].

1) CSIRO Chemical Physics, P.O. Box 160, Clayton, 3168, Australia.
2) AT&T Bell Laboratories, Murray Hill, N.J., U.S.A.
3) RMIT Microelectronics Technology Centre, Melbourne, Australia.

ABSTRACT

Diffusion, crystallization and phase separation processes in indium
implanted amorphous silicon are examined for low temperature annealing
(600°C). Both diffusion and crystallization are shown to be extremely
sensitive to the indium concentration. Diffusion coefficients more than
10 orders of magnitude higher than tracer diffusion coefficients in
crystalline silicon are measured, and amorphous to crystalline silicon
transitions at temperatures as low as 350°C are reported. Phase
separation is also observed.

INTRODUCTION

Despite the widespread technological interest in amorphous silicon,
little is known about atomic transport processes in this material. In
fact, it has generally been assumed that diffusion processes are similar
in both the amorphous and crystalline silicon phases. Recently, studies
have shown that fast diffusers in crystalline silicon such as Au and Cu
[1] and the alkali metals [2] also diffuse rapidly in amorphous silicon
at low temperatures. In contrast, high concentrations of conventional
slow diffusers in crystalline silicon such as In and Bi have been shown
to diffuse many orders of magnitude faster in amorphous silicon than
expected [2,3]. Surprisingly, in view of its technological importance,
diffusion in amorphous silicon has not been studied in nearly as much
detail as have the amorphous metallic systems [4]. It is, therefore,
important to fully characterize these intriguing rapid diffusion
processes exhibited by 'slow' diffusers.

In this study, we have further investigated the diffusion of indium
in amorphous silicon. We briefly report here on correlations between
indium diffusion and the concomitant processes of crystallization and
phase separation.

EXPERIMENTAL

Amorphous silicon layers were prepared on single crystal substrates
by ion implantation. Thick amorphous layers (2μm) were prepared by
sequential MeV Ar$^+$ ion bombardment at 77K according to the procedure
outlined by Donovan et al [5]. Thinner layers (200 nm) were prepared by
irradiation with 50 and 140 keV Ar ions to a fluence of 1.5×10^{14} cm^{-2} at
each energy. Indium was subsequently implanted at 80 keV into the near-
surface of these layers at a variety of doses. Samples were annealed in
a quartz tube furnace either in vacuum or flowing N_2. Diffusion,
crystallization and phase separation phenomena were investigated by
Rutherford backscattering/channeling (RBS) and transmission electron
microscopy (TEM) techniques.

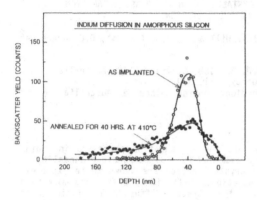

FIGURE 1: a) RBS spectra depicting the indium depth distribution as implanted with 80 kV In to a fluence of 3×10^{15} cm^{-2} (O) and following an anneal at 410°C for 40 hrs (●).

b) TEM diffraction pattern of sample after annealing.

RESULTS AND DISCUSSION

Figure 1 demonstrates rapid diffusion of indium in amorphous silicon. It depicts the indium depth distributions, for a 3×10^{15} In cm^{-2} implant into a 2μm amorphous silicon layer, both before and following an anneal at 410°C for 40 hrs. The inset TEM diffraction pattern confirms that the silicon remains amorphous during annealing. In this case, the estimated diffusion coefficient is 2×10^{-16} cm^{2}s^{-1}, which is over 10 orders of magnitude greater than extrapolated tracer diffusion measurements of indium in crystalline silicon at 410°C [6]. Such a comparison may not be particularly meaningful since our diffusion measurements in amorphous silicon have been carried out at very high indium concentrations (> 0.5 atomic percent). This is a regime which has not been investigated for crystalline silicon as a result of low indium solubility in the crystalline phase. It is significant to note, however, that within our detection limits, no phase separation was observed by TEM analysis, see Fig. 1. This indicates that we are most likely observing atomic diffusion of indium in amorphous silicon. These observations are intriguing in view of the fact that the peak indium concentration (1.5 atomic percent) is far in excess of the retrograde maximum solubility limit (0.02 atomic percent) of indium in crystalline silicon [7]. This may suggest a higher solubility limit of indium in amorphous compared with crystalline silicon.

Further studies to investigate the isothermal and isochronal annealing behaviour of the diffusion process indicated that the situation was decidedly more complex than the simple picture of rapid atomic diffusion of indium in amorphous silicon. For example, isothermal annealing studies have shown that this rapid indium diffusion does not follow a $\sqrt{time}$ dependence but, rather, slows down dramatically with time. At a given temperature, diffusing indium ultimately becomes immobile, and subsequent increases in temperature fail to reinitiate diffusion. One plausible explanation for this effect is phase separation which we have observed for high dose indium implants as shown in Figure 2.

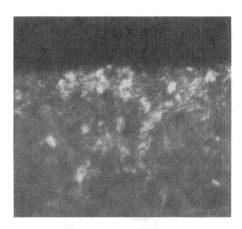 

40 nm.

FIGURE 2: Dark field TEM micrograph and diffraction pattern of a sample implanted with 80 kV In to a fluence of 5 x 10^{15} cm^{-2} and ramp annealed to 600°C in 10 minutes.

The dark field TEM micrograph and inset diffraction pattern in Figure 2 are taken from a 2μm thick amorphous layer implanted with 5x10^{15} In .cm^{-2} and ramped to 600°C in 10 minutes. The diffraction pattern indicates the existence of polycrystalline silicon and other precipitates. The micrograph highlights the precipitates, which from the diffraction pattern correspond to neither the normal tetragonal phase of indium nor to silicon. The precipitates are, therefore, either a metastable phase of indium or a silicon-indium intermediate phase. Experiments to determine the precise nature of these precipitates are currently underway. The important point to note from this example is that phase separation can occur concomitant with indium diffusion.

Figure 3 provides further insight into indium diffusion in amorphous silicon and related phenomena. Firstly, the use of identical annealing schedules (up to 535°C for 15 minutes) for two closely spaced indium doses clearly illustrates the dramatic concentration dependence of the indium diffusion process. The RBS profiles for the 2x10^{15} In cm^{-2} case, before and following annealing, are identical, indicating that little indium diffusion has taken place in amorphous silicon (see inset TEM diffraction pattern). In contrast, the 3x10^{15} In cm^{-2} case exhibits considerable indium diffusion. Furthermore, the indium concentration drops off dramatically at the amorphous silicon - single crystal silicon boundary, (indicated by the arrow at a depth of approximately 125nm). This is consistent with the assertion of an indium solubility difference in amorphous compared with crystalline silicon, and/or with a dramatic difference in the rate of diffusion in the different phases.

Finally, the most intriguing observation indicated in Figure 3 is contained in the inset diffraction pattern of Figure 3b, which shows polycrystalline silicon rings imaged from the previously amorphous layer. In view of the fact that the corresponding indium profile in Figure 3b has attained its final (immobilized) state, we suggest that the indium has first diffused in amorphous silicon, then polycrystalline silicon has nucleated and immobilized the diffusing indium. Furthermore,

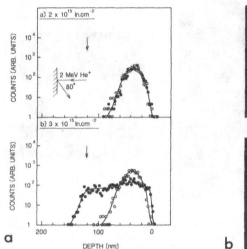
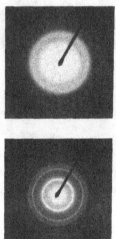

FIGURE 3: RBS spectra and TEM diffraction patterns for samples implanted with a) 2×10^{15} In.cm^{-2} and b) 3×10^{15} In.cm^{-2} and annealed up to 535°C for 15 minutes. The RBS spectra depict indium profiles as-implanted (O) and following annealing (●). TEM diffraction patterns are from samples after annealing.

it appears that indium significantly decreases the crystallization temperature of silicon. Indeed, we have observed an amorphous to crystalline silicon transition at temperatures as low as 350°C for very high doses (10^{16}cm^{-2}) of indium. Details of the concentration dependent indium diffusion and crystallization phenomena are given elsewhere [8].

In conclusion, the most significant features to note from this study are that the processes of indium diffusion in amorphous silicon, reduction in the silicon crystallization temperature and phase separation are intimately related. Our results show that indium diffuses rapidly in amorphous silicon, and further suggest that this is followed by the processes of silicon crystallization and phase separation.

ACKNOWLEDGEMENTS

The Australian Special Research Centres Scheme is acknowledged for partial support of this project. RGE also acknowledges the CSIRO Postdoctoral Award Scheme and the CSIRO Division of Chemical Physics for financial support.

REFERENCES

1. R.G. Elliman, J.M. Gibson, D.C. Jacobson, J.M. Poate and J.S. Williams. Appl. Phys. Lett., 46, 478 (1985).

2. M. Reinelt and S. Kalbitzer. J.De.Physique, 42, C4-843 (1981).

3. R.G. Elliman.
    Rad. Eff. Lett., 67, 77 (1981).

4. B. Cantor and R.W. Cahn.
    Chapter 25 in Amorphous Metallic Alloys, edited by F.E. Luborsky.
    (Butterworths, Sydney).

5. E.P. Donovan, F. Spaepen, D. Turnbull, J.M. Poate and D.C. Jacobson.
    J. Appl. Phys. 57, 1795 (1985).

6. See for example, D. Shaw in Atomic Diffusion in Semiconductors.
    Edited by D. Shaw. (Plenum Press, N.Y. 1973).

7. F. Trumbore, Bell System Tech. J., 39, 205 (1960).

8. R.G. Elliman, J.M. Poate, K.T. Short and J.S. Williams.
    To be published.

AMORPHIZATION AND RECRYSTALLIZATION PROCESSES IN
MONOCRYSTALINE BETA SILICON CARBIDE THIN FILMS

J.A. EDMOND, S.P. WITHROW*, H.S. KONG, AND R.F. DAVIS
Department of Materials Engineering, North Carolina State University,
Raleigh, NC 27695-7907
*Solid State Division, Oak Ridge National Laboratory, Oak Ridge, TN 37831

ABSTRACT

Individual, as well as multiple doses of $^{27}Al^+$, $^{31}P^+$, $^{28}Si^+$, and $^{28}Si^+$ plus $^{12}C^+$ were implanted into (100) oriented monocrystaline β-SiC films[+]. A critical energy of ≈16 eV/atom required for the amorphization of β-SiC via implantation of Al and P was determined using the TRIM84 computer program for calculation of damage-energy profiles coupled with results of RBS/ion channeling analyses. In order to recrystallize amorphized layers created by the individual implantation of all four ion species, thermal annealing at 1600, 1700, or 1800°C was employed. Characterization of the recrystallized layers was performed using XTEM. Examples of SPE regrown layers containing; 1) precipitates and dislocation loops, 2) highly faulted, microtwinned regions, and 3) random crystallites were observed.

INTRODUCTION

The processing steps leading to the development of selected electronic devices in cubic (beta) silicon carbide (E_g=2.3 eV) thin films involve ion implantation to introduce electrically active dopants. As the dose of the implanted specie is increased, the near surface region of this compound semiconductor becomes progressively damaged; atomic disorder and eventual amorphization of the structure occurs. Early work by Hart et al. [1] utilized Rutherford Backscattering (RBS)/channeling techniques in order to study both disorder production in monocrystalline 6H α-SiC (E_g=2.8 eV) by ion implantation and the subsequent thermal annealing of that damage. Williams et al. [2] have previously considered structural alteration in monocrystalline (0001) α-SiC as a result of Cr and N implantation. These authors have made direct comparisons of the theoretical damage profiles calculated using the computer codes E-DEP-1 [3] and TRIM [4] with those determined by RBS/channeling on experimentally implanted SiC samples. From the N implant results, it was determined that the critical energy density (CED) for randomization in their material was between 10 and 20 eV/atom (randomized regions refer to depths in the crystal where the aligned spectra coincide with the rotating random spectra).

The first objective in our investigation was to employ RBS ion channeling and Monte Carlo computer simulations together with the CED model [5] for implantation induced damage production in order to quantify, by a more novel approach, the disordering process during ion implantation of Al and P in β-SiC.

Solid-phase-epitaxial (SPE) regrowth during the thermal annealing of amorphous layers in compound semiconductors has been and continues to be the subject of a host of studies throughout the world [see, e.g.6] and also comprises our second objective in the present research. The quality of SPE regrown layers is generally very poor even when taking the utmost precautions. The two major problems are that nonstoichiometry results during implantation [7] and that dissociation of the constituent elements of the compound semiconductor generally occurs at different temperatures during thermal

annealing. We have studied SPE regrowth of β-SiC in order to better
characterize these two effects.

EXPERIMENTAL

Thin films of monocrystalline (100) β-SiC were epitaxially grown in-house
on (100) silicon wafers via chemical vapor deposition [8]. Each sample was
then mechanically polished, oxidized, and etched in HF in order to obtain a
clean, undamaged and smooth surface. After mounting in ultra high vacuum,
samples were implanted at room temperature with either P or Al at energies of
110 keV and 130 keV, respectively. The implants were made at an incident angle
7° off normal to avoid channeling effects. Following implantation, in situ He
backscattering analysis of the samples was obtained using 2.5 MeV $^4He^+$ ions
incident along the <110> axial direction. The dosimetry was stepwise increased
and additional ion scattering analyses made in order to measure the
incremental increase in lattice damage as a function of implantation dose.
Data were obtained until the backscattering yield from the damaged region was
the same as that expected from a sample amorphous to the surface. Theoretical
damage energy deposition profiles have been obtained for implantation of both
species using the TRIM84 code [9]. These calculations were executed using a
threshold displacement energy of 16 eV for SiC. However, it was determined
that this parameter could be increased as high as 65 eV for SiC and have
little effect on changing the TRIM84 profile and thus the CED value for
amorphization.

Amorphous layers were also produced in β-SiC by implanting Al, P, Si and
Si plus C in order to study SPE regrowth upon annealing. The former two
species were implanted in order to dope β-SiC p-type and n-type, respectively;
the latter two were used for preamorphization prior to subsequent dopant
introduction at concentrations below which amorphization occurs. A summary of
implant species and conditions is given in Table I. Solid-phase-epitaxial
regrowth of amorphized layers was achieved by thermally annealing samples in 1
atm. of Ar at 1600, 1700, or 1800°C for 300 s. Annealing in this temperature
range is necessary for optimizing electrical characteristics of implanted
p-type and n-type layers in β-SiC [10]. Residual lattice damage in the surface
implanted regions before and after annealing has been visually evaluated using
cross-sectional transmission electron microscopy (XTEM). The procedure for
XTEM sample preparation is discussed in Ref. 11.

Table I.   Summary of ion implantation conditions for SPE regrowth study.
All implants performed using an offset angle of 7°.

FIGURE NO	ION SPECIES	ENERGIES(keV)	DOSES (cm^{-2})	* PEAK CONC.(cm^{-3})	IMPLANT TEMP.
3	$^{27}Al^+$	110,190	6E14,9E14	1E20	RT
4	$^{31}P^+$	110,220	6E14,1E15	1E20	LN_2
5	$^{28}Si^+$	80,160,320	6E14,1E15,2E15	1E20	LN_2
6	$^{28}Si^+$	120,160,320	2.3E14,3.2E14,5.1E14	3E19	LN_2
	$^{12}C^+$	50,67,141	2.7E14,3.2E14,4.8E14	3E19	LN_2

* Peak concentration values determined by LSS calculation.

RESULTS AND DISCUSSION

RBS/Channeling and TRIM84 Analyses

The RBS spectra in Fig. 1a illustrate the accumulation of damage in β-SiC as a result of P implantation. Prior to implantation, both an aligned spectrum from the undamaged sample, labeled 'virgin', and a random spectrum, obtained by rotating the sample around its normal axis to approximate the spectrum expected from amorphous SiC, were measured. As can be seen in the figure, damage accumulated in the sample with increasing implant dose until the scattering yield from the damaged region of the aligned spectrum was coincident with the yield from random rotation and thus indicative of an amorphous crystal. This condition initially occured at a depth of approximately 90 nm for a dose between 3.0 (not shown) and 5.0 x 10^{14} cm^{-2}. At the latter dose a buried amorphous layer ranging in depth between 55 nm and 118 nm was observed. The width of the amorphous region increased with increasing implant dose, until at the highest dose measured, 3.0 x 10^{15}/cm^2, a surface amorphous layer 154 nm deep resulted, as indicated in the figure. The theoretical TRIM84 damage energy deposition profile for P implantation into SiC is shown in Figure 1b as a solid curve. The ordinate represents the energy deposition in eV/Å. In order to determine the CED value for amorphization in β-SiC, the experimental minimum and maximum depths over which the P implant amorphized the crystal were measured as a function of dose from backscattering spectra like shown in Fig. 1a. These depths are plotted in Fig. 1b with the

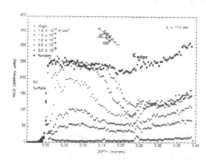

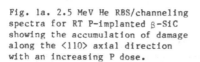

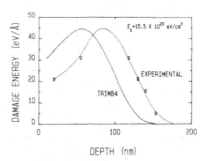

DEPTH (nm)

Fig. 1a. 2.5 MeV He RBS/channeling spectra for RT P-implanted β-SiC showing the accumulation of damage along the <110> axial direction with an increasing P dose.

Fig. 1b. Comparison between theoretical (TRIM84) and experimental damage-energy profiles for 110 keV P-implanted β-SiC.

corresponding ordinate representing k_{CED}/dose, where k_{CED} is a scaling factor that produces a curve through the data with the same maximum value and shape as the theoretical profile (see Ref. 12 for a more complete description of the procedure employed for determining k_{CED}). This curve is shown dashed in Fig. 1b. For P implantation at room temperature, k_{CED}, which refers to the CED value, is 15.5 x 10^{23} eV/cm^3, or 16 eV/atom.

398

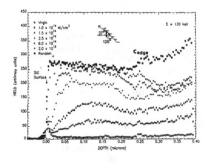

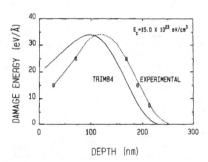

DEPTH (nm)

Fig. 2a. MeV He RBS/channeling spectra for RT Al-implanted β-SiC showing the accumulation of damage along the <110> axial direction with an increasing P dose.

Fig. 2b. Comparison between theoretical (TRIM 84) and experimental damage-energy profiles for 130 keV Al-implanted β-SiC.

Fig. 2a shows the RBS spectra for damage accumulation as a result of room temperature implantation of Al in β-SiC. The measured lattice damage increased rapidly for doses greater than 1×10^{14} cm^{-2}. The sample first became amorphous at a depth of approximately 120 nm for a dose between 4.0 and 6.0×10^{14} cm^{-2}. For the 6.0×10^{14} cm^{-2} implant, a buried amorphous layer ranging in depth from 70 to 168 nm resulted. A surface amorphous layer 213 nm in depth resulted from implanting with a dose of 2×10^{15} cm^{-2}.

The theoretical and experimental damage versus energy profiles for Al in β-SiC are compared in Fig. 2b using the same method as for the P implant described above. For Al, the CED value for amorphization is 15.0×10^{23} eV/cm^3, or 15.5 eV/atom. As expected, the amorphizing energy obtained from both analyses is nearly identical.

Deviation between the theoretical and experimental profile depths was very significant for both implanted species, possibly indicating a need for revision of parameters in the TRIM84 code for the implantation of SiC. However, a more likely source of deviation may be in the method in which the program computes damage production from recoils. The recoils are not individually followed in the program but rather the amount of damage they produce is approximated through the use of theoretical calculations. For that reason, the authors are presently investigating another Monte Carlo program that tracks damage production from recoils as well as from the primary ions.

Solid-Phase-Epitaxial Regrowth

Figure 3 shows an XTEM micrograph of an Al double implant region having a peak concentration of 1×10^{20} Al/cm^3. A buried amorphous layer having a crystalline cap of 10 nm resulted after implantation (Fig. 3a). The lower amorphous/crystalline (a/c) interface located at a depth of ≈170 nm is very diffuse as a result of implanting at room temperature. After annealing at 1600°C for 300 s in Ar, the amorphous layer had regrown (Fig. 3b) by SPE. However, a high concentration of defects was observed. Precipitates and/or dislocation loops formed where the upper and lower a/c interfaces were initially located. A broad band of defects (40 nm – 110 nm) resulted where the two a/c interfaces converged during SPE regrowth. In contrast, by annealing a

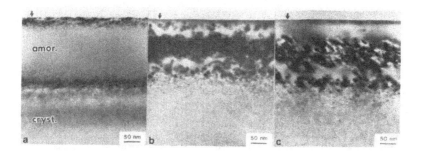

Figure 3. XTEM micrographs showing the surface of a sample which has been double implanted with Al to a peak concentration of 1×10^{20} Al/cm^3. (a) As-implanted; (b) annealed at 1600°C; (c) annealed at 1800°C for 300 s.

like sample at 1800°C (see Fig. 3c) many of the precipitates did not appear. In this instance, a virtually defect free surface region (0–50 nm) resulted. Additionally, a band containing loops and stacking faults formed at the regrowth convergence as well as small loops where the lower a/c interface was initially located.

Figure 4 illustrates the regrowth properties of a P double implant with a peak concentration (as calculated by LSS) of 1×10^{20} P/cm^3. The annealing temperature was 1700°C. Clearly, the supersaturation of P in the SiC matrix became sufficiently high to cause the layer to regrow in a polycrystalline condition after the first 100 nm of regrowth. However, within the first 100 nm, many small precipitates and loops formed. The regrowth properties of amorphous layers obtained using a lower atomic concentration of P is presently being investigated.

Figure 4. XTEM micrograph of a sample which has been double implanted with P to a peak concentration of 1×10^{20} P/cm^3 and subsequently annealed at 1700°C for 300 s. The surface appears rough as a result of polycrystalline regrowth.

In order to preamorphize β-SiC for subsequent dopant introduction, implantation of Si and Si plus C was conducted. Figure 5 shows an XTEM micrograph of a Si triple implant region prior to and after thermal annealing. The peak concentration of Si is 1×10^{20} Si/cm^3. After implantation, an amorphous surface layer 440 nm in depth was observed. After annealing at 1700°C for 300 s in Ar, the layer regrew epitaxially. The first 70 nm regrew moderately defect free. Thereafter, severe microtwinning and faulting occurred resulting in a polycrystalline layer (Fig. 5b) with a highly preferred orientation.

The XTEM micrographs in Fig. 6 directly compare the structural regrowth properties of implanted and amorphized layers created using Si and Si plus C. Figure 6a shows the amorphous layer which was formed by the Si triple implant with a peak concentration of 3×10^{19} Si/cm^3. The a/c interface is located 400 nm below the sample surface. After annealing at 1700°C for 300 s in Ar, the layer regrew epitaxially without severe faulting and/or microtwinning as was observed in Fig. 5b. However, a high concentration of precipitates and/or loops formed throughout the regrown bulk. In an attempt to eliminate these defects, a triple C implant was superimposed on the triple Si implant thus simulating implantation of SiC into SiC. The projected range peaks were

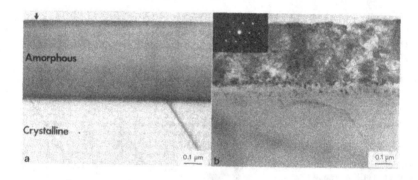

Figure 5. XTEM micrographs and diffraction pattern of (100) β-SiC which has been triple implanted with Si to a peak concentration of 1×10^{20} Si/cm^3. (a) As-implanted; (b) annealed at 1700°C for 300 s. (The diffraction pattern is near [011] and from the microtwinned and highly faulted layer).

matched (1:1) SiC in order to obtain the correct stoichiometry using LSS theory. Fig. 6c shows an XTEM micrograph of the regrown layer previously implanted and amorphized with SiC at a peak concentration of 3×10^{19} implanted SiC/cm^3. The annealing conditions were the same as described for the sample shown in Fig. 6b. Quite clearly, implanting Si plus C did not structurally improve the quality of the regrown layer. In fact, microfaulting and microtwinning occurred upon regrowth from 140 nm to the sample surface. However, a recent investigation using SIMS has revealed that implants of ^{30}Si$^+$ and ^{13}C$^+$ in SiC do not follow LSS theory and in fact, the above ^{28}Si$^+$ and ^{12}C$^+$ implant profiles may have deviated significantly. Therefore, the authors are further investigating the implantation of both ^{30}Si$^+$ and ^{13}C$^+$ in SiC in an attempt to subsequently improve the character of the regrown implanted layer.

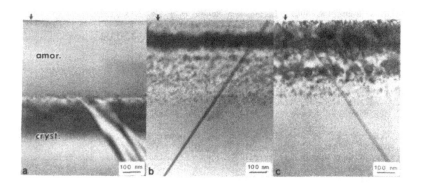

Figure 6. XTEM micrographs comparing the regrowth properties of samples implanted with equal atom concentrations ($3 \times 10^{19}/cm^3$) of Si (b, center) and Si + C (c, right). The as-implanted amorphous layer is also shown (a). Samples were annealed at 1700°C for 300 s.

CONCLUSIONS

Using RBS/channeling and the TRIM84 computer code it has been determined that the critical energy for amorphization of β-SiC at room temperature is $\cong 16$ eV/atom. Furthermore, it has been shown that amorphous SiC undergoes SPE regrowth upon thermal annealing at a temperature as low as 1600°C. For the case of amorphous layers created by P implants, polycrystalline regrowth was observed. Layers implanted with Al and Si regrew as single crystals but with residual line and planar defects. Initial attempts have shown that overlaying implants of C on Si does not improve the regrowth properties of amorphous SiC.

ACKNOWLEDGEMENT

The authors gratefully acknowledge the support of this program by the Office of Naval Research under contract N00014-82-K-0182 P005 and to the ONR Fellowship program for support of one of the authors (Edmond). Work at Oak Ridge was sponsored by U.S. D.O.E. Division of Materials Sciences under contract DE-AC05-840R21400 with Martin Marietta Energy Systems, Inc.

REFERENCES

[1]   R.R. Hart, H.L. Dunlap, and O.J. Marsh, Rad. Effects 9, 261 (1971).

[2]   J.M. Williams, C.J. McHargue, and B.R. Appleton, Nucl. Instr. and Meth. 209/210, 317 (1983).

[3]   I. Manning and G.P. Mueller, Comp. Phys. Comm. 7, 85 (1974).

[4]   J.P. Biersack and L. G. Haggmark, Nucl. Instr. and Meth. 174, 257 (1980).

402

[5]   J.R. Dennis and E.B. Hale, J. Appl. Phys. 49, 1119 (1978).

[6]   J.P. Donnelly, Nucl. Instr. and Meth. 182/183, 553 (1981).

[7]   L.A. Christel and J.F. Gibbons, J. Appl. Phys. 52, 5050 (1981).

[8]   H.P. Liaw and R.F. Davis, J. Electrochem. Soc. 132, 642 (1985).

[9]   J.P. Biersack and W. Eckstein, J. Appl. Phys. A34, 73 (1984).

[10]   J.A. Edmond, H.J. Kim, and R.F. Davis, to be published in 1985
       MRS Symposia Proceedings on Rapid Thermal Processing, Boston, MA 1985.

[11]   C.H. Carter, Jr., J.A. Edmond, J.W. Palmour, J. Ryu, H.J. Kim, and R.F.
       Davis, to be published in MRS Symposia Proceedings on Microscopic
       Identification of Electronic Defects in Semiconductors, San Francisco,
       CA, 1985.

[12]   W. Maszara, G.A. Rozgonyi, L. Simpson, and J.J. Wortman, to be published
       in 1985 MRS Symposia Proceedings on Beam-Solid Interactions and Phase
       Transformations, Boston, MA, 1985.

† These isotopes were used throughout this study unless otherwise noted.

PART VII

Ion Beam Mixing and Metastable Phase Formation

METASTABLE ALLOY FORMATION BY ION BEAM MIXING

F.W. SARIS, J.F.M. WESTENDORP AND A. VREDENBERG
FOM-Institute for Atomic and Molecular Physics, Kruislaan 407, 1098 SJ
Amsterdam, The Netherlands

ABSTRACT

In Ion Beam Mixing new surface alloys are produced by a combination of
vacuum deposition and ion irradiation. One may wonder what the advantages are
of this combination. Indeed one may ask:
- Why not just ion implantation?
- Why ion beams instead of laser or electron beams?
- Why ion mixing instead of evaporation only?
- What phases are formed and what is the stability of ion mixed phases?
In an attempt to answer these questions the role of ion beam mixing in modern
materials modification will de delineated. Areas of controversy and further
development will also be illustrated.

Fig. 1

Typical Experimental set-up for the combination of vacuum deposition and ion
beam mixing and in situ RBS and RHEED analysis (courtesy: High Voltage Engineering Europa).

1. INTRODUCTION

In ion mixing new surface alloys are produced by a combination of vacuum deposition and ion irradiation. A typical experimental set-up for ion mixing is shown in figure 1. In this system thin films of a large variety of materials can be evaporated under UHV conditions on different kinds of substrates, which can be loaded and unloaded through a vacuum-lock. The evaporated layers can be mixed in situ with one and other or with the substrate material using ion beams from a medium energy heavy ion accelerator. The samples can be heated in the manipulator for annealing prior to, during or after ion irradiation. Rutherford backscattering analysis of the sample surface layers can also be done in situ using He ions from the same accelerator.

One may wonder what the advantages are of this combination of ion irradiation with evaporation and thermal annealing. In fact one may ask:
- Why not just ion implantation?
- Why ion beams instead of laser or electron beams?
- Why ion mixing instead of evaporation only?
- What phases are formed and what is the stability of ion mixed phases?

In an attempt to answer these questions the role of ion beam mixing in modern materials modification will be delineated, areas of controversy and further development will also be illustrated. Although ion beam mixing was first applied successfully to silicide formation, this paper will be concerned mainly with metal alloys. For a recent and more elaborate review of this field one is referred to [1] and [2].

2. WHY NOT JUST ION IMPLANTATION?

Although ion implantation has become a standard tool in the semiconductor industry to dope semiconductor surfaces, it is not an accepted method yet for the modification of metal surfaces. The reason is simply that in order to alter the corrosion, friction, wear, optical or magnetic properties of metal surfaces by ion implantation requires a dose which is many order of magnitude larger than the dose needed to change the electronic properties of semiconductor surfaces. Significant modifications of metal surfaces are obtained only by alloying to high concentrations in the order of ten atomic percent and more. With ion implantation only this becomes very time consuming and uneconomical except for special applications.

The maximum concentration in the surface that can be reached by direct ion implantation is limited by the sputtering phenomenon. Roughly, if for every incoming ion S atoms are removed from the substrate then in steady state the surface concentration of dopants will not exceed 1/S. Moreover the thickness alloyed by ion implantation is limited to the ion range. Rather high beam energies of heavy ions are required if alloying beyond the thousand Ångstrom range is desired.

By ion mixing instead of pure ion implantation these limitations, in principle, are overcome. An incoming ion can mix several hundred atoms of a coating with hundreds of atoms of the substrate, if the ion energy is sufficient to penetrate at least to the interface between them. Therefore there is an efficiency enhancement factor which makes ion mixing for surface alloying a lot more attractive than pure ion implantation. Moreover, the sputtering limitation on the alloying concentration can be overcome and rather thick films can be built-up by repeated evaporation and ion irradiation.

The mixing efficiency has become one of the key parameters of investigation. A strong temperature dependence has been found. One can now make a distinction between a low temperature region in which ion mixing is relatively independent of the substrate temperature and a steep temperature dependent region. The origin of this peculiar temperature dependence is not yet clear. Some authors attribute the strong dependence to radiation enhanced diffusion [3] whereas others prefer an explanation in terms of spike effects [4]. Some authors believe that pure ballistic effects dominate in the lower temperature

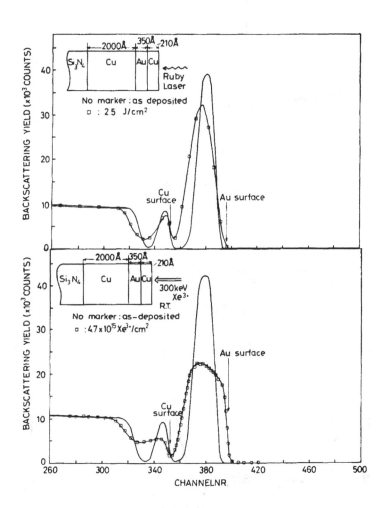

Fig. 2

RBS spectra of a Cu/Au/Cu sandwich structure before and after laser (above) and ion beam irradiation (below).

408

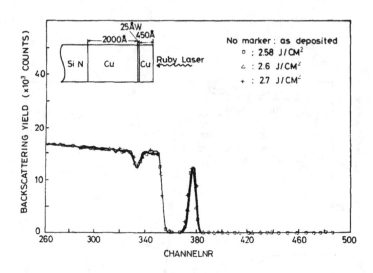

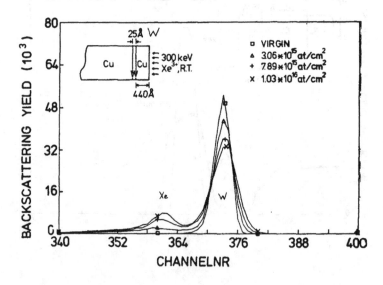

<u>Fig. 3</u>

RBS spectra of a Cu/W/Cu sandwich structure before and after laser (above)
and ion beam irradiation (below).

region [5] whereas others have shown that even below 77 K radiation enhanced diffusion and thermodynamics are still relevant to ion-induced atomic rearrangement in metals [6]. These fundamental problems will be discussed further in the following sections.

3. WHY ION BEAMS INSTEAD OF LASER OR ELECTRON BEAMS?

In surface alloying by laser or electron beams a predeposited film is controllably incorporated into the substrate by melting the film as well as part of the substrate. Mixing in the liquid state will be followed by rapid resolidification. The alloying element is thus incorporated in the near surface region of the substrate and the rapid quench rates fabricate novel metastable alloys. This kind of surface modification is relatively simple, rapid and easily applied over large surface areas. A possible advantage of ion irradiation over laser or electron beams must come from the fact that ions are massive particles that transfer kinetic energy to target atoms thus forcing them to mix in collision cascades. Ion mixing does not necessarily involve mixing in the liquid phase. In collision cascades, through ballistic effects and by radiation enhanced diffusion one may expect to form metastable alloys of systems which cannot be alloyed by laser or electron beams because of their limited or zero miscibility in the liquid phase.

To test this a comparison has been made between laser and ion mixing of Cu with Au and with W. The systems of Cu/Au and Cu/W are collisionally identical but thermodynamically very different. Cu/Au has a negative heat of mixing and shows complete solid solubility. Cu and W, on the other hand, are completely insoluble even in the liquid phase and there are no compounds known. Figure 2 shows that a Cu/Au/Cu sandwich structure can indeed be mixed by pulsed laser melting using 2.5 J/cm^2 from a Q-switched ruby laser and by 300 keV Xe ion irradiation. Figure 3, however, shows that pulsed laser irradiation has no effect on a Cu/W/Cu sandwich structure whereas after Xe irradiation the RBS spectra give clear evidence for ion mixing [7].

It is important to notice that ballistic mixing is only the first step in the description of ion mixing. The relative importance of purely collisional mixing, thermodynamic effects and diffusion due to defect generation and migration is still of great interest. For instance, figure 4 shows that collisionally identical systems may exhibit pronounced differences in mixing parameter [8]. Although for these systems the number of random walk steps per unit time induced by ion irradiation should be roughly the same, the results show a strong biasing by negative heat of mixing. For Cu-W, which according to Miedema [9] has a heat of mixing equal to +24 kJ/mole, the measured Xe-ion mixing parameter at room temperature is about half that of Cu-Ta which is also an immiscible system but with a calculated heat of mixing equal to 0 kJ/mole. Still larger mixing is observed for Cu-Au which has a negative heat of mixing of -10 kJ/mole. These effects persist also at low temperature (10 K) and for lower mass projectiles (Ne). Rather similar effects have been reported by the Caltech group [10] who have argued that the overwhelming majority of atomic mixing takes place after the initial ballistic regime, in the thermalizing regime during which the kinetic energy distribution among particles within the volume of a well developed cascade is described by a Maxwell-Boltzmann distribution. Cheng et al. [11] have devised a method for estimating the mean kinetic energy of atoms during the time wherein the majority of mixing occurs. From a fit to their data a value of k_BT_{eff} is obtained, the effective temperature at which mixing occured. Values of $k_BT_{eff} = 1-2$ eV were obtained.

The known scale of chemical biasing of the ion induced mixing process has thus been used to establish that the dominant contribution to the ion mixing occurs when particle kinetic energies are of order 1 eV. If this is true the implication is, according to the Caltech group [10], that ion mixing involves diffusion in the molten phase. If indeed ion mixing involves melting one should still not be able to mix systems which are immiscible in the liquid phase, consequently there should not be much advantage over laser or e-beam

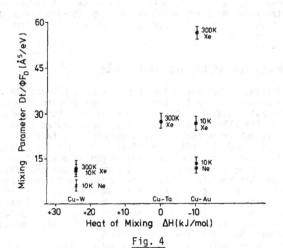

<u>Fig. 4</u>

Ion Beam Mixing Parameter measured for Cu-W, Cu-Ta and Cu-Au systems as function of their calculated heat of mixing.

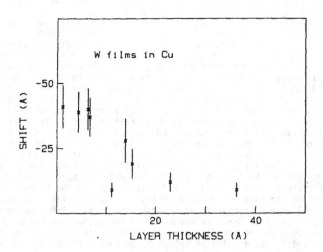

<u>Fig. 5</u>

Shift of thin W films in Cu due to irradiation with 460 keV Xe ions to a dose of $5-7 \times 10^{15}$ as function of the W film thickness. A negative shift means inward away from the surface in the direction of the Xe beam.

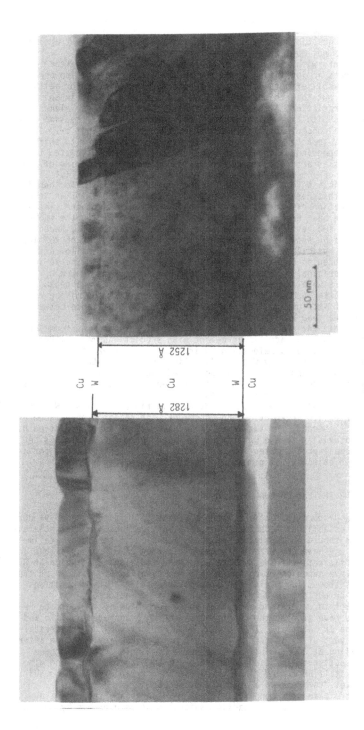

Fig. 6

Cross section transmission electron micrograph of Cu-W sandwich structures
before and after Xe ion irradiation. Before two thin W layers imbedded in Cu
at different depths are clearly visible. After ion irradiation the shallow W
layer is shifted in.

alloying. So, the question of the role of thermal spikes and phase transitions in the cascades is not just of academic interest.

Our experiments with thin films of W, Ta and Au in Cu give evidence of directional effects in ion mixing which may be explained by radiation enhanced diffusion or by recoil effects but certainly not by thermal spike or melting effects. It turns out that thin films of Au, Ta and W in a Cu matrix shift inward under ion irradiation. Although the degree of mixing of these elements with Cu correlates with the heat of mixing, the inward shift is the same within the experimental uncertainty. Figure 5 shows that for a dose of $5-7 \times 10^{15}$ Xe ions of 460 keV impacting on thin films of W in Cu causes them to shift inward by 10-50 Å depending on the layer thickness [8]. We have attempted to explain these shifts as artifacts in RBS analysis due to the clustering of the W film [12]. The here presented dependence of the shift on the initial film thickness contradicts with clustering, for the measured shift should then increase with film thickness. We have also tried to explain the shift in terms of a matrix relocation effect but then the shift must be towards the sample surface if the ion beam energy is decreased sufficiently [13]. In the whole energy range from 100 keV - 500 keV we always observe inward shifts. Recent X-TEM work, figure 6, shows that these shifts are not an artifact of RBS analysis but a real physical effect. A sandwich structure with a shallow and a deep W layer imbedded in Cu has been examined prior to and following Xe ion irradiation. The originally closed W layer near the surface is broken up into many small clusters with a diameter of 10-50 Å, distributed over a depth of 150-200 Å. This latter number is in excellent agreement with the mixing measured by RBS. Most intriguing, however, is the decrease in distance between the shallow and the deep W layers observed after ion irradiation.

We believe that the shift is due to preferential motion of W atoms inward during each individual cascade, in agreement with predictions made by Roush et al. [14] who calculated a relocation inward for the heavier atoms in a cascade of unlike atoms. The heavy W atoms recoiling in the Cu matrix will keep their momentum inwards and will be less deflected sideways than Cu recoils. Note that this bombardment induced shift of heavier impurities in lower mass substrates will distort SIMS profiles also.

Two other interesting features are observed in figure 6. First, W precipitates in Cu under ion irradiation, in agreement with recent observations by Nastasi c.s. [15]. Second, the top layer of Cu atoms after ion irradiation has been regrown epitaxially to the unterlying Cu layer. Once the W film has been broken up into small precipitates the underlying Cu layer acts as seed for irradiation induced epitaxial alignment of the top layer.

4. WHY ION MIXING INSTEAD OF EVAPORATION ONLY?

In order to produce coatings of amorphous metal alloys the use of vapour quenching is an established technique. For many applications, however, the as-evaporated films should be densified and their adhesion to the substrate improved. Heating may have the effect of crystallization of amorphous alloys and may not help sufficiently if the substrate is oxidized. Ion irradiation at RT is known to improve adhesion of coatings, and may even lead to epitaxy as the above result has shown. Ion mixing can also stimulate solid state reactions where diffusion or nucleation barriers act as limiting factors.

In an attempt to understand amorphous phase formation by solid state reactions [16], we have studied the reaction between an evaporated Ni film and a monocrystalline Zr substrate. In contrast with earlier work on polycrystalline sandwich layers of Ni and Zr [17], we did not observe any reaction after prolonged annealing at 300 K. This is true for a Zr single crystal surface which has been sputter-cleaned and annealed prior to Ni deposition, and also for a Ni-monocrystalline Zr interface which has been irradiated with 5×10^{14} Xe ions of 180 keV, to break-up a possible oxide barrier.

Apparently, in the case of a monocrystalline Zr-Ni interface there is a nucleation barrier for solid phase reaction which cannot be overcome by heating to 300 K (see figure 7a). After ion mixing of this interface to a dose of 5×10^{15} Xe ions the reaction proceeds rapidly, see figure 7b. Now heating to 300 K is sufficient to consume all of the evaporated Ni [8].

5. WHAT PHASES ARE FORMED AND WHAT IS THE STABILITY OF ION MIXED PHASED?

Although this question is fundamental to understanding the mechanism of ion mixing and its possible applications we shall not discuss it here. Phase formation and ion beam mixing is reviewed in this symposium by J.W. Mayer and M. Nastasi (paper A-D 2.4) and the stability of thin film amorphous metal alloys by F.W. Saris et al. in symposium E (paper E 3.1).

CONCLUSIONS

(1) There is an efficiency enhancement factor which makes ion mixing for surface alloying a lot more attractive than pure ion implantation. Moreover the sputtering limitation on the alloying concentration can be overcome and rather thick films can be built-up by repeated evaporation and ion irradiation.
(2) The ion mixing rate shows a temperature dependence the origin of which is still not clear.
(3) There is evidence that metastable alloys can be formed by ion mixing of systems that cannot be mixed by laser or electron irradiation. This must be due to ballistic effects by ion irradiation.
(4) The question of the role of thermal spikes and phase transitions in the cascades is not just of academic interest.
(5) A bombardment induced shift is observed of heavy impurities in lower mass substrates, which is a real physical effect and not an artifact of RBS analysis. It may distort SIMS profiles.
(6) Ion mixing can stimulate solid state reactions where diffusion or nucleation barriers act as limiting factors.

ACKNOWLEDGEMENTS

This work is part of the research program of the Stichting voor Fundamenteel Onderzoek der Materie (Foundation for Fundamental Research on Matter) and was made possible by financial support from the Nederlandse Organisatie voor Zuiver-Wetenschappelijk Onderzoek (Netherlands Organization for the Advancement of Pure Research).

REFERENCES

[1] Surface Modification and Alloying by Laser, Ion and Electron Beams, edited by J.M. Poate, G. Foti and D.C. Jacobson, Plenum Press, 1983.
[2] Ion Beam Modification of Materials, edited by B. Manfred Ullrich, North Holland - Amsterdam, 1985.
[3] A.D. Marwick in ref. 1 p.211.
[4] U. Shreter, F.C.T. So, B.M. Paine and M.A. Nicolet, MRS Symp.Proc.Vol. 27 (1984), edited by G.K. Huber et al. (Elsevier Science Publ.Co.Inc.) p.31.
[5] H.H. Andersen, Appl.Phys. 18 (1979) 131 and ref. 3.
[6] J. Bøttiger, S.K. Nielsen, H.J. Witlow and P. Wierdt, Nucl.Instr.and Meth. 218 (1983) 684.
[7] Wang Z.L.,.J.F.M. Westendorp and F.W. Saris, Nucl.Instr.and Meth. 209/210 (1983) 115.
[8] J.F.M. Westendorp, thesis Univ.Utrecht 1986.
[9] A.R. Miedema, Philips Techn.Rev. 36 (1976) 217.

414

[10] W.L. Johnson, Y.T. Cheng, M. van Rossum and M.A. Nicolet in ref. 2,
     p.657.
[11] Y.T. Cheng, M. van Rossum, M.A. Nicolet and W.L. Johnson, Appl.Phys.Lett.
     45 (1984) 185.
[12] J.F.M. Westendorp, P.K. Rol, J.B. Sanders and F.W. Saris, Nucl.Instr.
     and Meth. B 7/8 (1985) 616.
[13] J.F.M. Westendorp, F.W. Saris and U. Littmark, to be published, and
     U. Littmark, Nucl.Instr.and Meth. B 7/8 (1985) 684.
[14] M.L. Roush, F. Davarya, O.F. Goktepe and T.D. Andreadis, Nucl.Instr.
     and Meth. 209/210 (1983) 67.
[15] M. Nastasi, F.W. Saris, L.S. Hung and J.W. Mayer, J.Appl.Phys. 58 (1985)
     3052.
[16] R.B. Schwarz and W.L. Johnson, Phys.Rev.Lett. 51 (1983) 415.
[17] B.M. Clemens, W.L. Johnson and R.B. Schwarz, J.Non-Cryst.Solids 61-62
     (1984) 817.

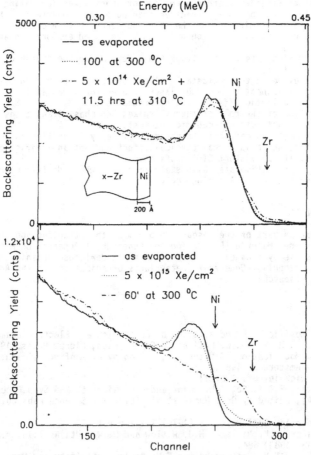

Fig. 7

RBS spectra of a thin film of Ni on top of a single crystal of Zr before and
after annealing to 300°C. Only after ion mixing with 5×10^{15} Xe ions does
subsequently annealing result in Ni reaction with Zr (below).

FORMATION OF SURFACE LAYERS OF ICOSAHEDRAL Al(Mn)*

J. A. KNAPP AND D. M. FOLLSTAEDT
Sandia National Laboratories, Albuquerque, NM 87185

ABSTRACT

   Surface layers of the icosahedral phase of Al(Mn) have been formed from
thin, alternating Al/Mn layers deposited on Al or Fe surfaces by rapid
electron-beam or laser melting, by ion beam mixing, and by solid-state
diffusion. The electron beam and laser treatments are similar to other liquid
quenching techniques used previously to form the phase, but have well defined
temperature histories which allow us to place limits on the melting point of
the icosahedral phase, the time needed for its nucleation from the melt, and
its growth velocity. Ion beam mixing is a way of forming the icosahedral
phase which is quite different from melt quenching; the phase is formed during
ion beam mixing at temperatures of 100-200°C. For mixing at <60C an amorphous
phase with icosahedral short-range order is formed; this phase can be
converted to the icosahedral phase by subsequent annealing. Formation of the
icosahedral phase by reacting the as-deposited layers in the solid state is a
new technique not previously reported. The results presented here place new
restrictions on proposed structural and thermodynamic models for the
icosahedral phase.

INTRODUCTION

   Recently, diffraction analysis of a melt-quenched alloy of Al-Mn produced
a very surprising result: a phase was found exhibiting sharp electron
diffraction spots, but with icosahedral orientational symmetry, which is
inconsistent with invariance under lattice translation.[1,2] This discovery
has inspired a large number of studies, both experimental and theoretical,
aimed toward understanding how this symmetry can be produced in an ordered
solid. One explanation which has been advanced is based on microtwinning of a
crystalline matrix [2-4]; however, a number of experimental observations
appear to argue against microtwinning.[1,5,6] Presently, many workers are
suggesting that the structure of the icosahedral phase is that of a
"quasicrystal", which uses two basic structural units to fill space in a non-
periodic manner [7] analogous to a Penrose tiling.[8] Such an atomic
arrangement would be a new class of ordered structure. Still others suggest
that the phase is a three dimensional analog of the hexatic phase [9], which
also shows long range orientational order and lacks translational symmetry,
but has more limited spatial ordering than a quasicrystal. Clearly,
determining the structure and thermodynamic properties of this novel
metastable phase and exploring other techniques which lead to its formation
are of considerable interest.
   We have been studying the formation of thin surface layers of the
icosahedral phase of Al(Mn) on metal substrates. Sample preparation begins
with the deposition of alternating Al and Mn layers on an Al or Fe substrate;
the layers are then converted to the icosahedral phase by melt quenching, ion
beam mixing or solid phase reaction. We reported earlier [10] that ion beam
mixing to a sufficient fluence with the substrate held at or above 100°C
converts the layers to fine grains of icosahedral Al(Mn). That work and
independent studies by Lilienfeld, et al.[11] were the first to show
icosahedral phase formation in the solid phase by ion beam mixing, as well as
the formation of an amorphous phase at 60°C or below. We also demonstrated

---

*This work performed at Sandia National Laboratories supported by the U.S.
Department of Energy under contract number DE-AC04-76DP00789.

that two different electron beam treatments of such layers on Al produced fine
grains of the icosahedral phase.[10]

More recently, even finer-grained (~2 nm) layers of icosahedral Al(Mn)
have been formed by using a 22 ns laser treatment, and larger-grained (0.5- 1
µm) layers containing the icosahedral and related phases by using electron
beam melting of the Al/Mn layers on Fe substrates. Although generally similar
to the rapid quenching methods which have been used previously to form the
icosahedral phase, our electron and laser beam treatments produce well-
defined, calculable temperature histories by which some of the thermodynamic
properties of the phase and the kinetic factors leading to its formation can
be deduced.

Finally, we have discovered that annealing thin deposited layers in the
solid state without any beam treatment also produces the icosahedral phase.
This transformation was observed during in situ annealing of the layers in the
transmission electron microscope (TEM). This result is analogous to the
formation of amorphous phases by solid-state diffusion which has recently been
reported.[12-14] This treatment may have important implications if the
icosahedral phase is found to have useful properties.

This paper will survey our methods for forming surface layers of the
icosahedral phase, and detail the inferences that may be drawn about the
properties of this novel metastable alloy.

RESULTS

Alternating Al and Mn layers were vapor deposited on electropolished
substrates of either Al or Fe to total thicknesses of 50-150 nm, typically
using 6-8 pairs of layers, in a vacuum of 5×10^{7} Torr. Deposition rates were
0.1-0.5 nm/s. Adjusting the relative layer thicknesses allowed samples with
Mn concentrations between 12 and 18 at.% to be prepared. The actual
concentration of Mn in the surface alloy was measured for each sample using
Rutherford backscattering spectrometry (RBS). Analysis using (d,p) nuclear
reactions was used to measure the total oxygen and carbon in the layers; for a
typical sample these were 3.5×10^{16} O/cm^2 and 1.7×10^{16} C/cm^2. The O is
believed to be mostly in surface oxides. Samples were analysed after
treatment by RBS and examined by TEM after backthinning by jet
electropolishing.

Electron-beam and laser treatments

Samples for the surface melting studies were given a preliminary ion beam

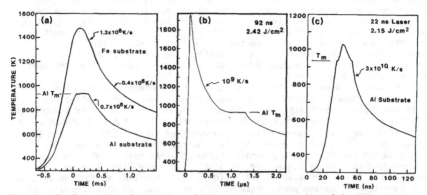

Fig. 1. Surface alloy temperature histories calculated for (a) LEBA treatment
on Al and Fe substrates; (b) PEBA treatment on Al; (c) laser treatment on Al.
Note the differences in time and temperature scales.

stayed there for ~200 µs. Figures 2(c) and (d) show results from such a layer with ~16 at.% Mn after treatment with the same beam conditions as for the Fe sample above. The diffraction pattern from the fine-grained layer is the ring pattern shown in Fig. 2(c). The d-spacings, excluding those of residual Al, all agree well with those of the most intense reflections of the icosahedral phase obtained with x-ray diffraction [18] as seen in Table I. The relative intensities of our observed rings are also in agreement. Figure 2(d) is a dark-field micrograph of the layer showing grains 10-50 nm in diameter which were imaged using the two brightest rings (d=0.217, 0.206 nm). The small grain size indicates a high nucleation density of $10^{16}-10^{17}/cm^3$. We infer that the alloy layer was highly undercooled with respect to the melting point of the icosahedral phase, and so the phase rapidly nucleated and solidified as a fine-grained layer on top of the still-molten Al substrate. Other evidence supporting this interpretation is given elsewhere.[19] The melting point of icosahedral Al(Mn) is thus believed to be above 660°C.

The other e-beam treatment was produced with a much faster, pulsed e-beam system (PEBA).[10] An alloy layer with ~16 at.% Mn on an Al substrate was again "stitched" with Xe and then treated with a 92 ns pulse of electrons, depositing a total energy of 2.4 J/cm². With this much faster energy deposition rate, the surface temperature was not constrained by the melting point of the substrate and rose to ~1700°C within 80 ns, while the the substrate melted to a depth of 4 µm. The temperature is shown in Fig. 1(b) to cool quickly (~0.9x10^9 K/s) down to the melting point of Al, where it remained for ~0.5 µs while the substrate resolidified. The resulting surface layer again consisted of very fine grains of the icosahedral phase, apparently because it was quenched so quickly to well below its melting point. Grains 10-30 nm in diameter are illuminated by dark-field imaging in Fig. 2(f), again using the two brightest rings in Fig. 2(e). Additional details of the e-beam treatments of these layers are presented elsewhere.[10,19]

An even faster quench and shorter melt time of the Al(Mn) surface alloy on Al has been achieved with a 22 ns, 2.15 J/cm² ruby laser pulse. The temperature history in Fig. 1(c) was calculated with an absorbed energy fraction of 0.135, obtained using the observed melt threshold of 2.05 J/cm². The temperature rose to ~760°C and remained above the melting point of Al for only 20 ns; after the substrate had solidified, the quench rate was 3x10^{10} K/s. The resulting surface alloy exhibits the smallest grains of the icosahedral phase that we have formed. Grains ~2 nm in diameter are illuminated with dark-field imaging in Fig. 3(b) using the bright ring in Fig. 3(a). Although the diffraction pattern does not show as many rings as are visible with larger grains, the rings match the most intense rings of the icosahedral phase, with the qualification that the bright ring corresponds to the bright doublet seen in Fig. 2(c). Also, there are more rings than observed for the amorphous phase (see below). The formation of the icosahedral phase on this very rapid time scale places significant restrictions on the kinetics of its nucleation and growth, as discussed later.

Ion beam mixing

Ion beam mixing of the deposited layers is a solid-state process quite different from melt quenching.[10,11] Bombardment of the deposited layers by a high energy ion beam produces cascades of displaced atoms which locally intermix the layers and produce considerable disorder. Depending on the temperature and composition of the sample, an amorphous phase may be retained, or more ordered phases may nucleate and grow. If the Al/Mn layers are irradiated at 60°C or less, we observe an amorphous phase. At 100-200°C, ion mixing to a sufficient fluence results in fine grains of the icosahedral phase. For one particular sample configuration, the crystalline phase Al$_6$Mn was also observed after mixing at elevated temperature.[20] All the ion beam mixing of alternating Al/Mn layers was done with Al substrates. The layers were irradiated with a Xe beam at 400 keV; a fluence of 1x10^{16} Xe/cm² was sufficient to complete the transformations of the deposited layers.

420

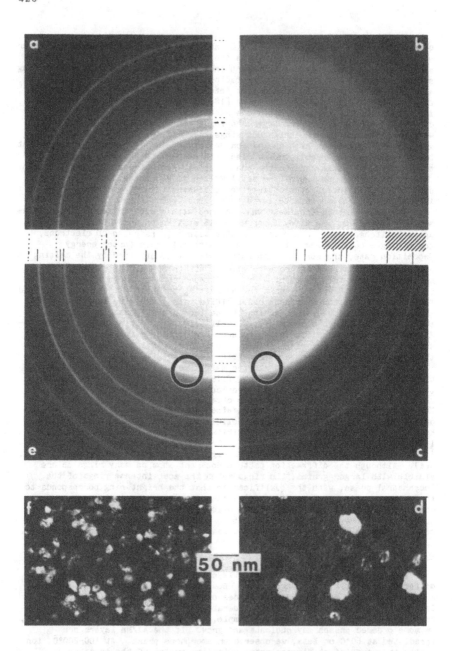

Fig. 4. Diffraction patterns and microstructures of Al/Mn layers before and after ion beam mixing with 1×10^{16} Xe/cm^2; (a) As-deposited layers. (b) Ion beam mixed at 60°C. (c) and (d) Ion beam mixed at 100°C. (e) and (f) Ion beam mixed at 200°C. Rings are marked as follows: icosahedral phase – solid lines, Al rings – dotted lines, Mn ring – dashed line, amorphous phase – hatched.

Figure 4 shows a series of TEM results illustrating the microstructures produced with different temperatures during ion beam mixing. The samples were attached to a Cu block during ion irradiation with a beam current less than 0.2 μa, and a thermocouple was used to monitor substrate temperatures. Figure 4(a) is an electron diffraction pattern from as-deposited layers with 16±1 at.% Mn. All the rings index to polycrystalline Al with the exception of the less intense ring at d=0.213 nm, which is near the position of the brightest ring of α-Mn (0.210 nm). Dark-field imaging showed 40 nm Al grains. After mixing a layer 50 nm thick of the same composition at 60°C, the surface was transformed to an amorphous phase, as shown by the diffuse rings in Fig. 4(b). The two outer rings have radii whose ratios to that of the bright inner ring (1.67, 1.95) are suggestive of local icosahedral coordination in the amorphous alloy.[21] The amorphous phase has also been produced in Al(Mn) alloys by high voltage electron irradiation of the icosahedral phase [11,22] and by very fast quenching of small droplets (~20 nm) of the molten alloy.[23]

When the layers were irradiated at 100°C, fine grains of the icosahedral phase resulted. Eight sharp rings are observed in the diffraction pattern in Fig. 4(c) which again agree in d-spacing and relative intensity with the most intense rings of the icosahedral phase. Dark-field imaging with the two brightest rings illuminated grains 25-45 nm in diameter, as shown in Fig. 4(d). When the same irradiation was done at 200°C, a similar microstructure resulted. The diffraction pattern, shown in Fig. 4(e), exhibits more rings

Table I. Diffraction Peaks from Icosahedral Al(Mn)

LEBA (a,b) d(nm)	PEBA (a,b) d(nm)	Ion Mixing (a,c) d(nm)	Intensity	Amor.+375°C d(nm) (a,b)	Melt Spinning(d) d(nm)	Intensity
		0.54(1)	w		0.542	
0.388	0.389	0.388(6)	s	0.391	0.385	22
0.336	0.335	0.335(5)	m	0.336	0.335	8
					0.314	
		0.297(4)	vw		0.286	1.5
0.254	0.254	0.254(3)	m (e)	0.257	0.252	3
0.218	0.217	0.217(2)	s	0.211 (f)	0.217	100
0.207	0.206	0.206(2)	s		0.2065	78
		0.193(2)	vw		0.1939	
		0.181(2)	w		0.1827	
		0.175(2)	vw		0.1757	1.5
0.150	0.150	0.150(1)	m	0.148	0.1496	11
0.146		0.146(1)	vw		0.1459	3
0.127	0.128	0.127(1)	m	0.128	0.1275	20
					0.1259	0.5
					0.1140	1
	0.110	0.110(1)	w		0.1101	5
0.109		0.1085(8)	vw	0.1085	0.1085	7
					0.1078	3
					0.1037	1
		0.0967(8)	vw		0.0962	1
		0.0955(8)	vw			
0.0904		0.0898(7)	vw			

(a) TEM, typical uncertainties in the last decimal place (in parenthesis) and relative intensities are given for one case. Intensities: strong (s), medium (m), weak (w) and very weak (vw).
(b) Obtained with 16±1 at.% Mn.
(c) Obtained with up to 17.5±1 at.% Mn.
(d) X-rays, with 14 at.% Mn [18]; includes all reflections with Intensity ≥ 1
(e) Possibly a pair of rings with spacings of 0.252(2) and 0.258(2) nm.
(f) Believed to be an unresolved doublet.

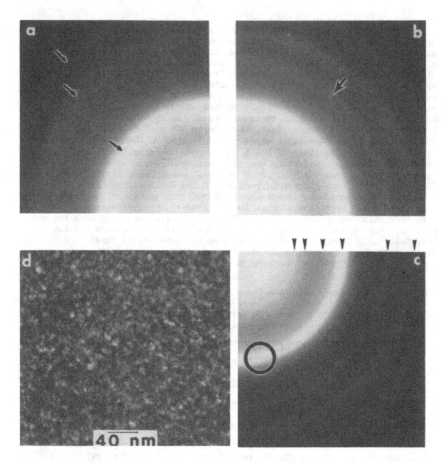

Fig. 5. Electron diffraction and microstructure obtained from an amorphous
Al(Mn) layer, produced by ion beam mixing at room temperature. (a) Amorphous
phase. (b) After annealing to 300°C. (c) and (d) After annealing to 375°C. The
rings due to the amorphous phase are marked in (a), then rings attributed to
the icosahedral phase are marked in (b) and (c).

than observed at 100°C, but the corresponding dark-field image in Fig. 4(f)
shows somewhat smaller grains, 10-25 nm in diameter. All of the rings again
fit those of the icosahedral phase.

When a sample with the icosahedral phase already present was ion
irradiated at room temperature to 1×10^{16} Xe/cm^2, the amorphous phase resulted.
Thus the icosahedral phase is not radiation stable at room temperature or
below, while it is obviously stable at its formation temperature of 100°C.
One sample configuration was examined which had a graded Mn concentration that
decreased from 17.5 at.% down to less than 8 at.% at the surface. When ion
beam mixed at an elevated temperature (not monitored), this sample exhibited
both the fine-grained icosahedral phase and circular grains of Al$_6$Mn 10 μm in
diameter.[20] This crystalline phase presumably nucleated and grew at the
depth for which the Mn concentration was appropriate (Al$_6$Mn is a line compound
at 14.3 at.% Mn.) The diffraction pattern from the icosahedral phase in this

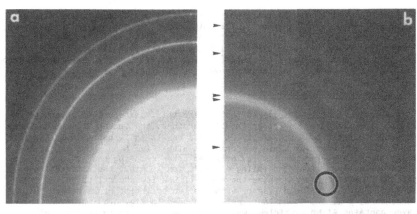

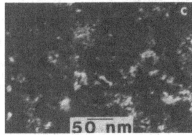

Fig. 6. Electron diffraction and microstructure obtained before and after annealing as-deposited Al/Mn layers _in situ_ in the electron microscope. (a) Before anneal. (b) After a 15 minute anneal at 350°C. Rings attributed to the icosahedral phase are marked.

sample exhibited a very complete set of 18 rings, all agreeing in d-spacing and relative intensity with values observed in x-ray examination of the icosahedral phase (see Table I). Subsequent irradiation of this sample at room temperature converted both the Al_6Mn and the icosahedral phase to an amorphous layer; thus Al_6Mn is also radiation stable at sufficiently elevated temperatures, but not at room temperature.

Annealing of the amorphous phase

If the amorphous phase formed by ion beam mixing at or below 60°C is subsequently annealed at increasing temperatures, both we [10] and others [11] observe that the next phase to form is the icosahedral phase; at still higher temperatures the crystalline phase Al_6Mn forms. Figure 5 shows TEM results from such an anneal sequence, performed _in situ_ in the microscope on a sample with an Al substrate which had been removed for examination of the amorphous layer. In Fig. 5(a) the diffraction pattern from the free-standing amorphous layer is seen to have a bright diffuse ring and two weak outer rings. Isochronal anneals of 15 min. duration were done at successively higher temperatures in 25°C increments. No change in the diffraction pattern was observed until 275°C, where an additional ring between the bright ring and the second ring was barely detectable. This new ring was more intense at 300°C, and is indicated by an arrow in Fig. 5(b); note that all the rings are somewhat less diffuse than in Fig. 5(a). At 300°C, a weak ring inside the bright ring could barely be distinguished; it grew in intensity and the rings continued to sharpen until the pattern in Fig. 5(c) was observed at 375°C. This pattern also shows a second inner, weaker ring, and an "intensity edge" just inside the bright ring. These features in Fig. 5(c) have d-spacings listed in Table I which are in reasonable agreement with the x-ray results for the icosahedral phase, with the qualification that the rings are too broad to resolve the doublet at the position of the bright ring. A dark-field image using the bright ring is shown in Fig. 5(d), where grains 3-6 nm are observed. When the anneal sequence reached 375°C, Al_6Mn began to nucleate where the surface layer was still in contact with the Al substrate, leaving the icosahedral phase in the free-standing film. At 400°C the Al_6Mn grew out into the rest of the layer, completely converting it within 4 minutes.

Annealing of as-deposited layers

We have recently discovered that if the thin, as-deposited layers of Al/Mn are annealed at an appropriate temperature, the icosahedral phase forms directly, without beam treatment.[24] This result is illustrated in Fig. 6. The anneal sequence was again performed in situ; Fig. 6(a) shows the Al and Mn rings of the as-deposited layers, obtained after removing the Al substrate by electropolishing. After annealing at 350°C for 15 min., the diffraction pattern from the layer exhibits the rings of the icosahedral phase, as shown in Fig. 6(b). Dark-field imaging with the two brightest rings showed icosahedral grains 5-30 nm in diameter, as seen in Fig. 6(c); however the layer contains Al_6Mn particles, which may have also been imaged in Fig. 6(c).

This process of solid-state conversion to the icosahedral phase is apparently dependent on the thickness of the alternating Al and Mn layers (typically we use 10-12 nm for Al and 2-3 nm for Mn). An attempt to observe interdiffusion between thicker Al and Mn layers at these temperatures, using RBS, was unsuccessful. It may be that the Al and Mn interdiffuse only over very short distances and form an interfacial layer of the icosahedral phase, which might then be a barrier to further diffusion. Further experiments on this process will be presented elsewhere.[24]

DISCUSSION

A number of deductions may be made about properties of the icosahedral phase from this work. With the e-beam and laser treatments the well-defined and calculable temperature histories permit limits to be placed on some of the thermodynamic properties of the material and kinetic factors leading to its formation. The observation of a fine-grained layer which solidifies on top of the molten Al substrate produced with LEBA implies two things. First, since the molten alloy was apparently highly undercooled (leading to the high density of nucleation sites with random orientation), the melting point of the icosahedral phase must be higher than 660°C. We can also take the melting point to be less than the liquidus in the equilibrium phase diagram [25] at 20 at.% Mn (which is now thought to be the central composition of the phase [26]). Thus limits of 660°C < T_m < 960°C are obtained for the icosahedral phase. The second implication is that the icosahedral phase is stable against transformation to other phases for at least 200 µs at 660°C.

The observation of very small icosahedral grains after the laser treatment places an upper limit on the time needed to nucleate the phase. Since the phase forms as small, randomly oriented grains, we infer that it again nucleated within the melt before the Al substrate solidified. With this interpretation, the time spent at or above 660°C gives an upper limit for the time required for nucleation: t_n < 20 ns. Even if the phase doesn't nucleate until a somewhat lower temperature in this case, the quench rate is so high that this limit would probably be increased by only 5-10 ns.

The LEBA result using an Fe substrate, which produced 0.5-1 µm grains of the icosahedral phase, also has two implications. Since many of the resulting grains are of the T phase, which is known to replace the icosahedral phase at slower quench rates [16], the quench rate for this sample must be near the limit for forming the icosahedral phase at this Mn concentration (~16 at.%). The calculation shows that the maximum quench rate was 1.3×10^6 K/s at 1100°C, and decreased to 0.7×10^6 K/s at 660°C. Thus we believe icosahedral phase formation requires a cooling rate $\dot{T} > \sim 1.0 \times 10^6$ K/s. The second inference is a limit on the growth rate for the phase. Since the nucleation density is much lower than for the LEBA-treated sample using an Al substrate, the layer must have finished solidifying before it cooled to 660°C. Knowledge of the time spent above this temperature, and the grain size, leads to a lower limit of ~1 cm/s for the growth rate of the icosahedral grains with this temperature history.

The ion beam mixing and annealing results provide information about radiation stability and nucleation of the icosahedral phase. First, as already mentioned, both the icosahedral phase and Al_6Mn are stable against radiation damage, but only at temperatures of 100°C and above. Either phase is converted to an amorphous phase under ion bombardment at room temperature. The nature of the Xe ion cascade should change little between 60° and 100°C, yet the microstructure produced changes from amorphous to icosahedral. This change implies that the icosahedral phase does not form within the ion cascades, but rather nucleates and grows during the subsequent defect evolution.

The transformations observed during annealing of the amorphous phase clearly indicate that it is the least stable of the phases examined here; it is also the most disordered. The icosahedral phase is apparently lower in free energy than the amorphous phase and has increased order. Since other studies show that the T phase becomes more dominant at slower quench rates [16], it is apparently somewhat more stable than the icosahedral phase, although the icosahedral phase transforms to Al_6Mn at 375°C with no evidence for T phase formation. The crystalline phase Al_6Mn is believed to be a stable equilibrium phase [25]; our results indicate that it has the lowest free energy, at least above 375°C. Thermodynamic modeling of this alloy system must include these relative stabilities.

The amorphous phase with its short-range icosahedral order has fundamental significance for a theoretical understanding of the icosahedral phase with its long-range order. Landau theories use this short-range order for liquids to deduce that the formation of a solid with long-range icosahedral order can be favored over other symmetries.[21] Since the amorphous phase has a quenched liquid structure, its observation supports the application of such theories to icosahedral Al(Mn). More generally, the observation of icosahedral short-range order in the amorphous phase complements the icosahedral symmetries of the other phases seen here. Of course, the icosahedral phase shows this symmetry in its long-range order, and perhaps even in its basic structural units.[27] The crystalline phase Al_6Mn contains icosahedral units within its orthorhombic unit cell.[28] The amorphous phase is thus the most disordered member of a set of Al(Mn) phases related to each other by icosahedral symmetry.

The formation of the icosahedral phase by solid-state interdiffusion of the deposited layers is an important new observation. This phenomenon is similar to the process of forming amorphous alloys by solid-phase annealing of alternating, thin layers of Au-La [12], Ni-Hf [13], or Ni-Zr [14]. The heat of mixing has been used to provide the thermodynamic driving force needed to explain the transformation of the two crystalline phases to an amorphous phase, although questions concerning the purity of the binary alloys have been raised.[29] In any case, the capability to form the icosahedral phase by such a relatively simple method may have important implications for other research as well as for potential application using powder metallurgy if such alloys are found to have useful properties.

Our observations of the icosahedral phase provide restrictions that must be accounted for by structural models. We have argued [10] that the observation of electron diffraction rings at the same spacings and relative intensities as those observed from alloys with much larger grains implies that each grain in our layers contains essentially the full icosahedral symmetry. In some microtwinning models [2,3], the icosahedral symmetry is accounted for by multiple diffraction between 2 or more of 20 possible microtwins in a single grain. For our fine-grained alloys, this suggests that several of the 20 microtwins would have to be present in each of the grains, some of which for the e-beam samples are 10 nm in diameter, and only 2 nm in diameter for

426

the laser. This requirement appears to be a serious restriction for such microtwinning models. Another model [4] accounts for the d-spacings of the icosahedral phase with a complicated crystalline structure whose unit cell is 2.67 nm on a side; since our smallest observed grains are only 2 nm in diameter this model also appears to be incorrect. Whatever structural models are proposed for the icosahedral phase must also account for nucleation within very short times (< 20 ns) from the liquid, as well as nucleation within Al/Mn layers during ion irradiation and thermal interdiffusion.

ACKNOWLEDGEMENTS

The authors wish to thank P. S. Peercy for his collaboration in the laser treatments, and whose results are included here. We also thank R. J. Birgeneau for valuable discussions. Technical assistance by G. L. Schuh and M. P. Moran is gratefully acknowledged.

REFERENCES

1. D. Shechtman, I. Blech, D. Gratias and J. W. Cahn, Phys. Rev. Lett. 53, 1951 (1984).
2. R. D. Field and H. L. Fraser, Mat. Sci. Engr. 68, L17 (1984).
3. M. J. Carr, submitted to J. Appl. Phys.
4. L. Pauling, Nature 317, 512 (1985).
5. R. Gronsky, K. H. Krishnan and L. Tanner, in Proceedings of the 43d Annual Meeting of the Electron Microscopy Society of America, ed. G. W. Bailey (San Francisco Press, San Francisco, 1985), p. 34.
6. D. Shechtman, D. Gratias and J. W. Cahn, C. R. Seances Acad. Sci., Ser. 2 300, 909 (1985).
7. D. Levine and P. J. Steinhardt, Phys. Rev. Lett. 53, 2477 (1984).
8. R. Penrose, Bull. Inst. Math. and Its Appl. 10, 266 (1974); see also M. Gardner, Sci. Amer. 236, 10 (1977).
9. D. R. Nelson and B. I. Halperin, Phys. Rev. B19, 2456 (1979).
10. J. A. Knapp and D. M. Follstaedt, Phys. Rev. Lett. 55, 1591 (1985).
11. D. A. Lilienfeld, M. Nastasi, H. H. Johnson, D. G. Ast, and J. W. Mayer, Phys. Rev. Lett. 55, 1587 (1985).
12. R. B. Schwarz and W. L. Johnson, Phys. Rev. Lett. 51, 415 (1983).
13. M. Van Rossum, M.-A. Nicolet, and W. L. Johnson, Phys. Rev. B29, 5498(1984).
14. Y.-T. Cheng, W. L. Johnson, and M.-A. Nicolet, Appl. Phys. Lett. 47, 800 (1985).
15. J. A. Knapp and S. T. Picraux, J. Appl. Phys. 53, 1492 (1982); J. A. Knapp, J. Appl. Phys. 58, 2584 (1985).
16. L. Bendersky, R. J. Schaefer, F. S. Biancaniello, W. J. Boettinger, M. J. Kaufman and D. Shechtman, Scr. Met. 19, 909 (1985).
17. L. Bendersky, Phys. Rev. Lett. 55, 1461 (1985).
18. P. A. Bancel, P. A. Heiney, P. W. Stephens, A. I. Goldman, and P. M.Horn, Phys. Rev. Lett. 54, 2422 (1985).
19. D. M. Follstaedt and J. A. Knapp, 1985 MRS symposium J proceedings.
20. D. M. Follstaedt and J. A. Knapp, to appear in J. Appl. Phys.
21. D. R. Nelson and B. I. Halperin, Science 229, 233 (1985).
22. K. Urban, N. Moser and H. Kronmuller, to appear in Phys. Stat. Sol.
23. S. D. Ridder, L. A. Bendersky and F. S. Biancaniello, presented at The Metallurgical Society fall meeting, Oct. 16, 1985, Toronto, Canada.
24. D. M. Follstaedt and J. A. Knapp, to be published.
25. Metals Handbook (American Society for Metals, Metals Park, Ohio, 1973), 8th ed. Vol. 8, p. 262.
26. R. J. Schaefer, L. A. Bendersky, D. Shechtman, W. J. Boettinger and F. S. Biancaniello, to be published.
27. P. Guyot and M. Audier, Phil. Mag. B52, L15 (1985).
28. W. B. Pearson, The Crystal Chemistry and Physics of Metal Alloys, (Wiley-Interscience, New York, 1972), p. 714.
29. J. J. Hauser, Phys. Rev. B32, 2887 (1985).

ION IRRADIATION INDUCED AMORPHOUS TO QUASICRYSTALLINE TRANSFORMATION: COMPOSITION DEPENDENCE IN THE ALMN SYSTEM

D. A. Lilienfeld, M. Nastasi[a], H. H. Johnson, D. G. Ast and J. W. Mayer
Department of Materials Science and Engineering
Cornell University, Ithaca, New York 14853

ABSTRACT

Multilayered $Al_{84}Mn_{16}$ films ion irradiated with Xe at room temperature transform into an amorphous phase. Subsequent Xe ion irradiation at 150C transforms the amorphous to the quasicrystalline phase. The quasicrystalline structure may be formed in the composition range of 91 atomic % Al to 80 atomic % Al.

INTRODUCTION

Since the discovery of the quasicrystalline state by Schectman et al.[1], the primary method for formation of the material has been melt spinning. Recent results have shown that ion beam mixing is effective in producing quasicrystalline material by a solid state reaction.[2,3] To date there has been no systematic investigation of the compositional dependence of the solid state quasicrystalline transformation.

We present results for the AlMn system for the compostion range of 95 atomic % Al to 31 atomic % Al and compare the results with structural models for quasicrystals.[4,5] These results are consistent with the current structural models.

The samples were first processed at room temperature (RT) to form the amorphous phase and then were processed at elevated temperature to form the quasicrystalline state. RT Xe^{++} ion irradiation (600 KeV $8x10^{15}$ ions/cm^2) was incapable of transforming the alloys into the quasicrystalline state. In the quasicrystal composition range, the alloys became amorphous: outside that range, the samples were amorphous or crystalline. Results of our previous work[2] show that 150°C is the optimum substrate temperature for the quasicrystalline transformation in the AlMn system.

EXPERIMENTAL

The samples used in this study consist of two types: the first is a compositionally varying sample which will be referred to as the wedge sample and the second is a fixed composition multilayer sample . The wedge samples are formed as a continuously varying thickness sample as shown in Figure 1. The substrate is rotated above an aperture which regulates the amount of material which is deposited on the substrate in such a way as to form a wedge of each material. This technique is described in detail in the paper by Nastasi et al.[3] The sample composition was checked by Rutherford backscattering (RBS) which showed that the composition gradient was approximately linear with a variation of 4 at %/mm but there was a definite non-linearity present. Experimentally this nonlinearity is not a problem for

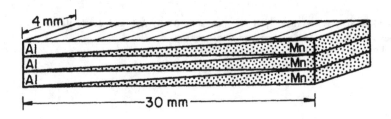

FIGURE 1. Schematic diagram of a wedge sample showing the linear composition gradient.

two different reasons. The first is that a wedge produces up to four sets of 2mm long specimens. Thus it is possible to perform 3 different experiments while retaining an unreacted set of samples and know the composition relationship of each transmission electron microscope (TEM) grid. Secondly all specimens were analyzed by energy dispersive x-ray spectrometry(EDS) individually which eliminates any error due to offsets between specimens.

Each substrate is then cleaved into four 1mm wide and 30mm long sections. Then each section is cleaved into 15 2mm long specimens and floated onto TEM grids. Single composition samples were prepared by depositing alternate layers of Mn and Al to form a multilayer sample approximately 500Å thick.[2] Samples were all given a room temperature(RT) 8×10^{15} 600keV Xe++ ions/cm^2 ion irradiation to render them amorphous prior to elevated temperature ion irradiations using 4×10^{15} Xe^{++} ions/cm^2.

Xenon ions, rather than Ne or Ar, were chosen for the irradiating species because ion mixing studies have shown this ion to be most efficient in producing homogeneous amorphous layers in RT irradiations. These ions also produce clear cut phase transformations in elevated temperature irradiations.[7]

The average composition of an unreacted AlMn multilayer film was determined by RBS while it was on the NaCl substrate. A sample of this film was placed on a TEM grid after the NaCl was dissolved. The sample was then examined by EDS and the intensity ratio of the x-rays was determined. EDS in the TEM was calibrated by the method of Barbour et al.[8]

The compositions and irradiation conditions are given in Table I.

Table 1.   Compostions and temperatures of ion irradiation for the AlMn wedge
and AlMn multilayer samples.
==

Sample	Al Composition	Xe ion irradiation temperature			
		RT	100	150	200
ALMn Wedge					
W1	not used	√		√	
W2	95 - 88	√		√	
W3	91 - 83	√		√	
W4	84 - 77	√		√	
W5	71	√		√	
W6	61 - 54	√		√	
W7	50 - 43	√		√	
W8	39 - 33	√		√	
W10	37 - 30	√		√	
W11	31	√		√	
Multilayer					
AlMn	84	√	√	√	√

RESULTS AND DISCUSSION

The focus of all the research has been on the amorphous to
quasicrystalline transformation. With the wedge samples, the composition
range for RT amorphization of AlMn can be investigated. In figure 2 are
presented electron diffraction patterns from three regions of different
composition. Pattern A in figure 2 is from a region with a composition of 95
at % Al and is not really amorphous but has crystalline rings which are not
due to fcc Al. Pattern B is amorphous and corresponds to a composition of 88
at % Al. Pattern C is not amorphous and is from an area with a composition of
78 at % Al. Below 78 at % Al, the RT ion irradiated samples were never
completely amorphous.

Samples that were RT ion irradiated and subsequently irradiated at 150°C,
yield the patterns seen in figure 3. Pattern A is from a region with 94 at %
Al and is not quasicrystalline. Patterns similar to this pattern exist down
to 91 at % Al. Below 91 at % Al, pattern B is found. This diffraction
pattern is quasicrystalline and persists to 80 at % Al. Below 80 at % Al, the
pattern that is found is shown in C of figure 3. This pattern is not
quasicrystalline and no quasicrystalline patterns are found below 80 at % Al.

The structural models that have been proposed by Guyot and Audier[5] and
Henley and Elser[4] can be used to predict a composition for the
quasicrystalline material. The Guyot and Audier model predicts a composition
of approximately $Al_{80}Mn_{20}$. The observation of a lower limit to the
quasicrystal formation is an indication that the material has a finite phase
field width and that the lowest composition of that field is 80 atomic % Al.

Figure 2. RT ion irradiation results for the AlMn wedge. The electron diffraction patterns are from regions of different compositions. The compositions are for pattern A, 95 atomic % Al, for pattern B, 88 atomic % and pattern C, 76 atomic % Al.

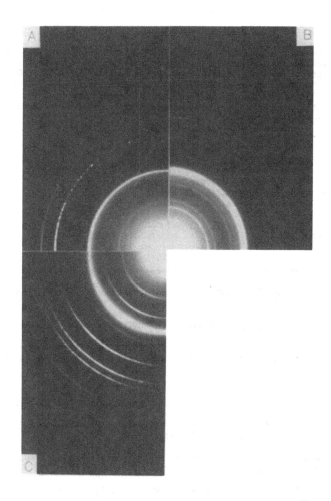

Figure 3. 150°C ion irradiation results for the AlMn wedge. The electron diffraction patterns are from regions of differing composition. The compositions are for pattern A, 95 atomic % Al, for pattern B, 88 atomic % Al and for pattern C, 76 atomic % Al.

CONCLUSIONS

There is a a lower Al composition below which it is impossible to form quasicrystals. The lower limit of 80 atomic % Al is consistent with two of the structural models.[4,5] For ion induced transformation at 150°C: above 91 atomic % Al, the samples transformed to a crystalline state. Between 91 atomic % Al and 80 atomic % Al, the samples transformed to the quasicrystalline state. Below 80 atomic % Al the samples were always crystalline.

ACKNOWLEDGEMENTS

Work was supported in part by NSF (L. Toth).

REFERENCES

[a] Present address: Los Alamos National Laboratory, Los Alamos, N. M.

[1] D. Schectman, I. Blech, D. Gratias, and J. W. Cahn, Phys. Rev. Lett. 53 1951 (1984)

[2] D. A. Lilienfeld, M. Nastasi, H. H. Johnson, D. G. Ast and J. W. Mayer, Phys. Rev. Lett. 55 1587 (1985)

[3] J. A. Knapp and D. M. Follstaedt, Phys. Rev. Lett. 55 1591 (1985)

[4] P. Guyot and M. Audier, Phil. Mag. B 52 L15 (1985)

[5] V. Elser and C. L. Henley, Phys. Rev. Lett. 55 2883 (1985)

[6] M. Nastasi, J. C. Barbour, J. Gyulai, L. S. Hung and J. W. Mayer, J. Vac. Sci. Technol. A 3 1903 (1985)

[7] M. Nastasi, D. A. Lilienfeld, H. H. Johnson and J. W. Mayer, submitted to J. Appl. Phys.

[8] J. C. Barbour, K. Sickafus and M. Nastasi, J. Vac. Sci. Technol. A 3 1895 (1985)

PHASE TRANSFORMATIONS IN NICKEL-ALUMINUM ALLOYS DURING ION BEAM MIXING

JAMES ERIDON*, LYNN REHN**, AND GARY WAS*
*University of Michigan, Dept. of Nuclear Engineering, Ann Arbor, Michigan 48109
**Argonne National Laboratory, Materials Science and Technology Division, Argonne, Illinois 60439

ABSTRACT

The effect of ion beam mixing of nickel-aluminum alloys with 500 keV krypton ions has been investigated over a range of temperature, composition, ion dose, and post-irradiation thermal treatments. Samples were formed by aternate evaporation of layers of aluminum and nickel. A portion of these samples was subsequently annealed to form intermetallic compunds. Irradiations were performed at both room temperature and 80 K using the 2 MV ion accelerator at Argonne National Laboratory. Phase transformations were observed during both in situ irradiations in the High Voltage Electron Microscope(HVEM) at Argonne and also in subsequent analysis of an array of irradiated samples. Electron diffraction indicates the presence of metastable crystalline structures not present in the conventional nickel-aluminum phase diagram. Transformations occur at doses as low as 5×10^{14} cm^{-2} and continue to develop as the irradiation progresses up to 2×10^{16} cm^{-2}. Layer mixing is followed through Rutherford Backscattering analysis. Samples are also checked with x-rays and Electron Energy Loss Spectroscopy (EELS). A thermodynamic argument is presented to explain the phase transformations in terms of movements on a free energy diagram. This analysis explains the interesting paradox concerning the radiation hardness of the NiAl[1] phase and the amorphous structure of mixed Ni-50% Al layers[2].

INTRODUCTION

Ion beam mixing of thin layers offers a method of preparing alloys with unique structures. This is due to the thermodynamically non-equilibrium nature of the alloys so formed. This same attribute - metastability - can be found in structures formed by the irradiation of equilibrium alloys. These two processes can be followed on a thermodynamic free energy diagram as shown in Figure 1.

Several alloys become amorphous when irradiated with heavy ions[3]. It has been noted that mixing at liquid nitrogen temperature of Ni/Al layers in a composition of 50% produces an amorphous structure[2]. However, irradiation of the intermetallic bcc phase NiAl at the same temperature produces no such transformation. Presumably, this result could be explained by the alternate paths which each process must follow on the free energy diagram of Figure 1. If the free energy of the amorphous phase lies between that of the pure elements and that of the intermetallic NiAl, then it should be easier to reach that point by lowering the free energy of the layers during mixing, rather than by raising the free energy of the intermetallic.

In order to test this supposition, and to investigate various theories of phase formation during ion beam mixing, samples of Ni/Al alloys were prepared in various compositions as both annealed intermetallics and elemental layers, and irradiated with heavy ions over a range of doses at both room temperature and 80 K. These samples were examined in a transmission electron microscope and analyzed for phases present with electron diffraction.

434

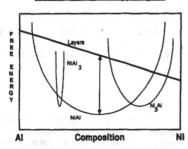

Schematic Free Energy Diagram

Figure 1. Schematic free energy diagram of the Ni/Al system showing energies of three intermetallic phases as well as the energy of a layered structure.

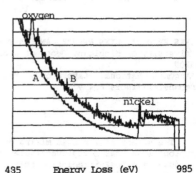

Figure 2. EELS plots showing the presence of oxygen in evaporated sample with outer Al layer (A), and lack of oxygen in similar sample with Ni outer layers (B).

EXPERIMENT

Samples were prepared by sequential electron-gun evaporation of nickel and aluminum layers on 300 mesh copper grids. This was done in a vacuum of better than 10^{-6} Torr. These grids were mounted in Crystalbond® and were soaked in acetone following evaporation to remove them from the mount. The resulting thin films smoothly spanned the holes in the mesh. All samples were prepared with nickel forming the outermost layers to impart oxidation resistance. The importance of this feature can be demonstrated by the EELS plot in Figure 2, which shows the presence of oxygen in a thin film in which aluminum forms the outermost layer, and the absence of oxygen in a similar film in which the aluminum is completely sandwiched between nickel layers. Despite the fact that some aluminum appears (from RBS) to migrate to the surface even in sandwiched samples, the nickel outer layers prevent the formation of any measurable oxidation. Impurities can have an important effect on the transformation kinetics and must therefore be carefully controlled.

Samples were prepared with three compositions corresponding to three intermetallics found in the Ni/Al phase diagram - $NiAl_3(\epsilon)$, $NiAl(\beta')$, and $Ni_3Al(\gamma')$. The NiAl samples were irradiated in the HVEM at Argonne National Laboratory. The other samples were irradiated in an ion pumped target chamber with a vacuum of better than 10^{-7} Torr at Argonne. Some samples were first annealed for 1 hour at 450°C in order to form the intermetallics mentioned previously. All samples were irradiated with 500 keV krypton ions which have a range of about 1200Å, far exceeding the thickness of the films (about 750Å). Doses ranged from 2×10^{14} cm^{-2} to 5×10^{16} cm^{-2}, and samples were monitored during irradiation with an infrared monitor to ensure that no samples suffered irradiation heating exceeding the detection limit of the monitor, which is 100°C. Samples were analyzed using electron diffraction to identify phases. Mixing was monitored by Rutherford Backscatering performed with 1.8 MeV helium ions on layers prepared and irradiated on glass slides. There was some concern that mixing of the glass substrate into the surface layers would affect the nickel/aluminum mixing efficiency, but similar experiments performed using polished nickel substrates instead of glass have shown the same results.

RESULTS

NiAl$_3$

In all cases, samples of NiAl$_3$ composition formed an amorphous phase with some residual elemental aluminum. Intermetallic NiAl$_3(\epsilon)$ became amorphous at the lowest dose of 2×10^{14} cm^{-2}, while the layers needed to be mixed to a dose of 2×10^{16} cm^{-2} before the amorphous phase appeared. At lower doses, the layered structure underwent different transformations at room temperature than at 80 K. At 80 K, the diffraction pattern indicated a smooth transition from elemental aluminum and nickel to less elemental aluminum and amorphous with increasing dose. At room temperature, the intermetallic NiAl formed at the lowest dose. As the mixing proceeded up to 1×10^{16} cm^{-2}, the β' rings faded and disappeared and a new ring appeared at a lattice spacing of 1.26Å and grew in intensity. Then, between 1 and 2×10^{16} cm^{-2}, the sample became amorphous with a residual amount of crystalline aluminum. The phase transformation can be followed in Figure 3, while the RBS spectra showing the mixing can be seen in Figure 4. The mixing of the layers is not complete, even at the high dose of 5×10^{16} cm^{-2}.

Ni$_3$Al

The intermetallic phase Ni$_3$Al(γ') irradiated at 80 K showed the simplest response, merely becoming disordered at 2×10^{14} cm^{-2} and remaining disordered up to 2×10^{16} cm^{-2}. The other three structures, including the layers irradiated at both room temperature and 80 K, and the γ' sample irradiated at room temperature, showed more interesting behavior. In all cases, the β' phase formed at low doses, and disappeared after a dose of 1×10^{16} cm^{-2} was reached. At this point, a very strong ring appeared at 2.21Å, along with several fainter rings. The lattice spacings of the faint rings vary with each set of samples, but the strong ring at 2.21Å is common to all three structures. This lattice spacing is not found in any Ni/Al intermetallic. In order to check for possible contamination as a cause of this strong ring, a mixed sample was annealed in a hot stage in a STEM at 400°C for 1/2 hour. The sample quickly reverted to a Ni$_3$Al ordered structure and the 2.21Å ring faded. This sequence of transformations in shown in Figure 5, along with a sequence of RBS spectra in Figure 6 showing the extent of mixing at different doses.

NiAl

Intermetallic NiAl(β') showed only some incomplete disordering after receiving a dose of 1×10^{16} cm^{-2} at -140°C. The Ni/Al layers irradiated at this temperature and composition showed more interesting behavior. The initial Ni/Al rings faded quickly, and at a dose of 1×10^{15} cm^{-2} the diffraction pattern showed an amorphous ring centered about 2.0Å. As the irradiation progressed, this hazy ring proceeded to sharpen, and other rings began to appear. At a dose of 2×10^{16} cm^{-2}, the amorphous ring was gone and a pattern showing disordered β' was present.

436

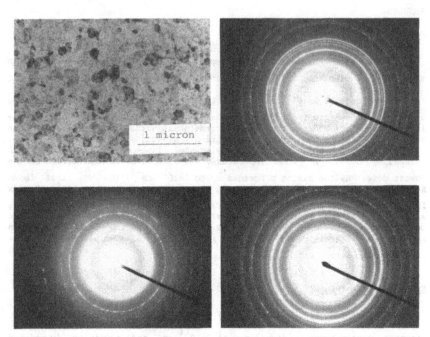

Figure 3. Micrograph and diffraction patterns from Ni-75%Al layers irradiated at room temperature showing (clockwise from upper left) initial evaporated fine grained structure; Al and NiAl rings formed at $2 \times 10(15)/cm^2$; Al rings and unknown bright ring at 1.26 A formed at $1 \times 10(16)/cm^2$; Al rings and amorphous halo formed after mixing to $5 \times 10(16)$.

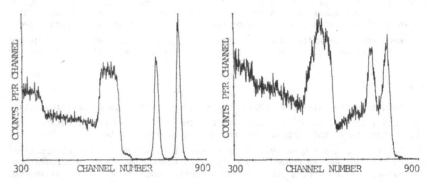

Figure 4. Rutherford Backscattering spectra showing initial layered structure of Ni-75%Al samples (left) and the same sample following mixing to a dose of $5 \times 10(16)$ ions/cm^2. Notice that the mixing is not complete.

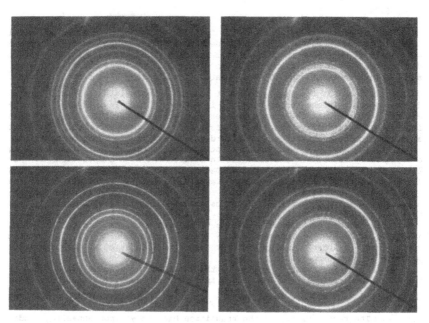

Figure 5. Sequence of diffraction patterns from Ni-25%Al layers irradiated at room temperature showing (clockwise from upper left) Ni and NiAl
rings at 2x10(15)ions/cm^2; Ni rings and two bright unknown rings around
the Ni (111) ring at 1x10(16); Ni rings and a bright unknown ring at
2.21 A; ordered Ni$_3$Al rings formed following annealing at 400 C for 1 hour.

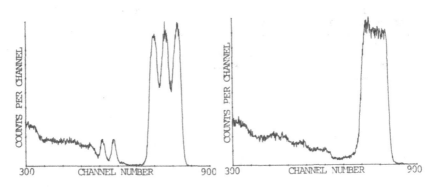

Figure 6. Rutherford Backscattering spectra showing initial layered
structure of Ni-25%Al samples (left) and the same sample following mixing
to a dose of 2x10(16)ions/cm^2. Notice that the mixing is complete.

DISCUSSION

When considering what phases will form during ion beam mixing of metal layers, thermodynamics will determine which phases are most likely, and kinetics will determine which ones will actually form. In the Ni/Al system, β' is the most stable intermetallic with a heat of formation of 13 kcal/g-atom, compared with 9 and 10 kcal/g-atom for ϵ and γ', respectively[4]. During mixing, β' frequently forms at low and intermediate (0.5–5×10^{15} cm^{-2}) doses, probably in the region near the interface of the Ni/Al layers. As the mixing proceeds and the sample becomes more homogeneous, the β' becomes less favorable energetically than other phases, such as the amorphous phase at aluminum-rich concentrations, and disordered γ' at nickel rich concentrations. Apparently, there is yet another structure, as yet unidentified, which is energetically favored in the irradiation environment at a composition of Ni-25%Al and which has a lattice plane spacing of 2.21Å in some direction. This phase can be formed at 80 K by mixing layers, but not by irradiation of the intermetallic γ' at this temperature. This illustrates the important interplay between thermodynamics and kinetics in determining phase formation. As the layers are mixed, the sample lowers its free energy and the new phase is inevitably favored energetically and may form even with the limited kinetics available at 80 K. The intermetallic γ' must, on the other hand, rise in free energy and re-order itself, and this cannot be done at 80 K, although it is possible at room temperature.

An interesting question raised by this work is the precise structure of the high dose Ni$_3$Al samples. The intensity of the 2.21Å ring indicates a large diffracting volume, while the absence of an entire set of new rings may mean that the new ring is the result of a minor modification of the basic fcc lattice. The reversibility of the transition back to γ' upon annealing indicates that it is not the result of an oxide or some other impurity. The further investigation of this structure should prove interesting.

REFERENCES

1. L.S. Hung, M. Nastasi, J. Gyulai, and J.W. Mayer, Appl. Phys. Lett. 42, 672 (1983).
2. J. Delafond, C. Jaouen, J.P. Riviere, and C. Fayoux, Materials Science and Engineering 69, 117 (1985).
3. J.L. Brimhall, H.E. Kissinger, and L.A. Charlot, Radiation Effects 77, 237 (1983).
4. R. Hultgren, R.L. Orr, P.D. Anderson, and K.K. Kelly, Selected Values of the Thermodynamic Properties of Metals and Binary Alloys. Wiley, New York (1963).

SILICIDE FORMATION BY HIGH DOSE TRANSITION METAL
IMPLANTS INTO Si.

F. H. SANCHEZ[a], F. NAMAVAR, J. I. BUDNICK, A. FASIHUDIN AND H. C. HAYDEN.
Department of Physics and Institute of Materials Science, University of
Connecticut, Storrs, CT 06268.

ABSTRACT

We report preliminary results of a study on silicide formation by means of
high dose transition metal implants into Si (100) single crystals.

100 keV Cr^+, Fe^+, Co^+ and Ni^+ were implanted at room temperature. For the
Cr^+, Fe^+ and Ni^+ implants, no silicide formation was observed after
implantation. However, both Rutherford Backscattering Spectrometry (RBS) and
X-Ray Diffraction (XRD) results clearly indicated the existence of $CrSi_2$
after the Cr-Si samples were annealed 4 hours at 550 °C. In the case of the
Fe^+ and Ni^+ implants, $FeSi_2$ and $NiSi_2$ were identified by XRD after annealing
the implanted samples half an hour at 400 °C. A layer of CoSi of about 1000 Å
was observed in the as implanted Co-Si samples by both RBS and XRD.

Ni^+ ions accelerated to 150 keV were implanted at 350 °C. A much broader
distribution and higher retention of Ni was obtained in this case, showing
evidence of long range atomic diffusion. NiSi and polycrystalline silicon
were observed by XRD in the as implanted samples.

The possibility of high dose ion implantation as a suitable technique for
producing transition metal silicides is discussed.

INTRODUCTION

Extensive studies on the formation of thin film transition metal silicides
have been carried out because of their technical applications (1) mainly in
Very Large Scale Integration (VLSI) technology. Usually, silicides are
produced by solid state reaction between a film of metal and the silicon
substrate. Conventional and rapid annealing processes are widely employed for
this purpose, and, to a lesser extent, ion beam mixing. By means of these
techniques, stable and metastable silicides of almost all transition metals
have been produced.

The question of whether ion implantation can be considered an adequate
technique to produce thin film contact silicides still remains unanswered. It
has been predicted (2) that sputtering of the target will impose a limit to
the achievable concentration of the implanted element, the maximum
concentration being inversely proportional to the sputtering coefficient. But
even within this limitation, and taking into account preferential sputtering,
it still could be possible to produce a wide range of transition metal
silicides using this technique. Furthermore, in recent works (3,4) we showed
that high concentration and retention of chromium and aluminum implanted into
silicon can be attained.

Here we present the preliminary results of an ongoing research program to
investigate fundamental aspects of silicide formation by means of ion

implantation. Even when it is not addressed to specific technological questions, we believe that the results of such investigation will also be of interest for that area. Many technologically important properties of silicide contacts, such as Schottky barrier height, contact resistance and corrosion stability are controlled by the silicon/silicide interface, and conventional silicide formation by thermal diffusion can lead to irreproducible electrical characteristics due to the native oxide present at that interface.

In this paper we will show the feasibility of producing transition metal silicide layers, either directly by implanting the metal into the silicon or by thermal annealing of the implanted samples.

EXPERIMENTAL PROCEDURE.

Polished Si single crystals with (100) orientation were uniformly implanted (scanned beam), at room temperature (RT) and at 350 °C, with 100 keV and 150 keV ions and current densities of 10-20 μA/cm$_2$ by means of the medium current implanter of the University of Connecticut. Implantation was performed at about 10^{-6} Torr, the samples being surrounded by a LN$_2$ cold trap. Samples were fastened to the heat sink with a conducting material.

A further description of the experimental set up is given in ref.4.

RESULTS.

1. Experimental ranges, retained doses and peak concentrations.

Peak concentrations and retained doses as obtained from RBS spectra for Ni$^+$ and Fe$^+$ implanted samples are shown in figure 1 (a and b). The results indicate that, at RT, the steady state concentration and retention of 100 keV implanted Ni would occur at a dose below 6x10^{17} Ni$^+$/cm$_2$ The results are quite similar to those found for the Cr$^+$ implants (3), where saturation under similar conditions of implantation occurred at a dose of 5x10^{17} Cr$^+$/cm$_2$ with a retention of 40 % of the implanted ions and a peak concentration of 42 at.%. A much higher retention is obtained in the case of 150 keV Ni$^+$ implants at 350 °C.

Figure 1c shows the experimental ranges for Ni$^+$ and Fe$^+$ implants. For 100 keV Ni$^+$ ions implanted at RT, the range starts decreasing for doses of about 5x10^{16} Ni$^+$/cm$_2$ where a peak concentration of 8.5 at. % is achieved. For the dose of 6x10^{17} Ni$^+$/cm$_2$ the range is considerably reduced because of the effect of sputtering (2). However, in the case of 150 keV Ni$^+$ ions implanted at 350 °C, the experimental range does not change for doses between 3 and 8x10^{17} Ni$^+$/cm$_2$

Figure 2 displays the depth concentration profile of the samples implanted with the highest doses of Ni$^+$ and Fe$^+$. The distribution of Ni implanted at 150 keV and 350 °C is much broader than the projected one indicating that long range atomic diffusion took place.

2. Silicide formation.

100 keV Cr$^+$, Fe$^+$, Co$^+$ and Ni$^+$ implants at RT.

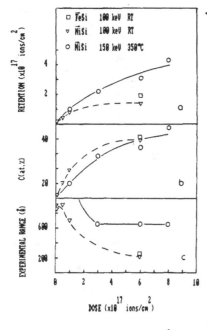

←FIGURE 1. Retained doses, peak atomic concentrations and experimental ranges (centroid of the implanted element distribution), for 100 keV Ni$^+$ and Fe$^+$ implants at room temperature and 150 keV Ni$^+$ implants at 350 °C, as a function of the implanted dose.

Figure 3 shows the RBS spectra from Fe$^+$ and Ni$^+$ implanted samples to a dose of 6×10^{17} ions/cm$_2$, before and after annealing half an hour at 400°C. Even when only a small redistribution effect can be seen in the spectra, XRD experiments revealed the formation of ꝺ-NiSi$_2$ and FeSi$_2$ (orthorhombic) after the thermal treatment.

Figure 4 demonstrates the RBS spectra of a Cr$_8^+$ implanted sample to a dose of 1×10^{18} Cr$^+$/cm$_2$, before and after annealing 4 hours at 550°C. The results show the silicide formation after the thermal treatment. XRD analyses identified this phase as γ-CrSi$_2$, in full agreement with the RBS data.

Figure 5 shows the RBS spectrum from a sample heavily implanted with Co$^+$ It can be seen that a CoSi layer (B20 type-clearly confirmed by XRD) was produced directly by implantation. The thickness of this layer is about 1000 Å. For a better understanding of this striking result, it should be mentioned that some temperature rise could have occurred during implantation because of the kind of sample-holder used in this particular case.

150 keV Ni$^+$ implants at 350 °C.

Figure 6 shows the RBS spectrum of a high dose implanted sample. Even when the RBS spectrum does not show clearly the compound formation, XRD analyses positively indicated the existence of η-NiSi as well as polycrystalline silicon.

DISCUSSION AND CONCLUSIONS.

In thin film systems, according to a proposed model (5), silicides are produced at high temperatures mainly by Si diffusion. The phonon vibrations weaken the Si-Si bonds allowing that process. At low temperatures near noble and group VIII transition metal silicides start forming by interstitial metal diffusion into Si, which again causes the above mentioned bond weakening, allowing Si to diffuse into the vacant metal sites. In our case, ion implantation produces a mixture of both components within the implantation range. Hence, only short range diffusion would be needed in the early stages of silicide formation. Afterwords, diffusion would take place in already formed silicide.

442

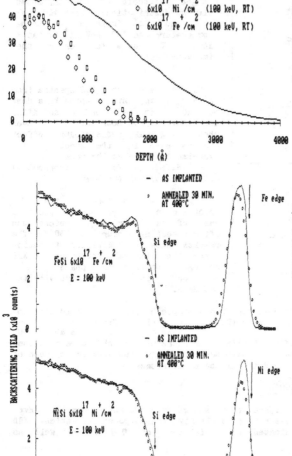

C(at.%)

- 8x10^{17} Ni$^+$/cm^2 (150 keV, 350°C)
◊ 6x10^{17} Ni$^+$/cm^2 (100 keV, RT)
□ 6x10^{17} Fe$^+$/cm^2 (100 keV, RT)

DEPTH (Å)

FIGURE 2. Depth concentration profile for the highest dose of 100 keV Ni$^+$ and Fe$^+$ implants at room temperature and 150 keV Ni$^+$ implants at 350°C. Profiles have not been corrected for depth resolution effects (~200 Å).

- AS IMPLANTED
◦ ANNEALED 30 MIN, AT 400°C

Fe edge

Si edge

FeSi 6x10^{17} Fe$^+$/cm^2

E = 100 keV

BACKSCATTERING YIELD (x10^3 counts)

- AS IMPLANTED
◦ ANNEALED 30 MIN, AT 400°C

Ni edge

Si edge

NiSi 6x10^{17} Ni$^+$/cm^2

E = 100 keV

ENERGY (MeV)

FIGURE 3. RBS spectra of 1.5 MeV He$^+$ ions from samples implanted with 100 keV Ni$^+$ and Fe$^+$ at RT before and after annealing 30 minutes at 400°C.

In three of the studied cases (Cr$^+$, Fe$^+$ and Ni$^+$ RT implants), silicide formation was achieved after post-implantation annealings. For Fe$^+$ and Ni$^+$ implants, XRD results show that a temperature of 400°C was enough to produce short range atomic rearrangement leading to disilicide formation. This is a temperature considerably lower than the one needed to form NiSi$_2$ (~ 800°C (6)) and FeSi$_2$ (~ 550°C (7)) by thermally induced reaction between a thin metal film with the Si substrate. However, this temperature was not still sufficient to produce a homogeneous silicide region. For the Cr$^+$ implants, RBS spectra indicate that a more homogeneous CrSi$_2$ layer was produced after

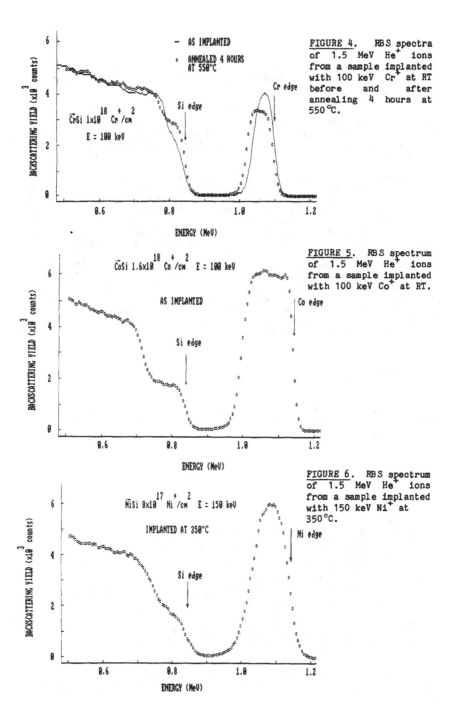

FIGURE 4. RBS spectra of 1.5 MeV He$^+$ ions from a sample implanted with 100 keV Cr$^+$ at RT before and after annealing 4 hours at 550 °C.

FIGURE 5. RBS spectrum of 1.5 MeV He$^+$ ions from a sample implanted with 100 keV Co$^+$ at RT.

FIGURE 6. RBS spectrum of 1.5 MeV He$^+$ ions from a sample implanted with 150 keV Ni$^+$ at 350 °C.

annealing at 550 °C.

The Ni$^+$ implants at 350 °C are a good example of how varying implantation conditions, much higher retained doses and different silicides phases can be obtained.

The production of a CoSi layer directly by implantation constitutes a very interesting case and quite unexpected result. It can be concluded that important long range diffusion effects occurred during implantation. This situation might be related to the suspected temperature rise during these implantations.

Summarizing, the present preliminary results indicate that ion implantation might be a suitable technique for the production of silicides. Concentrations of the implanted elements above 40 at.% were achieved in all the cases. This fact allows to us to predict that silicides of at least all the transition metals in the first row of the periodic table could be produced with the aid of this technique. The surprising result on the Co–Si system deserves further investigation as it suggests that under proper conditions, silicide layers can be produced directly by ion implantation. The results from higher temperature Ni$^+$ implants are perhaps indicating the direction to be followed in order to make ion implantation more efficient in producing silicides. By implanting at temperatures high enough a combination of compound formation plus atomic diffusion could provide a mechanism for this achievement. In this regard a systematic study is being carried out to investigate how temperature and other implantation parameters influence silicide formation by this technique.

ACKNOWLEDGEMENTS.

We are grateful to Dr. Quentin Kessel for his help with the Van de Graaff accelerator. We would like to thank C. H. Koch, J. Gianopoulous and Dr. P. Clapis for their technical assistance.

REFERENCES.

1. S.P.Murarka, Silicides for VLSI Applications, Academic Press, New York, 1983.
2. Z.L.Liau and J.W.Mayer, J. Vac. Sci. Technol., 15, 1629 (1978).
3. F.Namavar, J.I.Budnick, H.C.Hayden, F.A.Otter and V.Patarini, Mat. Res. Soc. Symp. Proc. 27, 341 (1984).
4. F.Namavar, J.I.Budnick, F.H.Sanchez and F.A.Otter, Nucl. Instr. Meth. Phys. Res. B7/8, 357 (1985).
5. K.N.Tu and J.W.Mayer in Thin Films-Interdiffusion and Reactions (J.M.Poate, K.N.Tu and J.W.Mayer, eds.) Wiley, New York, 1978, p.359.
6. K.N.Tu, E.Alessandrini, W.K.Chu, H.Krautle and J.Mayer, Jpn. J. Appl. Phys. Suppl. 2, Pt.1, 669 (1974).
7. S.S.Lau, J.S.-Y.Feng, J.O.Olowolafe and M.-A.Nicolet, Thin Solid Films, 25, 415 (1975).

* On a fellowship from CONICET, Republica Argentina. Permanent address: Departamento de Fisica, Universidad Nacional de La Plata, Republica Argentina.

PULSED LASER AND ION BEAM SURFACE MODIFICATION
OF SINTERED ALPHA-SiC

K. L. MORE[*], R. F. DAVIS[*], B. R. APPLETON[**], D. LOWNDES[**], AND P. SMITH[+]
* Department of Materials Engineering, North Carolina State University, Raleigh, NC 27695-7907
** Solid State Division, Oak Ridge National Laboratory, Oak Ridge, TN 37831
+ Microelectronics Center of North Carolina, Research Triangle Park, NC 27709

ABSTRACT

Pulsed laser annealing and ion beam mixing have been used as surface modification techniques to enhance the physical properties of polycrystalline α-SiC. Thin Ni overlayers (20 nm - 100 nm) were evaporated onto the SiC surface. The specimens were subsequently irradiated with pulses of a ruby or krypton fluoride (KrF) excimer laser or bombarded with high energy Xe^+ or Si^+ ions. Both processes are non-equilibrium methods and each has been shown to induce unique microstructural changes at the SiC surface which are not attainable by conventional thermal treatments. Under particular (and optimum) processing conditions, these changes considerably increased the mechanical properties of the SiC; following laser irradiation, the fracture strength of the SiC was increased by as much as 50%, but after ion beam mixing, no strength increase was observed.

High resolution cross-section transmission electron microscopy (X-TEM), scanning electron microscopy (SEM), and Rutherford backscattering techniques were used to characterize the extent of mixing between the Ni and the SiC as a result of the surface modification.

INTRODUCTION

Sintered α-SiC is a principal candidate material for high temperature and high strength applications. It is currently being considered for use, or is currently employed, in such areas as heat exchangers in fossil fuel systems, critical parts for uncooled gas turbine and adiabatic diesel engines, outlet channels on coal gasifiers, and high temperature bearings. However, during processing and subsequent machining, large surface flaws are created which are detrimental to and limit the overall strength of the ceramic. Modification of the near surface regions of the SiC can result in a considerable increase in the mechanical properties if either (1) the modification process changes the size and/or position of the critical flaws or (2) a compressive stress is induced at and/or near the specimen surface.

Two such surface modification processes, pulsed laser annealing and ion beam mixing, have been investigated in an attempt enhance the physical properties of Ni-coated, sintered α-SiC. Pulsed laser mixing ideally involves the melting of the coating and the near surface of the substrate material and the subsequent mixing of the two via liquid diffusion. If the melting threshold is not exceeded, the two materials may mix as a result of solid state mass transport. Ion beam mixing, on the other hand, involves the acceleration of high energy ions in a potential to a degree whereby they are implanted into the specimen surface and are used to mix layers of two different compositions. Each of the modification processes involves extremely rapid heating and cooling rates such that unique surface microstructures and non-equilibrium conditions can be created and preserved in rates approaching 10^9 K/s.

Ion beam mixing and laser annealing are relatively new processing techniques, but each has been extensively reviewed [1-4]. These techniques have most recently been applied to metal-semiconductor [5-9] and metal-insulator [10,11] systems. Investigations have also been made into the surface modification of single crystal β-SiC thin films grown in the authors' laboratory [12-15] as well as α-SiC ceramic systems [16,17].

The emphasis of this paper is on the microstructures formed as a result of the surface modification. The subsequent fracture strength results have been reported previously [18,19] but will be reviewed.

EXPERIMENTAL PROCEDURE

The material investigated in this study was an ≈98% dense α-SiC[+] produced by conventional sintering of submicron SiC particles at approximately 2373K and 0.1 MPa Ar. This SiC also contained 0.5 wt% free C and 0.42 wt% B in the sintered state. Boron is contained both in solid solution in the α-SiC and in separate B-containing grains within the α-SiC matrix. The sintered SiC plates were subsequently cut into standard flexure test bars (0.32 cm X 0.64 cm X 5.08 cm) and the surfaces sequentially ground using 100, 400, and 600 grit diamond wheels. The edges of each bar were bevelled at 45° to prevent corner flaws from acting as fracture origins. The SiC surfaces were sputter cleaned in a vacuum of approximately 10^{-7} Torr followed by the evaporation of 50 nm or 100 nm thick Ni overlayers. Each side of a bar was coated separately; the other sides were masked by glass slides to prevent additional deposition. The maximum specimen temperature reached during Ni deposition did not exceed 323K.

Ion beam mixing was achieved on the 50 nm Ni coated SiC using either 350 keV Xe^+ at a dose of 2.5 X 10^{16} Xe^+/cm^2 or 140 keV Si^+ at a dose of 1 X 10^{17} Si^+/cm^2. The diameter of the ion beam was ≈2.54 cm; thus, to ion beam mix the entire specimen, there was considerable overlap of the beam spots. Laser irradiation was conducted using one of two lasers: a pulsed ruby laser (λ = 693 nm, τ = 25 ns, nominal pulse energy density = 1.2 J/cm²) or a KrF excimer laser (λ = 248 nm, τ = 45 ns, pulse energy density = 1.85 J/cm²). The surfaces of each Ni-coated specimen were sequentially irradiated with laser pulses until all four sides were completely annealed; the size of the laser beam was ≈0.4 cm.

Each modified SiC bar was broken in 4-point flexure to determine the fracture strength and compared to the fracture strength values similarly obtained for the as-received SiC. Inner and outer span lengths of 1.27 cm and 2.54 cm, respectively, and a crosshead speed of 8.5 X 10^{-4} cm/s were used for the fracture research.

The modified surfaces of the laser irradiated and ion beam mixed SiC were examined using high resolution X-TEM specimens prepared by mechanical thinning to approximately 75 μm, dimpling, and subsequent low energy (5-6 kV), low angle (12°) Ar^+ ion beam thinning. The X-TEM results were compared to the chemical analysis derived from RBS. 1 MeV, 2 MeV, and 2.5 MeV He^+ beams were used for the RBS analyses.

RESULTS AND DISCUSSION

Starting Material

The microstructure of the α-SiC consisted of fairly equiaxed as well as elongated grains of SiC having an average grain size of ≈8.5 μm, much darker grains of the B-containing sintering aid (identified in previous research and confirmed by the manufacturer to be B_4C [20]), and particles of free C having a mottled surface texture. A polished surface of this material is shown in Figure 1. The average strength of the as-received SiC was found to

---

[+] HEXOLOY; SOHIO Engineered Materials Company, Niagara Falls, NY 14302 be approximately 310 MPa, as determined from 4-point bend experiments.

An ≈100 nm as-deposited Ni overlayer is shown in the X-TEM micrograph of Figure 2. From the X-TEM results, the Ni-coating was found to be fairly uniform in thickness across the SiC surface. Since the SiC surface was ground using 600 grit diamond as the final processing step, the resulting surface was fairly rough. The Ni coating was found to be non-conformal around extremely rough surface areas.

Ion Beam Mixing

For the ion beam mixing experiments, the SiC was coated with 50 nm of Ni. The coated SiC specimens were initially bombarded with 350 keV Xe^+ at a dose of 2.5 X 10^{16} Xe^+/cm^2. The projected range of Xe^+ in Ni is approximately 55 nm, which coincided with the Ni-SiC interface; thus, the maximum energy was deposited at this interface. The resulting microstructure is shown in the X-TEM micrograph of Figure 3. The new surface microstructure consisted of four distinct areas: (a) a residual polycrystalline Ni overlayer, (b) a Ni-SiC mixed region, (c) an amorphous SiC region caused by ion implantation damage, and (d) the underlying α-SiC substrate (which was highly faulted at the amorphous SiC - SiC substrate interface). The associated RBS spectra for the Xe^+ mixed SiC is shown in Figure 4. Analysis of the RBS results showed that the stoichiometric composition of the mixed region was Ni_2Si. Previous results obtained for similarly processed β-SiC thin films showed formation of the same silicide [12]. However, as noted by the arrows on the X-TEM micrograph of Figure 3, there was considerable bubble formation at the Ni - silicide interface as a result of the high dose of heavy Xe^+ ions. It was believed that this extensive bubble formation would be detrimental to the overall strength of the SiC. For this reason, and for the compatability of using a self-ion, the Ni coated SiC bars were subsequently ion beam mixed using Si^+ ions.

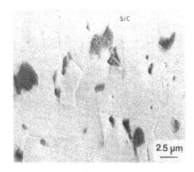

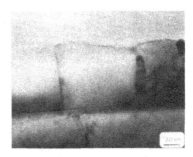

Figure 1. Scanning electron micrograph of a polished and etched α-SiC surface which also reveals the presence of free C, B_4C particles, and porosity.

Figure 2. A 100 nm as-deposited Ni overlayer on SiC.

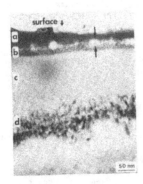

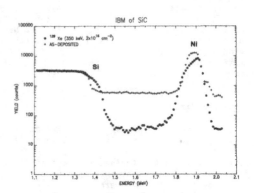

Figure 3. Cross-section TEM micrograph of 350 keV Xe$^+$ (2.5 X 10^{16} Xe$^+$/cm^2) ion beam mixed SiC. The arrows point to bubble formation at the interface.

Figure 4. 2.5 MeV He$^+$ RBS spectra comparing as-deposited SiC to 350 keV Xe$^+$ (2.5 X 10^{16} Xe$^+$/cm^2) ion beam mixed SiC.

The subsequent ion beam mixing experiments were conducted using 140 keV Si$^+$ at a dose of 1 X 10^{17} Si$^+$/cm^2. The resulting surface microstructure is shown in the X-TEM micrograph of Figure 5. The surface microstructure following Si$^+$ implantation was more uniform than that obtained after Xe$^+$ ion beam mixing and there was no interface bubble formation. The new microstructure consisted of: (a) a thin residual polycrystalline Ni overlayer and a Ni-rich amorphous SiC region, (b) a mixed Ni-SiC region, (c) an ≈150 nm amorphous SiC region caused by ion implantation damage, and (d) the SiC substrate. The RBS spectra for the ion beam mixed SiC using Si$^+$ is shown in Figure 6. From the RBS results, the stoichiometric composition of the mixed layer was found to be Ni$_3$Si$_2$. High resolution X-TEM was used to further examine the microstructure of the mixed region. Figure 7(a) shows an X-TEM micrograph of the mixed – amorphous – SiC substrate regions. The lattice fringes in the SiC substrate correspond to {0001} planes having a d-spacing of 1.51 nm. A high resolution X-TEM micrograph of the mixed region noted in Figure 7(a) is shown in Figure 7(b). The mixed region was found to consist of small crystalline islands in an amorphous matrix. The size of these islands was nominally 5 – 10 nm and the lattice fringes of the majority of the islands ran along the same crystallographic direction. The lattice spacing of the islands measured 0.69 nm, which corresponded to the c-axis spacing of orthorhombic Ni$_3$Si$_2$, as predicted from the RBS analysis.

Following ion beam mixing using either Xe$^+$ or Si$^+$, there was no observed change in the fracture strength of the SiC. After processing only four specimens, this approach was discontinued. At least within the limits created by the use of Ni coatings, the ion species, and parameters of bombardment used in this work, it was concluded that ion beam mixing was not a viable method for increasing the fracture strength of the α-SiC.

Pulsed Laser Mixing

Initial laser annealing experiments were conducted using pulsed ruby laser irradiation of 100 nm Ni-coated SiC at a nominal pulse energy density of 1.2 J/cm^2. Cross-sectional TEM micrographs of the resulting surface microstructure following ruby laser irradiation are shown in Figure 8.

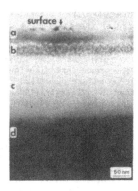

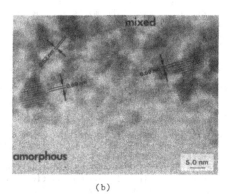

Figure 5. Cross-section TEM micrograph of 140 keV Si$^+$ (1 X 10^{17} Si$^+$/cm^2) ion beam mixed SiC.

Figure 6. 2.5 MeV He$^+$ RBS spectra comparing as-deposited SiC to 140 keV Si$^+$ (1 X 10^{17} Si$^+$/cm^2) ion beam mixed SiC

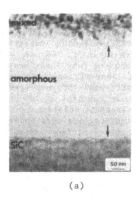

(a)

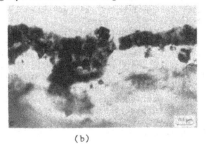

(b)

Figure 7. High resolution X-TEM micrographs of 140 keV Si$^+$ (1 X 10^{17} Si$^+$/cm^2) ion beam mixed SiC. (a) Micrograph showing the mixed region - amorphous SiC and the amorphous SiC - SiC substrate interfaces (indicated by arrows) and (b) a high resolution micrograph of the mixed region.

(a)

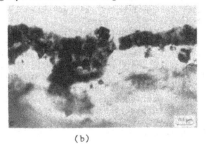

(b)

Figure 8. Cross-section TEM micrographs of pulsed ruby laser irradiated (1.2 J/cm^2) 100 nm Ni-coated SiC.

As noted from the X-TEM results, no mixing occurred between the Ni overlayer and the SiC substrate. Essentially, all of the energy of the laser beam was absorbed by the Ni coating and used to heat it to ever increasing temperatures. The Ni melted, flowed into the near-surface regions of the many pre-existing surface voids, and rapidly resolidified as a polycrystalline layer of non-uniform thickness across the specimen surface. Areas were also found where there was no Ni present. The main reason that no mixing occurred between the Ni and SiC appeared to be that the initial Ni overlayer (100 nm) was too thick. Fathy, et al. [13] observed mixing following pulsed ruby laser irradiation of 50 nm Ni-coated β-SiC thin films at the same pulse energy density (1.2 J/cm^2).

Following pulsed ruby laser irradiation, the fracture strength of the SiC was increased by approximately 50%. It is reasoned that the much larger remaining volume of each of the Ni-containing voids acted as a thermal insulator which mitigated heat loss from the Ni for a time sufficient to allow for a limited chemical reaction between the Ni and the walls of the voids. During rapid cooling, the greater contraction of the Ni induced a compressive stress in the near surface regions of the voids. Thus, the large surface flaws present in the starting material were prevented from acting as fracture origins; instead, subsurface flaws became the strength controlling flaws.

In an attempt to bring about mixing between the Ni overlayer and the SiC substrate, a KrF excimer laser was used for subsequent experiments. The excimer laser was chosen for three primary reasons: (1) the wavelength of the KrF laser (λ = 248 nm) is considerably shorter than that of the ruby laser (λ = 693 nm); thus, the photon energy of the KrF is significantly greater than the band gap of the α-SiC (E$_g$ = 3.023 eV) which can allow for absorbtion of the laser radiation by both the Ni coating and SiC substrate, (2) the pulse duration is twice as long for the KrF laser (τ = 45 ns), and (3) the KrF laser beam is much more homogeneous than the ruby laser beam.

The pulse energy density used for the excimer laser mixing experiments (1.85 J/cm^2) was higher than the pulse energy density used for ruby laser irradiation (1.2 J/cm^2), but the initial Ni thickness was kept at 100 nm. The new surface microstructure resulting from KrF laser irradiation is shown in the X-TEM micrographs of Figure 9. Three distinct regions were observed and are labelled in Figure 9(a): (a) a chemically mixed surface layer, (b) a chemically bonded interface, and (c) a highly faulted SiC substrate. During KrF laser irradiation, the Ni overlayer melted and reacted with the underlying SiC. Although the SiC itself did not melt, the temperature reached during irradiation was sufficiently high to allow for interdiffusion of Si and C into the Ni overlayer (the solubility of SiC in molten Ni is moderately high). Figure 9(b) shows a higher magnification micrograph of the interface. The "mixed" overlayer is bonded to the SiC and follows the surface contours of the substrate. The bubbles present at the interface were caused by the reaction at the interface; a small amount of C left the SiC, probably as CO_2, as a result of the reaction between the extremely hot SiC and air. Carbon lattice fringes were observed around the bubble edges indicating the presence of trapped C.

The RBS results obtained for a KrF laser irradiated specimen are compared to the RBS analysis of a ruby laser annealed specimen in Figure 10. The RBS spectrum for the ruby laser annealed SiC shows no mixing between the Ni coating and the SiC. However, the steps in the Si and C edges in the KrF laser irradiated spectrum indicate that limited mixing did occur. X-ray spectra taken from the mixed layer showed that the layer contained mostly Ni; Si was found at the interface region and, to a much lesser extent, at the specimen surface. These results are comparable to results obtained by Narayan, et al. [15] who found that KrF laser irradiation of 50 nm Ni-coated β-SiC thin films resulted in complete mixing of the Ni and SiC.

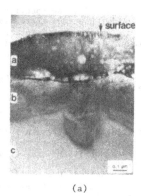 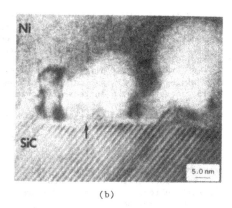

(a)   (b)

Figure 9. Cross-section TEM micrographs of KrF laser irradiated, 100 nm Ni-
coated SiC (1.85 J/cm²). (a) Micrograph showing entire surface
microstructure following irradiation and (b) higher magnification micrograph
of the interface (as indicated by the arrow). The lattice fringes in the
SiC substrate correspond to the {0001} planes having a d-spacing of 1.51 nm.

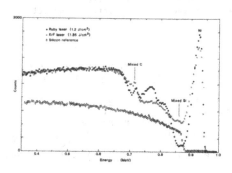

Figure 10. 1.0 MeV He⁺ RBS specta comparing ruby laser irradiated SiC
(1.2 J/cm²) to SiC irradiated with a KrF laser irradiated at 1.85 J/cm².

    Following KrF laser irradiation, the fracture strength of the SiC was
increased by approximately 20%. The significant increase in strength resulted from an induced compressive stress at the specimen surface. The compressive stress was caused by two interacting effects: (1) the enhanced
bonding at the Ni-SiC interface as a result of the interdiffusion of SiC
into the Ni overlayer during irradiation and (2) the difference in thermal
contraction between the Ni and the SiC; on cooling, the Ni contracts to a
greater extent than the SiC [18,19].

CONCLUSIONS

    Ion beam mixing of 50 nm Ni overlayers on sintered α-SiC using 350 keV

452

Xe^+ resulted in the formation of a mixed Ni–Si region below the specimen
surface having a stoichiometric composition of Ni_2Si. However, there was
considerable bubble formation at the mixed region – amorphous SiC interface
as a result of using a high dose of heavy Xe^+ ions. Ion beam mixing using
140 keV Si^+ resulted in the formation of a mixed layer having a stoichiometric composition of Ni_3Si_2. The new surface microstructure following ion
beam mixing using Si^+ was more uniform and there was no bubble formation.
There was no strength increase following ion beam mixing using either Si^+ or
Xe^+.

Ruby laser irradiation of 100 nm Ni-coated SiC resulted in no mixing
between the Ni overlayer and the SiC substrate. There was, however, a 50%
strength increase as a result of ruby laser annealing. The strength increase was caused by the flow of molten Ni during irradiation into the
surface regions of pre-existing flaws and the subsequent limited reaction
between the molten Ni and the walls of the voids.

KrF laser irradiation of 100 nm Ni-coated SiC resulted in a mixed Ni-
SiC surface region. The mixed surface was caused by the interdiffusion of
SiC into the molten Ni overlayer during irradiation. A significant 20%
increase in fracture strength occurred as a result of the surface processing.

ACKNOWLEDGEMENTS

The authors acknowledge the support of the U.S. Department of Energy
under ORNL Subcontract No. 19X-43377C and the Basic Energy Sciences Division, U.S. Department of Energy through the SHaRE program under Contract No.
DE-76-C-05-0033 with Oak Ridge Associated Universities and under Contract
No. DE-AC05-84OR-21400 with the Martin Marietta Corporation.

REFERENCES

1. S. D. Ferris, H. J. Leamy, and J. M. Poate, eds., Laser Solid
Interactions and Laser Processing – 1978, American Institute of
Physics, No. 50 (1979).
2. C. W. White and P. S Peercy, eds., Laser and Electron Beam Processing
of Materials (Academic Press, New York, 1980).
3. B. R. Appleton, J. Materials for Energy Systems, 6(3), p. 200 (1984).
4. B. R. Appleton, in Ion Implantation and Beam Processing, edited by J.
S. Williams and J. M. Poate (Academic Press, New York, 1984).
5. J. M. Poate and J. W. Mayer, eds., Laser Annealing of Semiconductors
(Academic Press, New York, 1984).
6. J. W. Mayer, B. Y. Tsaur, S. S. Lau, and L. S. Hung, Nucl. Inst. and
Meth., 182/183, p. 1-13 (1981).
7. B. Y. Tsaur, Z. L. Liau, and J. W. Mayer, Appl. Phys. Lettr., 34(2), p.
168 (1979).
8. D. Fathy, O. W. Holland, and J. Narayan, J. Appl. Phys., 58(1), p. 297
(1985).
9. S. S. Lau, B. Y. Tsaur, M. von Allmen, J. W. Mayer, B. Stritzker, C. W.
White, and B. R. Appleton, Nucl. Inst. and Meth., 182/183, p. 97
(1981).
10. G. C. Farlow, B. R. Appleton, L. A. Boatner, C. J. McHargue, C. W.
White, G. J. Clarke, and J. E. E. Baglin, Presented at the Fall Meeting
of the Materials Research Society, San Fransisco, CA, 1984
(unpublished).
11. C. W. White, G. Farlow, J. Narayan, G. J. Clarke, and J. E. E. Baglin,
Materials Letters, 2(5A), p. 367 (1984).
12. J. Narayan, D. Fathy, O. W. Holland, B. R. Appleton, R. F. Davis, and
P. F. Becher, J. Appl. Physics, 56(6), p. 1577 (1984).

13. D. Fathy, J. Narayan, O. W. Holland, B. R. Appleton, and R. F. Davis, Materials Letters, 2(4B), p. 324 (1984).
14. D. Fathy, O. W. Holland, J. Narayan, and B. R. Appleton, Nucl. Inst. and Meth., B 7/8, p. 571 (1985).
15. J. Narayan, D. Fathy, O. W. Holland, and B. R. Appleton, Materials Letters, 3(7/8), p. 261 (1985).
16. C. J. McHargue, G. C. Farlow, C. W. White, J. M. Williams, B. R. Appleton, and H. Naramoto, Mat. Sci. and Eng., 69, p. 123 (1985).
17. B. R. Appleton, H. Naramoto, C. W. White, O. W. Holland, C. J. McHargue, G. Farlow, J. Narayan, and J. M. Williams, Nucl. Inst. and Meth., B1, p. 167 (1984).
18. K. L. More and R. F. Davis, Presented at the Fourth International Symposium on The Fracture Mechanics of Ceramics, Blacksburg, VA, 1985, to be published in the conference proceedings.
19. K. L. More and R. F. Davis, Presented at the Conference on Tailoring Multiphase and Composite Ceramics, University Park, PA, 1985, to be published in conference proceedings.
20. R. F. Davis, J. E. Lane, C. H. Carter, Jr., J. Bentley, W. H. Wadlin, D. P. Griffis, R. W. Linton, and K. L. More, Scanning Electron Microscopy, III, p. 1161 (1984).

METASTABLE PHASES PRODUCED IN NICKEL BY HIGH DOSE La AND O
IMPLANTATION, AND PULSED LASER MELT QUENCHING

D.K. SOOD[1], A.P. POGANY[1], G. BATTAGLIN[2], A. CARNERA[2], G. DELLA MEA[2],
V.L. KULKARNI[2], P. MAZZOLDI[2] and J. CHAUMONT[3].

1. Microelectronics Technology Centre, Royal Melbourne Institute of
Technology, Melbourne 3000, Australia.
2. Departimento di Fisica, Universita di Padova, Via Mazzolo n.8, 35131
Padova, Italy.
3. Laboratoire Rene Bernas, Universite Paris XI, 91406 Orsay, France.

ABSTRACT

Nickel single crystals implanted with La, O and La + O, up to about 9 a/o
concentrations were irradiated with spatially homogenised ruby laser pulses (16
ns FWHM) up to 3.7 J/cm^2 in vacuum ($\sim 10^{-2}$ Torr). Phase changes, defect
structure and solute migration were studied before and after laser melt
quenching using RBS/channeling, TEM and SEM techniques.
Both La and La + O implants produced a buried amorphous layer on top of a
polycrystalline layer above the single crystal matrix. O implants did not
produce an amorphous phase, but a buried layer containing polycrystalline Ni,
metastable hcp Ni particles and a dense dislocation network was formed. Pulsed
laser melt quenching of these implanted metastable phases lead to a variety of
new phases - all crystalline.

INTRODUCTION

Ion implantation and pulsed laser melt quenching of metals have been
employed extensively [1,2] to produce metastable surface alloys in crystalline
or amorphous phases. Our previous study [3] by RBS/channeling of pulsed laser
treatments (PLT) of La implanted in Ni to doses of 1×10^{16} ions/cm^2 showed -
a) the nonsubstitutional crystalline phase of La in Ni as obtained by ion
implantation could not be transformed into a metastable solid solution after
PLT, b) evidence of enhanced defect trapping and impurity - defect interaction
and c) formation of La surface peak which was attributed to oxidation and
capture of La at the molten surface. In the present work, we extend this study
to much higher implant concentration of La (5×10^{16} ions/cm^2) in order to
produce an amorphous phase. In addition, the role of oxygen picked up from the
ambient during melt quench is explicitly studied by oxygen and La + O dual
implants in nickel.

EXPERIMENTAL

Electropolished Ni single crystals, <100> orientation, or well annealed
polycrystalline Ni samples, having a damage free surface and mirror finish have
been used in the present experiment for La and/or O implantations and
subsequent laser irradiations. Three types of implanted regions have been
produced: 1) implanted with La^{++} at 250 keV energy, 2) implanted first with
La^{++} at 250 keV and successively with O^{18} at 35 keV, 3) implanted with only
O^{18}. The implantation energy of 35 keV for the O^{18} implants has been chosen in
order to produce an O distribution which would approximately overlap the La
distribution. Each implant was to a dose of 5×10^{16} ions/cm^2. The implanted
crystals were irradiated under low pressure conditions (10^{-2} torr) at various
energy densities up to 3.7 J/cm^2 with ruby laser pulses of 16 nsec FWHM. From
the spatially homogenized large laser spot of 16 mm size, a region of 2.5 mm
diameter was carefully selected for irradiation which ensured energy uniformity

better than ±10% over the selected area. Implanted and irradiated regions have been analyzed by Rutherford backscattering (RBS) and channeling using 1.8 MeV He^+ beam and scattering angle of 135°. The O^{18} amount was determined by using the ^{18}O (p, α) ^{15}N nuclear reaction. Samples for transmission electron microscopy (TEM) were prepared by jet thinning from the backside, while protecting the irradiated face. Surface topography was studied using a scanning electron microscope (SEM).

RESULTS AND DISCUSSION

The projected range, R_p (straggling width ΔR_p) were calculated [4] as 31.3 (6.4) nm for La ions and as 29.4 (17.4) nm for O ions. A heat flow calculation [3] assuming reflectance of 0.68 for Ni yields the melt threshold at $1J/cm^2$, boil threshold at $2J/cm^2$; and melt depth and duration at 2.5 J/cm^2 of 420 nm and 49 ns respectively. In general, our RBS/channeling data lead to very limited and qualitative information whereas TEM results on similar samples gave more detailed results on the microstructure. The results obtained from a combined analysis of RBS and TEM data are now presented for each of the three systems:

1) La implanted Ni

La implantation in Ni produced an amorphous layer as suggested by the dechanneling yield, χ vs. depth curve (Fig.1a) exhibiting a damage peak with value of χ~1 in addition to identical depth profiles of La in aligned and random direction (Fig. 1b). The existence of a buried amorphous layer is confirmed by the TEM results (from a similar sample) in Fig. 2a, where the characteristic diffuse rings can be seen (inset) in the electron diffraction pattern (EDP). The amorphous region is seen as the darker part of Fig. 2a. In fact, the detailed TEM analysis of this sample has shown that different phases are produced at different depths depending on the implant concentration. There appear to be three distinct regimes with increasing implant concentration - i) metastable solid solution (MSS) of La in single crystal Ni at concentrations below a critical value above which ii) the MSS transforms to a polycrystalline phase containing metastable hcp nickel. Such an fcc to hcp transformation, as a precurser to amorphization in Ni during ion bombardment has been reported earlier [5] and is believed to result from a shear transformation stabilized by ion damage, including self ion damage; and iii) at a still higher La concentration, an amorphous phase is formed. Since the implant profile is somewhat skew (Fig. 1b), we observe the following sequence of phases with increasing depth: a very thin single crystal layer of Ni, followed by an amorphous La-Ni alloy (Fig. 2a), below which lies a polycrystalline layer containing MSS of La in fcc and hcp particles of Ni on a <110> Ni single crystal matrix.

After pulsed laser melt quench, the amorphous phase disappears and instead coarse grain, randomly oriented polycrystallites are formed (Fig. 2b) and extensive twinning is also seen (Fig. 2c). There is no sign of any precipitates of either La, or any of the six La-Ni intermetallic compounds [6] or of La_2O_3. Yet channeling (Fig. 1b) shows no substitutionality of La. This would be consistent with La existing as a metastable solid solution either interstitial or substitutional in each grain of polycrystalline Ni (as confirmed by our preliminary TEM results on hot stage annealing up to 600°C which do indeed show subsequent appearance of La and La-Ni precipitates). The ambient peak concentration of about 5 a/o La in the melt (Fig.1b) is clearly below the critical concentration required to stabilize an amorphous phase. However, it is high enough to block liquid phase epitaxy at the advancing liquid solid interface which no longer remains planar and the polycrystallites are produced. Each grain undergoes a rapid quench from a miscible liquid (solution) of La-Ni [6] and thus forms an MSS on solidification in accordance with Sood's criteria [7].

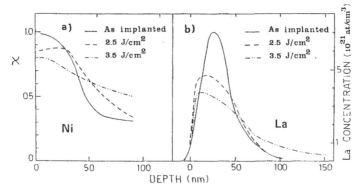

Fig.1.   RBS depth profiles for La implanted Ni. a) Aligned to random
ratios (χ) of Ni spectra, b) La profiles.   Each curve represents
random or <100> aligned yield.

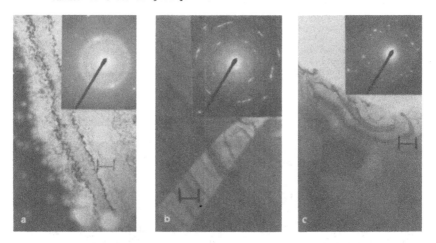

Fig.2.   TEM results on La implanted Ni a) As implanted, b) 2.5 J/cm^2 and
c) 3.5 J/cm^2.   Bar length 200 nm.

2) La + O dual implanted Ni

Additional bombardment of La implanted Ni with 5×10^{16} O ions/cm^2 produces
only a minor change (Fig. 3b) in the composition of amorphous phase (Fig. 4a)
and extends it right to the surface. Polycrystalline Ni layer lies underneath
as earlier. A careful search revealed no crystalline oxides at all indicating
formation of a ternary amorphous phase.

After pulsed laser melt quench, the as implanted amorphous phase completely
transforms to at least three crystalline phases - fcc Ni, NiO and La$_2$O$_3$.
However, no precipitates of either La or any intermetallic compounds are
observed. Fig. 4b shows randomly oriented fine NiO precipitates in the dark
field image obtained using the strongest NiO ring in SAD pattern, from a sample
treated at 2.5 J/cm^2. An example of La$_2$O$_3$ precipitates produced at 3.5J/cm^2 is
given in Fig. 4c which shows them lighting up in the DF image. The La$_2$O$_3$
particles can be seen to extend right to the surface - in agreement with the

formation of surface peak in the La depth profiles (Fig. 3b) after PLT. At higher laser fluence, La_2O_3 precipitates grow larger but NiO precipitates appear to be smaller and fewer indicating preferential growth of La_2O_3 at expense of NiO from and within the ternary liquid phase solution. These results are consistent with internal oxidation of Ni and La in the liquid phase which progressively depletes in La concentration till solidification is completed resulting in the observed microstructure containing precipitates of La_2O_3 and NiO distributed within the polycrystalline MSS of La in Ni. In addition, at $3.5J/cm^2$, we observe voids or gas bubbles showing clear black to white contrast change on underfocusing in TEM.

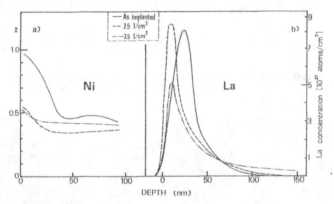

Fig.3. RBS depth profiles for La + O implanted Ni. a) Dechanneling yield for Ni, b) La profiles. Each curve represents random or <100> aligned yield.

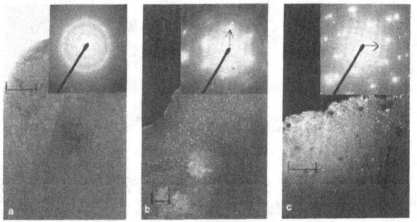

Fig.4. TEM results for La + O implanted Ni. a) As implanted, b) 2.5 J/cm^2, DF image showing NiO precipitates, c) 3.5 J/cm^2, DF image showing La_2O_3 precipitates. Bar length 200 nm.

3) O Implanted Ni

O^{18} implantation in Ni produced a distinct buried layer consisting of a dense dislocation network, polycrystalline Ni (Fig.5) and fine particles of a metastable hcp Ni phase (Fig. 6a, DF image on spot shown by arrow on SAD pattern in the inset). The matrix was (110) Ni and top surface had very thin ($\leq$5 nm) single crystal layer. These features are in qualitative agreement with calculated range profile of O^{18} in Ni, having projected range of 29 nm. Presence of high concentrations of implanted O appear to produce the buried layer. However no amorphous (Fig. 5) nor any oxide phases were formed indicating presence of a metastable solid solution of O in Ni.

Laser irradiation at 2.5J/cm^2 resulted in formation of fine, randomly oriented polycrystallites of NiO (Fig. 6b showing DF image using the indicated NiO ring) and a buried layer of much coarser polycrystallites of Ni (Fig. 6c showing DF image of same area as in Fig.6b but using the indicated Ni spot in EDP). Laser irradiation at 3.7J/cm^2 shows similar features but the polycrystalline Ni layer is now much thicker and consists of crystallites of larger size. The as-thinned sample after 3.7 J/cm^2 irradiation shows many holes in agreement with coarse topology seen with SEM. It may be noted that no similar oxide or polycrystallite formation was found on unimplanted control samples before or after laser irradiation.

These results described elsewhere in detail [8], are consistent with internal oxidation of Ni in the melt by implanted oxygen, based on Ni-O phase diagram [6]. Laser irradiation produces a molten region well beyond the O implant depth, with temperatures approaching boiling point (2730°C) of Ni. Before the solidification front arrives, NiO is precipitated out in liquid phase and solidifies at 1984°C well before Ni (at 1455°C). This phase separation could be associated with formation of polycrystalline Ni. However, we also observe a very curious and as yet unexplained formation of voids or bubbles (showing clear black to white contrast change on under-focusing in TEM) only at 3.7 J/cm^2 and in the presence of implanted oxygen.

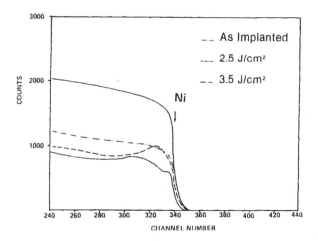

Fig. 5. RBS spectra for O implanted Ni.

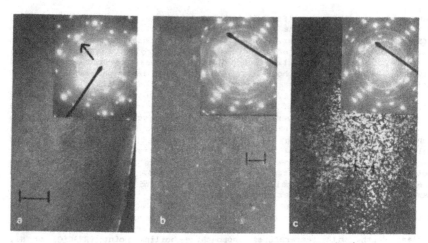

Fig. 6.   TEM results on 0 implanted Ni. a) As implanted, DF showing
          particles of hcp Ni, b) 2.5 J/cm^2, DF image of NiO, c) same areas as
          in b) but DF image of Ni. Bar length 200 nm.

CONCLUSIONS

a) <u>As implanted systems</u>: Both La and La + o implants produce amorphous
layers above a critical concentration. 0 implants produce a buried
polycrystalline layer of a metastable solid solution of 0 in Ni. In all three
cases, fcc to hcp shear transformation in the Ni matrix is observed.

b) <u>Laser melt quenching of implanted phases</u>: leads to a variety of new
phases - all crystalline. Precipitation of La or any of the six equilibrium
intermetallic phases is suppressed. Internal oxidation in the melt of both Ni
and La is observed. La appears to combine with oxygen preferentially (though
not exclusively) to Ni from within the ternary liquid phase solution.

REFERENCES

1. D.K. Sood, Radiation Effects <u>63</u>, 141 (1982).

2. S.T. Picraux and D.M. Follstaadt in <u>Surface Modification and Alloying by
    Laser, Ion and Electron Beams</u>, ed. J.M. Poate, G. Foti and D.C. Jacobson
    (Plennum Press, New York, 1983), p.287.

3. G. Battaglin, A. Carnera, G. Della Mea, L.F. Dona dalle Rose, V.N.
    Kulkarni, P. Mazzoldi, A. Miotells, E. Janitti, A.K. Jain, D.K. Sood and J.
    Chaumont, J. Appl. Phys. <u>55</u>, 3779 (1984).

4. J. P. Biersack and J.F. Ziegler in <u>Ion Implantation Techniques</u>, eds. H.
    Ryssel and H. Glawischnig (Springer-Verlag, Berlin, 1982), p.157.

5. Z.Y. Al-Tamimi, W.A. Grant and P.J. Grundy (to be published); Nucl.
    Instrum. Meth. <u>209/210</u>, 363 (1983).

6. M. Hansen, <u>Constitution of Binary Alloys</u> (McGraw Hill, New York, 1958).

7. D.K. Sood, Radiation Effects Lett. <u>67</u>, 13 (1981).

8. D.K. Sood and A.P. Pogany (to be published).

SOME CHARACTERISTICS OF $Al_{12}Mo$ IN ALUMINUM ANNEALED
AFTER IMPLANTATION WITH MOLYBDENUM

L. D. STEPHENSON,* J. BENTLEY,** R. B. BENSON, JR.,*** G. K. HUBLER,****
AND P. A. PARRISH*****
*U. S. Army Construction Engineering Research Laboratory, 2902 Newmark Drive,
Champaign, IL 61820
**Metals and Ceramics Division, Oak Ridge National Laboratory, P. O. Box X,
Oak Ridge National Laboratory, Oak Ridge, TN 37831
***North Carolina State University, Department of Materials Engineering,
229 Riddick Hall, Raleigh, NC 27695
****Condensed Matter and Radiation Sciences Division, Naval Research
Laboratory, 4555 Overlook Drive S. W., Washington, DC 20375
*****Defense Advanced Research Projects Agency, 1400 Wilson Boulevard,
Arlington, VA 22209

ABSTRACT

The characteristics of $Al_{12}Mo$ formed in aluminum annealed after implantation with selected maximum molybdenum concentrations were examined by analytical electron microscopy techniques. The $Al_{12}Mo$ was isolated as the only
precipitate in the microstructure for maximum as-implanted molybdenum concentrations up to 11 atomic percent. The morphology of the $Al_{12}Mo$ can be
selected by choosing the maximum as-implanted molybdenum level over the
same concentration range. A predominantly lamellar $Al_{12}Mo$ precipitate structure formed when aluminum was annealed at 550°C after implantation with maximum molybdenum concentrations in the range of 3.3 - 4.4 at.%. The orientation of the body centered cubic (bcc) $Al_{12}Mo$ precipitate with respect to the
face centered cubic (fcc) matrix can be expressed as $(\bar{1}23)_p \parallel (002)_m$ and
$[301]_p \parallel [310]_m$. An explanation for the experimentally observed orientation
relationship was developed based on the characteristic relationships between
the bcc $Al_{12}Mo$ precipitate and the fcc matrix. A continuous film of $Al_{12}Mo$
formed in the surface modified region when aluminum was annealed after
implantation with maximum molybdenum concentrations in the approximate range
of 8 - 11 at.%. The microstructure of the $Al_{12}Mo$ film was found to depend on
the annealing temperature. A granular film formed after annealing at 550°C
whereas a mottled film formed after annealing at 400°C. Sequential annealing
experiments revealed that the mottled film transforms to a granular film
which indicates the mottled film is metastable.

INTRODUCTION

Compounds of the type Al_xZ form, as has been shown by x-ray diffraction
analysis, in binary alloys of aluminum with selected transition metals Z,
where $12 \leq x \leq 3$ [1, 2]. In the aluminum-molybdenum and aluminum-tungsten systems a stable $Al_{12}Z$ phase forms, whereas metastable $Al_{12}Mn$ forms in the
aluminum-manganese binary alloy. These types of alloy systems are characterized by a series of peritectic reactions in the aluminum rich end of the
phase diagram and usually by a low solid solubility of the solute Z at lower
temperatures.

A program has been initiated recently to investigate the microstructures in selected binary aluminum-transition metal alloys in which compounds
of the type Al_xZ form where $12 \leq x \leq 3$. For the initial work, an alloy system
in which the phase $Al_{12}Z$ formed was desired, so the aluminum-molybdenum
system was selected.

The previously reported work for alloys containing an $Al_{12}Z$-type phase in the microstructure appears to be mostly with rapidly quenched, annealed alloys and mostly for aluminum-manganese alloys with manganese concentrations less than two weight percent [3-6]. In this work the $Al_{12}Z$ was not isolated as the only precipitate in the microstructure and the morphology of the $Al_{12}Z$ varied a great deal.

Ion implantation was selected as the initial alloy preparation technique in this work for several reasons. Ion implantation is a solid state alloy preparation technique rather than a liquid-solid one, such as rapid quenching, in which the as-quenched microstructure can play a dominant role in the morphology of the precipitate structure that is formed on annealing. The difference in the $Al_{12}Z$ phase morphologies between ion implanted, annealed aluminum-molybdenum alloys and the rapidly quenched, annealed alloys reported in the literature has been discussed recently in terms of the role of the as-quenched microstructure in the rapidly quenched alloys [7]. The solid solubility of molybdenum in aluminum is very low at all temperatures [1, 2]. It is well established that ion implantation can produce metastable primary solid solutions with very high degrees of supersaturation in many alloy systems [8]. The chemical composition profiles in the as-implanted layer can also be tailored by using multienergy implants, as was done in this work.

For aluminum annealed after implantation with selected maximum as-implanted molybdenum levels, there were several significant accomplishments in the results reported initially [9]. An $Al_{12}Z$-type phase, $Al_{12}Mo$, was isolated as the only precipitate in the microstructure. The morphology of the $Al_{12}Mo$ could be selected over a significant solute concentration range by choosing the maximum as-implanted molybdenum concentration. The results reported here deal with some of the characteristics of the $Al_{12}Mo$ phase for these morphologies.

EXPERIMENTAL PROCEDURE

Cold worked aluminum with a purity level of 99.999% was recrystallized by annealing at 350°C for thirty minutes which resulted in a grain size of the order of 0.5mm. The specimens were implanted to various maximum molybdenum concentrations using dual energy implants at the Naval Research Laboratory. The dual energy implants were at 50 and 110 keV for the various maximum as-implanted molybdenum concentration levels. Transmission electron microscopy specimens were prepared by electropolishing 3mm discs. Carbon extraction replicas from annealed specimens were used for x-ray microanalysis. The bulk specimens were annealed in a vacuum furnace at a pressure of approximately 10^{-4} Pa (10^{-6} torr). The annealing experiments in the electron microscope were carried out in a single tilt heating stage.

RESULTS AND DISCUSSION

As-Implanted Aluminum

The as-implanted maximum molybdenum concentrations of the alloys examined were determined by Rutherford backscattering spectrometry to be in the ranges of 3.3 - 4.4 atomic percent (at.%) and approximately 8 - 11 at.%. Selected area electron diffraction and electron microdiffraction analysis revealed that a metastable supersaturated solid solution had been retained in the as-implanted state for all the alloys considered. The observed as-implanted damage consisted of small dislocation loops and some tangled dislocations.

Aluminum Annealed after Implantation with 3.3 - 4.4 at.% Mo

As reported in the initial results when aluminum was annealed at 550°C for ninety minutes after implantation with maximum molybdenum concentrations in the range of 3.3 - 4.4 at.%, a predominantly coarse interlocking lamellar $Al_{12}Mo$ precipitate structure formed [9]. Subsequent results have shown that a fine lamellar $Al_{12}Mo$ structure forms when aluminum implanted with maximum molybdenum concentrations in the same range are annealed at 550°C for two minutes, and the coarser lamellar $Al_{12}Mo$ structure of Fig. 1 forms for a longer annealing time of thirty minutes at the same temperature. There appears to be both single growth ledges, as at A in Fig. 1, and double growth ledges, as at B in Fig. 1, associated with these $Al_{12}Mo$ lamellae. Elemental analysis of these discrete precipitates reported previously using energy dispersive x-ray microanalysis revealed a precipitate composition that was consistent with the composition of $Al_{12}Mo$ [9]. Convergent beam electron diffraction (CBED) analysis, also reported previously, revealed the space group and lattice parameter of the precipitate were consistent with those of $Al_{12}Mo$ as determined by x-ray diffraction analysis [9]. The Bravais lattice structure of the $Al_{12}Mo$ is body centered cubic (bcc).

The orientation relationship of the bcc $Al_{12}Mo$ precipitate with respect to the face centered cubic (fcc) matrix is presented in Fig. 2. As can be observed from Fig. 2, many planes from the two phases are close to being parallel. The orientation relationship of the bcc $Al_{12}Mo$ precipitate with respect to the fcc matrix, as can be observed from Fig. 2, can be expressed as $(\bar{1}23)_p \parallel (002)_m$ along with $[301]_p \parallel [310]_m$, where p and m represent the precipitate and matrix respectively. This orientation relationship appears to be consistent with both complete or partial reported orientation relationships for rapidly quenched, annealed aluminum-manganese alloys with low manganese compositions (i.e. less than 2 w/o Mn) where the $Al_{12}Mn$ precipitates exhibit various morphologies [3-5].

An explanation for the observed experimental orientation relationship can be developed based on the characteristic relationships between the bcc $Al_{12}Mo$ precipitate and the fcc matrix. The structure of the $Al_{12}Mo$ consists of the molybdenum atoms on a bcc Bravais lattice with each molybdenum atom surrounded by twelve aluminum atoms which form an icosahedron [10]. Eight faces of this icosahedron are parallel to {111} cubic planes. There is an equilateral triangular arrangement of aluminum atoms on $(111)_p$ that is almost

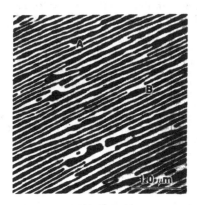

Fig. 1. Lamellar $Al_{12}Mo$ precipitate morphology in aluminum annealed at 550°C for 0.5 hours after implantation with maximum molybdenum concentrations in the range of 3.3 to 4.4 atomic percent.

464

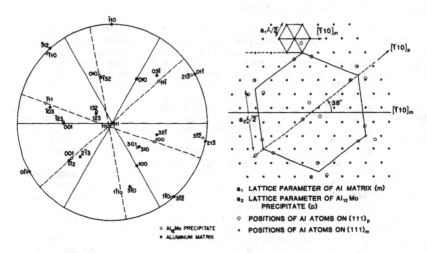

a₁ LATTICE PARAMETER OF Al MATRIX (m)
a₂ LATTICE PARAMETER OF Al₁₂Mo
     PRECIPITATE (p)
○ POSITIONS OF Al ATOMS ON (111)ₚ
● POSITIONS OF Al ATOMS ON (111)ₘ

○ Al₁₂Mo PRECIPITATE
● ALUMINUM MATRIX

Fig. 2. Orientation relationship for discrete $Al_{12}Mo$ precipitates in matrix.

Fig. 3. Matching of aluminum atom positions for $(1\bar{1}1)_p \parallel (111)_m$.

congruent with the equilateral triangular arrangement of aluminum atoms on $(111)_m$, as can be observed in Fig. 3. These arrangements of the atoms in the two phases will be discussed in more detail elsewhere. If the crystal axes of the matrix and precipitate have the same orientation, the triangles of aluminum atoms on their respective (111) planes are misoriented with respect to each other. A counterclockwise rotation of the precipitate by approximately 38° around a common (111) pole will bring one-third of the triangular configurations of aluminum atoms on $(111)_p$ into near-coincidence with aluminum atoms on $(111)_m$ and the remaining two-thirds into an almost closed-packed configuration between the matrix and precipitate aluminum atoms, as shown in Fig. 3. As can be observed in Fig. 2, this operation produces an orientation relationship very similar to the experimentally observed orientation relationship of this study. This arrangement of atoms on the $(111)_p$ with respect to $(111)_m$ could explain the close proximity of $(111)_p$ to $(111)_m$ in the experimental orientation relationship that is observed in the stereogram of Fig. 2. Trace analysis reveals that the lamellae interface planes are $(\bar{1}23)_p \parallel (002)_m$ which is consistent with the observed experimental orientation relationship. A factor influencing the selection of these particular interface planes could be their similar interplanar spacings [i.e. d $(\bar{1}23)_p$ = 0.2024nm and d $(002)_m$ = 0.2025nm]. If the geometrical orientation relationship based on the arrangement of atoms on (111) planes from both phases is shifted slightly to make the $(123)_p$ plane approximately parallel to $(002)_m$, as shown in Fig. 2, this orientation relationship then matches the experimental orientation relationship.

Aluminum Annealed after Implantation with 8 - 11 at.%Mo

A continuous film of $Al_{12}Mo$ has been observed in the surface modified region when aluminum was annealed after implantation with maximum molybdenum concentrations in the range of approximately 8 - 11 at.% [9]. The microstructure of the film was a function of the annealing temperature. A granular film was observed after annealing at 550°C for one and a half hours, Fig. 4a,

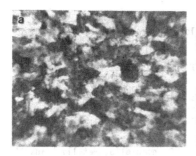

Fig. 4. Microstructure of continuous Al₁₂Mo films in aluminum implanted with maximum molybdenum concentrations in the range of 8 to 11 atomic percent: (a) granular film formed by annealing at 550°C for 1.5 hours, (b) mottled film formed by annealing at 400°C for 48 hours.

whereas a mottled film formed after annealing at 400 C°for forty eight hours, Fig. 4b. These results suggested that the mottled film may be a metastable microstructure. Sequential annealing experiments with bulk specimens revealed that the mottled film formed by an initial low temperature anneal at 400°C for forty-eight hours transformed to a granular film during a subsequent anneal at 550°C for four hours. Similar sequential in-situ annealing experiments with a heating stage in the transmission electron microscope produced results analogous to those for the bulk specimens. These results indicate that the mottled film is a metastable microstructure.

SUMMARY

A predominantly lamellar Al₁₂Mo precipitate structure was formed in aluminum annealed at 550°C for annealing times of up to thirty minutes after implantation with maximum molybdenum concentrations in the range of 3.3 - 4.4 at.%. The orientation relationship of discrete Al₁₂Mo precipitates can be expressed as $(\bar{1}23)_p \parallel (002)_m$ and $[301]_p \parallel [310]_m$. An explanation for the experimentally observed orientation relationship was developed based on characteristic relationships between the bcc Al₁₂Mo structure and the fcc matrix. A continuous Al₁₂Mo film formed in the surface modified region when aluminum was annealed after implantation with maximum molybdenum concentrations in the range of approximately 8 - 11 at.%. The microstructure of the film depended on the annealing temperature. A granular film formed during annealing at 550°C and a mottled film at 400°C. Sequential annealing revealed that a mottled film formed initially by annealing at 400°C for forty-eight hours transformed to a granular film on subsequent annealing at 550°C for four hours. This result indicates that the mottled film is a metastable microstructure.

ACKNOWLEDGEMENT

The authors are grateful for helpful discussions with H. H. Stadelmaier. The research of L. D. Stephenson was performed at ORNL under appointment to the laboratory graduate participation program administered by Oak Ridge Associated Universities for the U. S. Department of Energy. This work was sponsored by the U. S. Army Research Office under Contract DAAG-29-79-D-1003 and by the Division of Materials Sciences, U. S. Department of Energy under Contract DE-AC05-84OR21400 with Martin Marietta Energy Systems, Inc.

REFERENCES

1. M. Hansen, Constitution of Binary Alloys, 2nd ed. (McGraw-Hill, Inc., New York, 1958).

2. W. G. Moffatt, The Handbook of Binary Phase Diagrams (General Electric, Schenectady, NY, 1978).

3. E. Nes, S. E. Naess, and R. Hoier, Z. Metallkde., 63, 248 (1972).

4. A. K. Jena, D. P. Lahiri, T. R. Ramachandran, and M. C. Chaturvedi, J. Mater. Sci., 16, 2544 (1981).

5. D. Shechtman, R. J. Schaefer, and F. S. Biancaniello, Metall. Trans. A, 15, 1987 (1984).

6. A. Tonejc, J. Mater. Sci., 7, 1292 (1972).

7. R. B. Benson, Jr., J. Bentley, and L. D. Stephenson, in Mechanical Properties and Phase Transformations in Engineering Materials - Earl R. Parker Symposium on Structure Property Relationships, edited by S. D. Antolovich, R. O. Ritchie, and W. W. Gerberich, (Metallurgical Society of AIME, Warrendale, PA 1986) p. 355.

8. J. M. Poate and A. G. Cullis, in Treatise on Materials Science and Technology, Ion Implantation, edited by J. K. Hirvonen (Academic Press, New York, 1980), p. 85.

9. J. Bentley, L. D. Stephenson, R. B. Benson, Jr., P. A. Parrish and J. K. Hirvonen, in Ion Implantation and Ion Beam Processing of Materials, edited by G. K. Hubler, O. W. Holland, C. R. Clayton, and C. W. White. (Elsevier, New York, 1984), p. 151.

10. J. Adam and J. B. Rich, Acta Cryst., 7, 813 (1954).

ION IMPLANTATION AT ELEVATED TEMPERATURES*

NGHI Q. LAM and GARY K. LEAF
Argonne National Laboratory, 9700 S. Cass Ave., Argonne, IL   60439

ABSTRACT

A kinetic model has been developed to investigate the synergistic effects of radiation-enhanced diffusion, radiation-induced segregation and preferential sputtering on the spatial redistribution of implanted solutes during implantation at elevated temperatures. Sample calculations were performed for Al^+ and Si^+ ions implanted into Ni. With the present model, the influence of various implantation parameters on the evolution of implant concentration profiles could be examined in detail.

INTRODUCTION

During ion implantation, nonequilibrium point defects are produced in large numbers by the violent slowing-down of energetic ions in the host matrix. Spatial nonuniformity in the defect production and annihilation gives rise to persistent defect fluxes, e.g., towards the surface, or from the peak-damage region towards the mid-range and beyond-range regions where the defect-production rates are significantly lower. At elevated temperatures, these defect fluxes can preferentially transport certain alloying elements via defect-solute interactions, which causes a spatial redistribution of implanted species. This nonequilibrium phenomenon, called radiation-induced segregation (RIS), acts synergistically with radiation-enhanced diffusion (RED) and preferential sputtering (PS) during ion implantation, introducing a great complexity to the understanding of this process.

A comprehensive kinetic model has been developed recently to investigate the synergistic effects of all the processes mentioned above on the spatial redistribution of implanted atoms during implantation. The effects of spatially-nonuniform rates of damage and ion deposition, as well as the movement of the bombarded surface as a result of sputtering and introduction of foreign atoms into the system, were taken into account. The evolution of implant concentration profile in time and space was calculated for various temperatures, ion energies, and ion-target combinations. The results of the present work may be useful in elucidating the essential physics of the elevated-temperature ion implantation process.

BASIC KINETICS

We consider a metal substrate B into which A atoms are implanted at a flux ϕ (ions/cm^2·s). A number of A atoms will end up in interstitial sites, and the rest in substitutional sites of the host lattice. The respective rates of implantation into these sites are $f_i \phi$ and $f_s \phi$ (atom fraction/s). During implantation, point defects - vacancies and interstitials - are created by ion impact at a rate K (displacements per atom per second, dpa/s). Since the interstitial defects are distinguishable, the rate of interstitial production is partitioned into KC_{sA} and $KC_{sB} \equiv K(1-C_{sA})$ for A- and B-interstitials, respectively, with C_{sA} and C_{sB} being the respective concentrations of A and B atoms in substitution. The interstitials and vacancies annihilate by mutual recombination and/or diffusion to defect

---

*Work supported by the U. S. Department of Energy, BES-Materials Sciences, and BES-Applied Mathematical Sciences, under Contract W-31-109-Eng-38.

sinks. In addition, vacancies and B-interstitials can interact with free solute atoms, giving rise to the formation of vacancy-solute complexes and conversion of B-interstitials to A-interstitials, respectively.

The local concentrations (in atomic fractions) of vacancies (v), B-interstitials (iB), A-interstitials (iA), A-vacancy complexes (vA) and free substitutional solutes (sA) change with implantation time according to the following system of kinetic equations:

$$
\begin{aligned}
\frac{\partial C_v}{\partial t} &= -\nabla\cdot J_v && + K && - f_s\phi && + F_v \\
\frac{\partial C_{iB}}{\partial t} &= -\nabla\cdot J_{iB} && + K(1-C_{sA}) && && + F_{iB} \\
\frac{\partial C_{iA}}{\partial t} &= -\nabla\cdot J_{iA} && + KC_{sA} && + f_i\phi && + F_{iA} \\
\frac{\partial C_{vA}}{\partial t} &= -\nabla\cdot J_{vA} && && && + F_{vA} \\
\frac{\partial C_{sA}}{\partial t} &= -\nabla\cdot J_{sA} && - KC_{sA} && + f_s\phi && + F_{sA}
\end{aligned}
\tag{1}
$$

where the F's denote the local rates of creation and loss of species by chemical-type reactions (i.e. formation and dissociation of defect-solute complexes, and defect recombination and annihilation), and the J's are the fluxes of the mobile species, defined as [1]:

$$
J_v = -(1 + \sigma_v C_{sA})D_v \nabla C_v
\tag{2a}
$$

$$
J_{iB} = -(1 + \sigma_i C_{sA})D_{iB} \nabla C_{iB}
\tag{2b}
$$

$$
J_{iA} = -D_{iA} \nabla C_{iA}
\tag{2c}
$$

$$
J_{vA} = -D_{vA} \nabla C_{vA}
\tag{2d}
$$

$$
J_{sA} = -\sigma_i C_{sA} D_{iB} \nabla C_{iB} + \sigma_v C_{sA} D_v \nabla C_v .
\tag{2e}
$$

Here, the fluxes of defects and defect-solute complexes are proportional to their concentration gradients, the proportionality constants being the diffusion coefficients D_v, D_{iB}, D_{iA} and D_{vA}. The constants σ_v and σ_i are the capture factors for vacancy-solute and interstitial-solute encounters [1,2]. For strong defect-solute binding, $\sigma_v = 6$ and $\sigma_i = 6$, and for repulsion or no interaction, $\sigma_v = 2$ and $\sigma_i = 0$. The free solute flux arises from the coupling of A atoms with free B-interstitial and vacancy fluxes. Note that the vacancy flux-induced solute current, which is the last term of equation (2e), has the opposite sign of the other terms, because atom transport occurs in the direction opposite to that of the vacancy flux. The reader is referred to ref. [3] for more details of the model.

Concurrently with the buildup of solute concentration in the host matrix, the surface is subjected to displacements, due to the introduction of foreign atoms into the system and sputtering. The former gives rise to surface relaxation in the - x direction, at a rate δ_-, whereas the latter causes surface recession in the + x direction, at a rate δ_+. The rate δ_- can be calculated from the net atom flux towards the surface:

$$
\delta_- = J_{iA} + J_{iB} - J_v - J_{vA} ,
\tag{3}
$$

while the rate δ_+ is determined by the ion flux, the partial sputtering coefficients, S_A and S_B, and the surface concentrations of A and B atoms:

$$
\delta_+ = \phi\Omega[S_A C_A^s + S_B(1 - C_A^s)] ,
\tag{4}
$$

where Ω is the average atomic volume. The net surface displacement rate,

$$\dot{\delta} \equiv \frac{d\delta}{dt} = \dot{\delta}_- + \dot{\delta}_+ \tag{5}$$

is therefore controlled by the competition between the rates of ion collection and sputtering.

Equations (1) and (5) were solved numerically for a semi-infinite medium with the aid of the LSODE package of subroutines [4], starting from the thermodynamic equilibrium conditions. A moving boundary, which represents the receding surface, was accommodated by a transformation of the physical space x to a reduced space z:

$$z = 1 - \exp[-\beta(x - \delta)] \tag{6}$$

where β is a scaling factor. With this transformation, the region $\delta < x < \infty$ was mapped into the fixed region $0 < z < 1$.

Sample calculations were performed for two model systems: Al^+ and Si^+ implantations into Ni. The physical parameters used were tabulated in ref. [3]. The ions were implanted at 50 and 400 keV. The spatial dependence of the rates of defect production and ion deposition was calculated using the TRIM code [5], with a displacement energy of 30 eV. The most-probable damage and ion ranges for 50-keV Al^+ and Si^+ in Ni are ~13 and 27 nm, and ~12 and 24 nm, respectively. For 400-keV Al^+ and Si^+ in Ni, the respective most-probable ranges are ~175 and 235 nm, and ~160 and 215 nm. At 50 keV, sputtering is significant; the sputtering coefficients for the host and solute atoms, S_B and S_A, in Ni-Al and Ni-Si alloys were taken to be 3.5 and 2.0 atoms/ion, and 3.5 and 1.0 atoms/ion, respectively [6]. Here, we assumed that S_A and S_B were equal to the coefficients for the respective pure A and B elements. At 400 keV, sputtering is significantly reduced [6]; it was, however, assumed to be negligible in the present calculations. For each ion-target combination, the defect-production rate K was taken as $0.3K_0$, where K_0 is the calculated rate of defect production, and the factor 0.3 is the defect-production efficiency [7].

SAMPLE CALCULATIONS

Time and temperature dependence of implant redistribution during high-energy implantation

The time evolution of the Al redistribution during 400-keV implantation at 100, 500 and 800°C is shown in Fig. 1. The ion flux was 6.25×10^{13} ions/cm^2·s and the corresponding peak-damage rate was calculated to be $K_0^{max} = 4.92 \times 10^{-2}$ dpa/s. The most-probable damage and ion ranges are marked by R_d and R_i. The dislocation density was assumed to remain constant at 10^6 disl/cm^2 throughout the target during implantation. Since sputtering was assumed to be negligible, the surface relaxed in the - x direction; the calculated surface displacements, δ, are indicated in the figure for various times. During implantation, under the influence of damage-rate gradients, point defects flow out of the peak-damage region towards the surface and into the beyond-range region. Since Al solutes segregate in the direction opposite to the defect fluxes in irradiated Ni, there exists a net flux of Al atoms into the peak-damage region [8]. Consequently, there is no shift of the implant profile towards the sample interior; the ion distribution simply broadens with time. The effect of RIS on the implant distribution can also be demonstrated by the near-surface Al depletion on high-dose profiles, especially around 500°C where Al depletion is most severe. Long tails extending deep into the bulk, resulting from RED of Al atoms from the ion-peak region, are observed on the solute profiles for t > 10^2s. The contribution of RIS to the evolution of the Al tails is unimportant, because Al atoms and point defects migrate in opposite directions. The temperature dependence of the penetration

470

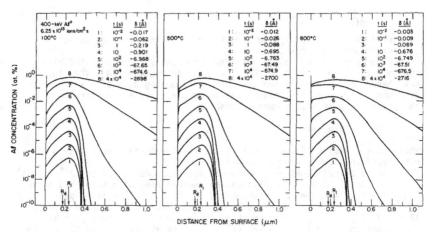

Fig. 1. Time evolution of the Al concentration profiles during 400-keV
implantation at 100, 500 and 800°C.

tails is, however, rather weak, because Al transport occurs via the
dominant interstitialcy mechanism. It should be pointed out that RED is
efficient during implantation because the effective rate of interstitial
production is always larger than the vacancy-production rate.

Figure 2 illustrates the time dependence of the Si redistribution
during implantation at 100, 500 and 900°C. The calculated peak-damage rate
was $K_0^{max} = 5.6 \times 10^{-2}$ dpa/s. The values of δ are indicated in the

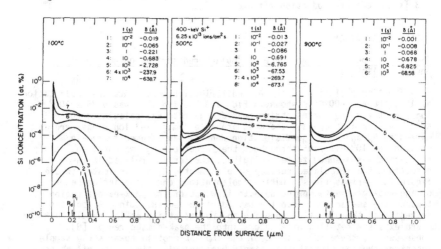

Fig. 2. Time evolution of the Si concentration profiles during 400-keV
implantation at 100, 500 and 900°C.

figure. During implantation at elevated temperatures, the preferential
association between Si atoms and Ni-interstitials induces a net flux of Si
in the same direction as the defect fluxes, out of the peak-damage region
[8,9]. As a result, after a short implantation time at, e.g., 500°C, a

depression in the Si concentration occurs in the peak-damage region, and a pronounced increase is seen at the surface and in the beyond-range region. The implant distribution peak moves inward, and eventually stops at a position near the end of the ion range. Unlike the Al-implantation case, the penetration tails on the Si profiles are caused by a combination of RIS and RED via an interstitialcy mechanism. For a given temperature and dose, these tails are significantly higher and longer than those on the Al concentration profiles shown in Fig. 1. At 100°C, the Si distribution peak is not shifted, only solute enrichment at the surface and long tails are observed. The tails on the 100 and 900°C profiles are, however, shorter than those at intermediate temperatures, e.g. 500°C, because of the decrease in the effectiveness of RIS at low ($\lesssim 0.2T_m$) and high ($\gtrsim 0.7T_m$) temperatures. The predicted shift of the Si distribution peak into the bulk is consistent with recent experimental measurements by Mayer et al. [10] in Si-implanted Ni at elevated temperatures.

Effect of preferential sputtering on implant distribution during low-energy implantation

Redistributions of Al and Si solutes in Ni during 50-keV implantation at 500°C are presented in Figs. 3 and 4, respectively. The rates of defect production and ion deposition, normalized to their peak values, are included in the top portions. For the ion flux used, 6.25×10^{13} ions/cm$^2 \cdot$s, the calculated peak-damage rates were $K_{max} = 7.92 \times 10^{-2}$ and 8.63×10^{-2} dpa/s for Al$^+$ and Si$^+$ bombardments of Ni, respectively. Since sputtering is significant at this energy, the bombarded surface recedes into the sample with time. The surface displacements are indicated in the figures.

In the Al-implantation case (Fig. 3), the Al concentration at the surface, C_{Al}^s, increases with time to a quasi-steady-state value, which is mainly determined by the sputtering coefficients and the sub-surface implant concentration. The time required to achieve quasi-steady-state is $\sim 10^4$s. The quasi-steady-state value of C_{Al}^s, 45.9 at.%, is significantly larger than that obtained in the high-energy implantation case, where C_{Al}^s is controlled by RIS. The total implant concentration remaining in the substrate is, however, smaller due to sputtering. The quasi-steady-state implant profile does not resemble the high-energy implantation profile (compare with Fig. 1); it decreases monotonically with depth from the value of C_{Al}^s.

The temporal evolution of the Si profile (Fig. 4) is quite different from that for Al, due to the different RIS behaviors. After a short time, e.g. 0.1 s, Si enrichment occurs at the surface, because of RIS and relatively small ion range, and the Si distribution peak starts moving into the sample interior. However, as the implantation goes on, the Si concentration at the surface, C_{Si}^s, decreases with time and eventually attains a quasi-steady-state value of $\sim 3.8 \times 10^{-4}$ at.% for $t > 10^4$s. This value is five-orders-of-magnitude smaller than C_{Al}^s, because, in this case, most of the implanted Si atoms segregate into the bulk. The Si profile always changes with time, even though C_{Si}^s does attain its quasi-steady-state value. The implant distribution continuously shifts towards the interior, and high-level tails extending deep into the sample are observed behind the distribution peak for $T \gtrsim 10^3$s.

ACKNOWLEDGEMENTS

The authors have greatly benefited from useful discussions with Prof. R. A. Johnson and Dr. H. Wiedersich. Mrs. Diane Livengood's contributions in skillfully formatting and typing the paper are also gratefully acknowledged.

472

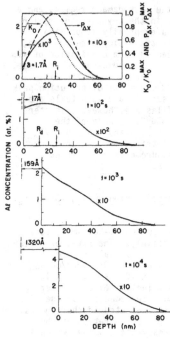

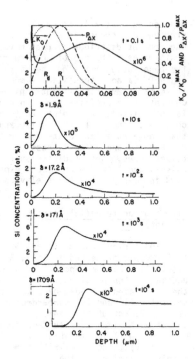

Fig. 3   Fig. 4

Fig. 3. Development of the Al profiles during 50-keV implantation at
         500°C. The normalized damage (K_0) and ion-deposition ($P_{\Delta x}$) rates
         are included in the top portion. The surface displacements, δ,
         resulting from sputtering are indicated.

Fig. 4. Development of the Si profiles during 50-keV implantation at
         500°C. The normalized damage and ion-deposition rates are shown
         in the top portion. Note that the depth scales in plots for
         t = 0.1 s and t > 10 s differ by a factor of 10.

REFERENCES

1. R. A. Johnson and N. Q. Lam, Phys. Rev. B13, 4364 (1976).
2. A. Barbu, Acta Metall. 28, 499 (1980).
3. N. Q. Lam and G. K. Leaf, submitted to J. Mater. Res.
4. A. C. Hindmarsh, in: Scientific Computing, eds. R. S. Stepleman et al.
   (North-Holland, Amsterdam, 1983) p. 55.
5. J. P. Biersack and L. G. Haggmark, Nucl. Instr. Methods 174, 257 (1980).
6. H. H. Andersen and H. L. Bay, in Sputtering by Particle Bombardment,
   ed. R. Behrisch (Springer, Heidelberg, 1981) p. 145.
7. P. R. Okamoto, L. E. Rehn and R. S. Averback, J. Nucl. Mater. 133 &
   134, 373 (1985).
8. N. Q. Lam, P. R. Okamoto and R. A. Johnson, J. Nucl. Mater. 74, 101
   (1978).
9. N. Q. Lam, K. Janghorban and A. J. Ardell, J. Nucl. Mater. 101, 314
   (1981).
10. S. G. B. Mayer, F. F. Mililllo and D. I. Potter, Mat. Res. Soc. Symp.
    Proc. 39, 521 (1985).

BINARY AND TERNARY AMORPHOUS ALLOYS OF ION-IMPLANTED Fe-Ti-C

D. M. FOLLSTAEDT AND J. A. KNAPP
Sandia National Laboratories, Albuquerque, NM 87185

ABSTRACT

The microstructure of Fe implanted with up to 50 at.% C was found to consist of hexagonal iron carbide precipitates oriented with respect to the Fe matrix. For higher C concentrations, an amorphous phase forms. This concentration dependence is explained in terms of the lattice structure of the iron carbide. In Ti-implanted Fe, substitutional Ti was found in the bcc Fe lattice for concentrations $\leq$ 15 at.% Ti. The work of others suggests that amorphous phases form for $\geq$ 33 at.% Ti. These results are discussed in terms of concentration boundaries of the ternary Fe(Ti,C) amorphous phase.

Ion beam alloying methods are currently being used to form metastable alloys [1], both for fundamental investigations of such alloys as well as for potential use to improve physical properties of components [2]. An important consideration in metal alloys is what phase will form upon implantation; one aspect of this question is to determine when amorphous phases will form. Rules are currently being advanced to predict alloy systems which will yield amorphous phases. By using ion irradiation and ion beam mixing as well as ion implantation, such rules can be evaluated over entire composition ranges.

To gain insight into amorphous phase formation, we have studied Fe alloys implanted with C, Ti and Ti + C. The Fe(C) alloys exhibit compound precipitation and amorphous phase formation; the precipitation and the concentrations at which the amorphous phase appears can be accounted for by considerations of the structure of the hexagonal carbide which forms. Based on conventional uses of the Fe(C) system, such alloys may be useful for improving mechanical properties by implanting C into ferrous components. Iron implanted with Ti is examined to a limited extent here, but by including ion irradiation studies by others [3], a more complete characterization of Fe(Ti) alloys is obtained. The microstructures of Fe(C) and Fe(Ti) are examined along with the known composition limits of amorphous Fe(Ti,C) alloys, which are important for their improved mechanical properties [2]. Taken together, a more complete determination of amorphous phase formation in this ternary system is obtained.

IRON IMPLANTED WITH CARBON

The microstructure of Fe implanted with C at room temperature has been examined previously [1,4] with transmission electron microscopy (TEM), and is illustrated in Fig. 1. For maximum concentrations of 38 and 49 at.% C, precipitates of the hexagonal phase Fe_2C were observed which were oriented with respect to the Fe matrix in which they formed. This carbide is usually written Fe_2C, but it exists over a wide composition range [5]; its lattice constants are a = 0.275 nm and c = 0.435 nm [6]. In Fig. 1a, a portion of the diffraction pattern from one of three symmetrically-equivalent variants of Fe_2C is outlined; it shows that the alignment of the precipitates to the bcc Fe matrix is [1]:

$(1\bar{2}10)$ || (111) - plane in Fig. 1

$[0001]$ || $[1\bar{1}0]$ ⎫
⎬ directions in that plane
$[10\bar{1}0]$ || $[11\bar{2}]$ ⎭

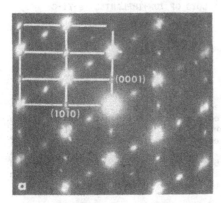

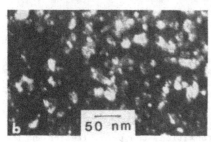

Fig. 1. a) Diffraction pattern and b) dark-field image of carbides in bcc Fe matrix obtained from Fe implanted with 2×10^{17} C/cm^2 at 30 keV.

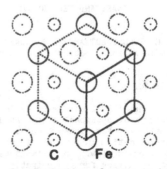

Fig. 2. Projection of atoms in iron carbide onto the (0001) plane, with unit cell outlined. Fe atoms are at 0 and 0.5 times c_o; C atoms are at 0.25 and 0.75 times c_o.

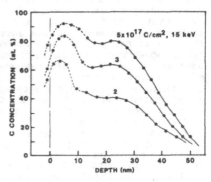

Fig. 3. Carbon depth profiles in Fe; the analysis uses a density of 7.86 g/cm^3. The dashed peak probably contains O also.

The carbide has a hcp lattice of Fe atoms with C atoms at octahedral interstitial positions between the Fe layers; the projected locations on the (0001) plane are shown in Fig. 2. Its precipitation in Fe can be interpreted as due to its structural simplicity and the above alignment [1]. Other work indicates that when compounds with simple structures are possible, they tend to form during ion implantation [7]. The precipitation may also be aided by the mobility of C in Fe at room temperature [8]. Conversely, when the possible compounds have complex structures, an amorphous phase is expected.

Hexagonal iron carbide has been found to exist over a wide composition range, but with $\leq$ 33 at.% C [5]. However the atomic arrangement shown in Fig. 2 suggests how this structure might exist with 49 at.% C. If each interstitial site is filled, a C:Fe ratio of 1 is achieved, thus allowing up to 50 at.%. With this hypothesis, a new phase would be required to accommodate additional C if it is implanted above this limit. It was suggested earlier [1] that an amorphous phase might form above 50 at.%, because no Fe carbides with higher C concentration are known and deposited C is amorphous. An amorphous phase had been reported for a fluence of

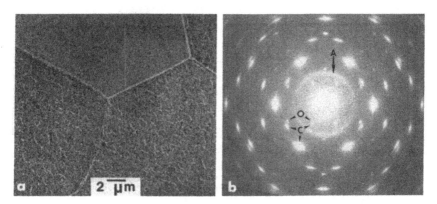

Fig. 4. a) Bright-field image of the Fe(C) surface alloy on Fe implanted with 5×10^{17} C/cm^2 at 15 keV, showing original grain boundaries and rough surface texture, and b) diffraction pattern from area within one grain.

10^{18} C/cm^2 at 30 or 50 keV, but only for a temperature of 77 K and/or lower purity of the Fe (99%) [7,9].

To look for the predicted amorphous phase formation above 50 at.%, we have examined high purity Fe (99.99%) implanted at room temperature with 2, 3 and 5×10^{17} C/cm^2 at 15 keV. Rutherford backscattering (RBS) was used to obtain the C concentration versus depth by examining the difference in backscattered yields from Fe between the implanted specimen and pure Fe. The resulting depth profiles are shown in Fig. 3. In each case, a peak is observed within 5 nm of the surface, followed by a plateau extending to 25 nm. The analysis in Fig. 3 assumes only C additions to the Fe; however nuclear reaction analysis shows an O content after implantation of ~ 1/5 the C content, i.e., 1.2×10^{17} O/cm^2 versus 6.7×10^{17} C/cm^2 for the highest fluence. Oxygen may have been incorporated into the near surface during the C implantation. Alternatively the sample may have oxidized upon subsequent air exposure; its surface is observed to be much rougher after the implantation, and appears brown. The O peak in the RBS spectrum shows that the O is within ~ 10 nm of the surface. Since the C profiles in Fig. 3 were derived from reductions in backscattered yield from Fe, to which O would contribute, the near-surface peaks are probably due to the presence of O and hence their profile is dashed.

For the low fluence with the plateau concentration of 42 at.%, the same microstructure as that seen in Fig. 1 was observed with TEM; there was no evidence of amorphous phase formation. For the intermediate fluence with the plateau at 63 at.%, the diffraction pattern showed a weak diffuse ring in addition to the carbide reflections; this ring was also seen with a maximum concentration of 56 at.% at a depth of 45 nm obtained with 30 keV C.

The microstructure of the high fluence sample in Fig. 3 is shown in Fig. 4. The surface alloy resisted electropolishing from the substrate side, and large areas of the alloy extending across many of the original Fe grains could be examined. Figure 4a shows several such grains, and an electron diffraction pattern from one grain is shown in Fig. 4b. The pattern contains the two variants of the carbide (spots labeled C) as seen in Fig. 1, plus a <110> diffraction pattern from γ-Fe$_2$O$_3$ (spots labeled O) and a diffuse ring (labeled A) at the same diameter as that observed with 63 at.%. The ring diameter corresponds to an atomic spacing of 0.34 nm, and does not match the most intense rings of surface iron oxides seen elsewhere in this sample. Examination of a tapered area of the alloy did not show the diffuse ring in the thinnest area near the surface, where O is present; thus

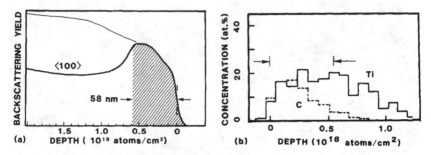

Fig. 5. a) Ion channeling spectrum and b) Ti and C concentration profiles for Fe implanted with 2 x 10^{17} Ti/cm^2 at 180 keV.

the ring and carbide reflections are taken to be due to material at the depth with the plateau in concentration at 80 at.%.

The diffuse ring in C-implanted Fe indicates the presence of the expected amorphous C-Fe phase; it may contain very little Fe since the iron carbide reflections from the same thickness suggest that the alloy is two-phased at 80 at.% C. The observation of the ring at 55-63 at.% C agrees with the expected amorphous phase formation above 50 at.%.

Some of the grain surfaces in Fig. 4a are rough and exhibit furrows, which apparently result from the high-fluence implantation. The surface roughness can be detected but not resolved with Nomarski optical microscopy; a difference between grains was observed at the lowest fluence and became more pronounced at the highest fluence. The TEM examination showed that individual grains were uniformly rough. Most grains have surface normals near <111>, but the sample was tilted by ~ 10° from normal incidence during implantation. We suggest that the variation between grains is due to small differences in grain orientations.

IRON IMPLANTED WITH Ti

The study of Ti-implanted Fe is difficult because C also enters the near-surface of the alloy from C-containing molecules in the vacuum system [2,10]. The resulting Fe(Ti,C) surface alloy is amorphous, but extends only to depths where there is sufficient C. In Fig. 5a, an RBS/channeling spectrum is shown from <100> Fe implanted with 2 x 10^{17} Ti/cm^2 at 180 keV. The disorder peak (cross-hatched) due to the amorphous phase extends only to 58 nm, and does not include all the implanted Ti as seen in the depth profiles in Fig. 5b, which were obtained as described in detail elsewhere [10]. The Ti within the amorphous layer will show no channeling effect with respect to the substrate lattice. However just behind this layer, the material has the bcc Fe structure, as shown by the electron diffraction pattern in Fig. 6 taken from a thickness containing both the amorphous layer (diffuse ring) and the substrate (spot pattern). Even though this depth contains as much as 15 at.% Ti, no Fe-Ti compound is observed, and thus the Ti may be substitutional in the bcc lattice.

The Ti K$_\alpha$ x-ray intensity was studied as a function of the incident angle of 2 MeV He ions about the <100> axis, and a small channeling dip to 78% of the yield at random orientation was observed. For comparison, the Fe backscattering yield behind the amorphous layer shows an ~ 50% channeling reduction, and the Fe K$_\alpha$ x-ray yield shows a dip to 57% of the yield at random orientation. The randomly oriented amorphous layer contains more than half the implanted Ti atoms, and thus even this modest channeling

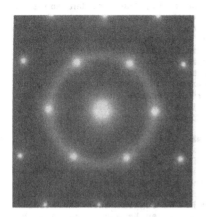

Fig. 6. Diffraction pattern from the amorphous layer (diffuse ring) plus the bcc substrate (spots).

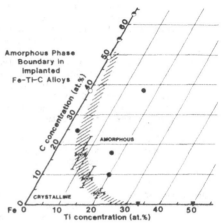

Fig. 7. Fe-Ti-C concentration triangle showing crystalline and amorphous regions.

reduction is consistent with most of the deeper Ti in the substrate being on substitutional bcc lattice sites.

The microstructures resulting from Ni^+ irradiation of the compounds Fe_2Ti and FeTi have been examined by others [3]; both were amorphized with doses producing ~ 1 dpa. Amorphization of these and other compounds was shown to correlate with their narrow composition ranges. Thus a narrow range was taken to indicate that a phase cannot survive disordering by irradiation. It is probable that Ti implantation into Fe to the concentration of Fe_2Ti would also produce amorphous Fe(Ti). Thus at concentrations up to 15 at.%, a significant fraction of implanted Ti is substitutional in the bcc Fe lattice, while for $\geq$ ~ 33 at.%, binary amorphous alloys are expected.

DISCUSSION

In the Fe-Ti-C composition triangle in Fig. 7, the minimum concentrations of Ti and C needed to stabilize the ternary amorphous phase are shown with error bars, and a cross-hatched boundary is drawn through them. These points were determined by comparing amorphous layer thicknesses with Ti and C depth profiles, as illustrated in Fig. 5 [10]. Also shown are C concentrations found to be crystalline (O) and crystalline + amorphous (◑), and Ti concentrations at which compounds were observed to be amorphized with ion irradiation (▼). Based on the previous work showing carbide precipitation in Fe, the boundary was not allowed to intersect the Fe-C binary axis below 50 at.% C. The absence of a binary amorphous phase at these concentrations is due to the apparent ease of formation of hexagonal iron carbide. The boundary might intersect the axis near or above 80 at.%; however, the C-rich amorphous phase perhaps should no longer be regarded as the same phase as amorphous Fe(Ti,C), which is composed principally of metallic atoms and has its strongest ring at a different atomic spacing (0.205 nm). On the Ti-rich side, we have extended our previously determined boundary [10] to intersect the Fe-Ti axis near 33 at.% Ti. Amorphous Fe(Ti) alloys are not formed at lower concentrations apparently because Ti is relatively soluble in bcc Fe (up to 10.5 at.% at high temperatures [11]).

478

When both C and Ti are present together in Fe, however, the mobile C atoms are thought to bond to Ti atoms due to their strong chemical affinity. This bonding is apparently inconsistent with the bcc lattice structure above the minimum concentrations noted in Fig. 7, and the amorphous phase forms. Bonding between neighboring Ti and C atoms has been inferred from Auger lineshapes obtained from this amorphous phase [12]. A two-phase crystalline alloy of bcc Fe and TiC is an alternative to the amorphous phase, and does form upon annealing to 600°C [13]. However, TiC formation does not occur during room temperature implantation, possibly because the Ti-C pairs are immobile. The Fe-Ti-C system may be just barely stable against such precipitation, since substituting N for C produces precipitation of TiN [14], which has the same structure as TiC. At high Ti and C concentrations, TiC may precipitate; the filled dots (●) in Fig. 7 indicate that the amorphous phase can be extended to at least these concentrations.

Our considerations of the interstitial structure in Fig. 2 can also explain observations on Fe alloys implanted with N, where it has been observed that above 50 at.%, blistering and loss of N occur [15,16]. At concentrations approaching 50 at.% N, the analogous nitride interstitial phase forms in Fe [17] and steels [16,18]. We suggest that the nitride phase can accomodate only up to one N atom per Fe atom, and thus the formation of a new phase, in this case gaseous N_2, occurs above 50 at.%. More generally, a number of metals implanted to high fluences of N showed blistering at saturated concentrations of 50-60 at.% [15]; since nitrides with N:metal atomic ratios significantly greater than 1 are not known for most metals, an upper limit of ~ 50 at.% is expected.

The authors wish to thank M. Moran and G. Schuh for their technical assistance. This work performed at Sandia National Laboratories supported by the U.S. Department of Energy under contract number DE-AC04-76DP00789.

REFERENCES

1. D. M. Follstaedt, Nucl. Inst. Meth. B7/8, 11 (1985).
2. D. M. Follstaedt, Nucl. Inst. Meth. B10/11, 549 (1985).
3. J. L. Brimhall, H. E. Kissinger and L. A. Charlot, Rad. Eff. 77, 273 (1983).
4. V. M. Drako and G. A. Gumanskij, Rad. Eff. 66, 101 (1982).
5. W. B. Pearson, Handbook of Lattice Spacings and Structures of Metals and Alloys, Vol. I and II (Pergamon Press, New York, 1958 and 1967), p. 919 and p. 1339, respectively.
6. Joint Committee on Powder Diffraction Standards, entry 6-0670.
7. B. Rauschenbach and K. Hohmuth, Phys. Stat. Sol. (a) 72, 667 (1982).
8. J. R. G. da Silva and R. B. McLellan, Mat. Sci. Engr. 26, 83 (1976).
9. B. Rauschenbach, private communication.
10. J. A. Knapp, D. M. Follstaedt and B. L. Doyle, Nucl. Inst. Meth. B7/8, 38 (1985).
11. Metals Handbook, Vol. 8, 8th Ed., (Am. Soc. Metals, Metals Park, Ohio, 1973), p. 307.
12. I. L. Singer, C. A. Carosella and J. R. Reed, Nucl. Inst. Meth. 182/183, 923 (1981).
13. D. M. Follstaedt, J. Appl. Phys. 51, 1001 (1980).
14. D. M. Follstaedt, J. A. Knapp, L. E. Pope and S. T. Picraux, Nucl. Inst. Meth. B12, 359 (1985).
15. A. Antilla, J. Keinonen, M. Uhrmacher and S. Vahvaselka, J. Appl. Phys. 57, 1423 (1985).
16. B. L. Doyle, D. M. Follstaedt, S. T. Picraux, F. G. Yost, L. E. Pope and J. A. Knapp, Nucl. Inst. Meth. B7/8, 166 (1985).
17. B. Rauschenbach and A. Kolitsch, Phys. Stat. Sol. (a) 80, 211 (1983).
18. F. G. Yost, S. T. Picraux, D. M. Follstaedt, L. E. Pope and J. A. Knapp, Thin Solid Films 107/108, 287 (1983).

ROLE OF CHEMICAL DISORDERING IN ELECTRON IRRADIATION INDUCED AMORPHISATION

D.E. LUZZI*†, H. MORI*, H. FUJITA*, and M. MESHII†
*Research Center for Ultra-High Voltage Electron Microscopy, Osaka University, Yamada-oka, Suita, Osaka 565 JAPAN
†Materials Research Center and Department of Materials Science and Engineering, Northwestern University, Evanston, IL 60201

ABSTRACT

The electron irradiation induced crystalline to amorphous (C-A) transition was investigated in the Cu-Ti alloy system. As the transition is found to be dose dependent, the energy accumulated during the irradiation must make the crystal lattice unstable with respect to the amorphous state. In this paper, two possibilities for this energy accumulation, chemical disordering and the accumulation of point defects, is examined using the intermetallic compound $CuTi_2$. Also, the effect of deviations from stoichiometry on the C-A transition is investigated in the compound CuTi. The Bragg-Williams long-rang order parameter, S, was found to decrease with electron dose. The rate of this decrease decreased with increasing temperature. At a critical temperature, the maximum obtainable degree of chemical disordering critically decreased. This critical temperature coincided with the critical temperature for the C-A transition, favoring the chemical disordering as the primary driving force. In CuTi, the C-A transition was hindered when there existed a Cu-rich deviation from stoichiometry.

INTRODUCTION

The occurrence of an electron irradiation induced crystalline to amorphous (C-A) transition has recently been reported [1,2]. The transition was found to be dose dependent [3] indicating a process in which the energy accumulated during the irradiation makes the crystal lattice unstable with respect to the amorphous state. This energy increase prior to amorphisation can be stored either as chemical disordering or as an increase in the density of point defects (no secondary defects are observed prior to amorphisation [4]). Recently, it has been reported that the energy increase due to irradiation induced chemical disordering is the major driving force for the C-A transition in Cu_4Ti_3 [5,6].
In this paper, the study will be expanded to the compound $CuTi_2$ in which a more in-depth experiment was conducted to determine the relationship between chemical and topological disordering. This relationship and the role of point defects in the C-A transition will be discussed under the conditions of stoichiometry and non-stoichiometry.

EXPERIMENTAL DETAILS: SPECIMEN PREPARATION

Alloy buttons composed of 99.99 percent pure Cu from Ishizu Pharmaceutical Co., Ltd. and 99.9 percent pure Ti from Vacom were formed by arc-melting in an Ar atmosphere in the proportion of 72, 45 and 52 atomic percent Ti. Bulk liquid phase quenchings were done using alumina crucibles for the 45 and 52 atomic percent Ti buttons. Various anneals were done as follows: 72 at.% Ti, 1223 K, 43.2 ks; 45 at.% Ti, 973 or 1153 K, 43.2 ks; 52 at.%Ti, 1153 K, 43.2 ks. Vacuum conditions for all of the anneals were better than 2.7×10^{-4} Pa. The buttons were sliced using a spark cutter and then mechanically polished to approximately 100 ^m thickness. Electron microscopy specimens were produced by standard jet polishing and electropolishing techniques [4].

EXPERIMENTAL DETAILS: ELECTRON MICROSCOPY: $CuTi_2$

The electron irradiation experiments were carried out in the Hitachi HU-3000 Ultra-High Voltage Electron Microscope (UHVEM) at Osaka University operating at an accelerating voltage of 2 MV. The dependence of the long-range order parameter, S, on the electron dosage and the final value of S was measured as a function of temperature over the temperature range of 183 to 363 K, paying special attention to the occurrence of the C-A transition. The final value of S is defined as the last measurable value of S prior to complete amorphisation at the temperatures at which the C-A transition occurs, or the steady-state value of S after long-term irradiation at the temperatures at which the C-A transition does not occur. Values of S were determined by the standard intensity ratio method used in various other experiments [7-11]. The temperature of the specimen was controlled using a liquid nitrogen cold stage with attached heater. Preliminary experiments were done to measure the degree of beam heating that occurs in the specimen at this voltage using the melting of vapor deposited In islands. The results indicated a temperature rise of approximately 30 K and all reported temperatures in this paper are corrected for this effect. All of the S experiments were done using fluxes in the range of 5×10^{23} $e/m^2 \cdot s$ and a beam diameter of 3 μm. The damage rate in displacements per atom per second (dpa/s) was calculated using Oen's Tables [12] for pure Cu with a threshold energy of 21 eV. The resulting cross-section value was multiplied by a factor of 0.6 to account for the directional anisotropy of the threshold energy [10,13]. Using this method, the resultant value was 3×10^{-3} dpa/s.

EXPERIMENTAL DETAILS: ELECTRON MICROSCOPY: CuTi

All irradiation experiments were done at 182 K with a flux of 1×10^{24} $e/m^2 \cdot s$. The occurrence of the C-A transition was monitored using standard BFI and SAD techniques.

RESULTS AND DISCUSSION: $CuTi_2$

In Figure 1, the temperature dependence of the amorphisation dosage of the compound $CuTi_2$ is presented. As found in CuTi [4] and Cu_4Ti_3 [5,6], the dosage is essentially independent of temperature at low temperatures (below 230 K). With increasing temperature, a slight dependence becomes evident at approximately 240 K (seen as an increase in the required dosage for amorphisation) and becomes more pronounced above 253 K. Finally, above some temperature called the critical temperature for amorphisation, the C-A transition does not occur. At all temperatures below the critical temperature for amorphisation, the end result of the irradiation is a complete transition to the amorphous state regardless of the dosage required. At temperatures above the critical temperature for amorphisation, the resulting microstructure is that of a large density of secondary defects [6].

If the electron irradiation of $CuTi_2$ is observed using a diffraction pattern containing superlattice reflections, some interesting observations can be made. In Figure 2, three series of SAD patterns are presented for the temperatures 183, 262 and 282 K. At all temperatures, the standard 00x systematic reflection of $CuTi_2$ can be seen at the start of irradiation consisting of four superlattice reflections (002,$00\bar{2}$, 004 and $00\bar{4}$) inside the first fundamental reflections (006 and $00\bar{6}$). The superlattice reflections are initially quite intense in this compound which is a great aid to the experiment. At 183 K, the occurrence of chemical disordering is evident after 0.12 dpa as a reduction in the intensities of the superlattice reflections and this continues until after 0.24 dpa they are quite weak. It is important to point out that during this early

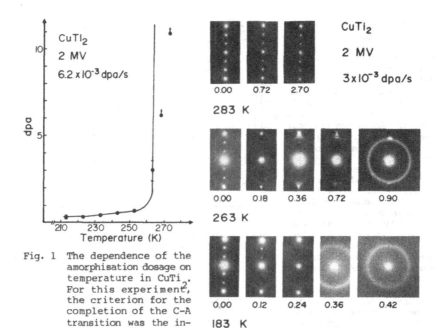

Fig. 1 The dependence of the amorphisation dosage on temperature in CuTi$_2$. For this experiment, the criterion for the completion of the C-A transition was the interruption of a bend contour.

Fig. 2 SAD patterns showing the evolution of CuTi$_2$ with electron irradiation at three different temperatures.

irradiation, the fundamental reflections remain strong indicating that amorphisation does not begin until significant chemical disordering occurs, and in fact, after 0.24 dpa, only the very beginning of amorphisation is indicated as a faint halo. Subsequently, the extent of the amorphous increases with the halo growing in intensity at the expense of the fundamental reflections (0.36 dpa). After 0.42 dpa at this temperature, the amorphisation is complete with only a halo present in the SAD pattern.

At 263 K, which is the temperature just below the critical temperature for amorphisation, a similar result as observed at 183 K can be seen although the rate of the process is reduced with amorphisation requiring 0.90 dpa. Also, at this temperature after 0.36 dpa, the fundamental reflections have become streaked in an "X" pattern and the beginnings of additional spots can be seen coincident with the first halo. This result is more obvious after 0.72 dpa and in the 0.90 dpa SAD pattern, it is clear that these previously absent, symmetrically positioned, diffuse spots remain after the fundamental reflections have completely disappeared. This SAD behavior was also seen in Cu$_4$Ti$_3$ [6]. At 283 K, which is slightly above the critical temperature for amorphisation, the superlattice reflections remain strong even after prolonged irradiation indicating that the attainable degree of chemical disordering is reduced.

In order to allow for more reliable conclusions to be drawn, in-situ quantitative measurements of the Bragg-Williams long-range order parameter, S, were made at eleven temperatures surrounding the critical temperature for amorphisation. The results are presented in Figure 3 as a function of electron dosage in dpa. In all cases, S decreased linearly with respect to the exponential of the dosage. At the temperatures below

482

the critical temperature for amorphisation, a – e, the value of S decreased rapidly to low values between 0.08 and 0.1 prior to a complete C-A transition. At 263 K, as previously mentioned in connection with Figure 2, the rate of decrease was reduced. Above the critical temperature for amorphisation, f – k, the rate of decrease of S was much reduced with final values of S after long-term irradiation being between 0.53 and 0.61. At these temperatures, no amorphisation occurred.

If the final values of S (as previously described) are plotted as a function of temperature, some important observations can be made. This is done in Figure 4. Here, the curve joining the low and high temperature values is arbitrarily drawn vertically. Two important conclusions can be made from this figure. The first is that there also exists a critical temperature for chemical disordering in $CuTi_2$ (268 K). The second conclusion is that the critical temperatures for chemical disordering and amorphisation are coincident (see Fig. 1). Below this critical temperature, the electron irradiation easily and rapidly produces significant chemical disordering which is always followed by a complete C-A transition. At temperatures above the critical temperature, both the rate and the final degree of chemical disordering are signifi-

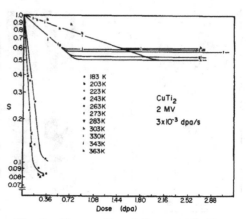

Fig. 3   The results of the quantitative measurements of the Bragg-Williams long-range order parameter as a function of electron dose in $CuTi_2$.

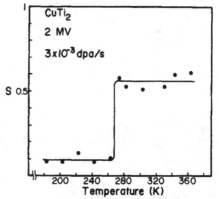

Fig. 4   The final values of the Bragg-Williams long-range order parameter as a function of temperature in $CuTi_2$. The presence of the critical temperature for chemical disordering is 268 K.

cantly reduced and the C-A transition does not occur. This comprises direct evidence that the irradiation induced chemical disordering plays a role in inducing the C-A transition in $CuTi_2$ (and Cu_4Ti_3 [5,6]) It can be expected that this will also be the case in other intermetallic compounds which possess a high chemical ordering energy.

It is unlikely that the electron irradiation produced Frenkel pairs could produce a sufficient energy increase in the crystal lattice and play a major role in inducing the C-A transition. This can be seen from a simple energy comparison which indicates that a defect density of 10^2 would be necessary to produce a sufficient energy increase. This density is higher than the value expected in crystals with typical spontaneous recombination volumes [6].

It is likely, however, that the point defects determine at what temperatures the C-A transition will occur and the rate at which it will occur. This is due to the competition between the irradiation induced chemical disordering and the point defect migration induced chemical reordering. At low temperatures, the migration of the point defects is inhibited permitting the observed rapid accumulation of chemical disordering and the subsequent amorphisation. As the temperature increases, the point defect migration increases which reduces the net rate of the chemical disordering and necessitates a higher amorphisation dosage. The critical temperature is the temperature at which the point defect migration induced reordering becomes sufficient to prevent the degree of chemical disordering required to induce the C-A transition.

In conclusion, it is proposed that the electron irradiation induced chemical disordering raises the energy of the crystal lattice to the point at which it becomes unstable with respect to the amorphous state; at this point, the C-A transition occurs. The temperatures at which the C-A transition occurs are controlled by the competition between the irradiation induced chemical disordering and the point defect migration induced chemical reordering. At the critical temperature and higher temperatures, the chemical reordering rate is sufficient to prevent the necessary chemical disordering accumulation for the C-A transition.

RESULTS AND DISCUSSION: CuTi

In order to determine the effect of non-stoichiometry on the C-A transition, an experiment was carried out in the CuTi ordered intermetallic compound. An expansion of the Cu-Ti phase diagram from 45 to 55 atomic percent Ti is presented as Figure 5. The 45 atomic percent Ti buttons were quenched from the liquid state to prevent composition modulations from occurring. Then, by controlling the annealing temperature, the composition of the CuTi in the two-phase specimens could be controlled. The three compositions investigated were 47.5, 48.7 and 52 atomic percent Ti. All of these compositions exhibited initial SAD patterns of the CuTi B11 structure. After irradiation to a dose of 1.4×10^{26} e/m^2 at 182 K with 2 MV electrons, the highly Cu-rich phase remained completely crystalline (Fig. 6a) with only secondary defects present in the BFI and crystalline spots in the SAD pattern. In the mildly Cu-rich specimen (Fig. 6b), the BFI exhibited typical C-A transition behavior [4] and the SAD produced a halo pattern. However, residual crystalline reflections persisted in this pattern. Since the Cu$_4$Ti$_3$ present in the same specimens easily became amorphous, it can be concluded that this was not an effect of the specimen preparation. The Ti-rich phase completely underwent the C-A transition as can be seen in Figure 6c.

This result indicates the importance of stoichiometry in the C-A

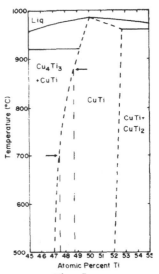

Fig. 5

Phase diagram of the Cu-Ti alloy system from 45 to 55 percent Ti showing the annealing temperatures and compositions of the Cu-rich phases.

484

transition. Both compounds to either side of CuTi (Cu$_4$Ti$_3$ and CuTi$_2$) exhibit the C-A transition, although CuTi$_2$ also exhibits a stoichiometric effect on its Cu-rich side similar to mildly Cu-rich CuTi [3] (the earlier experiments described in this paper were done using Ti-rich CuTi$_2$). At present, this effect of stoichiometry cannot be explained although preliminary qualitative experiments indicate that the accumulated chemical disorder is greatly reduced in highly Cu-rich CuTi [3].

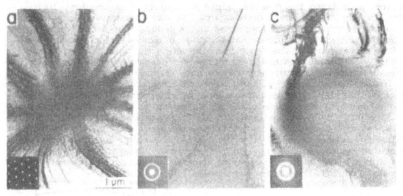

Fig. 6 The effect of non-stoichiometry on the C-A transition properties in CuTi. Irradiation was done with 2 MV electrons at 182 K to a dose of 1.4 x 10^{26} e/m^2. Compositions were a) 47.5 at.% Ti, b) 48.7% and C 52%.

REFERENCES

1. G. Thomas, H. Mori, H. Fujita and R. Sinclair, Scripta Metall. 16, 489 (1982).
2. A. Mogro-Campero, E. L. Hall, J. L. Walter and A. J. Ratkowski, in Metastable Materials Formation by Ion Implantation, edited by S. T. Picraux and W. J. Choyke (North-Holland, N.Y., 1982) p. 203.
3. D. E. Luzzi, H. Mori, H. Fujita and M. Meshii (unpublished).
4. D. E. Luzzi, H. Mori, H. Fujita and M. Meshii, Scripta Metall. 18, 957 (1984).
5. D. E. Luzzi, H. Mori, H. Fujita and M. Meshii, Scripta Metall. 19, 897 (1985).
6. D. E. Luzzi, H. Mori, H. Fujita and M. Meshii, Acta Metall. (in press).
7. G. J. C. Carpenter and E. M. Schulson, Scripta Metal. 15, 549 (1981).
8. M. Z. Hameed, R. E. Smallman and M. H. Loretto, Phil. Mag. A 46, 237 (1982).
9. H. C. Liu and T. E. Mitchell, Acta Metall. 31, 863 (1983).
10. S. Banerjee, K. Urban and M. Wilkens, Acta Metall. 32, 299 (1984).
11. G. J. C. Carpenter and E. M. Schulson, J. Nucl. Mater. 23, 180 (1978).
12. O. S. Oen, ORNL Report No. 4897, 1973).
13. P. Jung, R. L. Chaplin, H. J. Fenzl, K. Reichelt and P. Wombacher, Phys. Rev. B 8, 553 (1973).

ELECTRON BEAM-INDUCED PHASE TRANSFORMATION IN MgAl$_2$O$_4$ SPINEL

S.J. SHAIBANI* AND S.N. BUCKLEY**
* Department of Metallurgy, University of Oxford, Parks Road, Oxford, England
** Materials Development Division, AERE Harwell, Didcot, Oxfordshire, England

ABSTRACT

Bombarding stoichiometric MgAl$_2$O$_4$ spinel at elevated temperatures with high energy electron beams gradually converts the irradiated area into γ-alumina. This phase transformation, which involves the diffusion of magnesium out of the irradiated area, shrinks the oxygen ion sublattice, thereby generating internal stresses which cause localised crystal fractures.

Three categories of crack are observed : type 1, the majority, are straight and lie at the centre of the irradiated area; type 3 are circular and they are located near the edge of the irradiated area in the spinel matrix: and, type 2 are hybrids. A quantitative model which explains the development of all types of crack is presented.

The proportion of γ-alumina produced, the fractional loss of magnesium and the extent of cracking are found to depend on the electron beam energy, the electron dose, the beam profile and the irradiation temperature.

INTRODUCTION

The most commonly reported effects of an electron beam on oxygen-containing materials are decomposition and amorphisation, but the occurrence of a genuine phase transformation is rarely observed. With ceramic oxides, such as magnesia [1] and α-alumina [2-5], electron irradiation produces islands of metal and oxygen bubbles. Many natural silicates (forsterite, enstatite, anthophylite, talc and α-quartz) experience a crystalline-to-amorphous transition [6-8], although decomposition takes place in synthetic cordierite and enstatite [9]. The carbonate minerals, dolomite and calcite, are also decomposed by being reduced to oxides with an accompanying release of carbon dioxide gas [10]. In this paper we show that the above examples are contrasted by another ceramic oxide, stoichiometric magnesium aluminium oxide spinel, which undergoes a phase transformation when subjected to intense beams of electrons in a high voltage electron microscope (HVEM) at high temperatures. Our interest in spinel lies in it being one of a number of ceramic oxides which may have potential applications in nuclear fusion reactors, whose environment can be approximated to some extent by this type of experiment.

Previous work on spinel has noted the formation of metal islands after electron irradiation to high doses [11], which is similar to the response of its constituent oxides, magnesia and α-alumina. However, the results of the present research reveal that this behaviour is preceded by a phase transformation of spinel → γ-alumina. The transformation is identified by the presence of extra reflections in the diffraction patterns of irradiated areas and it is brought about by the replacement of magnesium ions with aluminium ions. A recent study [12] has identified a long-range ordering of the aluminium ions in γ-alumina, and this is accounted for by adjusting the tabulated lattice parameter of 7.90 Å [13,14] to 23.70 Å. This is supported by additional evidence obtained during this work although our understanding of the ordering is not complete.

The phase transformation involves the rearrangement of cations which decreases the original oxygen ion separation by up to 2.2 per cent, and such a contraction creates huge internal stresses which lead to microcracking.

The cracks fall into three distinct categories depending on the experimental conditions, and we propose a single quantitative model based on elasticity theory calculations to explain all types of crack. The rate of the spinel → γ-alumina transformation increases with irradiation temperature. Assessments of the dependences of the transformation, and the crack initiation, on electron beam energy are complicated by beam-energy dependent variations of both the total electron flux and the intensity profiles. The latter influences the reaction rate via a Laplacian flux profile term controlling the ionic drift mobilities. The rate is higher at lower electron energies, implying that ionisation processes dominate any displacement effects.

EXPERIMENTAL PROCEDURE

Single-crystal specimens of spinel suitable for examination in a transmission electron microscope (TEM) were prepared by the ion-beam thinning of 3 mm discs, which had been ground to a thickness of 50 μm and polished to an optical finish. The plane of the TEM foils was close to (111) and residual damage was removed in a 16-hour anneal at 1000 $^\circ$C, during which gold support rings were bonded to the specimens. The HVEM irradiations were conducted in an A.E.I. EM7 with accelerating voltages between 200 and 1000 kV, and at specimen temperatures of 750 - 850 $^\circ$C. Surface contamination of the specimens was minimised by using liquid nitrogen cooled anticontaminators, located immediately above and below the HVEM hot stage, and a vacuum better than 10^{-6} torr. Identical beam profile and peak intensity conditions cannot be obtained over the whole electron energy range and so a fixed condenser-one lens excitation was set for all irradiations. With fully-focussed beams, this produced peak electron fluxes and beam 'diameters' which varied from 1.56×10^{24} e m^{-2} s^{-1} and $\sim$1 μm at 1000 keV to 8.22×10^{22} e m^{-2} s^{-1} and $\sim$2 μm at 200 keV.

The amounts of magnesium and aluminium in the irradiated areas were measured with a Link Systems energy-dispersive x-ray analysis (EDX) facility in the HVEM. Values obtained for the Mg:Al ratio were normalised to the stoichiometric value of 0.5, which was assumed to prevail in the non-irradiated areas. This procedure compensates for minor variations in the counting efficiency of the lithium-drifted silicon detector, the presence of background counts and any systematic errors. All EDX data were obtained at room temperature so as to minimise any electron damage by the probe whose diameter at an energy of 400 keV was about 0.5 μm.

RESULTS

Diffraction pattern data

The presence of an irradiation-induced product was detected by comparing selected area diffraction patterns taken before and after irradiation in the HVEM. Figure 1 is a 111 pattern from an area irradiated at 750 $^\circ$C with 1000 keV electrons to a total peak dose of about 2×10^{27} e m^{-2}. The pattern contains : (i) fundamental reflections which were also observed prior to irradiation; (ii) six faint additional reflections 0.12 Å^{-1} from 000; (iii) additional reflections at A, B, C, D, E and F separated from the fundamental reflections next to them, with the ratio of the reciprocal space distances from 000 of the latter to the former being a constant 0.978 ± 0.003; and, (iv) evidence that some strongly-excited fundamental reflections, for example at G and H, obscure the additional reflections which could be expected to be found in close proximity to them.

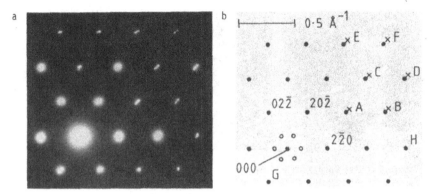

Figure 1. (a) Selected area diffraction pattern of electron-irradiated spinel; (b) Locations of additional reflections around 000 (open circles), and separation of fundamental reflections (filled circles) from neighbouring additional reflections (crosses).

EDX microanalysis data

A series of experiments with electron energy E and irradiation temperature T both constant showed that the Mg:Al ratio S in the irradiated zone decreased steadily with increasing electron dose D. The rate of this magnesium loss dS/dD decreased with increasing specimen thickness z up to a limiting value z_{max}, beyond which it was almost constant. Using material thick enough to minimise these surface effects ($z \sim 1.4$ μm > z_{max}), we found that dS/dD decreased with increasing divergence of the electron beam from its fully-focussed condition, i.e. as the intensity profile became flatter.

With fixed irradiation temperature (750 °C), the efficiency of the irradiation - expressed as the electron dose $D_{\frac{1}{2}}$ needed to halve the initial magnesium concentration - increased with decreasing energy and $D_{\frac{1}{2}}(E) = 5.06$, 3.37, 0.752 and ~ 0.6-0.9×10^{27} e m^{-2} at E = 1000, 600, 400 and 200 keV, respectively. The uncertainty in the low energy value arose from the difficulty in being able to locate precisely the centre of the irradiated area, which is larger at 200 keV than at 1000 keV. With fixed electron energy (1000 keV), the efficiency increased rapidly with T and $D_{\frac{1}{2}}(T) = 5.06$, ~ 2 and $\sim 0.5 \times 10^{27}$ e m^{-2} at T = 750, 800 and 850 °C. Data at higher temperatures are approximate because of the increased risk of microcracking which interferes with conduction from the irradiated area. This raises the indicated temperature by an unknown amount and so affects S further.

Microcracking and other phenomena

When the irradiation temperature was raised above ~ 800 °C or the material thickness was less than about 0.5 μm, the rate of magnesium loss became excessive and microcracks formed. Three distinct classes of crack geometry, illustrated in Figure 2, were observed. Type 1 cracks were the most frequently seen, being straight and confined to the centre of the irradiated area; they tended to be produced in thin material during irradiations at high energy and high temperature. Type 3 cracks were favoured by thick material, low energy and low temperature; they took the form of circular arcs in the spinel matrix around the periphery of the irradiated zone. Type 2 cracks were hybrids of types 1 and 3. The amounts of magnesium in the vicinity of all types of crack were found to be zero.

488

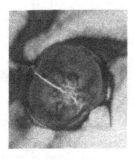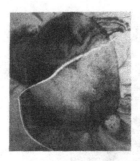

Figure 2. Microcracking in electron-irradiated spinel : (a) type 1; (b) type 2; (c) type 3. The cracks are 1-2 μm in length.

In some experiments where microcracking did not occur, the specimen became thinner producing small increases in electron transmission after appreciable irradiation times. The extent of thinning was assessed from stereo microscopy. Another observation was the creation of small (∿ 50 nm) faceted aluminium plates on the surface. These plates were usually associated with a defocussed beam profile whereas fully-focussed beams produced large (∿ 500 nm) irregular metal islands [11].

DISCUSSION

The phase transformation spinel → γ-alumina

The first-order fundamental reflections in Figure 1 are 0.35 Å^{-1} from 000 and x-ray measurements of our material give a lattice parameter of 8.08 Å, which is commensurate with the most commonly accepted value for spinel [15]. Consequently, the reflections are 220 type. The faint additional reflections occurring near 000 have also been observed in other experiments [12], but that work reported many more of these additional reflections distributed in a similar arrangement around other strongly-excited fundamental reflections. The material responsible was identified as γ-alumina, but if the first-order additional reflections are indexed as 220, their reciprocal space distance of 0.12 Å^{-1} from 000 corresponds to a lattice parameter of 23.70 Å [12]. This three-fold increase in the tabulated value of 7.90 Å for γ-alumina [13,14] was justified both from the experimental data and from basic crystallography. A consequence of this new lattice parameter is that the hkℓ reflections in γ-alumina based on the old value must be properly relabelled as 3h3k3ℓ.

The well-defined splitting of reflections illustrated in Figure 1 provides further evidence that γ-alumina is formed. This splitting is the separation of neighbouring fundamental and additional reflections and the reciprocal space ratio of 0.978 can be ascribed to d (hkℓ in spinel) : d (3h3k3ℓ in γ-alumina), which converts to a lattice parameter ratio of 8.08 Å : 23.70 Å. Hence, our data based on the splitting of reflections agrees exactly with the value derived from the distribution of additional spots [12], and we conclude that γ-alumina is formed as a product of electron-irradiated spinel. The continuity of the γ-alumina product with the parent spinel matrix implies that a phase transformation has been effected by the radial diffusion of magnesium ions out of the irradiated area and their replacement with aluminium ions. This process is characterised by a progressive decrease in S from 0.5 to 0 and a corresponding increase in the stoichiometry parameter n in $MgO.(Al_2O_3)_n$ from 1 to ∞. The interchange of cations is probably due to electron beam enhancement of the natural inversion which takes place in spinel above 750 °C [16]. The enhanced exchange arises from the intense electric fields which can be generated in ionic materials by a non-linear beam profile [11]. The observed increase in the effect of these fields at lower electron energies is consistent with an ionisation mechanism, because the cross-section for the latter increases with decreasing energy. An increase in temperature also accelerates the exchange by increasing thermal diffusion.

The model for microcracking

Transforming spinel to γ-alumina produces a contraction in the oxygen ion sublattice, witnessed by the reduction in the lattice parameters from 8.08 Å (spinel) to '7.90 Å' (spinel-related value of γ-alumina). The linear strain will be highest at the centre of the irradiated area and it will fall off radially to zero in the surrounding spinel matrix. We may represent the radial variation of this transformation strain by $\varepsilon_\gamma = - k \exp(-r^2/R^2)$, where k can reach a maximum linear value of 2.2 per cent for complete transformation, and R is a measure of the profile of the electron beam. Elasticity theory calculations for the case of plane stress show that this transformation strain leads to a radial stress of $\sigma_r = (kER^2/2r^2).\{1 - \exp(-r^2/R^2)\}$. The upper limit of this is one per cent of the Young's modulus, E, which is sufficient to break down atomic cohesion in a perfect crystal - no dislocations were detected in the vicinity of the heads of the cracks.

Figure 3. The effect of different transformation strain profiles on crack type : (a) straight type 1, (b) hybrid type 2, (c) circular type 3. The fracture strengths for γ-alumina and spinel are $\sigma_f(\gamma)$ and $\sigma_f(s)$, respectively, and ρ defines the size of the transformed region.

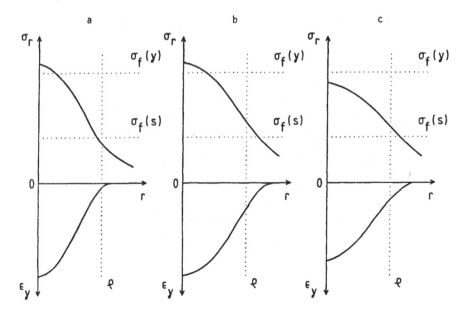

The conditions which give rise to the three types of crack depend on the nature of k and R. At high temperatures, the transformation rate is rapid and k is large, $k = k_1$ (say). For high electron energies, the beam size is smallest and we have the lowest value for $R = R_1$. These values combine to give a transformation strain which gives a type 1 crack, as depicted in Figure 3(a). This figure shows failure in the central irradiated area only, observed as a straight crack. Type 3 cracks occur at lower temperatures with $k = k_3$ ($< k_1$) and at lower energies when $R = R_3$ ($> R_1$), the consequence of which is shown in Figure 3(c). In this case, fracture occurs in the parent spinel matrix only in the form of circular arcs. The straight parts of type 2 cracks are explained by $k_2 = k_1$, and the circular component is due to $R_2 = R_3$ (Figure 3b).

Surface modification of beam-induced electric fields

The thinning of the irradiated area seen in some of our experiments may be explained by the surface becoming positively charged due to axial components of the electric fields caused by the beam profile [17,18]. Oxygen ions are attracted to the surface where they are neutralised and then lost in a continued erosion process. At the low damage rates found in defocused electron beam irradiations, the thinning is almost negligible and slight loss of oxygen may promote the aggregation of excess aluminium ions into the faceted plates observed.

REFERENCES

1. S.N. Buckley, unpublished work (1983).
2. G.P. Pells and D.C. Phillips, J. Nucl. Mat., 80, 207 (1979).
3. G.P. Pells and D.C. Phillips, J. Nucl. Mat., 80, 215 (1979).
4. T. Shikama and G.P. Pells, Phil. Mag. A, 47, 369 (1983).
5. G.P. Pells and T. Shikama, Phil. Mag. A, 48, 779 (1983).
6. D.R. Veblen and P.R. Buseck, Proc. 41st E.M.S.A., edited by G.W. Bailey (1983), p. 350.
7. P. Lorimer and P.E. Champness, in High Voltage Electron Microscopy, edited by P.R. Swann, C.J. Humphreys and M.J. Goringe (Academic Press, London, 1974), p. 301.
8. T.J. White and B.G. Hyde, Am. Min., 68, 1009 (1983).
9. W.E. Lee, T.E. Mitchell and A.H. Heuer, Radiat. Eff. (to be published, 1985).
10. E.D. Cater and P.R. Buseck, Ultramicroscopy (to be published, 1985).
11. S.N. Buckley, Proc. 13th Symp. on Fusion Technology, Varese, Italy, September 24-28, 1984 (Pergamon Press, Oxford, 1984), Volume 2, p. 1011.
12. S.J. Shaibani, S.N. Buckley and M.L. Jenkins, presented at 3rd Int. Conf. on Radiation Effects in Insulators, Guildford, England, July 15-19, 1985. To be published in Radiat. Eff.
13. E.J.W. Verwey, Z. Kristallogr., A91, 65 (1935).
14. E.J.W. Verwey, Z. Kristallogr., A91, 317 (1935).
15. Handbook of Chemistry and Physics, 55th ed., edited by R.C. Weast (Chemical Rubber Company, Cleveland, Ohio, 1974).
16. U. Schmocker and F. Waldner, J. Phys. C, 9, L235 (1976).
17. S.N. Buckley, Harwell Report AERE-R 11715, Chapter 3 (1985).
18. S.N. Buckley and S.J. Shaibani, Harwell Report (to be published, 1985).

ION BEAM MIXING IN BINARY AMORPHOUS METALLIC ALLOYS*

HORST HAHN**, R. S. AVERBACK**, T. DIAZ DE LA RUBIA***, and P. R. OKAMOTO**
** Materials Science and Technology Division, Argonne National Laboratory, Argonne, IL 60439
***Dept. of Physics, SUNY at Albany, Albany, NY 12222

ABSTRACT

Ion beam mixing (IM) was measured in homogeneous amorphous metallic alloys of Cu-Er and Ni-Ti as a function of temperature using tracer impurities, i.e., the so called "marker geometry". In Cu-Er, a strong temperature dependence in IM was observed between 80 K and 373 K, indicating that radiation-enhanced diffusion mechanisms are operative in this metallic glass. Phase separation of the Cu-Er alloy was also observed under irradiation as Er segregated to the vacuum and SiO_2 interfaces of the specimen. At low-temperatures, the amount of mixing in amorphous Ni-Ti is similar to that in pure Ni or Ti, but it is much greater in Cu-Er than in either Cu or Er.

INTRODUCTION

Diffusion in amorphous binary metallic alloys has been of increasing interest in the past few years. Some of this interest has been stimulated by the recent discovery that binary metallic systems can be rendered amorphous by low temperature annealing of constituent components, a process which requires diffusion through the amorphous phase [1,2]. Diffusion studies in metallic glasses are also of interest as they can provide insight into the structure of these glasses. A fundamental understanding of diffusion processes in amorphous binary metallic systems, however, is presently only in its rudimentary stages [3]. One of the methods that has proved quite fruitful in elucidating diffusion mechanisms in crystalline materials has been measurements of defect properties and diffusion during particle irradiation [4]. Irradiation enables the diffusion controlling defects, vacancies and interstitial atoms in crystalline material, to be introduced into the material in controlled ways and this isolates different possible diffusion parameters and mechanisms [5]. Working on the premise that irradiation can be similarly useful in ferreting out diffusion mechanisms in metallic glasses, we have initiated a program to investigate diffusion in metallic glasses during irradiation, which we will denote here simply as ion beam mixing (IM). In this paper, we report on the temperature dependence of IM using impurity tracer diffusion in the homogeneous amorphous alloys, $Cu_{50}Er_{50}$, $Ni_{65}Ti_{35}$, and $Ni_{41}Ti_{59}$.

The study of IM in amorphous alloys is also of interest for understanding IM itself. Nearly all quantitative studies of IM in metals have involved either tracer impurities in pure materials, or bilayer specimens of two pure components. One of the questions that has arisen in IM studies is why bilayer samples frequently yield much higher mixing yields than do either of the component elements. For example, the normalized mixing yields, i.e. $Dt/\Phi \cdot F_D$, of Cu and Al are $\approx$ 23 and 20 $Å^5/eV$, respectively [6], whereas in the bilayer system Cu-Al, it is $\approx$ 95 $Å^5/eV$ [7,8]. For the Au-Ag system, $Dt/\Phi \cdot F_D$ is $\approx$ 120 and 75 $Å^5/eV$ for Au and Ag, respectively [6], whereas in the bilayer is $\approx$ 265 $Å^5/eV$ [8,9]. In both of these systems, IM is not expected to be strongly influenced by

*Work supported by the U. S. Department of Energy, BES-Materials Sciences, under Contract W-31-109-Eng-38.

chemical interactions [10]. By measuring IM using a tracer impurity in a homogeneous alloy, this question can be addressed. Finally, by measuring IM in an amorphous alloy, the importance of atomic structure for IM can be determined. For example, it has been debated for many years whether long range focussed replacement sequences are an important diffusion mechanism in irradiated materials. Such sequences are suppressed in amorphous systems.

EXPERIMENTAL

The amorphous Cu-Er films were prepared by vapor deposition of a premelted $Cu_{50}Er_{50}$ alloy onto oxidized Si-wafer substrates using electron beam heating. An Ag marker was deposited from a W-boat by resistance heating. The switching time from Cu-Er to Ag evaporation was kept below 1 sec by using a manual shutter and thereby avoiding contamination problems at the interface. The thickness of the bottom and top layers were $\approx$ 25 and $\approx$ 35 nm, respectively. The marker thicknesses ranged from 1-2 nm for all specimens. The background pressure during deposition was less than $1x10^{-7}$ mbar. The structure of the evaporated films was examined by means of a Read camera using Cu K_α radiation. Only specimens which showed no traces of crystallinity were used for the irradiations. The compositions of the films were checked by He backscattering analysis prior to the irradiations; it was always found to be nearly the same as the bulk alloy composition. This is due to the similar activities of Cu and Er in the alloy. In some cases, thin layers of Ti were evaporated onto the substrate and onto the top of the sample to prevent oxidation of the sample during the irradiations. The Ni-Ti were similarly prepared by evaporation from a bulk Ni-Ti alloy; however, the background pressure during deposition for these specimens was $\approx 2x10^{-8}$ mbar.

The irradiations, which employed 1-MeV Kr^+, and the subsequent He backscattering analysis were performed at the Argonne Tandem accelerator. All backscattering spectra for irradiations below 295 K were acquired without warming the sample to room temperature. All other backscattering spectra were recorded at room temperature. To prevent possible influences of compositional or contamination differences between specimens, all irradiations were performed on samples cut from the same substrate. The marker spreading was determined by fitting a gaussian function to the experimental points prior to and after irradiation and using the expression,

$$Dt = 1/2 \cdot (\sigma^2_{irrad} - \sigma^2_{unirrad}). \qquad (1)$$

D is the effective diffusion coefficient, and σ^2 is the variance of the gaussian function.

All irradiations were performed well below the crystallization temperatures of the systems. The crystallization temperature for $Cu_{72}Er_{28}$ is $\approx$ 620 K [11] and for NiTi, $\approx$ 800 K [12].

RESULTS

Figure 1 shows three spectra corresponding to a $Cu_{51}Er_{49}$ sample. The top spectrum was acquired prior to any thermal or irradiation treatment. After irradiation with 1×10^{16} 1-MeV Kr^+-cm^{-2} at 10 K, appreciable spreading of the marker is observed, whereas no obvious change in the Cu or Er signals is detected. At the higher temperatures, 295 and 373 K, however, phase separation of the homogeneous Cu-Er alloy was observed (results shown for 295 K only). This was due, presumably, to the high affinity of the Er for oxygen at the specimen surface and in the SiO_2

substrate (the vapor pressure in the target chamber was 1×10^{-7} mbar). Correspondingly, Cu segregated to the center of the specimen. Without irradiation, no segregation effects were observed in the spectra. As the Ag marker behaves similar to the Cu, the marker spreading by IM was reduced by the segregation behavior. This caused deviations from linearity in plots of Dt versus ion dose for the higher temperature irradiations. Although the Er rearrangement is quite remarkable, and requires long-range diffusion, the x-ray Read film showed no evidence of crystalline compound formation, even at the highest irradiation temperature. This clearly demonstrates radiation-enhanced diffusion in this amorphous alloy.

In order to avoid the segregation effects of the Er for quantitative IM studies, subsequent Cu-Er samples were sandwiched between thin Ti films. These were prepared by covering the oxidized Si-substrate with 10 nm of Ti, evaporating the Cu-Er film with the marker, and finally adding 4 nm of Ti to the top of the sample. As Fig. 2 shows, no significant changes in the Cu or Er signals are observed for this sample after irradiation. This is true even for the irradiations at 373 K.

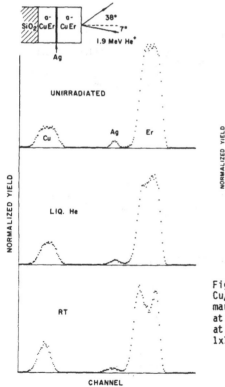

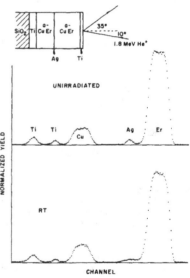

Fig. 2. Backscattering spectra of a $Cu_{48}Er_{52}$ thin film sample with an Ag marker and a protective Ti-coating at the substrate film interface and at the surface. Irradiation data: 1×10^{16}/cm^2 1-MeV Kr$^+$.

Fig. 1. Backscattering spectra of a $a_2Cu_{51}Er_{49}$ thin film sample with an Ag marker. Irradiation data: 1×10^{16}/cm^2 1-MeV Kr$^+$ at the indicated temperature. The yield from the substrate is not shown.

The results of the IM experiments in Cu-Er are summarized in Fig. 3, which shows the marker spreading, Dt, vs ion dose, Φ. No deviation from linearity could be observed at 80 or 373 K. There is a strong temperature dependence in the IM between 80 and 373 K, paralleling the segregation effect in the uncovered specimen. X-ray analysis of irradiated spots at RT and 373 K showed no crystalline phases. Thermal diffusion could not be detected during the times involved in the irradiation experiment. In fact, thermal annealing at 378 K for 80 hrs. also did not result in measurable broadening of the marker. For comparison, Dt values for a Ag-marker in pure Er and in a Cu-4.5 at% Er alloy are also included in Fig. 3.

To compare IM data for different systems, it is useful to scale values of Dt to the damage energy of the irradiation, $\Phi \cdot F_D$, where F_D is the damage

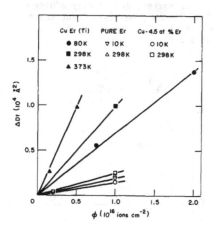

Fig. 3. Spreading of an Ag marker as a function of ion dose in a $Cu_{48}Er_{52}$ amorphous alloy protected at both surfaces with thin Ti films. IM data for pure Er and Cu-4.5 at.% Er are also shown for comparison.

energy deposition per unit length normal to the specimen surface. In Table 1, values of $Dt/\Phi \cdot F_D$ for the amorphous alloy, $Cu_{48}Er_{52}$, pure Er, a Cu-4.5 at.% Er alloy, and NiTi specimens are summarized. Values of F_D were calculated using the computer program TRIM [13].

TABLE 1: Values of IM

system	marker	F_D	$(Dt/\Phi \cdot F_D)_{10,80}$	$(Dt/\Phi \cdot F_D)_{295}$	$(Dt/\Phi \cdot F_D)_{373}$
a-$Cu_{48}Er_{52}$	Ag	120	59.7	83.0	157.0
Cu-4.5at%Er	Ag	146	10.9	16.6	--
Er	Ag	102	18.9	19.5	--
a-$Ni_{65}Ti_{35}$	Hf	113	12.1		
a-$Ni_{41}Ti_{59}$	Pd	113		13.1	
Ni[14]	Hf	220	11.4	14.4	
Ti[6]	Hf	120	11	--	

units of $Dt/\Phi \cdot F_D$ are $Å^5/eV$; units of F_D are eV/A.

DISCUSSION

These experiments clearly show that diffusion in irradiated Cu-Er amorphous alloys is temperature dependent. This is observed both in the Ag marker spreading and the phase separation in the unprotected Cu-Er specimen. We presently attribute this temperature dependence to a radiation-enhanced diffusion (RED) mechanism. In crystalline material, RED arises from the diffusion of point defects produced by the irradiation. The temperature dependence of RED arises from the temperature dependent point defect mobilities and concentrations. These quantities are related to the intrinsic properties of the point defects, such as their enthalpies of migration and formation. In metallic glasses, the origin of the RED effect that we observe here is not yet clear as the mechanism of diffusion in metallic glasses is unknown. The fact that we observe RED, however, suggests that diffusion mechanisms in metallic glasses may be similar to that in crystalline material. To carry the analogy of diffusion in metallic glasses with that in crystalline metals further, we have made a very preliminary analysis of the Cu-Er data based on normal RED theory [5]. First, we assume that the temperature dependence of IM has Arrhenius behavior; this yields an activation enthalpy of ≈ .16 eV for the diffusion process. Within a simple RED model this activation enthalpy would suggest that the "slowest moving defect" in the amorphous medium has a migration enthalpy of .32 eV. This value is approximately a factor of two smaller than the vacancy migration enthalpy in pure Cu. Further measurements will be necessary before the significance of these numbers can be determined. For example, measurements at other temperatures are necessary to determine if the process does indeed have Arrhenius behavior. In addition, the influence of dose-rate should be determined as a dose-rate dependence is predicted by RED theory.

The low temperature IM results show that the normalized IM parameter, $Dt/\Phi \cdot F_D$, can be greater in the alloys than in either of the pure components. It indicates that the underlying cause of the larger IM values in bilayer specimens than in pure specimens containing markers, cannot be attributed solely to chemical driving forces but requires an understanding of displacement processes in alloys. Presently, such an understanding is nearly completely lacking. It appears that the role of focussed collision sequences (RCS) in IM is small as IM in the crystalline material is smaller than in the amorphous material. The opposite behavior would be expected if RCSs were a dominant mechanism of IM. Finally, we point out the strong ion beam mixing effect between Cu-Er alloy and the SiO_2 substrate. The strong segregation effect demonstrates the importance of chemical driving forces in IM in the temperature range in which defects are mobile. In the present case, the segregation is driven, presumably, by the high heat of formation for Er_2O_3. The reaction of the amorphous film with the substrate, which only occurs in the presence of irradiation at temperatures below 373 K, very likely results in bonding of the evporated film to the underlying substrate. This could have interesting applications for thin film technology and we are currently testing for such bonding.

REFERENCES

1. R. B. Schwarz and W. L. Johnson, Phys. Rev. Lett. 51, 415 (1983).
2. Y.-T. Cheng, W. L. Johnson, M-A. Nicolet, Appl. Phys. Lett. 47, 800 (1985).
3. Y. Limoge, G. Brebec, Y. Adda, Diffusion and Defect Monographs Series Vol. 7, 185 (1983), Ed. F. J. Kedves, D. L. Beke.
4. See e.g., Properties of Atomic Defects in Metals, ed. N. L. Peterson and R. W. Siegel (North Holland, Amsterdam, 1978).
5. Y. Adda, M. Beyeler, G. Brebec, Thin Solid Films 25, 107 (1975).

496

6. S.-J. Kim, M-A. Nicolet, R. S. Averback, D. Peak, this conference.
7. F. Besenbacher, J. Bottiger, S. K. Nielsen, H. J. Whitlow, Appl. Phys. A 29, 141 (1982).
8. B. M. Paine and R. S. Averback, Nucl. Instr. and Meth. B 7/8, 666 (1985).
9. M. Van Rossum, Y.-T. Cheng, W. L. Johnson, M-A. Nicolet, Appl. Phys. Lett.
10. W. L. Johnson, Y.-T. Cheng, M. Van Rossum, M-A. Nicolet, Nucl. Instr. and Meth. B 7/8, 657 (1985).
11. M. Atzmon, K. M. Unruh, W. L. Johnson, J. Appl. Phys. 58, 3865 (1985).
12. J. L. Brimhall, Nucl. Instr. and Meth. B 7/8, 26 (1985).
13. J. Biersack, L. G. Haggmark, Nucl. Instr. and Meth. 174, 257 (1980).
14. R. S. Averback, D. Peak, Appl. Phys. A 38, 139 (1985).

A COMPARATIVE STUDY OF THE CRYSTALLIZATION OF Ni-P AMORPHOUS ALLOYS
PRODUCED BY ION-IMPLANTATION AND MELT SPINNING.

JAMES HAMLYN-HARRIS*, D. H. ST.JOHN* AND D. K. SOOD**
* Dept. Metallurgy & Mining, Royal Melbourne Institute of Technology,
124 La Trobe Street, Melbourne, 3000, Australia
** Microelectronics Technology Centre, Royal Melbourne Institute of
Technology, 124 La Trobe Street, Melbourne, 3000, Australia

ABSTRACT

Implantation of 40keV P+ ions into high purity Ni was employed at room
temperature to a dose of 3x10E17 ions/cm² to produce an 1100 A thick
amorphous surface alloy of Ni-14 wt% P. Commercially available melt spun
metallic glass ribbons of nominal composition, Ni-11 wt% P were used for
comparison of crystallization behaviour studied with TEM and RBS techniques.
The DTA analysis was employed to construct a TTT curve for the melt spun
glass, which was then used as a guide for selecting time and temperature of
crystallization of the implanted amorphous alloy. The melt implanted glass
is found to be less stable and crystallizes more readily than melt spun
glass of similar composition. A detailed study on nucleation and growth of
crystallites, mode of crystallization and effect of surface proximity will
be presented.

INTRODUCTION

Amorphous metal alloys can be made by a number of rapid solidification
methods including melt spinning, as well as by particle beam processes such
as ion implantation.
Compared to crystalline materials of the same composition, these alloys
exhibit improved corrosion resistance, higher hardness, better ductility,
and improved wear resistance. However, these properties are hard to take
advantage of because amorphous metals can not be made much thicker than
about 60 microns.
An alternative method is to use high dose ion implantation to amorphise
the surface of engineering components. The resultant intrinsic surface layer
is an integral part of the substrate, and is not susceptible to any kind of
de-lamination or spalling.
The thermal stability of this metastable amorphous layer is crucial to
the application of the processed component.
The aim of this work is to compare the thermal stability of melt spun
glass and ion implanted glass of similar composition. The Nickel-Phosphorus
system was chosen for study because of the relative ease of making amorphous
films by implantation [1], and the commercial availability of melt spun Ni-P
glass.
A number of basic differences exist between these materials. Amorphous
layers made by implantation are relatively thin (of the order of 100 to 1000
Angstroms) so glasses made this way are in close proximity to both a
crystalline substrate and to a free surface. Both of these regions may act
as nucleation sites for crystallization, and/or as solute sinks, thereby
modifying composition, and thus stability.

EXPERIMENTAL

Melt spun glass

The material used was a commercial brazing alloy 'MBF60', manufactured
by Nippon Amorphous Metals. This is an alloy of 11 wt.% P in Nickel, with

some 0.1 wt.%C. The glass was in the form of ribbon of thickness 40 microns and width 25 mm.

Thermal stability data for the glass was obtained using three methods, depending on crystallization time. For times ranging from 0.05 to 1.6 minutes, conventional Differential Thermal Analysis (DTA) was used. Times were taken as the inverse of heating rates. For times ranging from 10 to 240 minutes, DTA was done isothermally, until crystallization became to slow for detection by thermal changes. For times ranging from hours to weeks, crystallization was confirmed by optical microscopy.

Hot stage transmission electron microscopy (TEM) was performed on some specimens using a Jeol 100CX instrument.

Ion implanted glass

Annealed and electro-polished poly-crystalline nickel sheet (0.5 mm thick) was implanted with 80 keV P_2+ ions (being the equivalent of twice as many 40 keV P+ ions), to avoid any possibility of contamination by O_2+ ions (mass 32 amu). Most nominal doses were of $3 \times 10E17$ atoms/cm^2, with some at half that amount. Compositions were measured later using standard Rutherford Backscattering (2 MeV He++ ions).

Implanted specimens were electrochemically thinned from the crystalline side in a Tenupol apparatus using a solution of 40 % orthophosphoric acid, 40 % methanol, and 20 % water, at 20 Volts.

Crystallization was studied by hot stage TEM.

RESULTS AND DISCUSSION

Melt spun glass

A portion of the Time Temperature Transformation (TTT) curve based on experimental evidence was determined for the melt spun glass. This is shown in figure 1 along with a 'nose' for the same composition, as calculated from theory by Naka [2].

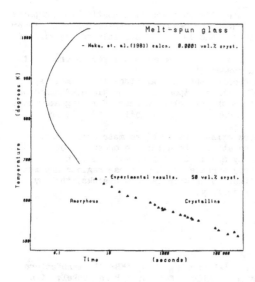

Figure 1. Predicted portion of Ni-P TTT curve (full line), with experimental data for melt spun Ni-P glass.

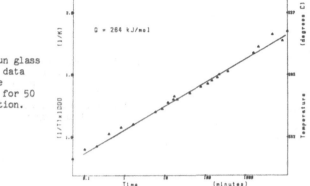

Figure 2. Melt spun glass
thermal stability data
re-plotted to give
activation energy for 50
vol.% crystallization.

If the same experimental data is re-plotted as 1/T vs log time
(figure 2), the overall activation energy for crystallization (50 vol.%) can
be calculated. Our result of 264 kJ/mol compares favorably
with the work of Von Heimendahl [3], who has used a similar method to
determine the actvation energies for crystallization of two Fe-B glasses,
2826 (440 kJ/mol) and 2826A (270 kJ/mol).

The glass was found to contain quenched-in crystals ranging in size
from <1 to 40 microns. The number density of these crystrals was 5x10E9
crystals/cm^3 +/- 2x10E9. Wherever crystallization was observed by hot stage
TEM, the melt spun glass crystallized from the surface of these crystals
(fig. 3). The crystallization product was identified as grains of Ni-Ni$_3$P
eutectic, with an inter-lamelar spacing of about 10 Angstroms.

The presence of quenched-in crystals implies that if this glass were
ideal, i.e. had no quenched-in crystals, it would be far more stable, as
homogeneous nuclei would have to form before crystallization.

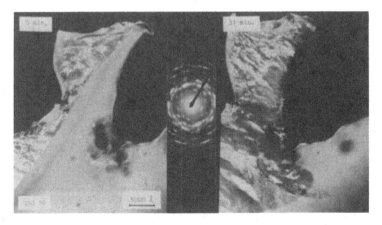

Figure 3. Melt spun glass crystallization. Note the quenched-in crystal
(left) acts as a nuclei for crystal growth. Dark field (Ni spot)
transmission electron micrograph.

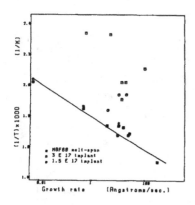

Figure 4. Temperature dependence of growth rate of the melt spun glass and the implanted specimens. Dotted points indicate values obtained by optical metallography, while all other points are obtained by hot stage TEM.

Figure 4 shows the dependance of growth rate on temperature. Note that the dotted points indicate data obtained from optical microscopy, whilst all other points were determined by hot stage TEM observations. The activation energy for crystal growth (movement of an atom from an amorphous site to a crystalline site) was determined from this figure using the assumptions of Koster [4]. The activation energy for growth was determined to be 170 kJ/mol, which compares well with Koster's determination of 194 kJ/mol for Fe-B glass [4], using the same method.

This figure differs from the activatation energy value determined from TTT data for crystallization.

Ion implanted glass

The composition of the high dose implants as measured by Rutherford Backscattering was 23 +/- 2 at. %.

Figure 5. Experimental TTT curve for melt spun glass (by DTA and optical microscopy), and data for high and low dose implanted amorphous layers (by hot stage TEM).

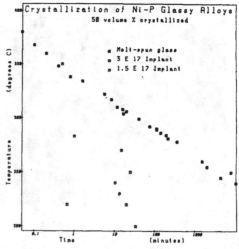

Though the crystallizing phases themselves were the same, a number of differences between the implanted glass and the melt spun glass were evident.

Firstly, the growth morphology was not that of a eutectic, as observed in the melt spun glass.

Secondly, nucleation did not occur at the amorphous-crystalline interface as was expected, but along the thinnest parts of the TEM foils.

Thirdly, crystallization was detected at much lower temperatures (i.e. 50 - 100°C lower) than was the case for the melt spun glass, indicating a less stable amorphous structure.

Figure 5 shows estimated 50 vol.% crystallized data points for both doses of implanted glass and points for the melt spun glass. The stability (fig. 5) and the growth rates (fig. 4) of the implanted glass show no systematic or Arrhenius temperature dependence, but it seems clear that the implanted glass is less stable than the melt spun glass.

Three growth regimes could be identified in implants of both doses, as can be seen in figure 6.

(a) Fast growth of perhaps single crystal Ni_3P occurs along the edge of the TEM wedge. These crystals contain long dislocations often dividing them into sub-grains of slightly different orientation. Growth rates in this regime range from 1000 to 10 000 A/s at around 240°C. This fast growth may be attributable to the dominance of the surface diffusion coefficient.

(b) Perpendicular to this, the crystals grow more slowly (1-100 A/s), and appear to decelerate as the TEM wedge becomes thicker.

(c) Eventually fine needle-like structures evolve out of the slow growth product and continue into thicker portions of the TEM wedge. A fine structure was detectable within each needle, and this may be a fine eutectic or a precipitate of nickel in Ni_3P, corresponding to a change in overall composition.

The transition from single crystal to needle like structures appears to be in agreement with the work of Bagley and Turnbull [5], who found that small changes in the composition of Ni-P glasses prepared by vapor deposition

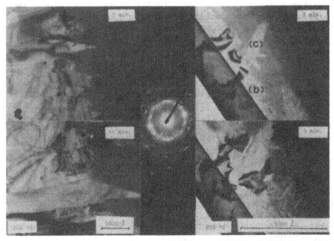

Figure 6. Growth of crystals into amorphous material for low dose implant (left) and high dose implant (right). Transmission Electron micrograph. The electron diffraction pattern is common to both specimens. Note the formation of needles (c) in both cases. The fast growth region tends to curl up during crystallization, and thus appears as a dark band.

resulted in a change in the interface shape of crystallizing Ni_3P from planar to dendritic (needle-like).

SUMMARY OF RESULTS AND DISCUSSION

	Melt spun	Implanted	Low dose imp.
Atomic % P	18.9	23 (by RBS)	not measured
Implantation dose (atoms/cm^2)	n.a.	3x10E17	1.5x10E17
Thickness	40 microns	1100 A (by RBS)	not measured
Cryst. product	Ni-Ni_3P eutectic	Ni_3P, Ni-Ni_3P eutectic	Ni_3P, Ni-Ni_3P eutectic
Activation energy	264 kJ/mol (cryst.) 170 kJ/mol (growth)	-	-

The difference in the two activation energies calculated above is not easily explained given that new crystal growth occurs on quenched-in crystals of the same phases (Ni, Ni_3P) already existing in the glass. Nucleation should not be necessary and growth of new crystals should begin immediately. However, there is an incubation time consistent with the need for nucleation and thus a correspondingly higher activation energy for crystallization than for subsequent growth is obtained. Further work is being carried out to determine the reason for this phenomenon.

The consistency of morphologies produced on crystallizing implanted glass with those observed by Bagley and Turnbull [5] confirms the RBS composition result which indicates that the implant compositions are richer in phosphorus than the eutectic of the metallic glass. Conclusions about stability differences between implanted and melt spun glasses must take composition differences into account. Work is currently in progress to implant phosphorus to eutectic composition.

CONCLUSIONS

1. Ni-P glass made by ion implantation is less stable than melt spun glass.
2. Crystallization phases are the same, but the morphology is different.
3. Changes in composition of implanted glass alter the morphology, without affecting the crystallization products.

REFERENCES

1. A. Ali, W. A. Grant, P. J. Grundy, Phil. Mag. B, 37, 353-376, (1978).

2. M. Naka, A. Inoue, T. Masumoto, Sci. Rep. Res. Inst. Tohoku Univ. A, 29, 184-194, (1981).

3. M. Von Heimendahl, J. Mat. Sci. Lett. 2, 1983, 796-800.

4. U. Koster, Z. Metallkde, 75, H.9, 691-697, (1984).

5. B. G. Bagley, D. Turnbull, Acta Met., 18, 857-862, (1970).

Author Index

Appleton, B.R., 369, 445
Ast, D.G., 427
Atwater, Harry, 337
Averback, R.S., 349, 491

Baily, A.C., 349
Bansal, Atul, 191
Barton, Roger, 303
Battaglin, G., 455
Benson, Jr., R.B., 461
Bentley, J., 461
Bijerk, Fred, 119
Bloembergen, N., 3, 201, 225, 245
Boggess, Thomas F., 213
Bose, Arijit, 191
Boyd, Ian W., 213
Braunstein, G., 233, 251, 263
Brown, W.L., 53, 319
Buckley, S.N., 485
Budnick, J.I., 439
Burcham, Jr., H.L., 161

Campisano, A., 329
Cannavo, S.U., 329
Carnera, A., 455
Chaumont, J., 455
Chen, M., 289, 309
Christie, W.H., 161
Culbertson, R.J., 357
Cuomo, J.J., 343

Davis, R.F., 395, 445
Della Mea, G., 455
Diaz de la Rubia, T., 491
Downer, Michael C., 15
Dresselhaus, G., 233, 251
Dresselhaus, M.S., 233, 251, 263

Edmond, J.A., 395
Elliman, R.G., 319, 389
Elman, B.S., 233, 251
Eridon, James, 433
Exarhos, Gregory, 179

Fabricius, N., 219
Fasihudin, A., 439
Fauchet, P.M., 149
Ferla, G., 329
Flytzanis, C., 25
Fogarassy, E., 173
Follstaedt, D.M., 297, 415, 473
Fröhlingsdorf, J., 271
Fujita, H., 479

Gandolfi, L., 329
Gerritsen, Hans C., 119
Gibson, J.M., 389

Häger, J., 25
Hahn, Horst, 491
Hamlyn-Harris, James, 497
Harper, J.M.E., 343
Hayden, H.C., 439
Headrick, R.L., 363
Herbots, N., 369
Hermes, P., 219
Hubler, G.K., 461
Huang, C.Y., 201, 245

Jacobson, D.C., 233, 251, 389
James, R.B., 161
Jellison, Jr., G.E., 131, 143
Jipson, V.B., 309
Johnson, H.H., 427
Joshi, N.V., 185

Kasuya, A., 257
Knapp, J.A., 415, 473
Knudson, Alvin R., 375
Kong, H.S., 395
Kulkarni, V.L., 455
Kurz, H., 201

La Ferla, A., 329
Lam, Nghi, 467
Larson, B.C., 113
Leaf, Gary K., 467
Lee, El-Hang, 155
Lehman, J., 185
Licoppe, C., 167
Lilienfeld, D.A., 427
Liu, J.M., 225, 245, 329
Lowndes, D.H., 131, 143, 173, 445
Luzzi, D.E., 479

MacDonald, C.A., 277
Maher, D.M., 319
Malvezzi, A.M., 201, 225, 245, 27'
Marinero, E.E., 289, 309
Mashburn, Jr., D.N., 131, 143
Maszara, W.P., 381
Mayer, J.W., 427
Mazzoldi, P., 455
Meshii, M., 479
Mills, B.E., 161
Mills, D.M., 113
More, K.L., 445
Mori, H., 479
Moss, Steven C., 213

504

Murakami, Kouichi, 119

Namavar, F., 439
Narayan, J., 173
Nastasi, M., 427
Nishina, Y., 257
Nissim, Y.I., 167
Noggle, T.S., 369

Okamoto, P.R., 349, 491

Pamler, W., 289, 309
Parrish, P.A., 461
Peercy, P.S., 99, 125, 283, 297
Pennycook, S.J., 131, 357, 369
Perepezko, J.H., 297
Picraux, S.T., 283
Pinizzotto, R.F., 213
Poate, J.M., 125, 389
Pogany, A.P., 455
Pospieszczyk, A., 219
Preston, John S., 137

Rehn, Lynn, 433
Richmond, Eliezer Dovid, 375
Rimini, E., 329
Rozgonyi, G.A., 381
Rubin, Kurt A., 303

St. John, D.H., 497
Sanchez, F.H., 439
Saris, Frans W., 119, 405
Seiberling, L.E., 363
Seidman, D.N., 349
Servidori, M., 329
Shaibani, S.J., 485
Shank, Charles V., 15
Shen, Y.R., 39
Simon, B., 25
Simpson, L., 381
Sipe, John E., 137
Smirl, Arthur L., 213
Smith, D.A., 343
Smith, Henry I., 337
Smith, P., 445
Sood, D.K., 389, 455, 497
Spaepen, F., 277
Speck, J.S., 263
Steinbeck, J., 233, 251, 263
Stephenson, L.D., 461
Strizker, B., 219, 271

Thompson, Carl V., 337
Thompson, Michael O., 99, 125, 283
Tischler, J.Z., 113

Tsao, J.Y., 125, 283
Turnbull, David, 71

Vach, H., 25
Van Brug, Hedser, 119
Van der Wiel, Marnix J., 119
Van Driel, Henry, 137
Venkatesan, J., 233, 251, 263
von der Linde, D., 219
Vredenberg, A., 405

Walther, H., 25
Was, Gary, 433
Westendorp, J.F.M., 405
White, C.W., 173
Wilkens, B., 233
Williams, J.S., 83, 319, 389
Withrow, S.P., 131, 395
Wood, R.F., 131, 143
Wortman, J.J., 381

Yu, L.S., 343

Zuhr, R.A., 369

Subject Index

Adsorbates, 39
AlAs, 185
Alloy phase formation (see also, ion beam mixing)
  [AlGa]As, 185
  Al/Ni, 433
  allotropic, 297
  icosahedral, 415, 427
  Si-Ge, 103
  silicide formation, 439
Al(Mn), 415, 427
$Al_{12}Mo$, 461
Aluminum, 5, 283, 461
Amorphization
  Cu-Ti, 479
  Fe-Ti-C, 473
  gallium, 271
  Ni, 455
  Ni/Al, 433
  Ni/P, 497
  silicon, 53, 71, 83, 213, 319, 329, 349, 357, 375, 381
  silicon carbide, 395
  tellurium, 303
Amorphous silicon
  crystallization, 71, 83, 99, 167, 319, 329, 349, 357, 375, 381
  diffusion, 91, 389
  melting dynamics, 99, 125, 131
  melting temperature, 71, 99
Annealing
  dynamics, 85
  ion-induced, 319, 329
  silicon, 83, 167
Auger heating, 15
Auger recombination, 202
$Au_x(TeO_2)_{1-x}$, 309

Bragg-Williams parameter, 479
Brass, yellow, 5

Carbon (see graphite)
Carbon monoxide, 44
Carrier densities, 204
Clusters
  carbon, 251
  ionic, 257
  gold, 309
Coherent optical processes, 39
Collision cascades, 53
Copper, 277
Crystallization
  collision-limited, 278

explosive, 71, 99, 125, 131
Ni-P, 497
radiation-enhanced, 319
SiC, 395
silicon, liquid phase, 53, 71, 99, 113, 125, 131
silicon, solid phase (see amorphous silicon)
Te-Ge, 306
Te-Se, 306
CVD, laser-assisted, 149
CW laser
  melting, 137, 303
  mixing, 185

Damage
  implantation, 58, 83, 349, 381
  mechanical, 155
Defects
  dislocation loops, 87, 357
  interstitials, 357
  point, 349
  radiation-induced, 319, 329, 371
  twins, 90
Desorption, 257, 363
Dielectric films
  TiO_2, 179
  ZrO_2, 179
Diffusion, 389, 467, 491
Distribution
  angular, 29
  rotational, 30
Drude, 17, 201

Electric dipole, 41
Electron-hole
  pairs, 3, 15, 62
  plasma, 17, 201, 225
Electron irradiation, 349, 479, 485
Electron-phonon interactions, 3, 201
Ellipsometry, time-resolved, 143
Evaporation, laser induced, 4, 225, 257
EXAFS, 119
Explosive crystallization (see crystallization)

Femtosecond
  dynamics, 15
  laser, 11, 15
Fe-Ti-C, 473

506

Fluorescence, laser-
  induced, 25
Fracture, 485
Fracture strength, 446
Furnace heating, 83, 319

Gallium, amorphous, 271
GaAs, 11, 161, 185, 219, 225
Germanium, 11, 71, 100, 143,
  149, 201, 257, 337, 369
Gold, 277
Gold clusters, 309
Grain growth, 337
Graphite
  carbon clusters, 251
  dielectric constant, 245
  laser melting, 233
  liquid carbon, 263
  plasma, 245, 251
  reflectivity, 11, 251
  surface scattering, 25
  surface structures, 263

Ice, 63
Icosahedral phase
  Al(Mn), 415, 427
  solid state diffusion, 424
Interface roughness, 167
Interface temperature, 117
Interface velocity
  liquid phase, 99, 113, 131,
    233, 279, 287, 296
  solid phase, 75, 83, 167, 319
Ion beam deposition, 369
Ion beam mixing
  Au/Cu, 407
  Cu/Er, 491
  icosahedral phase, 415, 427
  metastable alloys, 405
  Ni/Al, 433
  Ni/Ti, 491
  SiC, 445
  W/Cu, 408
Ion implantation
  into aluminum
    Mo, 461
  into carbon
    Ge, 251
  into iron
    C, Ti, 473
  into nickel
    Al, Si, 467
    La, O, 455
    P, 497

into silicon
  Ar, 88, 330
  As, 100, 173, 319, 358
  Au, 91
  Co, 439
  Cr, 439
  Cu, 131
  Fe, 439
  In, 92, 100, 125, 173, 389
  N, 100
  O, 100
  Ni, 439
  P, 100, 330
  Sb, 86, 173, 358
  Si, 100, 375, 381
  Sn, 100, 125
  Zn, 100, 125
into silicon carbide
  Al, P, Si, 395
Ion-induced epitaxy, 319, 329
Ion-solid interactions (see also
  ion implantation and ion beam
  mixing)
  annealing, 329, 375
  crystallization, 88, 319
  damage, 53, 375, 381, 395
  film alignment, 343
  film deposition, 369
  grain growth, 337
  mixing, 405, 445
  oxide, 363
  sputtering, 55, 343
Ionization
  laser-induced, 3, 25, 257
  spikes, 53

Lamellae, 137, 463
Laser
  CO_2, 138, 161
  dye, 16
  excimer, 27, 132, 143, 173,
    272, 290, 309, 445
  Kr, 303
  Nd:YAG, 180, 213, 220, 225
  ruby, 100, 251, 263, 298,
    409, 455
  spin-flip Raman, 27
Laser quenching, 271, 303
Laser-solid interactions
  femtosecond, 11, 15
  microsecond, 4
  millisecond, 4
  nanosecond, 4, 99, 113, 119,
    131, 143, 155, 161, 173,
    179, 219, 233, 251, 257, 263,

271, 283, 289, 298, 409, 415,
445, 455
picosecond, 7, 201, 213, 219,
225
Lattice disorder, 15

Manganese, 297
Marker, 493
Mass transport, 155
Maxwell Boltzmann, 6
Melting
allotropes, 297
aluminum, 283
brass, yellow, 5
copper, 278
CW laser-induced, 137
GaAs, 161, 219, 225
gallium, 271
germanium, 11, 76, 143
gold, 278, 309
graphite, 233, 263
heterogeneous, 78, 138
internal, 125, 131
silicon, 7, 71, 99, 113, 143,
213
tellurium, 289, 303, 309
Metal
aluminum, 5, 283, 461
copper, 277
gallium, amorphous, 271
gold, 277, 309
manganese, 297
nickel, 455, 467, 497
niobium, 343
tellurium, 292, 303
tin, 257
Metallic glass, 497
Metastable phase
gallium, 271
manganese, 297
Ni/Al, 433
silicon (see amorphous Si)
MgAl$_2$O$_4$, 485
Molecular beam scattering, 25
Morphological instabilities, 19
Morphology, 137, 149, 381
Multiphoton absorption, 3, 161,
213

Ni-Al alloys, 427, 433
Nickel, 455, 467, 497
Niobium, 343
Nitric oxide, 25
Nonlinear optics, 39, 161

Nucleation
internal melt, 99, 125
manganese, 297
silicon, 71, 83, 319
surface, 99

Optical recording, 303
Overheating (see superheating)
Oxidation, 173, 363

Pentadecanoic acid, 47
Periodic structures, 149
Phase separation, 389, 491
Picosecond laser, 7, 9, 201,
213, 225, 245, 277
Plasma
collisional, 3
density, 201
electron-hole, 15, 225
resonance, 201
Plasmon-aided recombination, 201
Polarization, 40
Precipitation, 83, 357, 455, 461
Pulsed electron beam, 416

Radiation-enhanced diffusion,
271, 491
Radiation-enhanced segregation,
467
Raman
microprobe, 149
spectroscopy, 179
stimulated, 42
surface, 42
Recrystallization (see
crystallization)
Reflectivity
aluminum, 287
copper, 280
GaAs, 220, 225
germanium, 143
gold, 280
graphite, 233, 245, 251
silicon, 9, 15, 125, 131,
143, 167, 201
tellurium, 292, 309
Refractive index, 3, 17, 143

Scattering
inelastic, 26
quasi-specular, 26
Raman, 42, 179

Second harmonic generation
GaAs, 11

508

surface, 43
Segregation
  coefficients, 128, 191
  graphite, 233
  interfacial instability, 191
  silicon, 128
Silicides, 439
Silicon
  amorphous (see amorphous
    silicon)
  ionization, 257
  liquid, 71, 99, 119
  melting, 7, 71, 99, 113, 131,
    143, 155, 201
  metallic, 10
  oxidation, 45
  precipitation, 83, 357
  polycrystalline, 90, 213
  surface reconstruction, 44
  thermodynamics, 71, 204
Silicon carbide, 395, 445
Solid phase epitaxy
  radiation-enhanced, 319
  SiC, 395
  thermal, 71, 83, 167, 357
Solubility, 99
Spectroscopy
  laser, 25
  molecular adsorbates, 49
  nonlinear optical, 39
Sputtering, 55, 343, 467
Stopping power, 53
Superheating
  GaAs, 219
  silicon, 78, 113
Superstructure, Si/Ge, 369
Surface
  disorder, 365
  GaAs, 161
  morphology, 149
  probe, 39
  Raman, 42
  reconstruction, 43
  scattering, 25
  symmetry, 40
  temperature, 5
Susceptibility, 40
Synchrotron, 113

Tellurium, amorphous, 289, 303
Thermal
  annealing, 329, 381, 389
  expansion coefficients, 113
  spike, 59

Time-of-flight measurements,
  220, 257
Tin, 220
Transient conductance
  aluminum, 283
  silicon, 100, 125
  tellurium, 290

Undercooling
  manganese, 297
  metals, 277
  silicon, 71, 99, 113, 125,
    131

Wave mixing, 39

X-ray absorption, 119
X-ray diffraction
  silicon, 113
  time-resolved, 113